Plasma-Assisted Nitrogen Fixation for Sustainable Process Industries

# Plasma-Assisted Nitrogen Fixation for Sustainable Process Industries

*Edited by*

*Volker Hessel*
*Adelaide University*
*Adelaide, Australia*
*and*
*University of Warwick*
*Coventry, UK*

*Annemie Bogaerts*
*University of Antwerp*
*Antwerp, Belgium*

*Gabriele Centi*
*University of Messina*
*Messina, Italy*

*Evgeny Rebrov*
*Eindhoven University of Technology*
*Eindhoven, The Netherlands*
*and*
*University of Warwick*
*Coventry, UK*

*Nguyen Van Duc Long*
*Adelaide University*
*Adelaide, Australia*
*and*
*University of Warwick*
*Coventry, UK*

WILEY

Published by John Wiley & Sons, Inc., Hoboken, New Jersey.



***Library of Congress Cataloging-in-Publication Data Applied for:***

Hardback ISBN : 9781394283019

Cover Design: Wiley
Cover Images: Courtesy of Nguyen Van Duc Long

Set in 9.5/12.5pt STIXTwoText by Straive, Pondicherry, India

# Contents

# About the Editors

**Prof. Volker Hessel** studied chemistry at Mainz University/D. 1994: Institut für Mikrotechnik Mainz/Germany (Director R&D); 2005: Professor at Eindhoven University of Technology/NL; 2018: Deputy Dean (Research), Professor at Adelaide University, Australia; and 2019: part-time professor at University of Warwick/UK. He is the author of over 800 peer-reviewed publications. His research is on process intensification by flow chemistry and plasma processing, waste-to-value, sustainability, and space resources. He is program lead in the ARC Centre of Excellence Plants for Space (P4S) and Research Director of the Andy Thomas Centre for Space Resources.

**Annemie Bogaerts** is a Professor of Physical Chemistry at the University of Antwerp, Belgium, and is well known for her research on plasmas for electrification of chemical reactions, including nitrogen fixation, by both experiments and modeling. She has published over 700 peer-reviewed papers since 1995 and has been invited *ca.* 300 times as an invited/plenary speaker at international conferences. She has supervised 65 finished Ph.D. students and is currently supervising 30 Ph.D. students (including joint Ph.D.'s). She has been the Vice-President and President of the International Plasma Chemistry Society (2020–2022; 2022–2023).

**Gabriele Centi** is a Professor of Industrial Chemistry at the University of Messina, Italy, and President of the European Research Institute of Catalysis, Brussels, Belgium. He is the author of over 600 publications in the areas of catalytic materials, low-carbon, and green processes and technologies. He has recently focused on the electrification of chemical processes, artificial leaf/tree devices, and photo-, electro-, and plasma catalysis. He is the past President of the International Association of Catalysis Societies and was also President of the European Federation of Catalysis Societies.

**Evgeny Rebrov** studied chemistry at Novosibirsk State University/Russia (1995), got Ph.D. in Physical Chemistry from Boreskov Institute of Catalysis (Novosibirsk, 1999) and then D.Sc. in Chemical Engineering (2012). He is a Professor of Energy Intensified Chemical Reaction Engineering at the University of Warwick (UK) and Eindhoven University of Technology (The Netherlands) and Director of Stoli Chem Ltd. (UK), an award-winning SME specializing in the design and manufacturing of chemical reactors. He is the author of numerous publications and patents on reactor design for process intensification using electricity-based energy fields (light, radio frequency, microwave, plasma). He focuses on the use of plasma for the efficient processing of chemicals and porous materials.

**Nguyen Van Duc Long** is a Senior Researcher with affiliations at Adelaide University, Australia, and the University of Warwick, UK. He received his Ph.D. in chemical engineering from Yeungnam University, South Korea, in 2012 and worked as a Senior Process Engineer at Samsung SDI, South Korea. His industrial experience includes process design, optimization, and scale-up, with a focus on implementing innovative solutions across manufacturing sectors, engineering, and equipment vendor companies. He has authored over 80 peer-reviewed articles, three book chapters, and two books published by Springer and Wiley. His research covers process intensification for separation and reaction processes, as well as digital twin and machine learning applications in chemical processes.

Four of the five editors have a prestigious joint Synergy Grant of the European Research Council (ERC) on plasma catalysis for chemical reactions.

# List of Contributors

***Muhammad Yousaf Arshad***
School of Chemical Engineering
Adelaide University
Adelaide
SA
Australia

***Susanta Bera***
Dutch Institute for Fundamental
Energy Research (DIFFER)
Eindhoven
The Netherlands

***Michal Bialkowski***
Research & Development Division
Casale SA
Lugano
Switzerland

***Pierdomenico Biasi***
Research & Development Division
Casale SA
Lugano
Switzerland

***Annemie Bogaerts***
Research Group PLASMANT
Department of Chemistry
University of Antwerp
Antwerp
Belgium

Electrification Institute
University of Antwerp
Antwerp
Belgium

***Filippo Buttignol***
Research & Development Division
Casale SA
Lugano
Switzerland

***Sarayute Chansai***
Department of Chemical Engineering
School of Engineering
University of Manchester
Manchester
UK

***Patrick Cullen***
School of Chemical and Biomolecular
Engineering
The University of Sydney
Sydney
New South Wales
Australia

***Alberto Garbujo***
Research & Development Division
Casale SA
Lugano
Switzerland

***Ana Gómez-Ramírez***
Departamento de Física Atómica
Molecular y Nuclear
Universidad de Sevilla
Seville
Spain

Laboratory of Nanotechnology on
Surfaces and Plasma
Instituto de Ciencia de Materiales de
Sevilla (CSIC-Universidad de Sevilla)
Seville
Spain

***Agustín R. González-Elipe***
Laboratory of Nanotechnology on
Surfaces and Plasma
Instituto de Ciencia de Materiales de
Sevilla (CSIC-Universidad de Sevilla)
Seville
Spain

***Yury Gorbanev***
Research Group PLASMANT
Department of Chemistry
University of Antwerp
Antwerp
Belgium

Electrification Institute
University of Antwerp
Antwerp
Belgium

***Christopher Hardacre***
Department of Chemical Engineering
School of Engineering
University of Manchester
Manchester
UK

***Volker Hessel***
School of Chemical Engineering
Adelaide University
Adelaide
SA
Australia

School of Engineering
University of Warwick
Coventry
UK

***Jungmi Hong***
School of Chemical and Biomolecular
Engineering
The University of Sydney
Sydney
New South Wales
Australia

***Rune Ingels***
Norwegian University of Life Sciences
Faculty of Environmental Sciences and
Natural Resource Management, Ås
Norway

***Haoxuan Jiang***
State Key Laboratory of Electrical
Insulation and Power Equipment
Center for Plasma Biomedicine
Xi'an Jiaotong University
Xi'an
PR China

***Leon Lefferts***
Catalytic Processes & Materials
MESA+ Institute for Nanotechnology
University of Twente
Enschede
The Netherlands

***Tianyu Li***
State Key Laboratory of Electrical
Insulation and Power Equipment
Center for Plasma Biomedicine
Xi'an Jiaotong University
Xi'an
PR China

***Liangliang Lin***
Key Laboratory of Synthetic and
Biological Colloids
Ministry of Education
School of Chemical and Material
Engineering
Jiangnan University
Wuxi
China

***Servando Marín-Meana***
Departamento de Física Atómica
Molecular y Nuclear
Universidad de Sevilla
Seville
Spain

Laboratory of Nanotechnology on
Surfaces and Plasma
Instituto de Ciencia de Materiales de
Sevilla (CSIC-Universidad de Sevilla)
Seville
Spain

***Adrián Megías-Sánchez***
Departamento de Física Atómica
Molecular y Nuclear
Universidad de Sevilla
Seville
Spain

Laboratory of Nanotechnology on
Surfaces and Plasma
Instituto de Ciencia de Materiales de
Sevilla (CSIC-Universidad de Sevilla)
Seville
Spain

***Anthony B. Murphy***
CSIRO
Manufacturing
Lindfield
NSW
Australia

***Magnus Nyvold***
Norwegian University of Life Sciences
Faculty of Environmental Sciences and
Natural Resource Management
Norwegian
University of Life Sciences, Ås
Norway

and

N2 Applied AS, Asker, Norway

***Manuel Oliva-Ramírez***
Departamento de Física Atómica
Molecular y Nuclear
Universidad de Sevilla
Seville
Spain

Laboratory of Nanotechnology on
Surfaces and Plasma
Instituto de Ciencia de Materiales de
Sevilla (CSIC-Universidad de Sevilla)
Seville
Spain

***Jose Luis Osorio Tejada***
School of Engineering
University of Warwick
Coventry
UK

***Raffaele Ostuni***
Research & Development Division
Casale SA
Lugano
Switzerland

***Evgeny Rebrov***
School of Engineering
University of Warwick
Coventry
UK

Department of Chemical Engineering and Chemistry
Eindhoven University of Technology
Eindhoven
The Netherlands

***Kevin Rouwenhorst***
Catalytic Processes & Materials
MESA+ Institute for Nanotechnology
University of Twente
Enschede
The Netherlands
Ammonia Energy Association
Ashburn, VA, USA

***Mateo Ruiz-Martín***
Departamento de Física Atómica
Molecular y Nuclear
Universidad de Sevilla
Seville
Spain

Laboratory of Nanotechnology on Surfaces and Plasma
Instituto de Ciencia de Materiales de Sevilla (CSIC-Universidad de Sevilla)
Seville
Spain

***Jing Sun***
State Key Laboratory of Electrical Insulation and Power Equipment
Center for Plasma Biomedicine
Xi'an Jiaotong University
Xi'an
PR China

***Nam Nghiep Tran***
School of Chemical Engineering
Adelaide University
Adelaide
SA
Australia

***Mihalis N. Tsampas***
Dutch Institute for Fundamental Energy Research (DIFFER)
Eindhoven
The Netherlands

***Xin Tu***
Department of Electrical Engineering and Electronics
University of Liverpool
Liverpool
UK

***Nguyen Van Duc Long***
School of Chemical Engineering
Adelaide University
Adelaide
SA
Australia

University of Warwick
Coventry, UK

***Weitao Wang***
Department of Electrical Engineering and Electronics
University of Liverpool
Liverpool
UK

***Shanshan Xu***
Department of Materials
School of Natural Sciences
University of Manchester
Manchester
UK

***Dimitrios Zagoraios***
Dutch Institute for Fundamental
Energy Research (DIFFER)
Eindhoven
The Netherlands

***Tianqi Zhang***
School of Chemical and Biomolecular
Engineering
The University of Sydney
Sydney
New South Wales
Australia

***Rusen Zhou***
State Key Laboratory of Electrical
Insulation and Power Equipment
Center for Plasma Biomedicine
Xi'an Jiaotong University
Xi'an
PR China

***Renwu Zhou***
State Key Laboratory of Electrical
Insulation and Power Equipment
Center for Plasma Biomedicine
Xi'an Jiaotong University
Xi'an
PR China

# Preface

The rapid development of industrialization and the increasing demand for nitrogen-based products, particularly in agriculture and chemical manufacturing, have made nitrogen fixation one of the most crucial industrial processes of our time. However, traditional processes, such as the Haber-Bosch process, remain highly energy-intensive and dependent on fossil fuels, contributing significantly to global carbon emissions. As industries move toward sustainable and energy-efficient processes, plasma-assisted nitrogen fixation technology has emerged as a promising alternative that aligns with global sustainability goals.

Plasma technology is a spearhead innovation among the eminent topic of "chemical electrification," besides electrolysis. Fueled by renewable energy expansion, today's industrial electrification moves far ahead into "mine-electrification," "electro agriculture," and more. An emerging (spearhead) technology can tell a lot not only about the future of one industry (nitrogen fixation) but also beyond about an ongoing technology step-change within chemical engineering. Plasma technology and nitrogen fixation allow to assess the global value of transitional innovation *via* revisiting and redesigning sustainability and energy efficiency like almost no other merger of market push-and-pull. Spear-headed viewpoints are able to crystallize the pros and cons and showcase the big opportunities, challenges, bottlenecks, and also the limits. From this vantage point, we hope this book advances plasma-assisted nitrogen fixation by promoting technical improvements, environmental impact and life cycle assessments as well as renewable energy integration for industrial applications.

This book, *Plasma-Assisted Nitrogen Fixation for Sustainable Process Industries*, explores how plasma technology can revolutionize ammonia and $NO_x$ production by improving efficiency, sustainability, and industrial resilience. It highlights the potential of plasma technology and provides a detailed overview of plasma-assisted nitrogen fixation, from the fundamental principles to practical applications, as well as cutting-edge solutions that enhance efficiency and sustainability. This book emphasizes reactor designs and discusses industrial feasibility, including decentralized production, making nitrogen fixation accessible to remote areas and smaller scale operations. In addition to covering current advances, the book explores future possibilities, making it a valuable resource for those interested in cutting-edge manufacturing methods for ammonia and $NO_x$ synthesis. As sustainability becomes a top priority, this book offers a pathway to greener and cleaner industrial processes.

The Haber-Bosch ammonia process is undoubtedly one of the five most important chemical processes for industry, mankind, globe, and technology history. It serves as the perfect "testbed" for any emerging technology. Plasma researchers have embraced this challenge and continue to do so, benefiting from 120 years of fundamental knowledge on this reaction, including reaction kinetics, catalyst, and sustainability studies. In a process that is so much energy driven, a large number of plasma technology studies have struggled to transition from fascinating laboratory studies to convincing industrial piloting. Ammonia may, therefore, not be the prime industrial translation example for plasma technology, yet better seen as of intellectual primacy. This is why this book also covers, on equal height, plasma-based $NO_x$ synthesis, which, in the very recent years, has become a great industrial showcase that currently demonstrates industrial translation of plasma technology in terms of founding companies, pilot-scale demonstrations, and commercialized products.

This book serves as a comprehensive resource for a wide range of readers, including scientists, engineers, industry professionals, and policymakers. It also fills a crucial knowledge gap in plasma-assisted nitrogen fixation, making this innovative technology more accessible and understood. More than just a technical guide, it connects the disciplines of chemistry, physics, and engineering, presenting complex concepts in an accessible way for researchers and practitioners. Additionally, entrepreneurs and innovators in clean technology can discover new opportunities in sustainable ammonia and $NO_x$ production. Last but not least, there is a stellar interest from academics, ecological activists, and stakeholders outside plasma and even outside chemistry: in soil science, agriculture, agrochemicals, molecular biology, and even in medicinal hospital disinfection (as it comprises high hydrogen peroxide loads when done under water). This book presents plasma-assisted nitrogen fixation in a clear and accessible way, making it easy for readers from different backgrounds to understand both the fundamentals and real-world applications. In general, readers can choose chapters of interest and read them independently, allowing them to focus on specific topics, such as plasma technology, reactor designs, or industrial implementation based on their needs.

This book consists of 15 chapters, organized into five parts:

- Part 1 covers the fundamentals of nitrogen fixation, introducing the Haber-Bosch process, its limitations, and alternatives (Chapter 1), followed by a discussion on $NO_x$ production and nitric acid synthesis (Chapter 2).
- Part 2 explores plasma technology, detailing plasma generation, ionization, and applications (Chapter 3) as well as the latest advances in plasma reactors for $NO_x$

synthesis (Chapter 4) and ammonia production (Chapter 5). Chapter 6 builds on Chapter 4, discussing plasma-assisted $NO_x$ synthesis with a focus on reaction pathways, energy efficiency, and the role of catalysts.

- Part 3 focuses on the fundamental reaction mechanisms and kinetics of plasma-driven nitrogen fixation for ammonia and $NO_x$ production (Chapters 7–8).
- Part 4 evaluates environmental sustainability (Chapter 9) and the economic feasibility (Chapter 10) of plasma-assisted nitrogen fixation, discussing its carbon footprint, renewable energy integration, and cost-effectiveness.
- Part 5 introduces cutting-edge innovations, such as microplasma technology (Chapter 11), plasma–liquid interactions for nitrogen fixation (Chapter 12), and plasma electrochemistry for nitrate-to-ammonia conversion (Chapter 13). It also covers analytical methods in plasma diagnostics as a tool to enable innovation (Chapter 14) and concludes with perspectives on plasma-based fertilizer applications (Chapter 15).

We acknowledge support and inspiration to this book from the ERC Synergy Grant "Surface-COnfined fast modulated Plasma for process and Energy intensification" (SCOPE) from the European Commission with the Grant No. 810182. We are truly grateful to Wiley Publishing, especially Michael Leventhal, Vishali Chandra Mohan, and Govindanagaraj Deenadayalu, for their support and smooth communication throughout the publication process of this edited book. Last but not least, our sincere gratitude is also expressed to all chapter authors for their expertise, dedication, and valuable, innovative contributions to this book.

10 November 2025
Adelaide, Antwerp, Coventry, Eindhoven, and Messina

The Editors
Volker Hessel
Annemie Bogaerts
Gabriele Centi
Evgeny Rebrov
Nguyen Van Duc Long

# Part I

# Fundamentals of Nitrogen Fixation

# 1

# Fundamentals of Ammonia Production

*Kevin Rouwenhorst*[1,2*] *and Leon Lefferts*[1]

[1] *Catalytic Processes & Materials, MESA+ Institute for Nanotechnology, University of Twente, Enschede, The Netherlands*
[2] *Ammonia Energy Association, Ashburn, VA, USA*

## 1.1 Introduction to Nitrogen Fixation

Ammonia ($NH_3$) is produced from nitrogen ($N_2$) and hydrogen ($H_2$), both naturally and synthetically (see Eq. 1.1) [1, 2]. While the synthetic ammonia production routes have developed over the past century, some organisms in nature fix nitrogen in the form of ammonia with nitrogenase [1–4]. Another source of fixed nitrogen is atmospheric deposition via lightning, while fixed nitrogen is recycled via crop residues and animal manures such as guano [5, 6]. Furthermore, ammonium sulfate ($(NH_4)_2SO_4$) was produced as a byproduct from coke and town gas production from coal at the end of the nineteenth century, via the reaction of ammonia with sulfuric acid ($H_2SO_4$, Eq. 1.2) [7]. However, only half of the required nitrogen fixation could be obtained from these sources around the beginning of the twentieth century [5].

$$3H_2 + N_2 \leftrightarrow 2NH_3 \left(\Delta H^0 = -91.8\,\text{kJ mol}^{-1}\right) \tag{1.1}$$

$$2NH_3 + H_2SO_4 \rightarrow (NH_4)_2SO_4 \tag{1.2}$$

Various industrial nitrogen fixation technologies, such as the Birkeland-Eyde process, the Frank-Caro process, and the Haber-Bosch process, were commercialized in the early twentieth century [8, 9]. The Birkeland-Eyde process is a plasma-based process for fixing nitrogen into nitrogen oxides ($NO_X$), and this is discussed in Chapter 2. Among these technologies, the Haber-Bosch process is the dominant technology

*Corresponding author

*Plasma-Assisted Nitrogen Fixation for Sustainable Process Industries*, First Edition. Edited by Volker Hessel, Annemie Bogaerts, Gabriele Centi, Evgeny Rebrov, and Nguyen Van Duc Long.

Published 2026 by John Wiley & Sons, Inc.

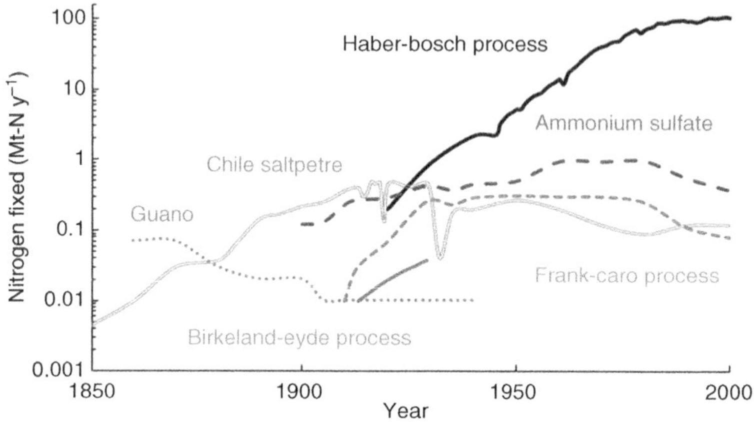


**Figure 1.1** Annual consumption of fixed nitrogen from various natural sources and from industrial nitrogen fixation technologies. Original sources for the data [6, 8, 10]. *Source:* Adapted from Ref. [11, 12].

today (see Figure 1.1), due to its higher energy efficiency for nitrogen fixation and the availability of low-cost natural gas for the production of hydrogen via steam methane reforming (SMR). Heat integration of exothermic ammonia synthesis with endothermic steam reforming has been essential to achieve a high energy efficiency.

## 1.2 The Haber-Bosch Process

The thermochemical ammonia synthesis process (Eq. 1.1) was published and patented by Haber and Le Rossignol in 1913 and 1916, which would be termed the Haber-Bosch process in the years to follow [13–15]. The feasibility of this process was demonstrated by Haber and Le Rossignol in 1908 with a tabletop system operating at 500–550 °C, 100–200 atm, and in the presence of an osmium catalyst, producing about 2 kg-$NH_3$ day$^{-1}$ [16]. Only one year before, Nernst, among others, had concluded that ammonia synthesis was not feasible during the 14th General Convention of the Bunsen Society [17]. However, Nernst's values for the thermodynamics proved erroneous, and synthetic ammonia synthesis was feasible after all. Mittasch and coworkers developed and patented the use of a multicomponent (*mehrstoff*) iron catalyst as a more abundant, safer alternative to toxic osmium for ammonia synthesis between 1909 and 1915 [18, 19]. Engineering challenges were dealt with by Bosch and co-workers in the 1910s [20], including the development of compressors, as well as steel materials resilient to operation at high hydrogen pressures, preventing hydrogen embrittlement. Up to this point, no industrial processes were operated at hundreds of bars [20, 21].

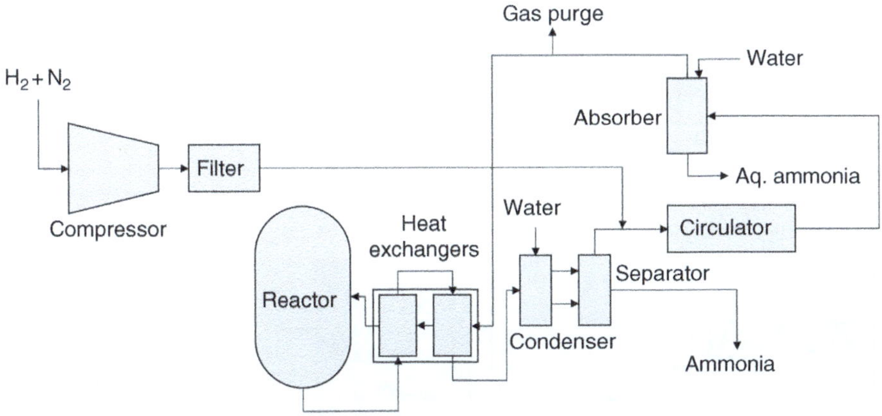


**Figure 1.2** Schematic overview of the Haber-Bosch ammonia synthesis loop as operated in Oppau (Germany). *Source:* Reproduced from Ref. [22] / with permission of American Institute of Chemical Engineers.

The first ammonia plant began operations in Oppau (Germany) in 1913. A schematic overview of a Haber-Bosch ammonia synthesis loop is shown in Figure 1.2. A modern Haber-Bosch process converts gaseous hydrogen ($H_2$) and atmospheric nitrogen ($N_2$) to ammonia ($NH_3$) at 350–500 °C and 100–400 bar using an iron-based catalyst [4, 23, 24].

## 1.3 Production Pathways to Ammonia

As given by Eq. (1.1), ammonia is produced from hydrogen and nitrogen. Typically, hydrogen production accounts for at least 90% of the energy requirements of ammonia production [25]. Ammonia can be produced from various fossil-based hydrogen sources, such as natural gas, coal, naphtha, and heavy fuel oil (HFO). Decarbonized hydrogen sources include biomass, biogas, and water splitting using renewable electricity. Nitrogen is purified from the air. Figure 1.3 shows conceptual pathways for the production of hydrogen as a feedstock for ammonia production.

Currently, ammonia is almost exclusively produced from fossil feedstocks, mainly natural gas and coal [26, 27]. About 72% of ammonia production uses natural gas, and about 22% of ammonia production uses coal, with HFO and naphtha accounting for about 4% and 1%, respectively, while 1% of ammonia is derived from other feedstocks [28]. Most ammonia production capacity using coal is located in China, with natural gas being the norm in other countries [29].

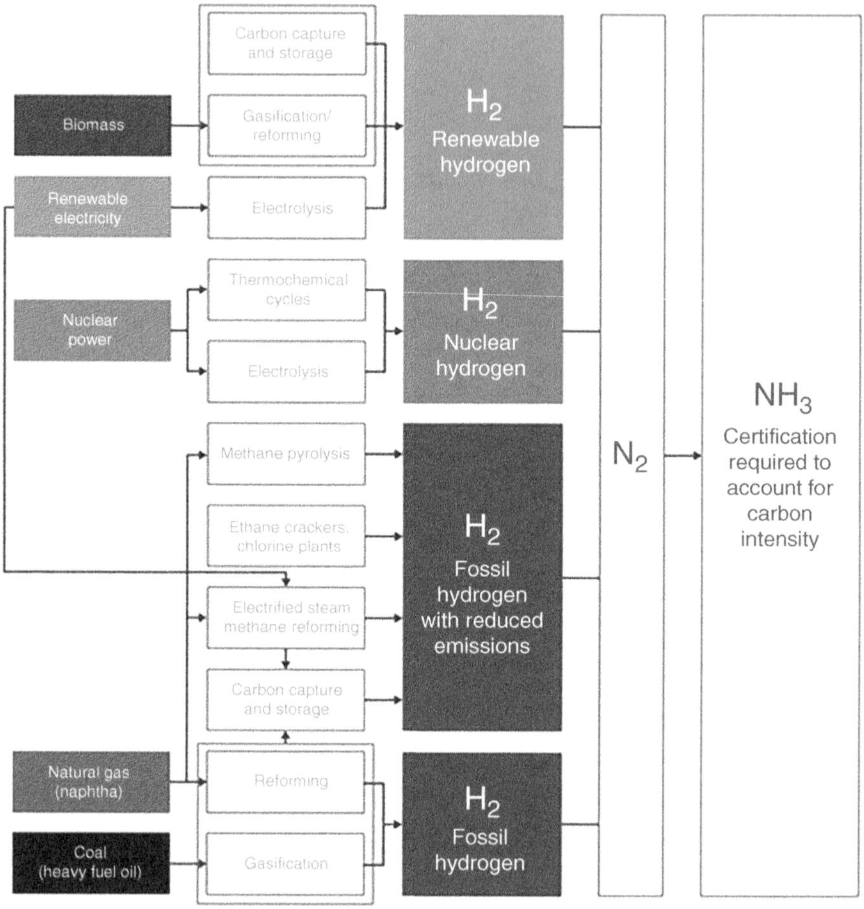


**Figure 1.3** Production pathways to ammonia from various feedstocks. *Source:* Reproduced from Ref. [25]/with permission of IRENA.

Renewable ammonia production accounted for just 0.1% of global production by the end of 2023 [30].

Global ammonia production consumes about 1–2% of the global energy supply, as well as 5% of the global natural gas supply, at the cost of 1–2% of global $CO_2$ equivalent emissions (0.5–0.6 Gt-$CO_{2,eq}$) [31–33].

Schematic overviews of a natural gas-based Haber-Bosch process and an electrolysis-based Haber-Bosch process are shown in Figure 1.4. A main difference between the processes is the method for hydrogen production and nitrogen purification. A natural gas-based Haber-Bosch process typically operates with a

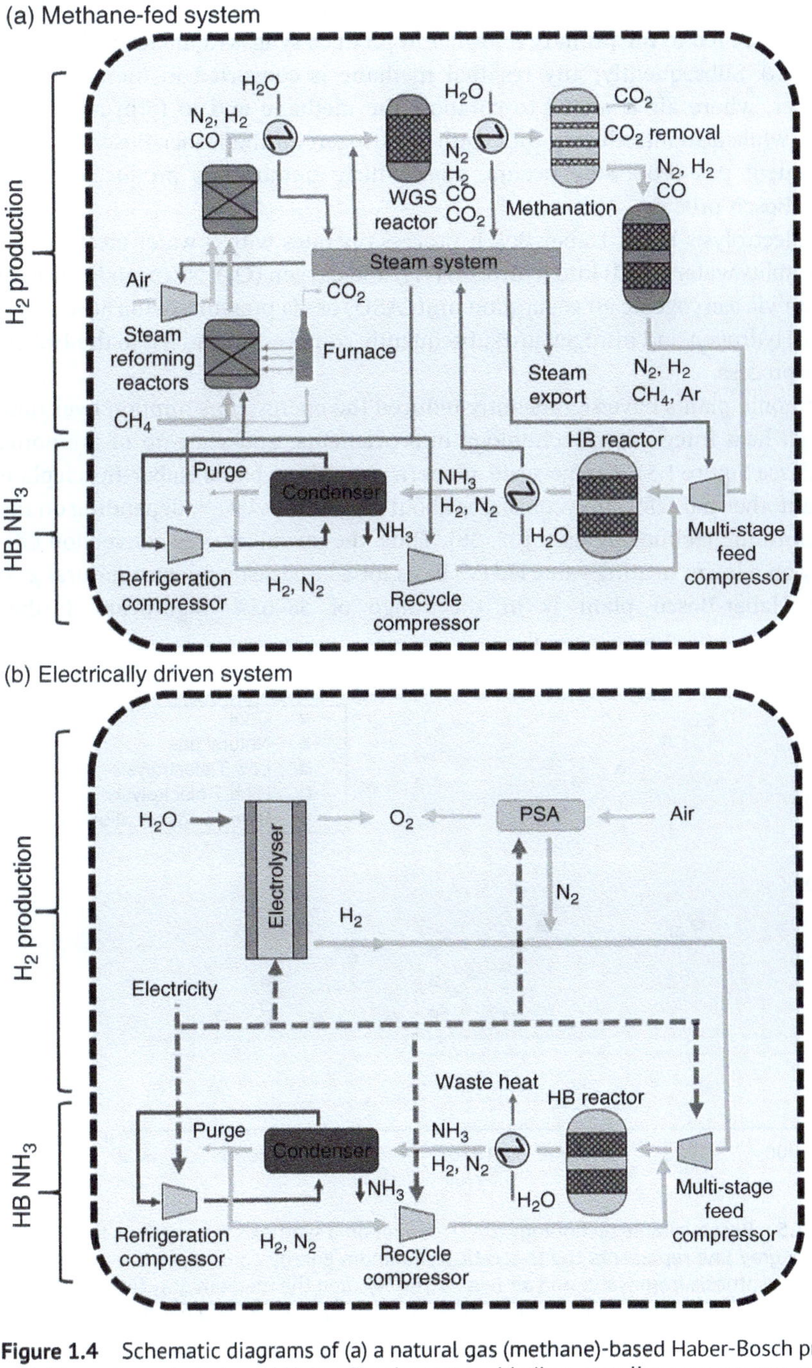


**Figure 1.4** Schematic diagrams of (a) a natural gas (methane)-based Haber-Bosch process and (b) an electrolysis-based Haber-Bosch process. Medium gray lines are process gas and water/steam, light gray lines are air, dark gray lines are ammonia, and dashed lines are electricity. *Source:* Ref. [34]/Royal Society of Chemistry/CC BY 3.0.

primary reformer and a secondary reformer. Steam ($H_2O$) and natural gas (primarily $CH_4$) are fed to the primary reformer to produce syngas (a mixture of $H_2$, CO, and $CO_2$). Subsequently, any residual methane is converted in the secondary reformer, where air is added to combust the methane and to form additional syngas while also introducing the required nitrogen for the Haber-Bosch process. Subsequent gas processing occurs, after which ammonia is produced in the Haber-Bosch process.

An electrolysis-based Haber-Bosch process operates with a water electrolyzer, which splits water ($H_2O$) into hydrogen ($H_2$) and oxygen ($O_2$). Nitrogen is purified from air via a cryogenic air separation unit (ASU) or via pressure swing adsorption (PSA). Hydrogen and nitrogen are subsequently compressed and fed to the Haber-Bosch process.

Ammonia plants have significantly reduced the energy consumption over time through heat integration, technology improvements, and scale-up of ammonia plants (see Figure 1.5). A large-scale, modern natural gas-based Haber-Bosch plant with autothermal reforming consumes about 29–32 GJ t-$NH_3^{-1}$, depending on the gas reforming technology used [35, 36]. Thus, the overall energy conversion efficiency on a lower heating value (LHV) basis for a large-scale modern natural gas-based Haber-Bosch plant is in the range of 58–65%. Significant further

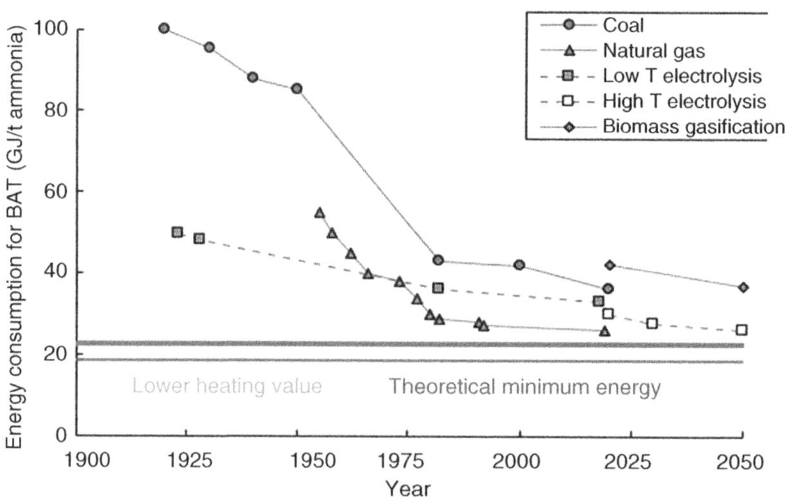


**Figure 1.5** Best available technology (BAT) for ammonia synthesis from various feedstocks. The dark gray line represents the theoretical minimum energy consumption required for ammonia synthesis from water and air (via $H_2$ and $N_2$), and the medium gray line represents the lower heating value of ammonia (18.646 GJ t-$NH_3^{-1}$). *Source:* Original data from Refs. [6, 8, 34–40].

optimization is not expected, as the process approaches a technological asymptote (see Figure 1.5). However, older plants can be optimized through revamps [41]. The addition of carbon capture and storage (CCS) technology for $CO_2$ emission abatement increases the energy consumption by around 1 GJ t-$NH_3^{-1}$ [26, 36].

Modern electrolysis-based Haber-Bosch plants using low-temperature electrolysis (alkaline electrolysis or proton exchange membrane (PEM) electrolysis) typically consume about 36 GJ t-$NH_3^{-1}$ [34, 36], equivalent to 52% energy conversion efficiency. Hydrogen production typically consumes most of the energy, with about 33 GJ t-$NH_3^{-1}$. Nitrogen purification from PSA for small-scale ammonia plants consumes about 0.6–0.9 GJ t-$NH_3^{-1}$, while nitrogen purification from cryogenic distillation for large-scale ammonia plants consumes about 0.3 GJ t-$NH_3^{-1}$ [42]. The ammonia synthesis loop typically consumes at least 1–2 GJ t-$NH_3^{-1}$ [34, 43].

Renewable ammonia synthesis from high-temperature electrolysis (solid oxide electrolysis) typically consumes about 30 GJ t-$NH_3^{-1}$ [34, 44] (62% efficiency) and is expected to decrease to 26 GJ t-$NH_3^{-1}$ in the long term [45] (72% efficiency). The lower energy consumption compared to low-temperature electrolysis is due to more efficient hydrogen production and greater heat integration across the process due to the utilization of excess steam from the ammonia converter as feedstock for the solid oxide electrolyzers [45, 46].

## 1.4 Novel Methods for Ammonia Synthesis

The Haber-Bosch process has been the dominant process for nitrogen fixation for over a century [6, 32, 47]. The Haber-Bosch process is highly optimized, and the energy efficiency of the natural gas-based ammonia production process is as high as 58–65% (LHV) [34]. This creates a high hurdle for new technologies.

Novel ammonia production technologies are especially relevant for decentralized applications, typically with a capacity below 10 t-$NH_3$ day$^{-1}$ [26]. At such small scales, the energy consumption of Haber-Bosch is typically high due to heat losses [42], and downscaling is costly due to the high pressures [48]. Heat losses for the Haber-Bosch ammonia synthesis loop can amount to 45–50 MJ kg-$NH_3^{-1}$ at the 10 kW scale [42, 49]. Thus, most novel technologies will initially focus on decentralized, niche applications for ammonia and its derivatives, especially if local production competes with imported ammonia via transport by truck.

A wide range of novel ammonia production technologies has been researched, such as electrochemical and photochemical processes, plasma-based processes (as discussed in the following chapters), chemical looping approaches, homogeneous synthesis, biological processes, and ammonia purification or electrochemical reduction from animal waste or wastewater [26, 50, 51]. Often, breaking the N≡N triple bond is the rate-limiting step in these processes, due to a high dissociation energy (945 kJ $mol^{-1}$). None of these technologies have been commercialized, with the current focus on research.

Electrochemical ammonia synthesis has received substantial research interest over the past decades [52–54]. This is due to the promise of the direct conversion of water and nitrogen to ammonia and oxygen, according to Eq. (1.3). However, this has remained a scientific challenge with low rates of ammonia formation [55], and false positives have been previously reported due to the presence of ammonia in the surroundings [56, 57]. Even with substantial optimization, the energy consumption of electrochemical ammonia synthesis will likely remain well above that of a Haber-Bosch plant fed with hydrogen produced via water electrolysis [58].

$$N_2 + 3H_2O \rightarrow 2NH_3 + \frac{3}{2}O_2 \tag{1.3}$$

Photochemical ammonia synthesis provides a direct pathway for converting water and nitrogen to ammonia using light [59]. The main advantage of this approach is the possibility of using water as a reducing agent for nitrogen under mild conditions. Instead of an electrocatalyst, a semiconductor is used for photocatalysis (typically titania, $TiO_2$) [60–63]. So far, no efficient photochemical system for ammonia synthesis has been developed. The poor stability of the photocatalysts and the low utilization factor of the light lead to a low solar-to-ammonia conversion [59, 64, 65].

Nitrogen oxides can be converted electrochemically to ammonia [66, 67]. Nitrite ($NO_2^-$) and nitrate ($NO_3^-$) can be electrochemically converted to ammonia. The conversion of nitrate to ammonia and water is given by Eq. (1.4). Polluted groundwater with nitrogen oxide concentrations around 1 mM or mmol/L, and low-level nuclear wastewater with nitrogen oxide concentrations up to 2 M can be used as a nitrogen oxide source [66]. If all nuclear wastewater were used for ammonia production, this would result in an ammonia production of less than 2 kt-$NH_3$ $year^{-1}$ [67]. Groundwater with 1 mM nitrogen oxide concentration would require about 55,000 t-water t-$NH_3^{-1}$, which is impractical from a water processing and purification perspective [67]. Therefore, electrochemical nitrogen oxide conversion to ammonia is not expected to play a significant role in decarbonizing ammonia production.

$$NO_3^- + 8e^- + 9H^+ \leftrightarrow NH_3 + 3H_2O \tag{1.4}$$

Chemical looping approaches involve separate thermodynamically stable reaction intermediates, which allow for controlling nitrogen activation and hydrogenation to ammonia independently [68]. Three main approaches for chemical looping have been proposed for ammonia synthesis, namely hydrogen chemical looping ($H_2$-CL), water chemical looping ($H_2O$-CL), and alkali or alkaline earth metal hydride chemical looping (AH-CL), sometimes in combination with electrocatalysis [51, 69, 70]. Chemical looping approaches have not found commercial applications for ammonia synthesis so far, in part due to the complex stepwise processing.

Homogeneous nitrogen reduction reactions have been reported since the 1960s [71, 72]. Homogeneous catalysis has not found practical application for ammonia synthesis due to low yields with up to 8–12 equivalents of ammonia per equivalent of catalyst [73, 74], fast catalyst deactivation, and expensive catalyst compounds [3, 50]. Homogeneous catalysts for ammonia synthesis are currently mostly useful for understanding nitrogenase, which fixes nitrogen in nature [75].

Three types of nitrogenases have been identified, with molybdenum (Mo) nitrogenase being the most common. The other types of nitrogenases are vanadium (V) nitrogenase and iron (Fe) nitrogenase [1]. The protein cycles of molybdenum nitrogenase are discussed in Refs. [1, 3, 76, 77]. The ideal reaction is given by Eq. (1.4). However, nitrogenase is not 100% selective toward ammonia, typically forming hydrogen in the process as well, according to Eq. (1.5).

$$\frac{1}{2}N_2 + \frac{3}{2}H_2O \rightarrow NH_3 + \frac{3}{4}O_2 \tag{1.5}$$

$$N_2 + 8H^+ + 8e^- \rightarrow 2NH_3 + H_2 \tag{1.6}$$

Theoretically, an energy input of 16 adenosine triphosphates (ATPs) is required for ammonia synthesis under ambient conditions, equivalent to 26–30 MJ kg-$NH_3^{-1}$ [76]. However, a maximum overall efficiency of about 10–15% is estimated for the enzyme nitrogenase (150–225 MJ kg-$NH_3^{-1}$), and even lower for the complete bacteria [4].

## 1.5 Applications of Ammonia

Ammonia is one of the basic chemicals used to produce chemical products and materials [78], next to ethylene ($C_2H_4$), propylene ($C_3H_6$), methanol ($CH_3OH$), BTX (benzene, toluene, and xylene) aromatics, and C4 compounds (butadiene, butene, isobutene, and butane).

About 183 Mt of ammonia was produced in 2020 [26, 27]. About 90% of all ammonia produced is consumed on-site as a feedstock for downstream processes, and the 18–20 Mt of ammonia is annually transported by ship [27, 79].

Nitrogen fertilizers account for about 80–85% of the current ammonia demand [25, 36]. Urea production ($CO(NH_2)_2$) accounts for about 55% of the global ammonia demand, and ammonium nitrate production ($NH_4NO_3$) accounts for about 15% of the global ammonia demand [25, 36]. Other nitrogen fertilizers include various nitrates, monoammonium phosphate and diammonium phosphate, ammonium sulfate, and mixtures of nitrogen fertilizers like urea ammonium nitrate (UAN) solution, calcium ammonium nitrate, and nitrogen, phosphorus, and potassium (NPK) fertilizers, which mix nitrogen with the other key nutrients, e.g. potassium and phosphate [80].

Other markets include the manufacturing of chemicals (e.g. amines and amides), pharmaceuticals, and polymers (e.g. polyamides, polyurethanes, acrylonitrile, and melamine). Ammonia is used in the mining industry (low-density ammonium nitrate explosives and metal brightening processes), refrigeration, waste treatment, and air treatment, such as nitrogen oxide ($NO_X$) abatement via selective reduction [25, 36].

Ammonia is also proposed as a carbon-free fuel and hydrogen carrier [25, 31, 81], as shown in Figure 1.6. As early as the nineteenth century, ammonia was proposed as a fuel [82]. Ammonia engines were used by Goldsworthy Gurney in the United Kingdom, and the ammoniacal gas engine was operated in New Orleans in the 1870s and 1880s [83]. Ammonia was used to fuel buses in Belgium during World War II due to the scarcity of other fuels [84]. NASA used ammonia to fuel its X-15 hypersonic rocket-powered aircraft in the 1960s [81].

Currently, ammonia is considered a fuel for power generation in thermal power plants and gas turbines. Ammonia can supplement fluctuating renewable electricity from solar PV and wind in countries that import energy [85]. In 2024, about 20% of ammonia was co-fired in a 1 GW coal-fired thermal power plant in Japan [86].

Ammonia is also considered a maritime fuel for the international transport of goods. The first ammonia-fueled vessel set sail in 2024, a tugboat with a dual-fuel ammonia-diesel engine [87]. A tugboat moves other vessels by pushing or pulling them, such as in harbors and canals. As of 2024, larger ammonia-fueled vessels with two-stroke engines are already on order, such as bulk carriers, gas carriers, tankers, and container ships. Bulk carriers transport metals, minerals, and coal. Gas carriers carry liquefied gases such as LNG (liquefied natural gas), LPG (liquefied propane gas), LEG (liquefied ethylene gas), hydrogen, $CO_2$, and ammonia. Tankers carry oil and petroleum products. Container ships transport goods in containers.

Finally, ammonia is considered a hydrogen carrier [88]. Liquid ammonia is transported more easily than hydrogen between continents, and ammonia infrastructure exists in major ports globally. Ammonia can be converted to hydrogen and nitrogen via the ammonia cracking reaction (Eq. 1.3), which is an endothermic reaction. Ammonia crackers might become relevant for energy import hubs, such as Belgium, the Netherlands, and Germany in Northern Europe, as well as

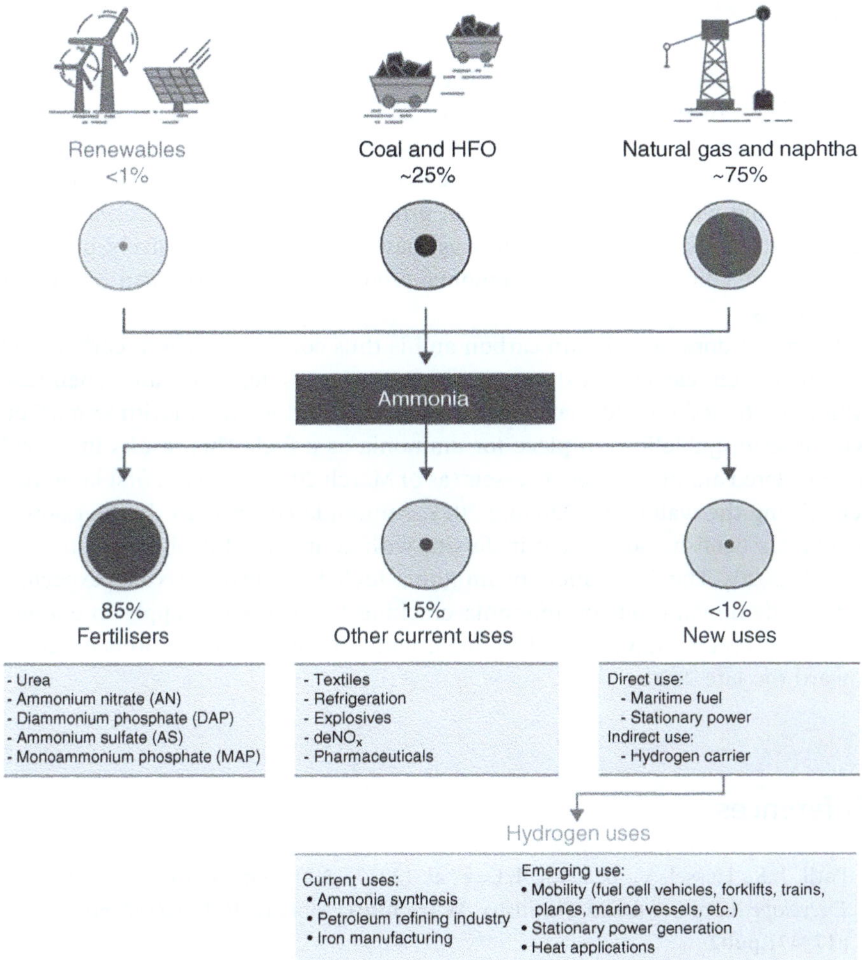


**Figure 1.6** Production methods and uses of ammonia. *Source:* Reproduced from Ref. [25]/with permission of IRENA.

Japan and South Korea in East Asia. Ammonia crackers typically operate at similar operational conditions to those of a steam methane reformer (600–900 °C and 10–40 bar), with a hydrogen yield of at least 75%, and sometimes above 80% [89, 90]. In recent years, plasma-based ammonia cracking into hydrogen is also gaining increasing interest, especially in warm plasmas [91–94].

$$2NH_3 \leftrightarrow 3H_2 + 1N_2 \left(\Delta H^0 = +91.8\,\text{kJ mol}^{-1}\right) \tag{1.7}$$

## 1.6 Conclusions

Ammonia has been produced via the Haber-Bosch process for over a century. Over time, the hydrogen production pathway has varied, with natural gas and coal as the current dominant feedstocks. Existing ammonia production capacity can be decarbonized via CCS. Alternatively, the emergence of renewable electricity from solar PV and wind allows for hydrogen production via water electrolysis. Novel electrified ammonia production pathways have been studied extensively, but these are not yet competitive with ammonia production with hydrogen via water electrolysis.

Ammonia does not contain carbon and is thus considered a zero-carbon fuel and hydrogen carrier, next to its current uses as fertilizer and chemical. Ammonia-fueled engines have been commercialized for the maritime market, with interim guidelines in place for ammonia as a fuel. This results in a total of 58 ordered ammonia-fueled vessels (as of March 2025), with the first large vessels hitting the waters in 2026 and 2027. Ammonia co-firing in thermal power plants has been demonstrated in Japan, with commercial deployment toward the late 2020s. The deployment of ammonia-fueled gas turbines is also expected. Finally, demonstrations of ammonia cracking for hydrogen supply to energy-importing regions have started in 2025, with commercial operations expected toward the late 2020s.

## References

1 Patil, B.S., Hessel, V., Seefeldt, L.C. et al. (2017). Nitrogen fixation. In: *Ullmann's Encyclopedia of Industrial Chemistry*. Wiley https://doi.org/10.1002/14356007.a17_471.pub2.
2 Appl, M. (2011). Ammonia, 1. Introduction. In: *Ullmann's Encyclopedia of Industrial Chemistry*. Wiley https://doi.org/10.1002/14356007.a02_143.pub3.
3 Foster, S.L., Bakovic, S.I.P., Duda, R.D. et al. (2018). Catalysts for nitrogen reduction to ammonia. *Nat. Catal.* 1: 490–500. https://doi.org/10.1038/s41929-018-0092-7.
4 Appl, M. (1999). *Ammonia: Principles and Industrial Practice*, 1e. Weinheim: Wiley-VCH Verlag GmbH https://doi.org/10.1002/9783527613885.
5 Smil, V. (1999). Detonator of the population explosion. *Nature* 400: 415. https://doi.org/10.1038/22672.
6 Smil, V. (2004). *Enriching the Earth: Fritz Haber, Carl Bosch, and the Transformation of World Food Production*. Cambridge, MA: MIT Press.
7 Nielsen, S.E. (2009). Ammonia synthesis: catalyst and technologies. *ACS Symp. Ser.* 1000: 15–39. https://doi.org/10.1021/bk-2009-1000.ch002.

**8** Ernst, F.A. (1928). *Industrial Chemical Monographs: Fixation of Atmospheric Nitrogen*. London: Chapman & Hall, Ltd.

**9** Travis, A.S. (2017). Globalising synthetic nitrogen: the interwar inauguration of a new industry. *Ambix* 64: 1–28. https://doi.org/10.1080/00026980.2017.1325585.

**10** Brightling, J.R. (2018). Ammonia and the fertiliser industry: the development of ammonia at Billingham, Johnson Matthey. *Technol. Rev.* 62: 32–47. https://doi.org/10.1595/205651318X696341.

**11** Rouwenhorst, K.H.R., Jardali, F., Bogaerts, A., and Lefferts, L. (2021). Plasma-based $NO_X$ synthesis: from the Birkeland-Eyde process towards energy-efficient and cost-effective plasma technology. *Energy Environ. Sci.* 14: 2520–2534. https://doi.org/10.1039/D0EE03763J.

**12** Rouwenhorst, K.H.R., Jardali, F., Bogaerts, A., and Lefferts, L. (2023). Correction: from the Birkeland–Eyde process towards energy-efficient plasma-based $NO_X$ synthesis: a techno-economic analysis. *Energy Environ. Sci.* 16: 6170–6173. https://doi.org/10.1039/D3EE90066E.

**13** Haber, F. and Le Rossignol, R. (1913). The production of synthetic ammonia. *Ind. Eng. Chem.* 5: 328–331. https://doi.org/10.1021/ie50052a022.

**14** F. Haber, R. Le Rossignol, Production of Ammonia, US 1202995, 1916.

**15** Sheppard, D. (2017). Robert Le Rossignol, 1884–1976: engineer of the "Haber" process. *Royal Soc.J. History Sci.* 71: 263–296. https://doi.org/10.1098/rsnr.2016.0019.

**16** Travis, A.S. (2018). *Nitrogen Capture: The Growth of an International Industry (1900–1940)*. Springer International Publishing https://doi.org/10.1007/978-3-319-68963-0.

**17** Nernst, W. (1907). Über das Ammoniakgleichgewicht. *Zeitschrift Elektrochem. Angew. Phys. Chem.* 13: 521–524. https://doi.org/10.1002/bbpc.19070133205.

**18** Farber, E. (1966). From chemistry to philosophy: the Way of Alwin Mittasch (1869–1953). *Chymia* 11: 157–178. https://doi.org/10.2307/27757266.

**19** Mittasch, A. and Frankenburg, W. (1950). Early studies of multicomponent catalysts. *Adv. Catal.* 2: 81–104. https://doi.org/10.1016/S0360-0564(08)60375-2.

**20** Prieto, G. and Schüth, F. (2015). The Yin and Yang in the development of catalytic processes: catalysis research and reaction engineering. *Angew. Chem. Int. Ed.* 54: 3222–3239. https://doi.org/10.1002/anie.201409885.

**21** C. Bosch, The Development of the Chemical High Pressure Method During the Establishment of the New Ammonia Industry, The Noble Prizes, 1932.

**22** Pattabathula, V. and Richardson, J. (2016). Introduction to ammonia production. *Chem. Eng. Prog.* 112 (9): 69–75.

**23** Liu, H. (2013). *Ammonia Synthesis Catalysts: Innovation and Practice*. World Scientific https://doi.org/10.1142/8199.

**24** Nielsen, A. (1995). *Ammonia: Catalysis and Manufacture*, 1e. Berlin Heidelberg: Springer-Verlag.

**25** IRENA (2022). *Innovation Outlook: Renewable Ammonia*. Abu Dhabi: Ammonia Energy Association https://doi.org/978-92-9260-423-3.

**26** Rouwenhorst, K.H.R., Krzywda, P.M., Benes, N.E. et al. (2020). Ammonia Production Technologies. In: *Techno-Economic Challenges of Green Ammonia as Energy Vector* (ed. R. Bañares-Alcántara and A. Valera-Medina), 41–84. Elsevier Science Publishing Co, Inc. https://doi.org/10.1016/B978-0-12-820560-0.00004-7.

**27** O. Hatfield, A review of global ammonia supply, in: NH3 Energy Conference, 2020.

**28** Bicer, Y., Dincer, I., Zamfirescu, C. et al. (2016). Comparative life cycle assessment of various ammonia production methods. *J. Clean. Prod.* 135: 1379–1395. https://doi.org/10.1016/j.jclepro.2016.07.023.

**29** Zhou, W., Zhu, B., Li, Q. et al. (2010). $CO_2$ emissions and mitigation potential in China's ammonia industry. *Energy Policy* 38: 3701–3709. https://doi.org/10.1016/j.enpol.2010.02.048.

**30** K. Rouwenhorst, Technology status: ammonia production from electrolysis-based hydrogen, Ammonia Energy, 2023. https://www.ammoniaenergy.org/articles/technology-status-ammonia-production-from-electrolysis-based-hydrogen/ (accessed December 24, 2023).

**31** The Royal Society, Ammonia: zero-carbon fertiliser, fuel and energy store, 2020.

**32** Erisman, J.W., Sutton, M.A., Galloway, J. et al. (2008). How a century of ammonia synthesis changed the world. *Nat. Geosci.* 1: 636–639. https://doi.org/10.1038/ngeo325.

**33** Mission Possible Partnership, Making Net-Zero Ammonia Possible: An Industry-Backed, 1.5°C-Aligned Transition Strategy, 2023.

**34** Smith, C., Hill, A.K., and Torrente-Murciano, L. (2020). Current and future role of Haber–Bosch ammonia in a carbon-free energy landscape. *Energy Environ. Sci.* 13: 331–344. https://doi.org/10.1039/C9EE02873K.

**35** CEFIC, European chemistry for growth: unlocking a competitive, low carbon and energy efficient future, 2013.

**36** International Energy Agency, Ammonia Technology Roadmap: Towards More Sustainable Nitrogen Fertiliser Production, 2021.

**37** European Fertilizer Manufacturers' Association (2000). *Production of Ammonia*. Brussels: European Fertilizer Manufacturers' Association.

**38** J.B. Hansen, P.A. Han, Roadmap to All Electric Ammonia Plants, in: NH3 Fuel Conference, Pittsburgh (PA), 2018.

**39** Grundt, T. and Christiansen, K. (1982). Hydrogen by water electrolysis as basis for small scale ammonia production. A comparison with hydrocarbon based technologies. *Int. J. Hydrog. Energy* 7: 247–257. https://doi.org/10.1016/0360-3199(82)90088-X.

**40** Zhang, H., Wang, L., Van Herle, J. et al. (2020). Techno-economic comparison of green ammonia production processes. *Appl. Energy* 259: 114135. https://doi.org/10.1016/j.apenergy.2019.114135.

**41** Katerina Kermeli, E. Worrell, W. Graus, M. Corsten, Energy Efficiency and Cost Saving Opportunities for Ammonia and Nitrogenous Fertilizer Production, 2017.

**42** Rouwenhorst, K.H.R., Van Der Ham, A.G.J., Mul, G., and Kersten, S.R.A. (2019). Islanded ammonia power systems: technology review & conceptual process design. *Renew. Sust. Energ. Rev.* 114: 109339. https://doi.org/10.1016/j.rser.2019.109339.

**43** Bañares-Alcántara, R., Dericks, G. III, Fiaschetti, M. et al. (2015). *Analysis of Islanded Ammonia-based Energy Storage Systems*. Oxford: University of Oxford.

**44** Cinti, G., Frattini, D., Jannelli, E. et al. (2017). Coupling solid oxide electrolyser (SOE) and ammonia production plant. *Appl. Energy* 192: 466–476. https://doi.org/10.1016/j.apenergy.2016.09.026.

**45** Hansen, J.B. (2015). Fuel processing for fuel cells and power to fuels as seen from an industrial perspective. *J. Catal.* 328: 280–296. https://doi.org/10.1016/j.jcat.2015.04.014.

**46** Hauch, A., Küngas, R., Blennow, P. et al. (2020). Recent advances in solid oxide cell technology for electrolysis. *Science* 370: eaba6118. https://doi.org/10.1126/science.aba6118.

**47** Liu, H. (2014). Ammonia synthesis catalyst 100 years: practice, enlightenment and challenge, Cuihua Xuebao/Chinese. *J. Catal.* 35: 1619–1640. https://doi.org/10.1016/S1872-2067(14)60118-2.

**48** Yoshida, M., Ogawa, T., Imamura, Y., and Ishihara, K.N. (2021). Economies of scale in ammonia synthesis loops embedded with iron- and ruthenium-based catalysts. *Int. J. Hydrog. Energy* 46 (57): 28840–28854. https://doi.org/10.1016/j.ijhydene.2020.12.081.

**49** Rouwenhorst, K.H.R. and Lefferts, L. (2020). Feasibility study of plasma-catalytic ammonia synthesis for energy storage applications. *Catalysts* 10: 999. https://doi.org/10.3390/catal10090999.

**50** Cherkasov, N., Ibhadon, A.O., and Fitzpatrick, P. (2015). A review of the existing and alternative methods for greener nitrogen fixation. *Chem. Eng. Process. Process Intensif.* 90: 24–33. https://doi.org/10.1016/j.cep.2015.02.004.

**51** Nørskov, J., Chen, J., Bullock, M. et al. (2016). *Sustainable Ammonia Synthesis Exploring the Scientific Challenges Associated with Discovering Alternative, Sustainable Processes for Ammonia Production*. Dulees, VA: US DOE Office of Science https://doi.org/10.2172/1283146.

**52** Giddey, S., Badwal, S.P.S., and Kulkarni, A. (2013). Review of electrochemical ammonia production technologies and materials. *Int. J. Hydrog. Energy* 38: 14576–14594. https://doi.org/10.1016/j.ijhydene.2013.09.054.

**53** McPherson, I.J., Sudmeier, T., Fellowes, J., and Tsang, S.C.E. (2019). Materials for electrochemical ammonia synthesis. *Dalton Trans.* 48: 1562–1568. https://doi.org/10.1039/c8dt04019b.

**54** MacFarlane, D.R., Cherepanov, P.V., Choi, J. et al. (2020). A roadmap to the ammonia economy. *Joule* 4: 1186–1205. https://doi.org/10.1016/j.joule.2020.04.004.

**55** Kibsgaard, J., Nørskov, J.K., and Chorkendorff, I. (2019). The difficulty of proving electrochemical ammonia synthesis. *ACS Energy Lett.* 4 (12): 2986–2988. https://doi.org/10.1021/acsenergylett.9b02286.

**56** Choi, J., Suryanto, B.H.R., Wang, D. et al. (2020). Electrochemical nitrogen reduction studies. *Nat. Commun.* 11: 5546. https://doi.org/10.1038/s41467-020-19130-z.

**57** Andersen, S.Z., Čolić, V., Yang, S. et al. (2019). A rigorous electrochemical ammonia synthesis protocol with quantitative isotope measurements. *Nature* 570: 504–508. https://doi.org/10.1038/s41586-019-1260-x.

**58** Izelaar, B., Ramdin, M., Vlierboom, A. et al. (2024). Techno-economic assessment of different small-scale electrochemical $NH_3$ production plants, Energy. *Environ. Sci.* 17: 7983–7998. https://doi.org/10.1039/D4EE03299C.

**59** Chen, X., Li, N., Kong, Z. et al. (2018). Photocatalytic fixation of nitrogen to ammonia: state-of-the-art advancements and future prospects. *Mater Horiz.* 5: 9–27. https://doi.org/10.1039/c7mh00557a.

**60** Schrauzer, G.N. and Guth, T.D. (1977). Photolysis of water and photoreduction of nitrogen on titanium dioxide1. *J. Am. Chem. Soc.* 99: 7189–7193. https://doi.org/10.1021/ja00464a015.

**61** Zhao, Y., Zhao, Y., Shi, R. et al. (2019). Tuning oxygen vacancies in ultrathin $TiO_2$ nanosheets to boost photocatalytic nitrogen fixation up to 700 nm. *Adv. Mater.* 31: 1–9. https://doi.org/10.1002/adma.201806482.

**62** Yang, J., Guo, Y., Jiang, R. et al. (2018). High-efficiency "working-in-Tandem" nitrogen photofixation achieved by assembling plasmonic gold nanocrystals on ultrathin Titania nanosheets. *J. Am. Chem. Soc.* 140: 8497–8508. https://doi.org/10.1021/jacs.8b03537.

**63** Liu, S., Wang, Y., Wang, S. et al. (2019). Photocatalytic fixation of nitrogen to ammonia by single Ru atom decorated $TiO_2$ nanosheets. *ACS Sustain. Chem. Eng.* 7: 6813–6820. https://doi.org/10.1021/acssuschemeng.8b06134.

**64** Zhang, S., Zhao, Y., Shi, R. et al. (2019). Photocatalytic ammonia synthesis: recent progress and future. *EnergyChem* 1: 100013. https://doi.org/10.1016/j.enchem.2019.100013.

**65** Guo, C., Ran, J., Vasileff, A., and Qiao, S. (2017). Rational design of electrocatalysts and photo(electro)catalysts for nitrogen reduction to ammonia ($NH_3$) under ambient conditions. *Energy Environ. Sci.* 11 (1): 45–56. https://doi.org/10.1039/C7EE02220D.

**66** van Langevelde, P.H., Katsounaros, I., and Koper, M.T.M. (2021). Electrocatalytic nitrate reduction for sustainable ammonia production. *Joule* 5: 290–294. https://doi.org/10.1016/j.joule.2020.12.025.

**67** John, J., MacFarlane, D.R., and Simonov, A.N. (2023). The why and how of $NO_x$ electroreduction to ammonia. *Nat. Catal.* 6: 1125–1130. https://doi.org/10.1038/s41929-023-01060-w.

**68** McEnaney, J.M., Singh, A.R., Schwalbe, J.A. et al. (2017). Ammonia synthesis from N2 and $H_2O$ using a lithium cycling electrification strategy at atmospheric pressure. *Energy Environ. Sci.* 10: 1621–1630. https://doi.org/10.1039/C7EE01126A.

**69** Swearer, D.F., Knowles, N.R., Everitt, H.O., and Halas, N.J. (2019). Light-driven chemical looping for ammonia synthesis. *ACS Energy Lett.* 4: 1505–1512. https://doi.org/10.1021/acsenergylett.9b00860.

**70** Wang, Q., Guo, J., and Chen, P. (2019). Recent progress towards mild-condition ammonia synthesis. *J. Energy Chem.* 36: 25–36. https://doi.org/10.1016/j.jechem.2019.01.027.

**71** Volpin, M.E., Shur, V.B., and Ilatovskaya, M.A. (1964). Fixation of nitrogen by systems based on dicyclopentadienyltitanium dichloride. *Bullet. Acad. Sci. USSR Division Chem. Sci.* 13: 1644. https://doi.org/10.1007/BF0084632.

**72** Volpin, M.E., Ilatovskaya, M.A., Kosyakova, L.V., and Shur, V.B. (1968). Catalytic fixation of nitrogen. *Chem. Commun.* (18): 1074–1075. https://doi.org/10.1039/C19680001074.

**73** Arashiba, K., Miyake, Y., and Nishibayashi, Y. (2010). A molybdenum complex bearing PNP-type pincer ligands leads to the catalytic reduction of dinitrogen into ammonia. *Nat. Chem.* 3: 120–125. https://doi.org/10.1038/nchem.906.

**74** Yandulov, D.V. and Schrock, R.R. (1979). Catalytic reduction of dinitrogen to ammonia at a single molybdenum center. *Science* 301 (2003): 76–78. https://doi.org/10.1126/science.1085326.

**75** Dance, I. (2010). Mimicking nitrogenase. *Dalton Trans.* 39: 2972–2983. https://doi.org/10.1039/b922606k.

**76** Burgess, B.K. and Lowe, D.J. (1996). Mechanism of molybdenum nitrogenase. *Chem. Rev.* 96: 2983–3011. https://doi.org/10.1021/cr950055x.

**77** Seefeldt, L.C., Hoffman, B.M., and Dean, D.R. (2012). Electron transfer in nitrogenase catalysis. *Curr. Opin. Chem. Biol.* 16: 19–25. https://doi.org/10.1016/j.cbpa.2012.02.012.

**78** Levi, P.G. and Cullen, J.M. (2018). Mapping global flows of chemicals: from fossil fuel feedstocks to chemical products. *Environ. Sci. Technol.* 52: 1725–1734. https://doi.org/10.1021/acs.est.7b04573.

**79** O. Hatfield, Country traded ammonia logistics and storage, present and future, in: Ammonia Energy Conference 2021, Boston, MA, 2021: pp. 1–16.

**80** Yara (2018). *Fertilizer Industry Handbook.* Yara.

**81** Valera-Medina, A., Xiao, H., Owen-Jones, M. et al. (2018). Ammonia for power. *Prog. Energy Combust. Sci.* 69: 63–102. https://doi.org/10.1016/j.pecs.2018.07.001.

**82** Sousa Cardoso, J., Silva, V., Rocha, R.C. et al. (2021). Ammonia as an energy vector: current and future prospects for low-carbon fuel applications in internal combustion engines. *J. Clean. Prod.* 296: 126562. https://doi.org/10.1016/j.jclepro.2021.126562.

**83** L.C. Bennick, E.H. Carlton, The Streetcars of New Orleans, 1965.

**84** Kroch, E. (1945). Ammonia – a fuel for motor buses. *J. Inst. Pet.* 31: 213–223.

**85** Palazzi, C., Nayak-Luke, R., Verschuur, J. et al. (2024). Green ammonia imports could supplement long-duration energy storage in the UK. *Environ. Res. Energy* 1 (4): 043001. https://doi.org/10.1088/2753-3751/ad785d.

**86** JERA, Start of Demonstration Testing of Fuel Ammonia Substitution at JERA's Hekinan Thermal Power Station: The World's First Demonstration Testing of 20% Ammonia Substitution at a Large-Scale Commercial Coal-Fired Thermal Power Plant, 2024. https://www.jera.co.jp/en/news/information/20240401_1863 (accessed May 11, 2024).

**87** A. Čučuk, Fortescue's ammonia-powered vessel completes set of sea trials, Offshore Energy, 2024. https://www.offshore-energy.biz/fortescues-ammonia-powered-vessel-completes-set-of-sea-trials/ (accessed October 6, 2024).

**88** IRENA, Global hydrogen trade to meet the 1.5°C climate goal, 2022.

**89** Makhloufi, C. and Kezibri, N. (2021). Large-scale decomposition of green ammonia for pure hydrogen production. *Int. J. Hydrog. Energy* 46: 34777–34787. https://doi.org/10.1016/j.ijhydene.2021.07.188.

**90** Trangwachirachai, K., Rouwenhorst, K., Lefferts, L., and Faria Albanese, J.A. (2024). Recent progress on ammonia cracking technologies for scalable hydrogen production. *Curr Opin Green Sustain Chem.* 49: 100945. https://doi.org/10.1016/j.cogsc.2024.100945.

**91** Fedirchyk, I., Tsonev, I., Quiroz Marnef, R., and Bogaerts, A. (2024). Plasma-assisted $NH_3$ cracking in warm plasma reactors for green $H_2$ production. *Chem. Eng. J.* 499: 155946. https://doi.org/10.1016/j.cej.2024.155946.

**92** L. Wang, C. Liu, Y. Zhao, H. Guo, X. Wang, W. Gong, On-site generation of cox-free hydrogen for fuel cell from ammonia decomposition by plasma-catalysis method, in: ACS National Meeting Book of Abstracts, New Orleans, 2008.

**93** Yi, Y., Wang, L., Guo, Y. et al. (2019). Plasma-assisted ammonia decomposition over Fe–Ni alloy catalysts for $CO_x$-free hydrogen. *AIChE J* 65: 691–701. https://doi.org/10.1002/aic.16479.

**94** Akiyama, M., Aihara, K., Sawaguchi, T. et al. (2018). Ammonia decomposition to clean hydrogen using non-thermal atmospheric-pressure plasma. *Int. J. Hydrog. Energy* 43: 14493–14497. https://doi.org/10.1016n2018.06.022.

# 2

# Fundamentals of $NO_x$ Production

*Filippo Buttignol, Alberto Garbujo, Raffaele Ostuni, Michal Bialkowski, and Pierdomenico Biasi*

*Research & Development Division, Casale SA, Lugano, Switzerland*

## 2.1 Introduction

In strong interconnection with ammonia production ($NH_3$), nitric acid ($HNO_3$) is among the most crucial inorganic acids in the chemical industry. Its relevance lies especially in the production of nitrogen (N)-based fertilizer, ammonium nitrate ($NH_4NO_3$), and nitrophosphates, among others, ensuring a readily available source of nitrogen for crops, boosting yields, and helping meet global food demands. Simultaneously, the oxidizing power of $HNO_3$ is fundamental in the polymer industry. Ensuring effective oxidation of aliphatic hydrocarbons, a considerable fraction of the worldwide production of $HNO_3$ is devoted to the oxidation of cyclohexanol and cyclohexanone for the synthesis of adipic acid, which, in turn, is further utilized for the manufacture of nylon-6,6. While these productions can be carried out using the so-called weak nitric acid (i.e. a diluted $HNO_3$ solution with a concentration in the 55–65% range), other processes require concentrated $HNO_3$ possessing higher purity. In this view, concentrated $HNO_3$ has found relevance in esterification and nitrification processes, which ultimately sustains the production of explosive agents (e.g. nitroglycerin), nitrocellulose lacquers, polyurethanes (via nitro- and dinitrotoluene), as well as dyes and insecticides (by chloronitrobenzene) [1].

Historically speaking, the large-scale production of $HNO_3$ began with the so-called Chile saltpeter (a mixture of $NaNO_3$, $KNO_3$, and NaCl). Once refined, $NaNO_3$ was mixed and heated together with concentrated $H_2SO_4$, yielding diluted vapors of $HNO_3$, which are ultimately condensed. However, when the natural

*Plasma-Assisted Nitrogen Fixation for Sustainable Process Industries*, First Edition. Edited by Volker Hessel, Annemie Bogaerts, Gabriele Centi, Evgeny Rebrov, and Nguyen Van Duc Long.

Published 2026 by John Wiley & Sons, Inc.

reserves of Chile saltpeter were considered to be close to exhaustion, an incredible effort was made to substitute this nitrogen source with atmospheric nitrogen. In this climate, the Birkeland-Eyde process was developed. Using nitrogen and oxygen from air and employing electric arcs, nitrogen oxides (NO, $NO_2$) were obtained and subsequently transformed into $HNO_3$ by absorption of $NO_2$ in $H_2O$ [2]. However, given the extremely high temperatures ($T > 2{,}000\,°C$) required for direct transformation of nitrogen and oxygen into NO and the consequent energy expenditures, this production pathway was soon abandoned. Starting from the beginning of the twentieth century and following the development and implementation of the Haber-Bosch process, the primary industrial pathway for $HNO_3$ production became the Ostwald process. Herein, $NH_3$ becomes the N-source for $HNO_3$ formation via an indirect process in which it is first catalytically combusted to form nitrogen oxides. Thus, the profound relationship between $NH_3$ and $HNO_3$ does not lie only in the use of the former to neutralize the latter and synthesize $NH_4NO_3$ but also because $NH_3$ is the feedstock itself for the large-scale production of $HNO_3$ [3].

## 2.2 The Ostwald Process

The Ostwald process, developed by Wilhelm Ostwald in the early 1900s, leverages three main steps: (i) the catalytic oxidation of $NH_3$ in the presence of oxygen to form nitrogen oxides (particularly NO), (ii) the homogeneous gas-phase oxidation of NO into $NO_2$ and (iii) the ultimate heterogeneous reaction between $NO_{2(g)}$ and $H_2O_{(l)}$ to yield $HNO_3$.

### 2.2.1 Catalytic Oxidation of Ammonia

In the first step of the Ostwald process, $NH_3$ is vaporized and preferentially oxidized to NO at high temperature ($T$ = 800–950 °C) and high contact time ($10^{-3}$–$10^{-4}\,s^{-1}$) generally over $Pt_{90}$-$Rh_{10}$ gauze catalysts according to Eq. (2.1) [4]:

$$4NH_3 + 5O_2 \longrightarrow 4NO + 6H_2O \quad \Delta H = -904\,kJ\,mol^{-1} \tag{2.1}$$

About 3–50 gauzes can be arranged in the burner (the reactor) possessing diameters up to 6 m. While the optimal operating temperature range is quite narrow, various pressure settings exist depending on the plant configuration. Pressures of about 1 to 12 bar are common when $NH_3$ combustion and NO absorption occur at the same pressure (mono-pressure configuration). Differently, if the absorption is carried out at higher pressure compared to the oxidation of $NH_3$ (dual-pressure configuration), lower pressures ranging from 1 to 6 bar can be employed [5].

Under appropriate conditions of operating variables and catalyst, a selectivity to NO of about 93–98% can be achieved. However, nitrogen and nitrous oxide ($N_2O$) can also be produced as by-products (Eqs. 2.2 and 2.3) [5]:

$$4NH_3 + 3O_2 \longrightarrow 2N_2 + 6H_2O \qquad \Delta H = -1{,}268\ \text{kJ mol}^{-1} \tag{2.2}$$

$$4NH_3 + 4O_2 \longrightarrow 2N_2O + 6H_2O \quad \Delta H = -1{,}105\ \text{kJ mol}^{-1} \tag{2.3}$$

Depending on the reaction temperature, ammonia oxidation may proceed through multiple pathways, producing nitrogen-containing species in ascending degrees of oxidation: nitrogen, $N_2O$, and NO. Of these, nitrogen is thermodynamically the most favorable product. Indeed, equilibrium calculations suggested that the non-catalytic combustion of ammonia would yield only nitrogen in the 200–1,000 °C temperature range. However, in the presence of a catalyst, the spectra of possible products change [4]. At lower temperatures (150–200 °C), nitrogen remains the sole nitrogen-containing product. As temperature rises, $N_2O$ formation begins and reaches its peak near 400 °C. This scenario is undesirable not only because of lower $NH_3$ conversion to NO but also because of the production of $N_2O$, a highly polluting agent. At these low temperatures, the reaction is kinetically controlled. The formation of NO – highly valued in nitric acid production – starts at approximately 300 °C and increases steadily as the temperature is increased. Under these conditions, the reaction rate starts to be controlled by mass transfer phenomena related to the diffusion of molecules of $NH_3$ onto the catalyst surface [5]. Despite the existence of various pressure configurations (vide infra), lower oxidation temperatures at equal pressure tend to favor the formation of undesirable nitrogen and $N_2O$. The selectivity in ammonia oxidation is largely governed by the competition between $NH_3$ and $O_2$ for active sites on the catalyst, as well as by the concentrations of surface intermediates [4]. At temperatures around 200 °C, the catalyst surface is predominantly N-covered, favoring the formation of nitrogen. As the temperature increases, the dissociative adsorption of oxygen raises the relative oxygen coverage, yielding progressively more oxidized products. Under typical industrial conditions, the catalyst surface becomes mainly oxygen-covered, promoting the production of NO. This is also ensured by the use of an appropriate composition of the $NH_3$-air mixture as will be discussed later.

#### 2.2.1.1 Reaction Mechanism and Kinetics

As mentioned above, the optimal operating conditions for $NH_3$ to NO transformation involve very high temperatures and low pressures and arise in a mass transfer-controlled reaction characterized by a considerable rate ($10^{-11}$ s). The substantial rate of the reaction under real operating conditions implies that the possible reaction intermediates most likely cannot be detected because they are very short-lived and present in very low concentrations. Furthermore, because of the issues with performing mechanistic studies under pressure, most investigations [6–8] have been conducted in vacuum or under reduced pressure, thus leading to high uncertainty when extrapolating results to obtain conclusions regarding the real process.

Because of these challenges, performing a proper study of $NH_3$ oxidation is not trivial, and a foolproof mechanistic interpretation is yet to be derived, to the best of our knowledge. Nevertheless, historically, three main mechanistic proposals based on a Langmuir-Hinshelwood model have been put out. These differ by the type of intermediate species, which, in multiple stages, lead to the final products ($N_2$, $N_2O$, NO, and $H_2O$). A nitroxyl (HNO)-, hydroxylamine ($NH_2OH$)-, and imine (NH)-based interpretation has been formulated [4]. In general, and since a very detailed mechanistic analysis is beyond the scope of this chapter, one might consider the following steps:

- Ammonia adsorption and dehydrogenation: $NH_3$ first adsorbs on the Pt surface and is successively dehydrogenated through intermediates such as $NH_2^*$, $NH^*$, and atomic $N^*$ (asterisks denoting adsorbed species).
- Oxygen adsorption and activation: simultaneously, oxygen molecules adsorb dissociatively, forming $O^*$ ad-atoms on the Pt surface.
- Formation of NO: the adsorbed nitrogen intermediates then react with $O^*$ to produce NO, which subsequently desorbs.

#### 2.2.1.2 Reaction Engineering and Parameter Effect

Together with the high selectivity, the catalytic oxidation of $NH_3$ under the conditions of the Ostwald process can also achieve very high theoretical conversion of about 98%. However, the effect of many parameters must be considered. The most important are (i) air concentration and consequent composition of the $NH_3$-air mixture, (ii) working pressure, (iii) working temperature, (iv) space velocity and number of Pt-Rh gauzes, and (v) the catalyst and its deactivation [5].

Considering Eq. (2.1), the $NH_3$-air mixture entering the burner should contain about 14.4% of $NH_3$. Nevertheless, since at higher $NH_3$ concentrations the conversion to NO drops, the usually employed ratio is lower and ranges from 10 to 13.5%, depending on the operating pressure. Conversion to NO might decrease because of local accumulation of $NH_3$ on the active site leading to an increase in production of the undesirable by-product $N_2O$, especially. Another reason why the $NH_3$-air mixture composition must be controlled is the possible formation of an explosive mixture if the $NH_3$ concentration exceeds the explosion limit. However, since at higher pressure the explosion limit decreases, $NH_3$-air mixture at higher $NH_3$ concentrations can be employed in high-pressure burners. Because of these two phenomena, it is also essential to ensure a proper mixing between $NH_3$ and air as well as to evenly distribute the mixture across the catalyst gauzes.

Despite the explosion limit declining with rising pressure, the $NH_3$ conversion decreases in the opposite order following Le Chatelier's principle because of the increase in volume dictated by Eq. (2.1). Thus, one might think of operating the burner at low pressures. However, together with the high NO yield (allowed by thermodynamics), the $HNO_3$ production would decrease dramatically, thus affecting the economics of the chemical plant. An accepted compromise is to work at an intermediate pressure of about 4–6 bar.

Together with the working pressure, and as mentioned above, a sufficiently high temperature of the burner is also critical in order to ensure effective $NH_3$ conversion to NO. However, despite an increase in temperature being beneficial, above ca. 900 °C, the $NH_3$ conversion to NO starts decreasing for a variety of reasons [5].

$$2NO \longrightarrow N_2 + O_2 \quad \Delta H = -90\ \text{kJ mol}^{-1} \tag{2.4}$$

$$2NH_3 \longrightarrow N_2 + 3H_2 \quad \Delta H = +46\ \text{kJ mol}^{-1} \tag{2.5}$$

The *in-situ* formed NO (metastable under operating conditions) might decompose to nitrogen and oxygen (Eq. 2.4). Simultaneously, decomposition of $NH_3$ can also take place (Eq. 2.5). Finally, Pt (and Rh) will tend to vaporize from the catalyst gauzes above ca. 800 °C, decreasing catalytic activity and reducing the economic sustainability of the chemical plant. Additionally, when the noble metals are lost, the gauze's mechanical strength deteriorates, shortening its service life. Meanwhile, the rhodium content rises on the catalyst surface. This phenomenon might result in lower activity and elevated formation of $N_2O$ and $N_2$, ultimately reducing the overall yield of NO if the surface concentration of Rh increases excessively [9].

To conclude, the conversion of $NH_3$ to NO also decreases depending on space velocity and number of catalyst gauzes [5]. If the gas velocities are too low (or an excessive number of gauzes is employed), NO decomposition into nitrogen and oxygen is favored (Eq. 2.4). On the opposite, if the gas velocities are too high (or an insufficient number of gauzes is employed), $NH_3$ will tend to slip and react with NO, decreasing its production.

#### 2.2.1.3 Catalyst: Pt and New Development

The state-of-the-art catalyst for $NH_3$ combustion in the Ostwald process comprises Pt-Rh gauzes (generally 95–90% Pt and 5–10% Rh). Rhodium is alloyed together with Pt in order to enhance the strength, thus reducing catalyst loss. The 5% to 10% of Rh is driven both by its higher cost compared to Pt but also by observations that the burner efficiency decreases at Rh contents higher than ca. 20%. Despite these notions being generally true, Rh should be avoided in burners operating at low temperatures ($T < 800$ °C) because its 3+ oxidized state would form and accumulate over the catalyst surface, decreasing the overall activity [1]. Ternary catalysts possessing about 5% of Pd (together with Pt and Rh) have been proven by lab-scale investigations to maintain the required yield of NO while decreasing capital expenditures given the lower cost of Pd [10].

Generally, about 3–50 closely stacked gauzes can be arranged in the burner. Suppliers provide wires possessing various diameters, as well as organized in different manners. Historically, the manufacture of the catalyst was carried out in the form of woven gauzes. However, in an attempt to minimize heat gradients, thus enhancing catalyst lifetime by decreasing the extent of metal loss, and in order to maximize selectivity to NO, novel, three-dimensional geometries possessing high porosity (the so-called knitted gauzes) were developed and very often implemented. Alternative three-dimensional designs have also been introduced, and these offer greater effectiveness compared to traditional knitted gauzes. In these configurations, two dense mesh layers, situated at the top and bottom, are joined by pile threads. The low porosity of the mesh layers promotes higher reaction rates compared to conventional knitted gauzes, while enhanced heat conduction through the pile threads mitigates large temperature gradients between layers, thus curbing precious metal losses and limiting catalyst deactivation. Furthermore, these weft threads placed in the central region create a third dense, highly active site for the reaction. As a result, these structures combine the high activity characteristic of woven gauzes with the reduced precious metal losses typical of knitted gauzes [11].

Irrespective of the design, and upon manufacturing, the surface of the gauzes is often quite smooth. This morphology exerts minimal resistance to gas-phase mass transfer limitations and yields lower initial levels of NO. After a brief period in service (ca. a few days), however, the catalyst surface area increases substantially due to microstructural changes and interactions with volatile components. As this surface growth intensifies, mass-transfer limitations decrease, and the production of NO consequently rises. In order to limit this period of activation, the growth of the crystal is initialized by the catalysts' suppliers by pretreating the gauzes in hydrogen flame [12].

While offering high activity and selectivity, Pt-Rh gauze catalysts are subjected to deactivation after prolonged periods of operation [9]. Deactivation is primarily a consequence of two concurring phenomena, i.e. platinum loss and catalyst poisoning. Platinum loss consists, mainly, in the vaporization of metallic platinum passing through an intermediate $PtO_2$ gas-phase species and onsets at about 750–800 °C. This phenomenon leads to lower NO yield and to a higher production of undesired by-products (e.g. $N_2O$). Simultaneously, losses due to mechanical forces (e.g. by abrasion) can occur if the $NH_3$-air mixture contains impurities, especially. The poisoning of the catalyst is, in turn, triggered by the formation of rust and/or by the presence of dust in the process air. In general, whatever material that can deposit over the catalyst surface acts as poison. This is minimized by the installation of a multistage filter prior to the burner at the air intake, which should remove 99.9% of all particles larger than 50 μm [5].

Noble metal losses greatly impact the operating costs in $HNO_3$ production plants. Accordingly, different techniques aimed at recovery Pt have been developed, and various types of systems are installed in most cases. In general, two families of systems exist: mechanical filters and recovery gauzes. In the former case, filters made up of ceramic fibers, mineral wool, or glass are installed in a zone where the gas temperature is low (about 400 °C). These systems possess low capital cost but high operational cost since their installation drives the occurrence of considerable pressure drops across the filter. For this reason, they are generally installed only in mono-pressure plants operating at high pressure. Different from such a filter, the recovery gauze approach consists of installing a material (Au-Pd-Ni alloys) stable at high temperature and which can react with $PtO_2$ vapors, forming an alloy. By diffusing into the Pd (i.e. the core component), the $PtO_2$ vapors react with it and are captured [1].

Non-noble metal-based catalysts in the oxide form can be used as well, offering a lower-cost alternative. Examples, either reported in literature and/or patented, are single metal oxides or mixed metal oxides of the perovskites and spinel types, based on the presence of Co, Bi, Mn, Cr, Ni, and Fe as active components. Among these, $Co_3O_4$-based systems have been investigated more extensively [13, 14]. However, the main limitation toward their practical implementation remains the lower conversion rates and, generally, the tendency to deactivate more rapidly and/or irreversibly. While Pt-Rh gauzes can achieve $NH_3$-to-NO conversion efficiency of about 95–98%, $Co_3O_4$-based systems perform in the 88–92% conversion range. Simultaneously, the sintering of the $Co_3O_4$ phase, which decreases catalyst porosity, as well as the permanent deactivation of the active $Co_3O_4$ component because of its reduction into the CoO phase by $NH_3$, limit the implementation of these catalytic systems [15].

### 2.2.2 NO Oxidation to $NO_2$

After the $NH_3$ burner, the process stream contains about 10% NO, 6% $O_2$ and 15% $H_2O$ and is further treated to oxidize NO to $NO_2$, which is required for further $HNO_3$ synthesis by absorption in $H_2O$. At the current stage, the industrial oxidation of NO proceeds as a homogeneous reaction (Eq. 2.6), i.e. without the involvement of a catalyst, and is carried out at ca. 40–60 °C and 4–10 bar with the aid of process air [5].

$$2NO + O_2 \rightleftharpoons 2NO_2 \quad \Delta H = -114\,\text{kJ mol}^{-1} \tag{2.6}$$

In practice, $NO_2$ is not the sole product since, once formed, it is in equilibrium with gas-phase $N_2O_4$ (Eq. 2.7). The equilibrium is shifted to the right at low temperature and high pressure.

$$2NO_2 \rightleftharpoons N_2O_4 \quad \Delta H = -58\ \text{kJ mol}^{-1} \tag{2.7}$$

Furthermore, also $N_2O_3$ can be obtained by the reaction between $NO_2$ and NO according to Eq. (2.8):

$$NO_2 + NO \rightleftharpoons N_2O_3 \quad \Delta H = -40\ \text{kJ mol}^{-1} \tag{2.8}$$

#### 2.2.2.1 Reaction Mechanism and Kinetics

NO oxidation exhibits a considerable temperature- and pressure-dependence. Due to the net reduction in the number of moles dictated by Eq. (2.6), the rate of NO conversion to $NO_2$ rises with increasing pressure, exhibiting a third-order rate equation function of the partial pressures of nitric oxide and oxygen [16]. Simultaneously, $NO_2$ formation is favored at low reaction temperatures. Practically, this is achieved through an interconnected system of heat exchangers, which also recovers a portion of the energy generated and provides sufficient residence time for the reaction occurrence. The dependence of $NO_2$ formation on the reaction temperature is a consequence of both thermodynamic and kinetic considerations. In the former case, the exothermicity of the oxidation process dictates that the equilibrium is shifted toward the products side of Eq. (2.6) (i.e. $NO_2$) at lower temperatures. In the latter case, it must be considered that the reaction exhibits an atypical kinetic behavior, in which its rate is higher at lower temperatures. This is believed to occur because of the two-step radical mechanism followed by the reaction. Since NO is a free radical with an unpaired electron, the first reaction step has been proposed to be an instantaneous dimerization (Eq. 2.6a), governed by an equilibrium constant ($K_{dim}$), which increases at higher temperatures following Arrhenius.

$$2NO \longrightarrow (NO)_2 \qquad [\text{fast}] \tag{2.6a}$$

$$(NO)_2 + O_2 \rightleftharpoons 2NO_2\ [\text{slow and rate} - \text{determining}] \tag{2.6b}$$

Upon dimerization, the dimer is transformed into $NO_2$ by consuming oxygen ($r_{ox\text{-}2}$) through Eq. (2.6b). This second, slower step governs the rate of NO oxidation. Accordingly, the rate of the overall process ($r_{tot}$) can be described as:

$$r_{tot} = \frac{r_{ox-2}}{K_{dim}} \tag{2.6c}$$

Thus, $K_{dim}$ increases at higher reaction temperatures, driving a lower, overall, reaction rate and thus explaining the inverse Arrhenius behavior exhibited by the oxidation of NO [17].

#### 2.2.2.2 Process Implementation Exploiting the Catalytic Oxidation of NO

As mentioned above, industrial NO oxidation is carried out without a catalyst in the gas phase and at medium pressures and low temperatures to enhance $NO_2$ formation. Despite these efforts, considerable optimization of the process is still possible. As an example, one might consider that at 300–400 °C and 4 bar, equilibrium calculation estimated an extent of NO oxidation of about 55–90%. However, under the same experimental conditions, the degree of oxidation achieved is generally only 20% [18]. This observation highlights a largely untapped potential to implement NO to $NO_2$ transformation, which, at least in principle, could be filled with the use of an appropriate catalyst. Furthermore, an increase in the oxidation temperature to ca. 300 °C would result in additional recovery of the heat of reaction at temperatures where high-pressure steam can be produced, thus decreasing the overall industrial footprint.

The catalytic oxidation of NO is far from an unexplored research territory, having been extensively examined using various catalyst formulations [19]. These include noble metals, metal oxides, perovskites, and C-based catalysts. With regard to noble metal-based systems, (i) Pt-, Ru-, and Pd-supported catalysts, (ii) their form modified with metal oxides (e.g. $MoO_3$, $Wo_3$, $V_2O_5$, $Ga_2O_3$, and $MnO_x$-$CeO_2$), and (iii) bimetallic systems, Pt-Pd, especially, have been extensively studied. Single metal oxides, based on the use of $MnO_x$ supported over $Al_2O_3$, $TiO_2$, and $ZrO_2$, and even multi-metal oxides in which the activity of MnOx species was modified by the introduction of Ce, Co, or La, also received considerable attention. Similarly, approaches based on the synthesis of perovskites took advantage of similar elements, thus resulting in tri-metallic structures based on Mn, Co, La, or Fe. Finally, carbon-based catalysts relying on activated carbons or activated carbon fibers were developed on the basis of their satisfactory adsorption capabilities, catalytic activity, and redox properties [20].

Irrespective of the formulation under consideration, appropriate design of the catalyst for NO oxidation remains challenging for two main reasons: (i) the gas-phase conversion of NO to $NO_2$ might limit the oxygen available for the catalytic reaction and (ii) strong oxidizers (NO, $O_2$, $NO_2$, $HNO_2$, and $HNO_3$) can either oxidize the catalyst or cause metal leaching. Another crucial consideration in developing catalysts for industrial oxidation of NO is the exact location of the reactor. If placed near the ammonia combustor exit (and not in the vicinity of the absorption tower), the lower $NO_2$ and higher $O_2$ concentrations together with a considerable temperature might enhance both catalytic activity and heat recovery. However, if the temperature is excessive, the thermodynamic might limit the kinetic profile of the reaction and thus prevent achieving sufficient conversion. The opposite is expected if the reactor were to be placed in positions closer to the absorption column, characterized by high $NO_2$ and lower $O_2$ concentrations and considerably lower temperatures. Consequently, the choice of the location and the related operating temperature for the catalyst bed must be a trade-off between thermodynamic constraints, kinetic considerations, and the extent of heat recovery targeted.

### 2.2.3 Nitrogen Oxides Absorption and $HNO_3$ Production

After cooling and compression, the process stream containing the mixture of nitrogen oxides (NO, $NO_2$, $N_2O_4$, and $N_2O_3$) is pumped at the bottom of the absorption tower, while deionized water (or dilute nitric acid, depending on the type of plant) is introduced from the top in countercurrent mode. The column can be as high as 80 m, with a diameter up to 6 m and possessing 50 trays. Diluted $HNO_3$, which is collected from the bottom of the tower at start-up, is pumped back within the tower in order to reach steady-state conditions in shorter times.

In short, $HNO_3$ synthesis can be described as the absorption of $NO_2$ in water following Eq. (2.) [5]:

$$3NO_2 + H_2O \rightleftharpoons 2HNO_3 + NO \quad \Delta H = -73\,\text{kJ mol}^{-1} \tag{2.9}$$

The maximum $HNO_3$ concentration at a certain composition of inlet gas stream can be determined by equilibrium considerations. The equilibrium constant ($K_p$) can be defined as:

$$K_p = \frac{p^2_{HNO3}p_{NO}}{p^3_{NO2}p_{H_2O}} = K_1K_2 \text{ with } K_1 = \frac{p_{NO}}{p^3_{NO2}} \text{ and } K_2 = \frac{p^2_{HNO3}}{p_{H_2O}}$$

In this rationale, the partial pressure of NO and $NO_2$ and the vapor pressure of $HNO_3$ and $H_2O$ are used. $K_2$ depends on the temperature and on the $HNO_3$ concentration. $K_1$ is in turn a function of the composition of the inlet gas entering the absorption tower. Following this, high partial pressures of $NO_2$ and low concentrations of NO are fundamental parameters that control the strength (i.e. concentration) of the $HNO_3$ retrieved at the bottom of the column [21].

According to Eq. (2.), NO is continuously reformed during absorption. In order to both promote re-oxidation of $NO_2$ from NO and to remove traces of $NO_2$ from the produced $HNO_3$, a stream of air is introduced in the tower. A plate-like column is generally used in which the absorption trays are of the sieve or bubble cap type. As mentioned above, both oxidation and absorption processes occur within the column, and, given the many components present in both gas and liquid phases, various reactions must be considered.

Oxidation processes take place in the space between the trays and involve three main reactions, namely, NO oxidation, $NO_2$ dimerization, and $N_2O_3$ formation, Eqs. 2.6–2.8. Secondary reaction between such nitrogen oxides and steam can also lead to $HNO_2$ and $HNO_3$ formation. However, these processes have very limited impact.

Simultaneously, absorption processes in the liquid phase occur over the trays. It is generally accepted that $HNO_3$ formation in the liquid phase can take place in three ways:

i) By reaction between $N_2O_4$ (or 2 $NO_2$) and water:

$$N_2O_4 + H_2O \longrightarrow HNO_3 + HNO_2 \quad \Delta H = -87\,kJ\,mol^{-1} \tag{2.9a}$$

ii) By reaction between $N_2O_3$ and water:

$$N_2O_3 + H_2O \longrightarrow 2HNO_2 \quad \Delta H = -50\,kJ\,mol^{-1} \tag{2.9b}$$

iii) By $HNO_2$ dissociation:

$$3HNO_2 \longrightarrow HNO_3 + H_2O + 2NO \quad \Delta H = -15\,kJ\,mol^{-1} \tag{2.9c}$$

Nitric oxide formed in this way is transported back into the bulk of the gas phase and, there, is further transformed into $NO_2$, $N_2O_4$, and $N_2O_3$ by the oxidation reactions mentioned previously (Figure 2.1). Within this rationale, the transport of $N_2O_4$ from the bulk of the gas phase to the gas–liquid interphase is considered the rate-limiting step [5].

**Figure 2.1** The new 1500 MTD capacity nitric acid production plant supplied by Casale SA in Navoiy (Uzbekistan).

Thanks to these reactions, a nitric acid solution of approximately 55–65% is typically taken from the base of the column, although concentrations can range from 30% to 70%. The final $HNO_3$ concentration is determined by factors including temperature, pressure, the number of absorption stages, and the nitrogen oxides concentration entering the absorber. Temperature affects both gas solubility and reaction kinetics. In the former case, lower temperatures favor the absorption of $NO_2$ into the liquid phase and drive the equilibrium toward $HNO_3$ formation. Following the same trend, lower temperatures also favor the kinetics of the oxidation reaction, as mentioned in the previous paragraph. For these reasons, the absorption tower is operated at reasonably low temperatures (30–70 °C), which are maintained by cooling coils fed with water and installed on the absorption trays. This cooling design allows to remove the heat generated by the oxidation of NO to $NO_2$ as well as by the successive nitric acid formation by absorption. Note, however, that the stability of $HNO_2$ increases at lower temperatures. Thus, excessive cooling in the bottom of the absorption tower might favor production of $HNO_2$ rather than $HNO_3$, which will have to be stripped, increasing the operation in the bleacher. Considering the working pressure of the absorber, increasing the total pressure raises the partial pressure of nitrogen oxides, improving their dissolution into the aqueous phase and thereby facilitating higher $HNO_3$ concentrations. This explains why medium- (ca. 5 bar) and high-pressure (about 10 bar) plants typically achieve higher $HNO_3$ concentrations than low-pressure systems. Finally, higher nitrogen oxide concentrations accelerate the rate of oxidation. This allows us to achieve higher $HNO_3$ production at the same design of absorption tower or to decrease the spacing between the plates while maintaining the same $HNO_3$ productivity [22].

At the very bottom of the tower, on the first tray, $HNO_3$ at the highest concentration is obtained. This product acid is pumped in the so-called bleacher, allowing stripping of the dissolved $NO_x$ with secondary air, and further stored in tanks. As mentioned above, $HNO_3$ production plants operating in this way can achieve a final $HNO_3$ concentration of about 55–65% (the so-called weak $HNO_3$). Higher concentrations are possible, but, as will be discussed later, they require further unit operations.

## 2.3 Type of Processes

### 2.3.1 Weak Nitric Acid

Early industrial facilities for producing weak nitric acid consisted of burning $NH_3$ at atmospheric pressure and performing $NO_x$ absorption at low pressure. This type of configuration is no longer constructed, and, in turn, it has been substituted by more efficient designs operating at higher pressures. At first, atmospheric combustion was combined with medium-pressure absorption (up to 5–6 bar). This approach allowed lowering the absorption expenses and offsetting the cost of the compressor since the improvement in absorption efficiency allows the equipment to be downsized. From this alternative, and in order to achieve a second decrease in plant volume, industrial settings operating also the $NH_3$ combustion unit at medium pressures were developed. Mono-medium-pressure plants consisting of a single $NH_3$ burner and $NO_x$ absorption tower can produce up to 700 tons of $HNO_3$/day and, generally, offer significant energy efficiency. Indeed, the moderate operating costs related, for instance, to air compression at medium pressures are more easily offset. However, with a rising need of $HNO_3$ and related products, as well as stricter regulations regarding $NO_x$ emissions, industrial alternatives proceeding with a high-pressure (up to 8–12 bar) absorption were also developed, in the United States and Russia especially. This is due to the generally lower cost of both energy and feedstock compared to the European Union. Differently from medium-pressure designs, mono high-pressure plants have been designed in order to optimize the last step of the Ostwald process, i.e. $NO_x$ absorption into water and consequent $HNO_3$ production. Indeed, mono high-pressure plants consisting of a single absorption tower can produce up to 1,000 tons of $HNO_3$/day. Since the higher operating pressure allows shrinking the size of the unit operations and of the footprint of the chemical plant in general, the higher operating costs (higher consumption of $NH_3$ at higher pressures) are partially counterbalanced by the lower initial capital investment. Irrespective of the medium- or high-pressure designs, if both combustion and absorption are performed at the same pressure (mono-pressure configuration), the so-called primary air (which is required for $NH_3$ oxidation and $NO_x$ absorption) is compressed at the desired pressure, mixed with superheated $NH_3$ and forced into the burner. A simplified scheme of a mono-pressure plant configuration owned by Casale SA (NA1000™ nitric acid technology) is reported in Figure 2.2.

Differently from these mono-pressure designs, dual-pressure alternatives have also been developed with the aim of attaining both the economic benefits of medium-pressure combustion and the high effectiveness of high-pressure absorption. This approach incorporates a compressor between the $NH_3$ oxidizer and the absorption tower. Dual-pressure processes entail an $NH_3$ oxidation unit running at about 3–6 bar and an $NO_x$ absorption unit operating at 10–13 bar. However, in older plants, burning of $NH_3$ might still be carried out at atmospheric pressure prior to a medium-pressure absorption tower. An example of a dual-pressure plant configuration (NA2000 nitric acid technology) is reported in Figure 2.3. Dual-pressure plants equipped with a single combustion unit can produce up to 1,500 tons of $HNO_3$/day making these last alternatives particularly attractive for large plants, especially when built in Europe. Compared to mono-pressure design, initial investments run higher in dual-pressure plants. However, the variable costs are

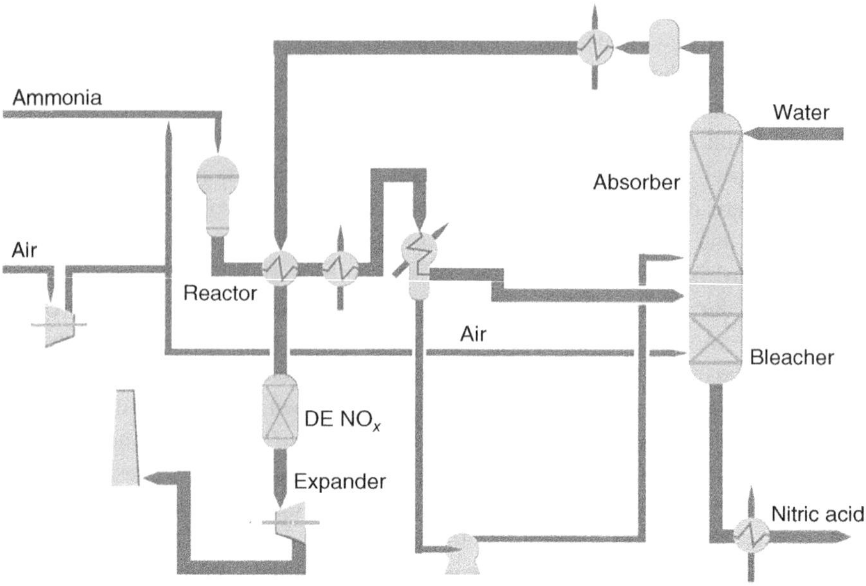


**Figure 2.2** Simplified diagram of the NA1000™ nitric acid technology, mono-pressure plant, owned by Casale SA.

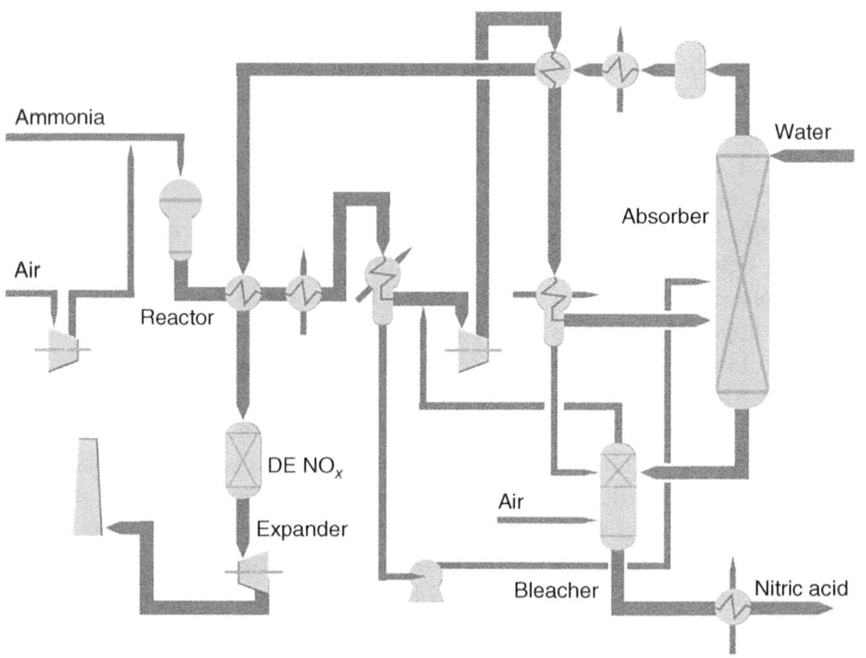


**Figure 2.3** Diagram of the NA2000 nitric acid technology, dual-pressure plant, owned by Casale SA.

considerably lower. Indeed, reduced operating pressures in the combustion unit permit improved $NH_3$ yield and catalyst performance as well as decreased Pt loss, leading to superior catalyst lifetime. These phenomena, together with a higher achievable production capacity, make the initial investment easily recoverable.

#### 2.3.1.1 General Process Description

Ahead of the $NH_3$ combustion unit, the primary air is filtered and supplied by a compressor that is driven by two turbines, the expander and a makeup driver. The expander recovers energy from the tail gases, which remain after the absorption column once they are decompressed to atmospheric pressure. Differently, the driver is usually a steam turbine (or an electric motor). Simultaneously, the inlet $NH_3$ is evaporated, superheated, and filtered. After mixing of the two process streams in the $NH_3$-air mixer and achieving the desired $NH_3$-air ratio, the mix is directed into the combustion unit and passed over the Pt-Rh gauze catalyst, producing NO, steam, and considerable heat. This stream of hot gases is used to heat up tail gases as well as to produce steam that can be employed to drive the steam turbine. Such cooling stages, together with enough residence time, allow oxidation of NO to $NO_2$ (and $N_2O_4$). Further cooling in one or more of the so-called cooler-condenser units occurs. A condensate of $HNO_3$ is produced and pumped in specific locations of the absorption column. In mono-pressure plants, only one cooler-condenser is present. In dual-pressure designs, the output gases coming from the first condensation are further pressurized, cooled, and directed to an additional cooler-condenser. From this unit, additional $HNO_3$ is obtained and mixed with the output bottom $HNO_3$ stream generated in the absorption tower [1].

Irrespective of the type of plant, the mixture of nitrogen oxides enters the column from the bottom. By absorption with water fed into the tower in countercurrent mode with respect to the stream of gases, liquid $HNO_3$ is obtained. Such a product is treated in a bleacher with secondary air to strip the dissolved nitrogen oxides. This nitrogen oxides-rich air stream is then fed to the column. As previously mentioned, this is driven by the need to oxidize the remaining NO as well as the quantities of NO produced within the tower. Purified $HNO_3$ is stored or pumped to downstream operations. On the contrary, the tail gases generally contain excessive quantities of $NO_x$ and $N_2O$ to be discharged directly into the environment. Thus, the stream is treated to remove such impurities. The final emission into the atmosphere occurs using an expander that allows the recovery of mechanical energy.

### 2.3.2 Concentrated Nitric Acid

While weak $HNO_3$ is perfectly suitable to be further employed for fertilizer production, nitration processes in industrial organic chemistry require $HNO_3$ at higher purity. Accordingly, the $HNO_3/H_2O$ mixture requires further treatment. Standard distillation fails in the objective since $HNO_3$ and $H_2O$ form an azeotrope (68% $HNO_3$ at $P = 1$ bar). Thus, appropriate strategies have been developed. These are divided into direct or indirect processes. In the former case, concentrated $N_2O_4$ in liquid form is reacted at high pressure (ca. 50 bar) with oxygen and water (or weak $HNO_3$) to yield concentrated $HNO_3$ according to Eq. (2.10):

$$2N_2O_4 + O_2 + H_2O \longrightarrow 4HNO_3 \tag{2.10}$$

In the latter case, industry takes advantage of extractive distillation. Starting from weak nitric acid utilizing the processes described earlier, a third component (generally sulfuric acid or magnesium nitrate) is then added as a dehydrating agent to increase the volatility of $HNO_3$ allowing its separation with a purity up to 99% (from the top of the column as vapors).

The direct processes for high-purity $HNO_3$ production are rather complex and consist of adding processing units to the plant for weak $HNO_3$ production. Upon $NH_3$ combustion, following the specifications mentioned in previous paragraphs, the output stream is treated in a highly efficient cooler, which is designed to remove as much water as possible, minimizing $NO_x$ absorption [1]. This liquid product, called condensate ($HNO_3$ concentration of about 2–3% under atmospheric combustion conditions), will be required later in the final absorption stage. Differently, the stream of gas containing $NO_x$ undergoes two oxidation steps in order to maximize $NO_2$ concentration. First, and after further cooling, the $NO_x$ stream is mixed with secondary air and treated in an oxidation tower in which $HNO_3$ is continuously recirculated from the bottom to the top of the column. This step allows us to achieve about 95% oxidation efficiency. In the second step, such a stream of concentrated $NO_x$ moves countercurrently to concentrated $HNO_3$ in the so-called postoxidizer, which allows us to obtain a fully oxidized stream of $NO_2$ saturated with $HNO_3$ vapors, as well as $HNO_3$ at a concentration of about 70% from the bottom of the postoxidizer. After these two oxidation steps, the process stream is cooled to −10 °C. This allows complete dimerization of $NO_2$ into $N_2O_4$. The process stream is further treated by washing in absorption towers with concentrated $HNO_3$ (first) and dilute $HNO_3$ (later). A last stripping step allows us to obtain the stream of concentrated $N_2O_4$, which is finally liquefied and fed to a stirred tank. At a pressure of about 50 bar and temperatures around 60–80 °C concentrated $N_2O_4$ reacts with oxygen and water (or weak $HNO_3$) to produce high-purity $HNO_3$. Herein, the most important parameter is to have an equimolar ratio between $N_2O_4$ and water. This ratio does not only affect process performance but also allows monitoring of the operations of the reactor. For instance, $HNO_3$ at very high purity ($\% \geq 98\%$) requires higher ratios [5].

These types of plants are rather uncommon nowadays, and a more preferred approach is based on combining weak $HNO_3$ production with extractive distillation. Following weak $HNO_3$ production (*vide supra*), a solution of $H_2SO_4$

(concentration up to 85%) or $Mg(NO_3)_2$ (60–80%) is employed in order to interact and extract water from weak $HNO_3$ and thus allows production of concentrated $HNO_3$ vapors (% up to 99%) from the head of the distillation tower.

In the case of $H_2SO_4$, preheated weak $HNO_3$ (concentrated up to ca. 68%) is fed into the distillation column while the $H_2SO_4$ solution is introduced from the head. The section of the column above the inlet of $HNO_3$ (irrigated with $H_2SO_4$) is the rectifying section, while the bottom one is the stripping part. More diluted $H_2SO_4$ is retrieved from the bottom, while concentrated $HNO_3$ vapors are obtained in the top. The former component undergoes re-concentration in a separate unit, and it is further pumped back into the head of the distillation column. Concentrated vapors of $HNO_3$ are condensed, de-aerated in a bleaching unit, and stored for further utilization.

In a similar manner, a solution of magnesium nitrate can also be used to extract water from diluted $HNO_3$ by extractive distillation. The bottom solution comprising diluted $Mg(NO_3)_2$ (40–65%) is easily re-concentrated in a single stage and is recycled back into the column. Traces of diluted $HNO_3$ are generally obtained by such an evaporator and are further recovered to be recycled into the distillation column. Concentrated vapors of $HNO_3$ are retrieved from the head of the tower and cooled in multiple stages to force condensation. A minor fraction is recycled back into the head of the tower, while the rest is treated as in the process utilizing $H_2SO_4$ [23].

Among these two routes, the magnesium nitrate process is generally preferred since (i) it ensures an easier re-concentration of the solution retrieved at the bottom of the column, (ii) it allows the use of cheaper material, thus saving capital investments, and (iii) magnesium nitrate is easier to handle compared to sulfuric acid.

## 2.4 Environmental Protection and Treatment of Exhaust Gases

The main environmental concerns related to $HNO_3$ production regard the emission of excessive quantities of oxides of nitrogen (NO, $NO_2$, and $N_2O$). Being promoters of several noxious processes, $NO_x$ are considered to be mainly responsible for air pollution. They can cause (i) photochemical smog by reacting with volatile organic compounds in the presence of sunlight, (ii) acid rains and eutrophication upon interaction with water in the atmosphere or in aquatic ecosystems, and (iii) a multitude of health and respiratory issues by diffusing in the alveolar cells and disrupting lung functions. Simultaneously, although not involved in the legal definition of $NO_x$, $N_2O$ is also a noxious pollutant [24]. According to its tremendous global warming potential (GWP = 270) and long atmospheric lifetime (ca. 120 years), $N_2O$ has progressively become the third major greenhouse gas responsible for ozone layer depletion and climate change [25]. Since different strategies can be implemented in order to meet the limits imposed by environmental regulations, the control of $NO_x$ (NO and $NO_2$) and $N_2O$ emissions will be treated separately.

### 2.4.1 Control of $NO_x$ Emissions

One of the most useful strategies to minimize the concentration of $NO_x$ in the tail gases to be discharged in the atmosphere is to optimize the performance of the absorption tower. For instance, high-pressure operations and an increased number of trays allow superior $NO_x$ absorption, thus decreasing the concentration of $NO_x$ in the headstream leaving the column, as discussed previously. By full optimization of the absorber, an $NO_x$ concentration as low as 150 ppmv can be achieved [1]. It is not surprising that industry took advantage of this approach to meet the legal requirements imposed by environmental legislation until the beginning of the present century. However, with the introduction of more stringent emission limits (e.g. 5–90 ppmv as of 2011 in the EU and 30–40 ppmv as of 2012 in the United States), the proper optimization of the absorber and of its operations was no longer a suitable strategy, alone. This statement is particularly valid since these limits comprise all operations, including start-up, shutdown, and malfunction.

Despite it depending on the type of plant considered, modern installations are characterized by an $NO_x$ concentration from the absorber of about 200–400 ppmv. Thus, the legal limits are met by implementing specific technologies that have been developed in the last 30 years. In this chapter, we will briefly cover three strategies: (i) the use of $H_2O_2$ in the absorber, (ii) the nonselective catalytic reduction (NSCR) approach, and (iii) the selective catalytic reduction (SCR) strategy.

$H_2O_2$ can be added at the top of the absorption tower in order to increase $NO_x$ absorption efficiency and thus lower the $NO_x$ concentration of the tail gases. Despite this strategy being mainly implemented in order to limit $NO_x$ emission at start-up of the plant, it is also used during standard operations if the temperature of the tail-gases is so low (e.g. $T < 80\,°C$) that undesired formation of ammonium nitrate might occur under SCR conditions. The $H_2O_2$ route is based on the following fast reactions:

$$NO + NO_2 + 2H_2O_2 \longrightarrow 2HNO_3 + H_2O \tag{2.11}$$

$$2NO_2 + 2H_2O_2 \longrightarrow 2HNO_3 \tag{2.12}$$

The main disadvantage of this approach is the high cost related to the use of $H_2O_2$.

Nonselective catalytic reduction (NSCR) involves (i) the reduction of $NO_x$ to a mixture of $H_2O$ and $N_2$ and (ii) the combustion of all the remaining $O_2$ in the tail gas, using an appropriate reducing agent (generally $H_2$, $CH_4$, or natural gas) over a suitable catalyst (supported noble metals, $V_2O_5$, or $Fe_2O_3$). The reaction is carried out at about 200–300 °C in the presence of $H_2$ and 450–575 °C when hydrocarbons are employed. Because of the high exothermicity of the combustion reactions, the temperature increases substantially and can reach even 800 °C. In these conditions, intermediate cooling between two stages is likely required because such a high temperature surpasses the conditions in which the final expansion turbine can operate. The advantage of this process is that it allows converting not only $NO_x$ but also $N_2O$ [4]. Despite this, the SCR strategy has largely substituted the NSCR approach because of numerous drawbacks. First, the occurrence of combustion reactions implies that not only $CO_2$ but also considerable quantities of CO (% < 1,000 ppmv) and highly toxic HCN are formed. Furthermore, a large excess of reducing agent is required since the reaction between $H_2$ or hydrocarbons with $O_2$ is preferential. Finally, the need for expensive $H_2$ and supported noble metal catalysts posed economic limitations [26].

Due to these serious limitations, modern installations operate following the SCR strategy using $NH_3$ as a reducing agent for the conversion of $NO_x$ ($NH_3$-SCR) and suitable catalysts [1]. Compared to NSCR, $NH_3$-SCR offers selectivity, which means that $NH_3$ preferentially reacts with $NO_x$ (i.e. lower consumption of reducing agent) and also produces only environmentally benign $N_2$ and water products (i.e. a C-free output stream). $NH_3$ is generally dosed in excess ($NH_3/NO_x = 1$–$1.5$) since under nonoptimal operations ($T > 400$ °C) it can undergo a secondary oxidation reaction with $O_2$ producing $N_2$ and $N_2O$ (in minor quantities) [27]. SCR also runs at a lower reaction temperature compared to NSCR, i.e. about 300–400 °C, depending on the catalyst employed. This allows us to use cheaper construction materials. The main drawback of this approach is that SCR requires the use of $NH_3$. Thus, its dosing and consumption must be optimized.

Depending on the composition of the tail gases, $NH_3$-SCR develops according to different schemes: the Standard SCR, Fast SCR, and $NO_2$-SCR reactions [28]. If NO accounts for the majority of the total $NO_x$ in the exhaust, the chemical equation corresponding to its reduction is known as the Standard SCR reaction (Eq. 2.1):

$$4NH_3 + 4NO + O_2 \longrightarrow 4N_2 + 6H_2O \tag{2.13}$$

The reaction also requires oxygen, which is available in excess in the exhaust gases. The presence of $NO_2$ can accelerate the $NO_x$ conversion through a different reaction pathway, known as the Fast SCR reaction (Eq. 2.1):

$$2NO + 2NO_2 + 4NH_3 \longrightarrow 4N_2 + 6H_2O \tag{2.14}$$

However, excessive $NO_2$ in the tail gases is not beneficial. When $NO_2$ is present above the equimolar amount of NO, the SCR process is hindered by the development of the $NO_2$-SCR reaction (Eq. 2.1):

$$6NO_2 + 8NH_3 \longrightarrow 7N_2 + 12H_2O \tag{2.15}$$

Among the materials active for $NO_x$ conversion through $NH_3$-SCR, mostly vanadia-supported catalysts ($V_2O_5/TiO_2$ promoted with $WO_3$) and Cu- and Fe-exchanged zeolites have successfully met the requirements to find industrial applicability [28]. Despite offering considerable activity and selectivity, at low temperatures especially, Cu-exchanged zeolites are very rarely used in the exhaust after-treatment of $HNO_3$ production plants because of their high costs. On the contrary, both V- and Fe-based materials are commercially employed. Among these two classes, although they are more expensive, Fe-exchanged zeolites have established themselves for several reasons. First, they exhibit superior SCR activity and selectivity, at $T > 350\,°C$, especially. Furthermore, Fe-exchanged zeolites are moderately active in $NH_3$ oxidation and very selective toward $N_2$. This allows us to control possible undesired emissions of unreacted $NH_3$ according to the $NH_3$-SCR chemistry. Finally, this class of material catalyzes $N_2O$ abatement through decomposition and also reduction reactions [29].

Instead, V-based catalysts generally lack sufficient $N_2O$ conversion activity and possess a pronounced tendency toward detrimental $N_2O$ formation via non-selective $NH_3$ oxidation above 400 °C. Finally, $V_2O_5/TiO_2$ generally has poor thermal stability and can release carcinogenic vanadium oxide compounds [30]. Because of the features mentioned above, Fe-exchanged zeolites also bear promises for a successful concomitant conversion of $N_2O$ and NO under $NH_3$-SCR conditions. This innovative abatement strategy has been recently investigated by some of us [31]. Irrespective of the type of catalyst, the material is shaped in the form of pellets or washcoated on monoliths, forming honeycomb-like structures [32]. The former case is mostly implemented when the reactor operates under pressure. Differently, the honeycomb morphology is preferential when only limited pressure drops are accepted, such as after the tail gas expander.

The main safety concerns to be considered when the SCR strategy is implemented are (i) the release of unreacted $NH_3$ in the output stream to be discharged (so-called $NH_3$-slip) and (ii) the formation of solid ammonium nitrate ($NH_4NO_3$) residues [27]. $NH_3$-slip does not have to be controlled only because it implies higher costs related to an inefficient usage of $NH_3$ in the reaction chamber, but also to meet legal limits especially. In order to minimize $NH_3$ slip, an optimized contact between the $NO_x$-$NH_3$ mixture and the catalyst is required. This is ensured by appropriately mixing $NO_x$ and $NH_3$ prior to their introduction in the abatement unit as well as by proper design of catalytic reactors ensuring correct flow dynamic conditions. Simultaneously with the possible release of toxic gaseous $NH_3$ in the atmosphere, $NH_3$-slip is also dangerous because, together with unreacted $NO_x$, it can lead to the formation of solid residues of $NH_4NO_3$. $NH_4NO_3$ formation occurs at relatively low temperature (i.e. $T < 100\,°C$), for instance, in cold sections downstream of the abatement unit, and poses danger because it can explode upon heating.

### 2.4.2 Control of $N_2O$ Emissions

Together with $NO_x$, $N_2O$ emissions also must be controlled and minimized. Despite this statement appearing obvious, there was no obligation to limit $N_2O$ emissions before the ratification of the Kyoto Protocol in 2005, especially, and, indeed, $HNO_3$ production plants represented the largest source of industrial $N_2O$ emission worldwide [33]. Following the implementation of more stringent environmental legislation, various strategies for the mitigation of $N_2O$ emissions have been developed. Three groups of strategy can be defined depending on the position in which the $N_2O$ controlling unit is installed: primary, secondary, and ternary [4, 5, 33]. Primary abatement involves minimizing $N_2O$ formation from its source, i.e. $NH_3$ combustion. This approach involves the optimization of the catalyst, of the conditions, and of the design of the burner. Secondary abatement consists in removing $N_2O$ from the $NO_x$ stream between the combustion unit and the absorber. Because high temperatures are required for $N_2O$ conversion, the unit is generally installed immediately after the burner. Finally, the ternary method comprises catalytic abatement of $N_2O$ by treatment of the tail gases leaving the absorption column. In this case, the reactor is often installed ahead of the expander turbine.

#### 2.4.2.1 Primary Measures

The improvement of $NH_3$ combustion to NO is a good preliminary strategy to minimize $N_2O$ production and its further emission. Both process conditions and catalysts can be optimized to reduce the yield of $N_2O$. From the process side, (i) the $NH_3$-air ratio and their mixing/distribution, (ii) the working temperature, and (iii) the working pressure can be tuned [4]. An $NH_3$-air ratio of about 1:9 together with efficient mixing of the two components and their homogeneous distribution across the catalyst is fundamental to achieving superior catalytic behavior. Considerably high temperatures ($T > 900\,°C$) and a reasonably fresh catalyst also lead to decreased production of $N_2O$. However, such high temperatures will promote catalyst deactivation by Pt losses, thus increasing $N_2O$ selectivity and productivity. Low $N_2O$ yield and high NO production can be achieved by working at lower pressure. However, this works at the expense of low $HNO_3$ productivity. In addition to these parameters, as has been described earlier in Section 2.2.1.2, the geometry of the gauzes can also be optimized to maximize $NH_3$-to-NO transformation, thus reducing $N_2O$ yield. In the view of primary measures, the development of novel catalysts (oxide-based systems) is probably the most appealing strategy, which, however, nowadays did not reach the level of maturity required for industrial implementation. Indeed, while ensuring lower $N_2O$ productivity compared to Pt-Rh gauze catalysts, metal oxides (e.g. promoted or unpromoted $Co_3O_4$) suffer decreased $NH_3$-to-NO activity, permanent deactivation by reduction in $NH_3$ and a higher tendency to be poisoned. Furthermore, oxide catalysts generally require lower reaction temperatures, thus complicating steam balance and the possibility of plant revamping [15].

#### 2.4.2.2 Secondary Measures

The second strategy to reduce $N_2O$ emissions consists in the thermal or catalytic decomposition of this pollutant inside or immediately after the burner in an appropriate unit [4]. In the case of thermal decomposition, the measure can be implemented either (i) by increasing the residence time of the process gas in the burner (ca. 1–3 seconds higher) or (ii) by installing an empty chamber between the burner and the first heat exchanger (about 3.5 m in length). In both cases, retrofitting is highly complicated since the investments would not be outweighed by lower operational costs. Because such homogeneous decomposition is a rather slow reaction, a more appealing approach is to promote conversion using a catalyst, which can be located downstream or inside the burner. In the former case, and similarly to the ternary abatement strategies, a unit comprising Fe-exchanged zeolites as a catalyst can be installed between the burner and the absorption tower [34]. While such a catalytic system works effectively, the main problems are represented by the intricate system of heat exchangers that follow the combustion unit (limiting revamping) and the chances of catalyst aging (leading to poorer performance) driven by the high $H_2O$ concentration present in the process stream (ca. 13%). Compared to the above-mentioned routes, the secondary measure that is most commonly implemented consists of installing a catalyst devoted to $N_2O$ decomposition beneath the Pt-Rh gauzes and the Pt recovery system. This strategy can allow for achieving $N_2O$ concentrations as low as 130–300 ppmv and offers the advantage of being, in principle, reasonably easy to implement in already existing plants. However, additional pressure drops are introduced, which might decrease the NO yield. Different examples of formulations have been developed by various companies, generally in pellet form. Catalysts based on spinels of Cu-Zn-Al, CuO/$Al_2O_3$, Fe/ZrO, and $Co_2AlO_4/CeO_2$, among others, are at different phases of industrialization [1].

#### 2.4.2.3 Ternary Measures

The last family of strategies to control $N_2O$ emissions consists of the catalytic decomposition or reduction of $N_2O$ in a downstream unit located before the expander at a sufficiently high temperature. Ternary measures were particularly useful in the view of revamping since they do not affect the core of the $HNO_3$ production plant. Also, the tail gases leaving the absorber undergo multiple steps of heat exchange on their way to the exhaust after-treatment unit, which allow, in most cases, the temperature required for the conversion process to be matched. In general, the choice of conversion system is determined by the composition, pressure, and, especially, temperature of the tail gases. This latter parameter is mainly a function of the type of plant. More specifically, the operative conditions of the $NH_3$ burner determine the temperature of the exhaust stream. As a rule of thumb, the higher the working pressure of the combustion unit, the higher the tail gas temperature. Unless old low-mono-pressure plants are considered, modern dual-pressure or high-mono-pressure configurations generate tail gases with a temperature in the 300–425 °C range. The most common position in which a ternary abatement unit gets installed is, generally, upstream of the expander. However, the unit can also be installed downstream to the final expander. This situation occurs when the pressure drops induced by the presence of the catalytic unit must be minimized or because of requirements driven by the temperature of the exhaust stream after the ternary unit. In the following, we will cover three main types of ternary abatement measures: $N_2O$ conversion via NSCR, by catalytic decomposition, and through SCR [4].

NSCR relies on using reducing agents based on $H_2$, natural gas, or $CH_4$ to promote total reduction of $N_2O$ (together with $NO_x$ species) and combust all $O_2$ present in the tail gases. The reader is addressed to Section 2.4.1 for further information.

In contrast to NSCR, the high-temperature $N_2O$ abatement in the absence ($N_2O$ decomposition) or presence of reducing agents (SCR of $N_2O$), especially, has found tremendous applicability. $N_2O$ decomposition refers to the catalytic transformation of $N_2O$ into environmentally benign $N_2$ and $O_2$ products at $T > 300$ °C following Eq. (2.16):

$$2N_2O \longrightarrow 2N_2 + O_2 \tag{2.16}$$

Despite a multitude of catalyst alternatives has been developed and investigated in the last 20 years, the reaction remains very sluggish, especially under real operating conditions. Catalysts based on Fe- or Cu-exchanged zeolites, transition (Fe, Cu, and Ni), and noble metals (Pd, Ru, and Rh) supported on $Al_2O_3$, $TiO_2$, and $CeO_2$, as well as mixed-metal oxides (spinels and perovskites) have received considerable scientific attention [35]. The presence of steam, especially, hampers the rate of $N_2O$ decomposition considerably [36], lowers the onset temperature, and thus limited the industrialization of this process to $HNO_3$ acid production plants characterized by high-temperature tail gases ($T > 500$ °C). However, in the presence of reducing agents, such as hydrocarbons (HCs), $NH_3$, or NO, especially, most of the activity is actually recovered [4]. This is the principle at the basis of the SCR of $N_2O$. To introduce a third component, which, by undergoing oxidation, promotes the rate of $N_2O$ conversion. One might look at the process as the heterogeneously catalyzed transfer of the O-atom from $N_2O$ to the reducing agent. In doing so, the reducing agent oxidizes. However, two fundamental differences exist between HCs/$NH_3$ and NO. The former reducing agents enhance $N_2O$ conversion in a stoichiometric manner, i.e. higher concentrations deliver higher $N_2O$ conversion activity. NO, instead, it increases the rate of $N_2O$ conversion catalytically. The second main difference is that while HCs/$NH_3$ have to be supplied to the ternary unit, NO (in conjunction with $NO_2$) is already present in the tail gases undergoing treatment. Among the various catalysts that received considerable interest, Fe-exchanged zeolites are the most used nowadays.

Considering HCs or $NH_3$, the onset temperature of the reduction reaction varies in the 200–400 °C range depending on the operating pressure, on the catalyst used, and on the type of reductant [37, 38]. In general, HCs require lower temperatures for ignition, which might not even require preheating of the tail gases. For this reason, one might think that this is a more competitive strategy compared to the SCR of $N_2O$ using $NH_3$. Nevertheless, one should also consider that the use of HCs implies the production of $CO_2$ and, possibly, CO. In addition, SCR of $N_2O$ by HCs is strongly inhibited by NO [39], which is present in the tail gases as well. For this reason, $NO_x$ must be first completely reduced with ammonia in a preliminary stage if HCs are employed as a reducing agent. On the contrary, the SCR of $N_2O$ by $NH_3$ is promoted by NO. In general, the rate of $N_2O$ reduction increases with the temperature, at higher working pressure, as the concentration of reductant increases, and using higher residence time [4]. In the case of HCs, however, higher working pressure also favors undesired combustion reactions between HCs and $O_2$. This increases HCs consumption, leading to higher costs, and raises $CO_x$ formation. Despite these drawbacks, the SCR of $N_2O$ by HCs is practically implemented in industry to treat tail gases at $300\,°C < T < 500\,°C$, especially. The HCs of choice is selected on the basis of several variables: (i) desired $N_2O$ conversion activity and operating temperature, (ii) the hydrocarbon's ability to selectively target $N_2O$ over $O_2$, (iii) extent of CO and $CO_2$ emissions, (iv) susceptibility to the interference of NO and $NH_3$, and (v) overall cost. Alkanes tend to function more efficiently compared to unsaturated hydrocarbons, with alkynes showing higher ignition temperatures and demonstrating lower selectivity due to their propensity to react with $O_2$. Among the alkanes, ethane stands out for its robust $N_2O$ conversion activity, high selectivity, and reduced susceptibility to NO inhibition compared to $CH_4$ [1].

Differently from HCs or $NH_3$, which have to be supplied to the ternary unit, increasing overall costs, a more appealing and economical strategy to promote the rate of $N_2O$ conversion is to exploit the quantities of NO already present in the tail gases, promoting the so-called NO-mediated $N_2O$ decomposition reaction (Eq. 2.17) [40].

$$N_2O + NO \longrightarrow N_2 + NO_2 \tag{2.17}$$

During the reaction, NO acts as a reducing agent (being transformed to $NO_2$) and promotes $N_2O$ decomposition in a catalytic manner. NO lowers the on-set temperature of $N_2O$ decomposition, allows recovery of the loss of activity triggered by the presence of steam, and thus permits achievement of full $N_2O$ conversion under practical conditions of temperature, pressure, space velocity, and real tail gas composition [40].

Alternatively, another promising approach is the concurrent abatement of $N_2O$ and $NO_x$ in a single catalytic bed, taking advantage of what we called $N_2O$-NO-SCR reaction [31]. In principle, $N_2O$-NO-SCR involves NO conversion via $NH_3$-SCR and $N_2O$ abatement following a combination of $N_2O$ decomposition, NO-mediated $N_2O$ decomposition, and SCR of $N_2O$ with $NH_3$. However, the performance delivered is not given by a simple sum of the individual processes. Rather, $N_2O$-NO-SCR is an appealing reaction scheme in which NO and $N_2O$ (and their corresponding abatement routes) mutually cooperate under reaction conditions, leading to simultaneous enhanced abatement performances.

NO-mediated $N_2O$ decomposition, together with $NO_x$ abatement via $NH_3$-SCR, are the two most implemented technologies for the control of $N_2O$ and $NO_x$ emission in $HNO_3$ production plants generating tail gases at $400\,°C < T < 525\,°C$. As previously mentioned, Fe-exchanged zeolites (based on FER, ZSM-5, and BEA frameworks) are the first catalytic choice. However, for $NO_x$ conversion via $NH_3$-SCR, $V_2O_5/TiO_2$ is still used because of its lower cost. At the same time, in the view of $N_2O$ conversion via SCR strategies, single- or mixed-metal oxides are also employed for the treatment of tail gases at high temperature ($T > 550\,°C$). Considering Fe-exchanged zeolites, their activity is often exploited, taking advantage of dual-bed configurations installed upstream of the expander [1].

Treatment of tail gases at $400\,°C < T < 525\,°C$ is generally carried out using a reactor equipped with two catalyst beds filled with Fe-exchanged zeolites (the same or different type can be utilized) and a manifold allowing the supply of $NH_3$ between the first and second beds. The first bed is designed to maximize $N_2O$ conversion via NO-mediated $N_2O$ decomposition and, indeed, operates at a full load of $NO_x$ species. At the output of the first bed, the exhaust stream contains $NO_x$ (now possessing a lower $NO/NO_2$ ratio), remaining traces of $N_2O$, steam, nitrogen, and oxygen. $NH_3$ is introduced, mixed with the process stream in static mixers, and the stream is distributed to the second bed. Here, the main aims are to promote $NO_x$ conversion via $NH_3$-SCR as well as to reduce the remaining quantities of $N_2O$. This design allows us to achieve 98–99% conversion under appropriate conditions.

If the tail gas temperature is in the 300–525 °C range, thus a lower bottom limit, a different dual-bed configuration (always relying on Fe-exchanged zeolites) aimed at promoting $N_2O$ conversion via HCs-SCR and $NO_x$ abatement using $NH_3$-SCR can be installed. Because of the detrimental effect of $NO_x$ on HCs-SCR, $NO_x$ is converted before $N_2O$ in this design. The process stream leaving the absorption tower is mixed with $NH_3$ and distributed in a first catalytic bed, which maximizes $NO_x$ conversion. Afterward, the HCs are injected. In the second catalytic stage, $N_2O$ is converted with an efficiency around 99% according to HCs-SCR.

## 2.5 Future Trends

The production of nitric acid is a cornerstone of the global chemical industry, primarily driven by the demand for fertilizers. Despite the well-established Ostwald process dominating nitric acid synthesis for over a century, there is a great interest in research and development to integrate new, sustainable feedstocks, to reduce costs and emissions, and to enhance process efficiency.

Green ammonia, produced by electrolyzing water for hydrogen and using nitrogen from air under renewable energy inputs, represents a crucial advancement toward decarbonizing the chemical sector. Traditional ammonia synthesis relies on the steam reforming of natural gas, with negative environmental impact. In contrast, green ammonia production offers a low-carbon pathway, especially if leveraging renewable electricity sources such as wind or solar. When used as a feedstock for nitric acid production, green ammonia can significantly reduce the overall carbon footprint of the process. Research efforts are currently focused on enhancing electrolyzer technologies to improve energy efficiency and reduce capital costs.

The efficiency of the Ostwald process and a reduction of costs can be in principle achieved by the development of novel catalytic systems. In the former case, industrial and academic research focused on the development of catalysts for the oxidation of NO between $NH_3$ combustion and $NO_2$ absorption. Individual and multi-metallic systems based on noble metals (Pt, Ru, or Pd) as well as transition metal oxide-supported catalysts, together with perovskites and C-based materials, are currently being investigated. The aim here is to increase the rate of transformation of NO-to-$NO_2$ in the 300–400 °C temperature range in order to enhance the fraction of recoverable heat and increment process intensification.

On a similar note, considerable cost reduction would be achieved by the development of suitable and cheaper alternatives to the traditional Pt-Rh $NH_3$ oxidation catalyst. Single metal oxides or mixed metal oxides of the perovskite- and spinel-type, based on the presence of Co, Bi, Mn, Cr, Ni, and Fe as active components, are currently actively studied. Among these, $Co_3O_4$-based systems have been investigated more extensively. Yet, considerable effort is still required in order to meet the standards required by industry.

An additional strategy under development aimed at achieving reduced power consumption and a better overall ecological footprint is the so-called plasma-assisted oxidation of $NH_3$ [41]. The principle involves the use of plasma discharges to dissociate reactant molecules, thereby inducing chemical reactions over suitable catalysts, e.g. metal oxides ($Co_3O_4$, $MoO_3$, and NiO among others) supported on $\gamma$-$Al_2O_3$. The main advantage of this approach would be to operate a non-thermal plasma reactor at ambient conditions, i.e. low temperature and atmospheric pressure, which is fed with streams of atmospheric air and pure oxygen, thus providing a fossil-free nitrogen fixation route [42].

## 2.6 Conclusions

The industrial production of nitric acid remains an essential pillar of the modern chemical industry, driven primarily by the demand for fertilizers and many chemical intermediates. While being over a century old, the Ostwald process has undergone significant advancements in terms of catalyst development, reaction optimization, and emission control. The use of Pt-Rh gauze catalysts for $NH_3$ oxidation remains highly effective in achieving remarkable activity and high selectivity toward NO. However, challenges related to catalyst costs and deactivation upon poisoning and noble metal losses necessitate continuous innovation. The implementation of advanced catalytic gauzes, optimized reactor configurations, and pressure management strategies has significantly improved process efficiency and economic feasibility. Additionally, environmental concerns have led to the widespread adoption of tail-gas treatment technologies, which have helped mitigate $NO_x$ and $N_2O$ emissions to comply with more stringent environmental regulations. Moving forward, the integration of green ammonia as a feedstock, the development of cost-effective alternative catalysts, and the intensification of NO oxidation processes represent promising avenues for further reducing the carbon footprint and enhancing sustainability in nitric acid production. Similarly, novel technologies based on the plasma-assisted synthesis of both $NH_3$ and $HNO_3$ are under development. In principle, these new strategies might offer the possibility of using milder process conditions, thus achieving a reduced power consumption and a better overall ecological footprint. Through ongoing research and technological advancements, the industry is poised to meet future global demands while adhering to increasingly strict environmental standards.

## References


1 Groves, M.C.E. (2020). *Kirk-Othmer Encyclopedia of Chemical Technology*, 1–37. Wiley https://doi.org/10.1002/0471238961.

2 Parsons, C.L. (1927). Nitric acid from ammonia. *Ind. Eng. Chem.* 19: 789–794. https://doi.org/10.1021/ie50211a013.

3 Brunborg, J. and Holmesland, P.B. (1985). *Nitric Acid and Fertilizer Nitrates*. New York: M. Dekker.

4 Pérez-Ramŕez, J., Kapteijn, F., Schöffel, K., and Moulijn, J.A. (2003). Formation and control of $N_2O$ in nitric acid production. *Appl Catal B* 44: 117–151. https://doi.org/10.1016/S0926-3373(03)00026-2.

5 Thiemann, M., Scheibler, E., and Wiegand, W. (2003). Nitric Acid, Nitrous Acid, and Nitrogen Oxides. In: *Ullmann's Encyclopedia of Industrial Chemistry* (ed. M. Bohnet), 177. Weinheim Germany: Wiley.

6 Nutt, C.W. and Kapur, S. (1969). Oxidation of ammonia on platinum. *Nature* 224: 169–169. https://doi.org/10.1038/224169a0.

7 Pignet, T. (1975). Kinetics of $NH_3$ oxidation on Pt, Rh, and Pd. *J. Catal.* 40: 212–225. https://doi.org/10.1016/0021-9517(75)90249-3.

8 Imbihl, R., Scheibe, A., Zeng, Y.F. et al. (2007). Catalytic ammonia oxidation on platinum: mechanism and catalyst restructuring at high and low pressure. *Phys. Chem. Chem. Phys.* 9: 3522–3540. https://doi.org/10.1039/B700866J.

9 Connor, H. (1967). The manufacture of nitric acid. *Platin. Met. Rev.* 11: 60–69. https://doi.org/10.1595/003214067X11129.

10 Heck, R.M., Bonacci, J.C., Hatfield, W.R., and Hsiung, T.H. (1982). A new research pilot plant unit for ammonia oxidation processes and some gauze data comparisons for nitric acid process. *Ind. Eng. Chem. Process. Des. Dev.* 21: 73–79. https://doi.org/10.1021/i200016a014.

11 Horner, B.T. (1993). Knitted Platinum alloy gauzes: catalyst development and industrial application. *Platin. Met. Rev.* 37: 76–85. https://doi.org/10.1595/003214093X3727685.

12 A. Bazhenov, G. Gushin, Proceedings of international nitrogen and syngas conference, 2017, 22–25.

13 Schmidt-Szałowski, K., Krawczyk, K., and Petryk, J. (1998). The properties of cobalt oxide catalyst for ammonia oxidation. *Appl. Catal. A* 175: 147–157. https://doi.org/10.1016/S0926-860X(98)00206-3.

14 Petryk, J. and Kołakowska, E. (2000). Cobalt oxide catalysts for ammonia oxidation activated with cerium and lanthanum. *Appl Catal B* 24: 121–128. https://doi.org/10.1016/S0926-3373(99)00099-5.

15 Sadykov, V.A., Isupova, L.A., Zolotarskii, I.A. et al. (2000). Oxide catalysts for ammonia oxidation in nitric acid production: properties and perspectives. *Appl. Catal. A* 204: 59–87. https://doi.org/10.1016/S0926-860X(00)00506-8.

**16** Tsukahara, H., Ishida, T., and Mayumi, M. (1999). Gas-phase oxidation of nitric oxide: chemical kinetics and rate constant. *Nitric Oxide Biol. Chem.* 3: 191–198. https://doi.org/10.1006/niox.1999.0232.

**17** Gopakumar, J., Myrstad, R., Borresen Anda, R. et al. (2025). Ostwald process intensification by catalytic oxidation of nitric oxide. *ACS Omega* 10: 2197–2211. https://doi.org/10.1021/acsomega.4c09111.

**18** Grande, C.A., Andreassen, K.A., Cavka, J.H. et al. (2018). Process intensification in nitric acid plants by catalytic oxidation of nitric oxide. *Ind. Eng. Chem. Res.* 57: 10180–10186. https://doi.org/10.1021/acs.iecr.8b01483.

**19** Hong, Z., Wang, Z., and Li, X. (2017). Catalytic oxidation of nitric oxide (NO) over different catalysts: an overview. *Catal. Sci. Tech.* 7: 3440–3452. https://doi.org/10.1039/C7CY00760D.

**20** Shen, Y., Ge, X., and Chen, M. (2016). Catalytic oxidation of nitric oxide (NO) with carbonaceous materials. *RSC Adv.* 6: 8469–8482. https://doi.org/10.1039/C5RA24148K.

**21** Carberry, J.J. (1959). Some remarks on chemical equilibrium and kinetics in the nitrogen oxides-water system. *Chem. Eng. Sci.* 9: 189–194. https://doi.org/10.1016/0009-2509(59)85001-6.

**22** Ingale, N.D., Chatterjee, I.B., and Joshi, J.B. (2009). Role of nitrous acid decomposition in absorber and bleacher in nitric acid plant. *Chem. Eng. J.* 155: 851–858. https://doi.org/10.1016/j.cej.2009.09.017.

**23** Winterbauer, H. (2005). Improved process for high concentration of nitric acid using magnesium nitrate. *Chem. Eng. Technol.* 28: 709–711. https://doi.org/10.1002/ceat.200500026.

**24** Peel, J.L., Haeuber, R., Garcia, V. et al. (2013). Impact of nitrogen and climate change interactions on ambient air pollution and human health. *Biogeochemistry* 114: 121–134. https://doi.org/10.1007/s10533-012-9782-4.

**25** Prather, M.J., Hsu, J., DeLuca, N.M. et al. (2015). Measuring and modeling the lifetime of nitrous oxide including its variability. *J. Geophys. Res. Atmos.* 120: 5693–5705. https://doi.org/10.1002/2015JD023267.

**26** Costa, C.N., Savva, P.G., Fierro, J.L.G., and Efstathiou, A.M. (2007). Industrial $H_2$-SCR of NO on a novel Pt/MgO–$CeO_2$ catalyst. *Appl Catal B* 75: 147–156. https://doi.org/10.1016/j.apcatb.2007.04.018.

**27** Buttignol, F., Rentsch, D., Alxneit, I. et al. (2022). Aging of industrial Fe-zeolite based catalysts for nitrous oxide abatement in nitric acid production plants. *Catal. Sci. Tech.* 12: 7308–7321. https://doi.org/10.1039/D2CY01486F.

**28** Nuguid, R.J.G., Buttignol, F., Marberger, A., and Kröcher, O. (2021). Selective Catalytic Reduction of $NO_x$ Emissions. In: *Petrodiesel Fuels* (ed. O. Konur). CRC Press https://doi.org/10.1201/9780367456252.

**29** Buttignol, F. (2024). *The Simultaneous Selective Catalytic Reduction of $N_2O$ and NO over Fe-Exchanged Zeolites. Ecole Polytechnique Fédérale de Lausanne (EPFL).*

**30** Elsener, M., Nuguid, R.J.G., Kröcher, O., and Ferri, D. (2021). HCN production from formaldehyde during the selective catalytic reduction of $NO_x$ with $NH_3$ over $V_2O_5/WO_3$-$TiO_2$. *Appl Catal B* 281: 119462. https://doi.org/10.1016/j.apcatb.2020.119462.

**31** Buttignol, F., Fischer, J.W.A., Clark, A.H. et al. (2024). Iron-catalysed cooperative redox mechanism for the simultaneous conversion of nitrous oxide and nitric oxide. *Nat. Catal.* 7: 1305–1315. https://doi.org/10.1038/s41929-024-01231-3.

**32** Buttignol, F., Garbujo, A., Biasi, P. et al. (2022). Effect of an $Al_2O_3$-based binder on the structure of extruded Fe-ZSM-5. *Catal. Today* 387: 207–215. https://doi.org/10.1016/j.cattod.2021.09.020.

**33** Isupova, L.A. and Ivanova, Y.A. (2019). Removal of nitrous oxide in nitric acid production. *Kinet. Catal.* 60: 744–760. https://doi.org/10.1134/S0023158419060041.

**34** W. Boll, Prolonged campaign length of high pressure nitric acid plant by optimization of secondary catalyst design for $N_2O$ abatement, Proceedings of International Nitrogen and Syngas Conference, 2017.

**35** Konsolakis, M. (2015). Recent Advances on nitrous oxide ($N_2O$) Decomposition over non-noble-metal oxide catalysts: catalytic performance, mechanistic considerations, and surface chemistry aspects. *ACS Catal.* 5: 6397–6421. https://doi.org/10.1021/acscatal.5b01605.

**36** Xia, H., Sun, K., Feng, Z., and Li, C. (2011). Effect of water on active iron sites for $N_2O$ decomposition over Fe/ZSM-5 catalyst. *J. Phys. Chem. C* 115: 542–548. https://doi.org/10.1021/jp1094917.

**37** Coq, B., Mauvezin, M., Delahay, G., and Kieger, S. (2000). Kinetics and mechanism of the $N_2O$ reduction by $NH_3$ on a Fe-zeolite-beta catalyst. *J. Catal.* 195: 298–303. https://doi.org/10.1006/jcat.2000.2980.

**38** van den Brink, R.W., Booneveld, S., Pels, J.R. et al. (2001). Catalytic removal of $N_2O$ in model flue gases of a nitric acid plant using a promoted Fe zeolite. *Appl Catal B* 32: 73–81. https://doi.org/10.1016/S0926-3373(00)00294-0.

**39** Pérez-Raḿrez, J. and Kapteijn, F. (2004). Effect of NO on the SCR of $N_2O$ with propane over Fe-zeolites. *Appl Catal B* 47: 177–187.

**40** Pérez-Raḿrez, J., Kapteijn, F., Mul, G., and Moulijn, J.A. (2002). No-assisted $N_2O$ decomposition over Fe-based catalysts: effects of gas-phase composition and catalyst constitution. *J. Catal.* 208: 211–223. https://doi.org/10.1006/jcat.2002.3559.

**41** Anastasopoulou, A., Keijzer, R., Butala, S. et al. (2020). Eco-efficiency analysis of plasma-assisted nitrogen fixation. *J. Phys. D. Appl. Phys.* 53: 234001. https://doi.org/10.1088/1361-6463/ab71a8.

**42** Patil, B.S., Cherkasov, N., Lang, J. et al. (2016). Low temperature plasma-catalytic $NO_x$ synthesis in a packed DBD reactor: effect of support materials and supported active metal oxides. *Appl Catal B* 194: 123–133. https://doi.org/10.1016/j.apcatb.2016.04.055.

# Part II

# Plasma-Assisted Nitrogen Fixation

# 3

# Introduction to Plasma Technology

*Anthony B. Murphy*

*CSIRO, Manufacturing, Lindfield, NSW, Australia*

## 3.1 Fundamental Concepts

Plasmas are often described as the fourth state of matter; as energy is added, a substance is transformed from a solid to a liquid, then to a gas, and finally to a plasma. The transformation that occurs when a plasma is formed is that some of the gas molecules are ionized to form ions and electrons. Hence, the term "ionized gas" is often used to describe a plasma. The term "plasma" was first introduced by Langmuir [1] to describe the regions in an ionized gas in which the electron and ion charges were balanced. Plasmas are generally composed of a range of species, including molecules, radicals, atoms, molecular and atomic ions (collectively termed the heavy species or heavy particles), electrons, and photons.

### 3.1.1 Types of Plasma

Most of the visible matter in the universe (at least 99.9%) is in the plasma state. Stars and the interstellar medium are composed of plasma, as is the Earth's ionosphere. On Earth, plasmas are much less common, but lightning arcs, sparks, flames, and the polar aurora are all examples of naturally occurring plasmas. As will be discussed in Section 3.4, plasmas are also of immense practical importance, so a wide range of man-made plasmas are routinely produced and applied.

These examples indicate that plasmas exist across a wide range of parameters. Plasmas can be classified according to their properties, including the pressure, the electron density, the degree of ionization, and the temperature. Several different temperatures can be defined, as will be discussed below; the most important is

*Plasma-Assisted Nitrogen Fixation for Sustainable Process Industries*, First Edition. Edited by Volker Hessel, Annemie Bogaerts, Gabriele Centi, Evgeny Rebrov, and Nguyen Van Duc Long.

Published 2026 by John Wiley & Sons, Inc.

the electron temperature, which describes the translational temperature of the electrons. There is a strong interdependence of the different properties; e.g. the electron density depends on the pressure and the degree of ionization, which in turn depends on the electron and possible other temperatures.

A widely used classification, shown in Figure 3.1, considers the electron density $n_e$ and the electron temperature $T_e$. The range of electron densities varies from around $10^8$ to $10^{24}$ $m^{-3}$ and depends on the pressure and the degree of ionization. Pressures range from high vacuum to above atmospheric, and the degree of ionization ranges from very low (as low as $10^{-10}$ in flames) to full or multiple ionization in thermal and thermonuclear fusion plasmas.

Of the man-made plasmas, thermonuclear fusion plasmas have high electron temperatures and densities, like those in the solar core, which is consistent with the goal of thermonuclear fusion research – to drive nuclear fusion reactions like those occurring in the sun.

Plasmas of industrial interest – glow discharges, non-equilibrium atmospheric pressure discharges such as dielectric barrier discharges (DBDs), warm plasmas such as microwave and spark discharges, and thermal arc and RF discharges – have approximately similar electron temperatures, typically in the range of 1–20 eV (1 eV ~ 11,600 K), but span a range of electron densities. The characteristics of such plasmas will be discussed in Section 3.4.

### 3.1.2 Scaling Parameters

While a conventional description of a plasma is as an ionized gas, most definitions are more restrictive. Typical definitions require quasineutrality, i.e. that the negatively and positively charged particles balance each other. Quasineutrality is driven by the electric field that is formed due to any separation between the electrons and ions, which counteracts their separation. Deviations from quasineutrality can occur within a characteristic length scale known as the Debye length, which describes the smallest length scale of a plasma, given by

$$\lambda_\mathrm{D} = \left(\frac{\varepsilon_0 k_B T_e}{n_e e^2}\right)^{1/2} \tag{3.1}$$

where $\varepsilon_0$ is the permittivity of free space, $e$ is the electronic charge, and $k_B$ is Boltzmann's constant. The dependence of Debye length on electron temperature and electron density is shown in Figure 3.2a. Deviations from quasineutrality occur in the sheath regions at the boundary between electrodes and the plasma, whose thickness is typically several $\lambda_\mathrm{D}$, and within the heads of streamers, which are discussed in Section 3.2. Another requirement for a plasma is that at least several electrons are present within a sphere of radius $\lambda_\mathrm{D}$; i.e.

$$\frac{4}{3}\pi n_e \lambda_\mathrm{D}^3 > 1 \tag{3.2}$$

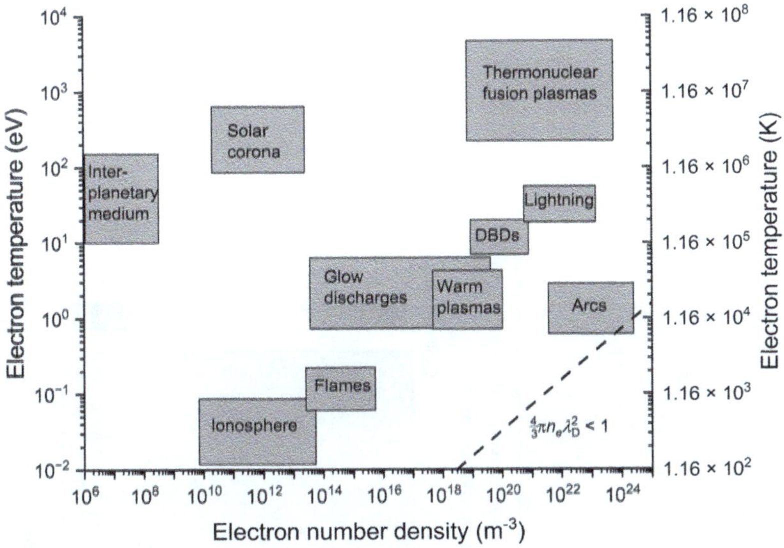


**Figure 3.1** Plasmas classified by electron temperature and electron density.

The range of electron temperatures and densities over which this criterion applies is indicated in Figure 3.1.

Two important frequencies are the electron and ion plasma frequencies, given by

$$\omega_{pj} = \left(\frac{n_j Z_j^2 e^2}{\varepsilon_0 m_j}\right)^{1/2} \tag{3.3}$$

where $n_j$ is the number density, $Z_j$ is the charge number, and $m_j$ is the mass of species $j$, which can be the electron or an ion. The plasma frequency determines whether a species can follow an applied electric field; if the frequency of the applied field is higher than $\omega_{pj}$, then species $j$ cannot follow the field. The dependence of $\omega_{pj}$ on electron density is shown for electrons and atomic hydrogen and argon ions in Figure 3.2b.

The concept of temperature is more complex for a plasma than a non-ionized gas. Many processes contribute to the energy of the plasma, including the translational energy of each species, the rotational and vibrational excitation of molecules and molecular ions, and the electronic excitation of molecules, atoms, and ions. Each of these can be described by a distribution function and temperature, with thermal equilibrium of a given species leading to the Maxwell–Boltzmann distribution:

$$f(\boldsymbol{v}_j) = \left(\frac{m_j}{2\pi k_B T_j}\right)^{3/2} \exp\left(\frac{\frac{1}{2} m_j v_j^2}{k_B T_j}\right) \tag{3.4}$$

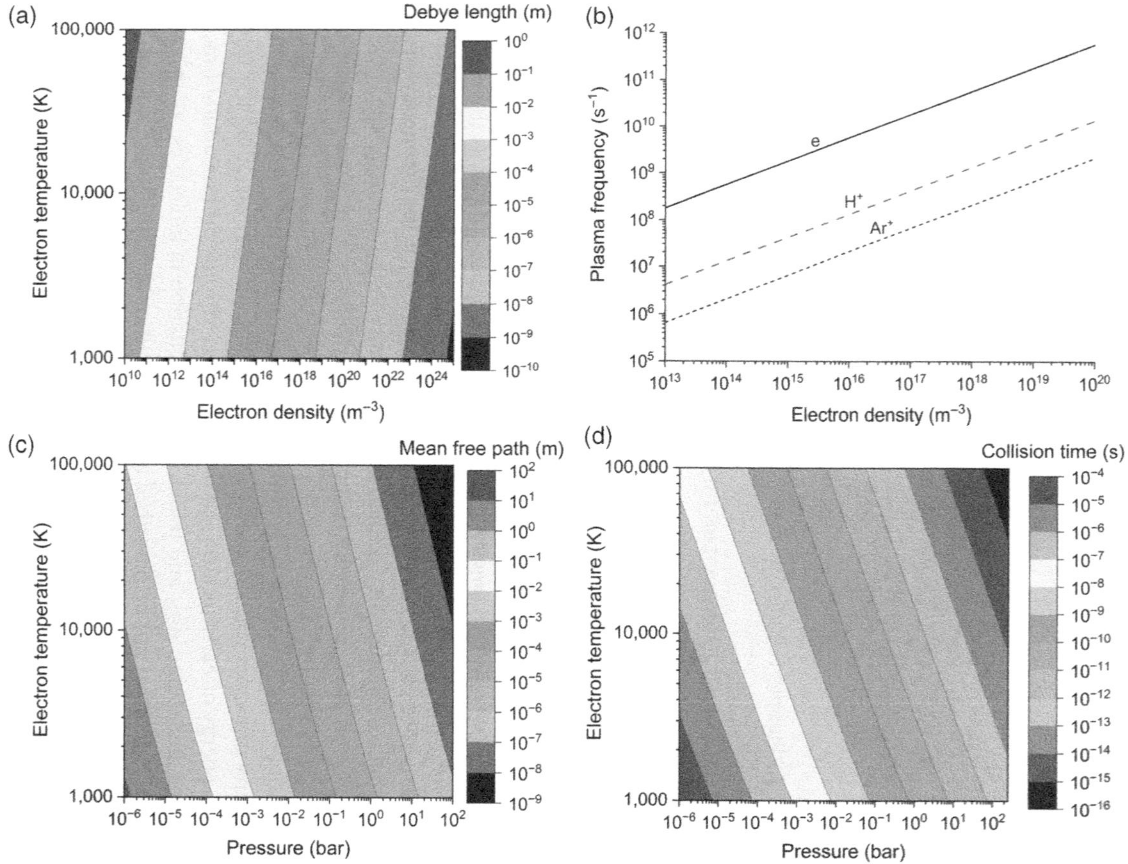


**Figure 3.2** (a) Debye length; (b) electron, hydrogen ion, and argon ion plasma frequency; (c) mean free path; (d) mean time between collisions; (c) and (d) are for electrons in a weakly ionized argon plasma with heavy particle temperature of 300 K.

where $\mathbf{v}_j$, $m_j$, and $T_j$ are the velocity, mass, and temperature of the $j$th species. The mean energy is given by $3k_BT_j/2$, and the average velocity is

$$\overline{v_j} = \left(\frac{8k_BT_j}{\pi m_j}\right)^{1/2} \tag{3.5}$$

The average kinetic energy exchange during an elastic binary condition between particles of mass $m_j$ and $m_k$ is given by [2]:

$$\Delta E_{\text{kin}} = \frac{2m_jm_k}{(m_j + m_k)^2}\left|E_{\text{kin}\,j} - E_{\text{kin}\,k}\right| \tag{3.6}$$

where $\left|E_{\text{kin}\,j} - E_{\text{kin}\,k}\right|$ is the difference in kinetic energy of the particles. For particles of the same mass, $\Delta E_{\text{kin}} = 1/2\left|E_{\text{kin}\,j} - E_{\text{kin}\,k}\right|$, and only around 10 successive collisions are required to eliminate distortions in the Maxwell–Boltzmann distribution. However, particularly at low pressures for which the time between collisions is relatively large, deviations from the Maxwell–Boltzmann distribution can occur, e.g. due to fluxes of energetic species from the wall (e.g. electrons released by ion impact) [3].

For the case of a collision between an electron and a heavy particle of masses $m_e$ and $m_h$, respectively, Eq. (3.6) gives $\Delta E_{\text{kin}} = 2(m_e/m_h)\left|E_{\text{kin}\,j} - E_{\text{kin}\,k}\right|$. Since $m_e \ll m_h$, a large number of collisions, at least 10,000, are required to equilibrate the temperatures of electrons and heavy species.

Energy is usually supplied to a plasma through an electric or electromagnetic field. While collisions act to equilibrate the species temperatures, the field accelerates the electrons, with the electron energy acquired between collisions given by

$$\Delta E_{\text{kin}\,e} = eE\overline{v_{de}}\,\overline{\tau_e} \tag{3.7}$$

where $E$ is the electric field strength and $\overline{v_{de}}$ is the mean electron drift velocity. The drift velocity is given by

$$\overline{v_{de}} = \frac{\overline{\tau_e}}{m_e}eE \tag{3.8}$$

where $\overline{\tau_e} = \overline{l_e}/\overline{v_{de}}$ is the mean time between electron collisions with $\overline{l_e}$ denoting the electron mean free path. For a plasma that is weakly ionized ($n_e/n \ll 1$, $n$ being the total or gas number density), the mean free path is given by

$$\overline{l_e} = 1/(nQ_{ea}) \tag{3.9}$$

where the momentum transfer cross-section for electron–atom collisions is $Q_{ea} \approx (3.6 \times 10^{-4}T_e - 0.1) \times 10^{-20}\,\text{m}^2$ for an argon plasma [4, 5]. The value of $Q_{ea}$ depends on the type of plasma. Its dependence on the electron temperature and pressure for argon is given in Figure 3.2c. Data for different plasmas can be found in Refs. [6, 7].

Equating $\Delta E_{\text{kin}\,e}$ with the energy exchange in a collision between an electron and heavy species,

$$\Delta E_{\text{kin}\,e} = (2m_e/m_h)3k_B(T_e - T_h)/2 \tag{3.10}$$

gives an estimate for the relative temperature difference [8].

$$\frac{T_e - T_h}{T_e} = \frac{3\pi}{32}\frac{m_h}{m_e}\left(\frac{e\overline{l_e}E}{\frac{3}{2}k_B T_e}\right)^2 \tag{3.11}$$

where $T_h$ is the heavy-particle temperature. The inverse dependence of $\overline{l_e}$ on $n$ (Eq. (3.9)) indicates that the ratio $E/n$, known as the reduced electric field, plays an important role in determining the kinetic equilibrium between electrons and heavy particles.

Figure 3.2d shows the dependence of $\overline{\tau_e}$ on the pressure and the electron temperature for a heavy-particle temperature $T_h$ of 300 K in a weakly ionized argon plasma. $\overline{\tau_e}$ is inversely proportional to $T_h$, so results for higher heavy-particle temperatures are easily determined from the figure.

For an atmospheric-pressure plasma with $T_e$ = 10,000 K and $T_h$ = 300 K, $\overline{\tau_e} \approx 1.7$ ps, indicating that equilibration between electrons and heavy particles will require at least tens to hundreds of nanoseconds. At low pressures, $\overline{\tau_e}$ is much larger.

A further scaling parameter, relevant to high-frequency excitation of the plasma, is $\omega\overline{\tau_e}$, where $\omega$ is the angular frequency of the applied voltage. Power transfer from the electric field is maximized for $\omega\overline{\tau_e} \approx 1$. This is achievable at low pressures using RF and microwave sources, since, as shown in Figure 3.2d, $\overline{\tau_e}$ is in the micro- to nanosecond range. However, at atmospheric pressure, the picosecond values of $\overline{\tau_e}$ would require excitation in the terahertz range, indicating that the power transfer is far from optimal, even for microwave sources.

## 3.2 Plasma Generation

Plasmas are generated by electrical breakdown – the formation of electrons and ions under the influence of an electric field. Breakdown occurs when the production of these charged species (i.e. ionization) overcomes their tendency to recombine to form neutral molecules or attach to molecules to form negative ions (i.e. recombination). Figure 3.3 summarizes the transitions that can occur that lead to the formation of different types of discharges.

A low density of seed electrons, of order $10^9$–$10^{10}$ m$^{-3}$, is always present in a gas because of the action of cosmic rays and background radioactivity [10]. An electric field accelerates these electrons, which can then attain the energy required to

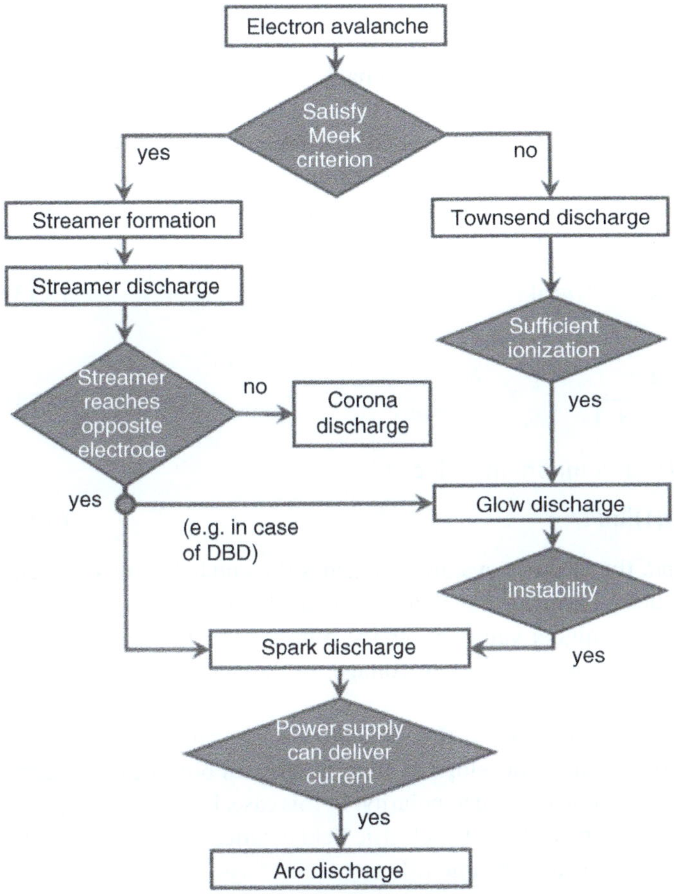


**Figure 3.3** Summary of the transitions that can occur in the formation of discharges. *Source:* Reproduced from Ref. [9]/with permission of IOP Publishing.

ionize surrounding molecules. This leads to exponential growth of the electron density, forming an avalanche. The ionization coefficient (also known as the first Townsend coefficient), which gives the number of electrons produced per unit length, can be approximated as [3, 11].

$$\alpha = Ap\exp(-Bp/E) \tag{3.12}$$

where $A$ and $B$ are constants that depend on the gas. Recombination is described by a second Townsend coefficient $\beta$, and a third Townsend coefficient $\gamma$ describes the production of electrons by the impact of ions on surfaces.

In a DC field between two electrodes separated by a distance $d$, electrons drift toward the anode (the positive electrode) and are lost upon reaching it. Maintaining the discharge (known as a Townsend discharge) requires enough secondary electrons to be produced by the interaction of the ions with the cathode (the negative electrode). The Paschen criterion for breakdown requires that sufficient ions are produced by ionization for the secondary electrons produced at the cathode to replace those lost at the anode. It provides a relation between the breakdown voltage $V_b$ and the product $pd$ [3, 11].

$$V_b = B(pd)/[C + \ln(pd)] \tag{3.13}$$

where

$$C = \ln \frac{A}{\ln(1/\gamma) + 1} \tag{3.14}$$

The function $V_b(pd)$ has a minimum value at

$$(pd)_{\min} = (e/A)[\ln(1/\gamma) + 1] \tag{3.15}$$

At lower values of $pd$, the electron avalanche reaches the anode before sufficient ions have been produced, requiring a higher electric field (and, therefore, voltage) to increase ionization. At higher values of $pd$, the number of collisions increases, and the electric field decreases for a given voltage, so the voltage again must be increased.

For high-frequency electric fields, the breakdown mechanism is different because the electron avalanche develops as time progresses but does not reach the anode because of the rapid change in polarity. In this case, breakdown requires that the energy gain of electrons from the electric field balance losses by collisions with the neutral gas, averaged over time [3].

With increasing pressure, the mean free path $\overline{l_e}$ (Eq. (3.9)) decreases because of the increased number density $n$. At atmospheric pressure, its value is around 10 μ m, much less than the distance between electrodes, so the influence of secondary electrons vanishes. At high pressure, breakdown proceeds by the formation of streamers. A streamer head containing a high charge density is formed, generating a high electric field, which causes further ionization, leading to the propagation of the streamer head toward the electrode of opposite polarity. The propagation mechanism depends on the polarity of the electrode toward which the streamer is propagating [9]. The streamer head is trailed by a streamer channel, which is slowly neutralized by recombination.

Two main criteria have been found in the literature for the transition of an electron avalanche to a streamer [12]. The first is that the background field must be high enough for electron-impact ionization to dominate recombination reactions. In dry air, the primary recombination mechanism is the attachment of electrons to molecules to form negative ions. Figure 3.4 shows that the ionization

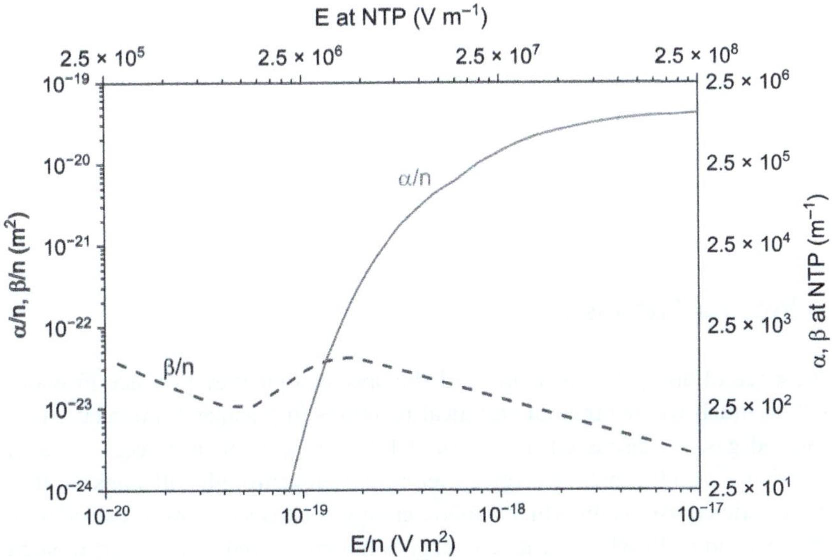


**Figure 3.4** Values of the ionization and recombination coefficients as a function of the electric field, all normalized to the gas density for dry air. Unnormalized values at NTP (normal temperature and pressure, 293.15 K and 1 atm) are also shown. *Source:* Adapted from [13].

coefficient $\alpha$ increases rapidly with the $E/n$, as per Eq. (3.12), while the recombination coefficient $\beta$ has a relatively weak dependence on $E/n$. Initial breakdown requires about 2 MV m$^{-1}$ at 1 bar. Once breakdown has occurred, this is reduced to about 0.5 MV m$^{-1}$ due to the presence of excited molecules, which can detach electrons from negative ions [13]. The second criterion to be satisfied is known as the Meek or Raether–Meek condition and requires that the number of electrons must reach $10^8$–$10^9$ for significant space charge effects to occur (i.e. for the charge to modify the electric field); if this occurs by exponential multiplication of an initial electron in the background electric field, it requires

$$\exp[(\alpha - \beta)d_{\text{ion}}] \approx 10^8 \text{ to } 10^9 \tag{3.16}$$

i.e. $\alpha d_{\text{ion}} \approx 18$ to 21, where $d_{\text{ion}}$ is the ionization length of the local avalanche (for which $\alpha - \beta > 0$). The Meek condition is an approximate criterion and has been modified to take into account many other factors (e.g. Ref. [12]).

A corona discharge occurs when $\alpha d_{\text{ion}}$ is at least around 9 [14] but does not meet the Meek condition and does not cross the full distance to the other electrode. If, however, the streamer crosses the full distance between the electrodes, a conductive channel is formed. Gas heating occurs through collisions on a time scale of

~100 ns at 1 bar, as discussed in Section 3.1.2, and a spark is produced. This can progress to an arc discharge on a time scale of ~100 μs if the power supply can deliver the required current (in the range of amperes to kiloamperes or more) [9]. If, however, the time scale of the discharge is limited to substantially less than required for gas heating, e.g. by the introduction of a dielectric barrier between the electrodes, as in a DBD, or by removing the voltage, as in a nanosecond-pulsed power supply, a glow discharge is produced.

## 3.3 Plasma Chemistry

The presence of energetic electrons and the species that they produce provides access to a much wider range of chemical reactions in plasmas than occurs in a non-ionized gas. As discussed in Section 3.1.2, the electrons are accelerated in the electric field and transfer energy to heavy particles through collisions. In that section, elastic collisions, in which kinetic energy is conserved, were considered. Inelastic collisions, in which kinetic energy is not conserved, also occur; in such cases, the composition or internal energy of at least one of the particles involved in the collision is changed. The collision partners can be an electron and a heavy particle or a pair of heavy particles; three-body collisions can also be important. Plasma chemistry is mainly driven by these inelastic collisions. Chemistry can also be driven by photons (e.g. photoionization is important in the propagation of streamers propagating toward the cathode) and by reactions occurring on surfaces.

### 3.3.1 Inelastic Collisions Between Electrons and Heavy Particles

Inelastic collisions with electrons can transfer the kinetic energy of electrons into the internal energy of atoms and molecules via rotational, vibrational, and electronic excitation of molecules and electronic excitation of atoms. Further, inelastic collisions can cause dissociation of molecules and ionization of atoms and molecules. Superelastic collisions, in which the internal energy of excited molecules or atoms is transferred back into the kinetic energy of electrons, can also occur.

The reactions depend strongly on the energy of the electrons, which in turn depends on the reduced electric field $E/n$. The mean electron energy for a Maxwell–Boltzmann distribution of electron energies (Eq. (3.4)) is $(3/2)k_BT_e$. The distribution has a high-energy tail so that reactions requiring well over the mean electron energy can occur.

Electron-impact vibrational excitation of molecules is an important process in non-equilibrium atmospheric-pressure plasmas. As shown in Figure 3.5, it dominates the energy transfer from electrons to molecules at low electron energies, up to a few eV. While the optimum electron energy range and the cross-section

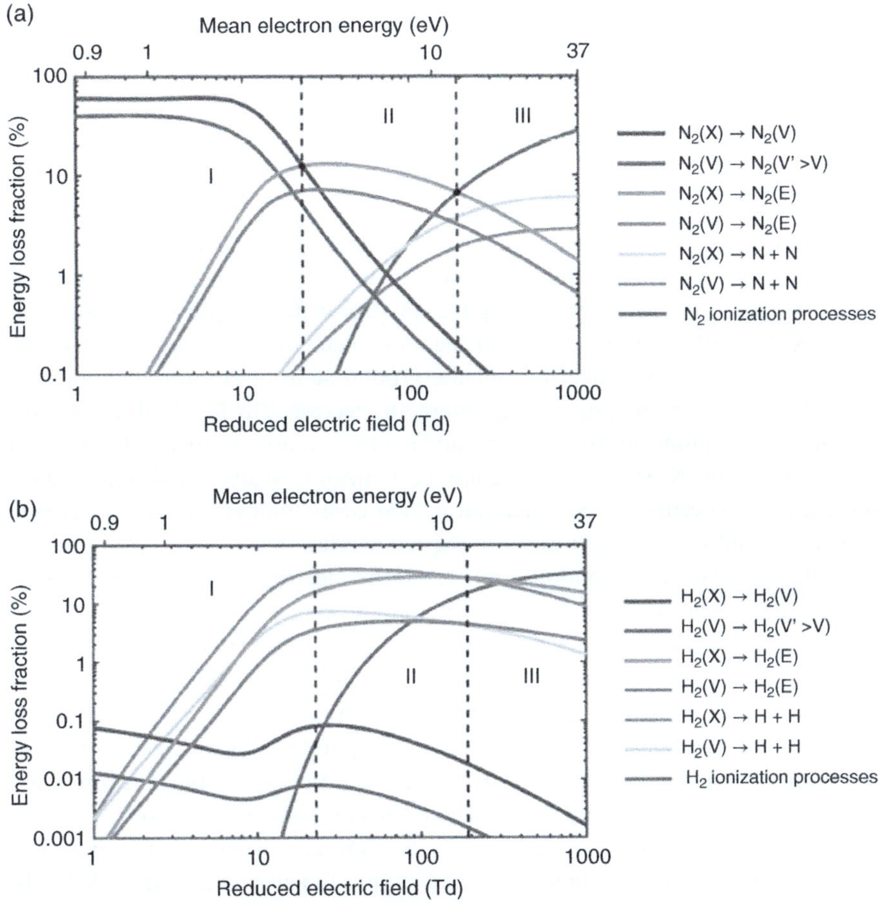


**Figure 3.5** Fraction of electron energy transferred to (a) nitrogen and (b) hydrogen species in a 3:1 mixture of $N_2$ and $H_2$ with $T_g$ = 400 K and $T_{vib}$ = 300 K. (X), (V), and (E) denote ground state, vibrationally excited, and electronically excited levels, respectively. 1 Td = $10^{-21}$ V m$^2$. *Source:* Ref. [15]/Royal Society of Chemistry/CC BY 3.0.

depend on the gas, vibrational excitation is typically most effective in the range $T_e \sim$ 1–3 eV, with almost every electron–molecule collision leading to vibrational excitation at 2 eV for $N_2$, $CO_2$, and CO [16]. As shown in the figure, the electron energy is related to the reduced electric field $E/n$, which is often given in units of townsend; 1 Td = $10^{-21}$ V m$^2$. At 1 bar, 1 Td corresponds to an electric field of $2.5 \times 10^4$ V m$^{-1}$.

Rotational excitation requires far less energy than vibrational excitation and typically proceeds by "quasi-elastic" energy transfer, with each collision transferring

$m_e/m_h$ of the electron energy. This typically ensures that the rotational temperature $T_{rot} \approx T_h$ at or near 1 bar.

Electronic excitation of atoms and molecules requires higher electron energies, usually of the order of 5 eV or more, as shown in Figure 3.5. For example, the first excited level (A $^3\Sigma_{u+}$) of a nitrogen molecule is about 8.0 eV, and the first excited level of an argon atom is 11.55 eV.

Dissociation through electronic excitation can be caused by a single collision via several mechanisms; those requiring the lowest energy are excitation to a repulsive molecular state and excitation to an attractive state with energy exceeding the dissociation threshold. As shown in Figure 3.5, dissociation requires high electron energies, particularly in the case of strong bonds such as the nitrogen triple bond.

Ionization can occur directly from the ground state of an atom or molecule through a collision with a sufficiently energetic electron. The first ionization energies of an argon atom, a nitrogen atom, and a nitrogen molecule are 15.8, 14.5, and 15.6 eV, respectively. Stepwise ionization, i.e. ionization of an excited atom or molecule, can also occur. Direct ionization is favored by high electron energies (see Figure 3.5) and low concentrations of excited species.

In addition to the above endothermic reactions, electron interactions with energetic species, such as electron–ion recombination, can release significant energy. In a molecular gas, electron–ion recombination can lead to dissociation (dissociative electron–ion recombination – $e + \mathbf{AB}^+ \rightarrow \mathbf{AB}^* \rightarrow \mathbf{A} + \mathbf{B}^*$). In atomic gases, three-body electron–ion recombination ($e + e + \mathrm{A}^+ \rightarrow \mathrm{A}^* + e$), with the excess energy being transferred to an electron, is important. Radiative electron–ion recombination ($e + \mathrm{A}^+ \rightarrow \mathrm{A}^* \rightarrow \mathrm{A} + h\upsilon$), in which the excess energy is converted into radiation, can also be important when the electron density is low.

A range of reactions that form negative ions can also occur. In non-equilibrium atmospheric-pressure discharges, three-body electron attachment ($e + \mathbf{A} + \mathbf{B} \rightarrow \mathbf{A}^- + \mathbf{B}$) can be important, e.g. in the formation of $O_2^-$ ions. At lower pressures, when three-body reactions are less likely, dissociative electron attachment ($e + \mathbf{AB} \rightarrow (\mathbf{AB}^-)^* \rightarrow \mathbf{A} + \mathbf{B}^-$) can be important; the cross-section depends strongly on the electron energy and has a peak at relatively high energies, around 6 eV for $O_2$ and 10 eV for CO, except for halogen molecules [16]. Electron impact detachment ($e + \mathbf{A}^- \rightarrow \mathbf{A} + e + e$) is an important detachment mechanism for high degrees of ionization; associative detachment (see Section 3.3.2) dominates in non-equilibrium discharges.

### 3.3.2 Inelastic Collisions Between Heavy Particles

Collisions between pairs of ground-state atoms or molecules are generally elastic. However, if one or more of the particles are excited, a wide range of inelastic collisions can occur. These include VT relaxation, in which energy is transferred from vibrational excitation to translational kinetic energy; VV relaxation, in which vibrational energy is transferred from one vibrationally excited molecule to another; the analogous RT and RR relaxation processes that involve rotationally excited molecules; and relaxation of electronic excitation to vibrational and rotational excitation and translational kinetic energy.

VT relaxation is slow unless chemical reactions occur. This can lead to high levels of vibrational excitation in gases such as $N_2$, $H_2$, CO, and $CO_2$ for non-equilibrium discharges. VV collisions can lead to "ladder-climbing," with vibrational energy transferred from one excited molecule to another.

In contrast, RT collisions are relatively fast, and rotational and translational temperatures are generally similar in atmospheric-pressure plasmas, as noted in Section 3.3.1.

Relaxation of electronic excitation to translational energy is very slow, while transfer to vibrational energy can occur through formation of intermediate complexes and can be relatively fast. When the degree of ionization exceeds $10^{-6}$, transfer of electronic energy is dominated by superelastic collisions with electrons [16].

Vibrational and electronic excitations increase the rate of endothermic reactions. The reaction rate coefficient in the case of vibrational excitation can be expressed as

$$k_R(E_{\mathrm{vib}}, T_h) = k_{R0} \exp\left(-\frac{E_a - \alpha_{\mathrm{FM}} E_{\mathrm{vib}}}{k_B T_h}\right) \tag{3.17}$$

when the activation energy $E_a > \alpha_{\mathrm{FM}} E_{\mathrm{vib}}$, where $k_{R0}$ is the pre-exponential term for ground-state molecules, $E_{\mathrm{vib}}$ is the vibrational excitation energy, and $\alpha_{\mathrm{FM}}$ is the efficiency of the excitation energy in overcoming the reaction barrier. For exchange-type reactions **A** + **BC** → **AB** + **C**, the Fridman–Macheret $\alpha$-model gives an approximate expression:

$$\alpha_{\mathrm{FM}} = E_{af} / \left(E_{af} + E_{ar}\right) \tag{3.18}$$

where *f* and *r* denote the forward and reverse reactions, respectively. The value of $\alpha_{\mathrm{FM}}$ is close to 1 for strongly endothermic reactions with activation energy close to the reaction enthalpy and close to 0 for exothermic reactions without activation energy [16]. Electronic excitation is harder to analyze, but the same general trends apply.

Dissociation of molecules through heavy-particle collisions requires at least one of the molecules to be excited. In the absence of electronic excitation, dissociation is a multi-step process requiring collisions between vibrationally excited molecules to transfer enough energy to one of the molecules to dissociate; this process is effective for a limited group of gases that includes $N_2$, $H_2$, CO, and $CO_2$.

While electron-impact ionization requires an electron with kinetic energy only slightly more than the ionization potential, heavy-particle collisions transfer much less energy to bound electrons. Ionization requires the colliding heavy particle to have a kinetic energy of 10–100 keV, which is rare in plasmas of industrial interest. However, the internal energy of one or more of the heavy particles can promote ionization. Penning ionization occurs when a metastable atom with excitation energy $E_m$ collides with another atom with ionization energy $E_{\text{ion}} < E_m$. This is particularly important for metastable helium atoms ($2^3$S), with an excitation energy of 19.8 eV, which is sufficient to ionize $N_2$, Ar, CO, etc. Associative ionization, in which the heavy particles bind to each other to form a molecular ion, can also occur.

Charge transfer reactions ($\mathbf{A}^+ + \mathbf{B} \rightarrow \mathbf{A} + \mathbf{B}^+$) can be resonant (typically if A = B) or non-resonant. The cross-section is much larger in the resonant case. Associative detachment ($\mathbf{A}^- + \mathbf{B}^+ \rightarrow (\mathbf{AB}^-)^* \rightarrow \mathbf{AB} + e$) is the most important electron detachment mechanism in non-equilibrium discharges. Collisions with excited particles ($\mathbf{A}^- + \mathbf{B}^* \rightarrow \mathbf{A} + \mathbf{B} + e$) can also lead to detachment if the excitation energy of collision partner B* exceeds the electron affinity of particle A; the process is important in warm and thermal plasmas.

### 3.3.3 Equilibrium in Plasmas

The properties of plasmas can be described in terms of equilibrium statistical distributions. The Maxwell–Boltzmann distribution of kinetic energies was given by Eq. (3.4). The distribution of excited states can be described by a Boltzmann distribution:

$$n_j = n_0 \frac{g_j}{g_0} \exp\left[ - \frac{(E_j - E_0)}{k_B T} \right] \tag{3.19}$$

where $n_j$ and $n_0$ are the number densities of particles in the state $j$ and the ground state, respectively, $g_j$ and $g_0$ are the statistical weights of the two states, and $E_j$ and $E_0$ are the energies of the two states.

Ionization equilibrium (for the reaction $\mathbf{A} \leftrightarrow \mathbf{A}^+ + e$) is described by the Saha equation:

$$\frac{n_e n_i}{n_a} = \frac{g_e g_i}{g_a} \left( \frac{m_a T}{h} \right)^{3/2} \exp\left( - \frac{E_{\text{ion}}}{k_B T} \right) \tag{3.20}$$

where subscripts $e$, $i$, and $a$ refer to electrons, ions, and atoms, respectively, and $h$ is Planck's constant. A similar expression can be used to describe dissociation equilibrium.

Plasmas in which all species follow the Maxwell–Boltzmann distribution (3.4) for translational energy distribution, the Boltzmann distribution (3.19) for all excited states, the Saha Eq. (3.20) for ionization equilibrium, and chemical equilibrium equations for all dissociation and other reactions, with the same temperature applying in each case, are said to be in LTE. Complete thermodynamic equilibrium (CTE), which also requires the spectral density to follow a Planck distribution, is never realized in laboratory plasmas. However, most regions of thermal plasmas can usually be described by LTE because of the high collision rate at atmospheric pressure. However, the large gradients that occur in the edge regions and close to the electrodes in thermal plasmas can lead to departures from LTE. Such regions can often be described as two-temperature plasmas, with $T_e > T_h$. In this case, electrons and heavy particles follow separate Maxwell–Boltzmann distributions, and the Boltzmann distributions of excited electronic and vibrational states are described by $T_e$. The equilibrium of chemical reactions driven by electrons is largely determined by the electron temperature; as a consequence, the Saha equation is modified [8].

As discussed in Section 3.1.2, equilibration between translational energies of electrons and heavy particles takes tens to hundreds of nanoseconds, and shorter-lived discharges do not reach LTE. In most cases, separate Maxwell–Boltzmann distributions can still be applied to electron and heavy-particle translational energies, and the rotational excitation temperature $T_{\text{rot}} \approx T_h$. However, since vibrational excitation can be much faster than VT relaxation, the vibrational temperature $T_{\text{vib}}$ is often higher than $T_h$. Further, higher vibrational states can have higher densities than given by the Boltzmann distribution. This accelerates the rate of chemical reactions in which the vibrational states take part in accordance with Eq. (3.17).

The differences in $T_e$, $T_h$, and $T_{\text{vib}}$ that occur in non-equilibrium plasmas lead to departures from chemical equilibrium. Departures can also occur when rates of particle transport (diffusion and convective flow) exceed rates of chemical reactions. For example, recombination reactions between atoms require three-body interactions to conserve momentum and kinetic energy and are therefore relatively slow; even in plasmas in thermal equilibrium, the rapid diffusion driven by steep temperature gradients can lead to more atoms being present at low temperatures than predicted by chemical equilibrium [17].

Departures from non-equilibrium chemistry facilitate many plasma applications. Nevertheless, in gas production applications, chemical equilibrium considerations often determine the optimum plasma type for a given reaction, with higher temperatures being favored for reactions with negative enthalpy $\Delta H$ and positive entropy $\Delta S$, since the Gibbs free energy $\Delta G = \Delta H - T\Delta S$ decreases with increasing temperature.

## 3.4 Plasma Technology

Over the past 150 years, plasmas have been increasingly used for a huge range of applications including lighting, welding, cutting, deposition, water disinfection, mineral processing, semiconductor etching, and gas cleaning. The applications take advantage of many different plasma capabilities, including providing high temperatures and high fluxes of energy, reactive species, energetic ions, and radiation, which promote physical changes and chemical reactions that are inaccessible in the non-ionized gas phase. Most of these capabilities are only accessible with particular types of plasma. Accordingly, a wide range of plasma reactors has been developed to produce plasmas with the required properties. Reactor design is the subject of Chapter 4. This section considers the range of plasma types and properties that are used in applications and the principles underlying the production of the corresponding plasmas.

The plasma types of industrial interest in Figure 3.1 are labeled as glow discharges, DBDs, and warm plasmas (examples of non-equilibrium atmospheric-pressure discharges), and arcs (an example of thermal plasmas). The range of electron temperatures, approximately 1–10 eV, is relatively narrow, while the range of electron densities, from $10^{16}$ to $10^{25}$ $m^{-3}$, is much greater. The latter reflects wide ranges of both pressure and fractional ionization $n_e/n$. Pressure is determined by the filling pressure of the vessel containing the plasma (or the surrounding atmosphere if there is no vessel) and the energy supplied to the plasma. The fractional ionization depends on these factors, as well as the lifetime and dimensions of the plasma.

Energy can be supplied by different methods. The great majority of plasmas of industrial interest are generated by electric fields with frequencies ranging from DC to microwave. DC plasmas are generated between two electrodes, with the current being carried by electrons and ions. AC plasmas are excited by AC electric fields in the range from 50 Hz to around 100 kHz; the frequency is low enough for the ions to follow the oscillation of the field, so both electrons and ions contribute to the current. In both DC and AC plasmas, Joule heating dominates. For RF plasmas (in the MHz range), the excitation frequency exceeds the ion plasma frequency for the lower ion densities that occur at the low pressures typically used for such plasmas (see Section 3.1.2), so the ions remain almost at rest. Further, the displacement current rather than conduction carries the current. As well as Joule heating of the bulk plasma, stochastic heating (in which the electrons from the bulk plasma collide with the sheath regions) is significant. Finally, in microwave plasmas (in the GHz range), while the electrons can still (just) follow the field, the electromagnetic wave cannot fully penetrate the plasma because the skin depth of the wave is smaller than the plasma radius.

### 3.4.1 Low-Pressure Plasma Applications

Low-pressure plasmas, which are typically glow discharges, are used for the etching of metals and semiconductors, deposition and growth of thin film coatings, functionalization of surfaces, and generation of light. Typical pressures are in the range of 0.1–100 Pa ($10^{-6}$–$10^{-3}$ bar), and the value of $pd$ is usually less than $pd_{\mathrm{min}}$ (see Eq. (3.15)).

Plasma etching typically uses an RF discharge; coupling to the plasma can be capacitive, inductive, or through excitation of a plasma wave. The aim of the process is to transfer patterns, usually defined by lithography, into a thin film, meeting specifications for etching rate, material selectivity, feature size, and aspect ratio. A critical advantage of plasma etching over liquid- or vapor-phase techniques is that plasmas can etch a substrate anisotropically. Anisotropy is critical in integrated circuit manufacturing, where the spacing between features is now in the 5–10 nm range, and is achieved by the production of energetic ions directed toward the substrate. The ion energy distribution function depends on sheath potential and width and the ion mean free path, mass, and transit time compared to the RF period $2\pi/\omega$ [9]. Large mean free paths require low pressure, for which the time between collisions is large (see Figure 3.2d); however, this also decreases the plasma density and etching rate. This is no longer a critical problem since modern etching equipment can control both plasma density and ion bombardment characteristics independently [18].

Plasma deposition is achieved using many techniques, which are based on chemical vapor deposition (CVD) or physical vapor deposition (PVD). CVD uses high temperatures to promote gas-phase chemical reactions that produce reactive species that are deposited on a substrate, while PVD uses physical means, such as heating or ion sputtering, to produce the gas-phase species from a target.

The application of a plasma to CVD and PVD has several advantages. The process can take place at a lower temperature, allowing heat-sensitive substrates to be used. Applying a bias voltage to the substrate can be used to accelerate ionized precursors; these energetic ions transfer energy to the film, changing the film's composition and structure. For example, the top layer of the film can be annealed, improving and stabilizing the film properties without the need for a separate post-annealing process [18].

Plasma-enhanced CVD (PECVD) typically uses an RF plasma. The plasma makes available a greater range of species, including radicals and molecular fragments, which are mainly produced by electron-impact dissociation of gas-phase species. Higher deposition rates (1–10 nm $s^{-1}$ or more) can be achieved than by CVD [19]. A wide range of pressures is used, depending on the required film composition and properties. PECVD is used for the deposition of many materials, including hydrogenated amorphous silicon (a-Si:H), mono- and nanocrystalline silicon (mc/nc-Si), hydrogenated silica ($SiO_2$:H) and silicon nitride ($SiN_x$:H), diamond-like carbon, TiN, and $TiO_2$, which are used, e.g. in solar cells, integrated circuits, and as wear-resistant and low-friction coatings.

Plasma-enhanced PVD methods include magnetron sputtering and cathodic arc deposition [20]. Magnetron sputtering is a plasma-enhanced PVD process in which permanent magnets are placed behind the target to produce a magnetic field that restricts electron motion, increasing the plasma density and, therefore, the flux of ions to the target. The plasma is produced using a DC discharge if the target is metallic or an RF discharge for an insulating target. HiPiMS (high-intensity impulse magnetron sputtering) is a recently developed process that uses a pulsed high-power plasma to increase the ionization of sputtered atoms and create high fluxes of depositing atoms.

Cathodic arcs use the action of short-lived DC arcs between the cathode (the material to be evaporated) and an anode to produce a high flux of ions. The ions are produced from "cathode spots," which are small transient molten regions, and are accelerated up to $\sim 10^5\ \mathrm{m\,s^{-1}}$ by the strong local pressure gradient and electric field of the expanding plasma. Cathodic arcs can be operated at very low pressures, or a gas can be added to react with the ions. The ions can be steered by a magnetic field to eliminate macroparticles before deposition onto a substrate. Applications include the deposition of wear-resistant tool coatings such as TiN and TiAlN and diamond-like carbon [18].

Low-pressure lamps, such as the fluorescent lamps used for domestic and industrial lighting, are gradually being replaced by LED lamps. In most discharge lamps, an AC current between two electrodes at either end of a glass tube is used to produce a glow discharge in low-pressure argon, with a small amount of mercury vapor used to emit UV radiation at 254 nm. For visible lighting, a phosphor is deposited on the inside of the glass tube. For UV lighting, used for disinfection and ozone production, no phosphor is used, and the glass is replaced by quartz to ensure good transmission. RF inductive coupling is also used, particularly for more powerful lamps [21].

While, in principle, low-pressure plasmas could be used for ammonia and $NO_x$ synthesis, the production rate would be strongly limited by the low gas density, so such a system would be unlikely to be of practical interest.

### 3.4.2 Non-equilibrium Atmospheric-Pressure Plasma Applications

Non-equilibrium atmospheric-pressure plasmas have been used for over a century for the production of ozone and for many decades for surface functionalization and electrostatic precipitation [22]. In the past 10–20 years, a much wider range of applications, such as gas cleaning and production, human medicine including disinfection, wound treatment and cancer treatment, food disinfection, and

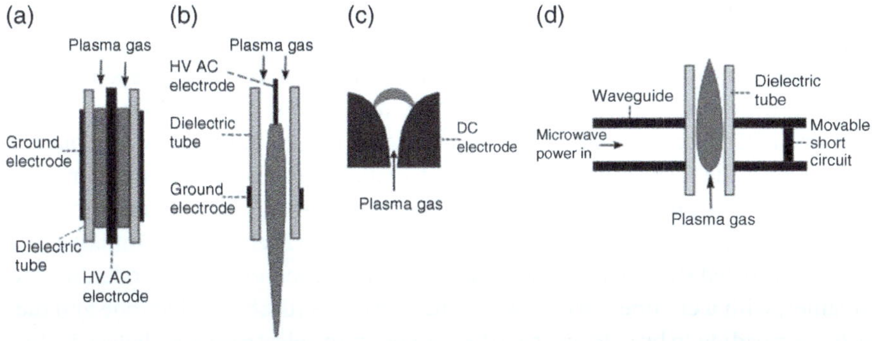


**Figure 3.6** Schematics illustrating (a) cylindrical dielectric barrier discharge, (b) atmospheric-pressure plasma jet, (c) gliding arc discharge, and (d) microwave discharge.

promotion of seed germination and plant growth, have been developed and are, in some cases, in commercial use.

We can differentiate between non-equilibrium cold plasmas, where the gas temperature ranges from room temperature to several hundred kelvin, and non-equilibrium warm plasmas, where the gas temperature may be as high as a few thousand kelvin. In both cases, the electron temperature is at least 10,000 K.

Cold plasmas are generally glow discharges or corona discharges. An important consideration in generating a cold plasma is preventing the transition from a streamer to a spark or arc discharge, as noted in Section 3.2. Common methods are the imposition of one or more dielectric barriers between the electrodes (the DBD), shown schematically in Figure 3.6a, and limiting the discharge duration by using a nanosecond-pulsed power supply. In both cases, the discharge duration is typically less than 100 ns. The dielectric barrier is effective because charge builds up on the dielectric, reducing the electric field to below that required to sustain the discharge.

Using a high gas flow or a noble gas, for which vibrational excitation does not occur, also reduces the heating of the heavy particles and delays the transition to a spark. This approach has been applied to atmospheric-pressure plasma jets (see Figure 3.6b), in which gas is passed through a narrow (~1 mm radius) tube and ionized using a high-frequency AC voltage. In most cases, a dielectric barrier is used between the electrodes, but this is not always required. Microplasmas, in which an array of small (~100 μm or less) plasmas is produced, are stabilized using a ballast resistor to limit the current and therefore heating [23].

DBDs are particularly widely used. Many configurations have been developed [24]. Figure 3.6a shows a cylindrical DBD with the dielectric barrier adjacent to the ground electrode. The dielectric barrier can also be adjacent to the high-voltage electrode, or two dielectric barriers can be used. Parallel flat-plate electrodes are an alternative to a cylindrical configuration. The discharge gap is in the range of 0.1–10 mm, and the voltage required is in the range of 1–100 kV, depending on the thickness of the discharge gap and the dielectric(s). Such DBDs are called "volume DBDs" because the discharge occurs in the gas gap volume. The plasma is usually in the form of large numbers of spatially localized microdischarges of radius about 100 μm, which are initiated shortly before the peak voltage is reached on each half cycle of the applied AC voltage. The microdischarges are initiated by a cathode-directed streamer, followed after a few nanoseconds by an anode-directed streamer, with a channel formed when the streamers reach the electrode and dielectric, spreading to broader surface discharges on the electrode and dielectric. In a 1 mm air discharge gap, the duration of the individual microdischarges is between 10 and 100 ns; the duration can be longer in argon [9]. Typical electron densities in the microdischarges are $10^{20}$ to $10^{21}$ $m^{-3}$ [22]. The reduced electric field is large, usually 100 Td or more, leading to high electron energies (at least 10 eV), which couple to dissociation and ionization reactions rather than vibrational excitation (see Figure 3.5). As noted above, the microdischarges extinguish because the charge transferred to the dielectric barrier reduces the electric field across the gas gap.

It is possible, e.g. by lowering the pressure, using helium or helium-rich gases, or choosing particular electrode arrangements, to obtain diffuse non-filamentary discharges [22, 25]. Another option is surface DBDs, in which the electrodes are placed on either side of the dielectric, with the discharge appearing on the exposed dielectric surface next to the electrode that does not cover the whole dielectric surface.

Packed-bed DBDs are usually cylindrical, with the discharge region filled with beads of a dielectric material. Packed-bed DBDs combine both volume and surface discharges since microdischarges occur both along the surface of the beads and in the gaps between the beads. They are a widely used arrangement for plasma catalysis and plasma gas cleaning since they allow close contact between the plasma and the catalytic or reactive surface. Plasma catalysis is an intensively researched application of non-equilibrium atmospheric-pressure plasmas. The plasma provides energy to the reactants, allowing catalytic surface reactions to occur at lower gas temperatures and pressures than in thermal catalysis [26].

Atmospheric-pressure plasma jets are usually DBDs, as noted above. Different electrode arrangements are used, including the pin high-voltage electrode and ring ground electrode shown in Figure 3.6b. Such jets allow treatment of a surface by the non-equilibrium plasma, which can have a heavy-particle temperature close to ambient if the plasma gas is a noble gas. This suits the jet to the treatment of heat-sensitive surfaces, facilitating applications such as sterilization of foods and other materials, direct application to skin for wound healing, and treatment of dermatological conditions [27]. Floating electrode DBDs are also used for such applications [28]. FE-DBDs usually have a flat plate high-voltage electrode covered by a dielectric, with the biological material serving as the ground electrode.

The interactions of atmospheric-pressure plasmas with liquids are critical in many applications. The discharge can occur within the liquid, above the liquid surface (possibly with the liquid serving as an electrode), in bubbles inside the liquid, and in a dispersed liquid (an aerosol) [29]. Applications such as wound and food treatment rely on the complex chemical reactions occurring at the boundary between the plasma and the biological material, particularly those involving reactive oxygen and nitrogen species. Discharges within the liquid can take the form of streamers or pulsed spark-type discharges. Bubble-type discharges can be formed by feeding gas through an electrode or through a separate gas diffuser. An important application of plasma discharges in liquids and bubbles is water treatment, with the degradation of many organic contaminants, including PFAS (per- and poly-fluoroalkyl substances), having been demonstrated [30].

Warm plasmas can be produced using sparks, microwave discharges, and gliding arc discharges. Spark discharges occur between two electrodes with a DC or AC voltage applied, but with the current limited to a few amperes. Gliding arc discharges use two electrodes whose separation increases with height, as shown in Figure 3.6c. A DC arc is struck at the bottom, where the separation is small, and convection moves the arc upward until it is too long to be supported by the power supply voltage, increasing the degree of non-equilibrium and eventually causing the arc to extinguish. Rotating gliding arc discharges apply swirl to the gas flow, allowing a larger proportion of the gas to pass through the discharge zone. Microwave discharges can be produced by many means; the most common arrangement at or near atmospheric pressure is for the plasma to be ignited in a dielectric tube inserted through a waveguide (Figure 3.6d). Microwave discharges are often operated somewhat below atmospheric pressure, which gives a broader plasma region.

Typical warm plasma electron densities at 1 bar are around $10^{18}$ to $10^{19}\,m^{-3}$, electron temperatures are in the range of 1–2 eV, and heavy-particle temperatures are at least 1,000 K. Warm plasmas are characterized by high levels of vibrational excitation, which is facilitated by the relatively low $E/N$ values, typically around 10 Td, as illustrated in Figure 3.5.

Both cold and warm atmospheric-pressure plasmas are widely used in the research and development of gas conversion processes. As discussed in Section 3.3.3, the choice is, to a significant extent, governed by the thermodynamics of the reaction. Processes such as ozone production from oxygen, ammonia production from nitrogen and hydrogen, and carbon dioxide methanation are exothermic and decrease entropy, so chemical equilibrium favors low temperatures, i.e. cold plasmas. Production of nitric oxide from nitrogen and oxygen, carbon dioxide splitting, and dry reforming of methane are endothermic and maintain or increase enthalpy and are therefore favored by high temperatures, i.e. warm plasmas. The presence of a catalyst is important if selectivity is required (e.g. in carbon dioxide hydrogenation reactions) and can also assist in reducing energy barriers.

Results bear out these points. For example, extensive research on carbon dioxide splitting and dry reforming of methane has been carried out using dielectric barrier, gliding arc, and microwave discharges, with the best conversion and energy efficiencies obtained in the warm plasmas [31]. Similarly, warm plasmas outperform cold plasmas in nitric oxide production [32]. Commercial ozone production uses DBDs, with water cooling used to keep the temperature as low as possible [22]. Ammonia production has been shown to be favored by DBDs and the presence of a catalyst [15], as discussed in Section 3.5. Carbon dioxide methanation requires an appropriate catalyst and is again performed successfully in a DBD [33].

### 3.4.3 Thermal Plasma Applications

Thermal plasmas are those in which the electron temperature and heavy-particle temperature are approximately equal. Large regions of thermal plasmas are typically in LTE. As discussed in Section 3.3.3, this requires a high collision frequency, which in turn requires a pressure close to or above 1 bar (see Figure 3.2d). The level of ionization is high – at least 5% single ionization, and often 100% single ionization and significant double ionization. At atmospheric pressure, once full single ionization is reached, $n_e \approx 2 \times 10^{23}\ \mathrm{m}^{-3}$.

The typical mechanism of thermal plasma formation, discussed in Section 3.2, involves the initiation of a streamer discharge that spans the gap between two electrodes to form a spark discharge, with a power supply that can provide the current necessary to maintain the discharge. The high electron density leads to a relatively low resistance, so the voltage is generally low – corresponding to $E/n \sim 2\ \mathrm{kV\ m}^{-1}$ in the arc column plus about 10–30 V across the sheath regions. This is well below the electric field required to break down air (see Figure 3.4), so a thermal plasma is generally initiated by short-circuiting and then separating the electrodes or with an RF pulse to reduce the breakdown voltage. Thermal plasmas can also be produced by an inductively coupled RF field.

We can divide thermal plasmas into three main classes, based on their excitation mechanism and electrode configuration, as shown in Figure 3.7 [8, 34].

Transferred arcs are formed between two electrodes, one of which is the "workpiece" – the material being processed by the plasma. Transferred arcs are usually DC but can also be low-frequency AC. Transferred arcs can reach temperatures of 20,000 K or more and are characterized by high current densities (up to about $10^8\ \mathrm{A\ m}^{-2}$), which heat the plasma by Joule heating. The high temperature

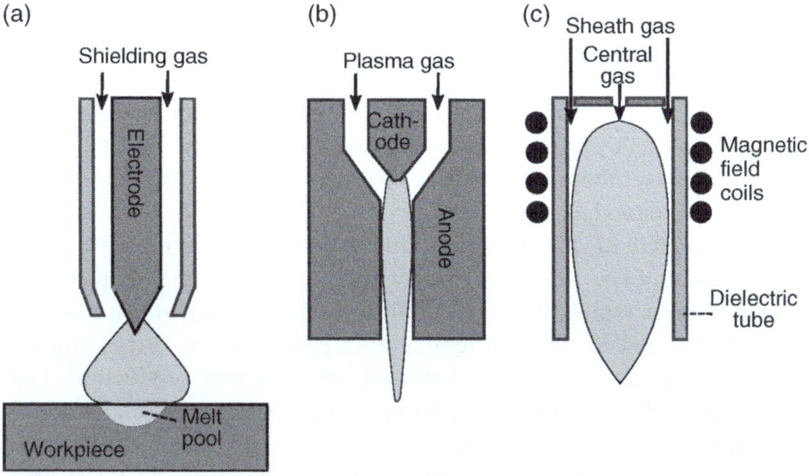


**Figure 3.7** Schematics illustrating (a) transferred arc, (b) non-transferred arc, and (c) inductively coupled RF thermal plasma. *Source:* Ref. [34]/with permission of IOP Publishing.

leads to intense radiative emission. The high heat fluxes associated with the high temperatures and current densities allow metals and other materials to be melted; this is applied in arc welding, plasma cutting, plasma waste treatment, electric arc furnaces for metallurgical applications, and chemical applications. Arc currents for most such applications are in the range from 50 to 500 A but can be much higher in metallurgical applications. For such currents, the interaction between the current and the self-induced magnetic field leads to a strong magnetic pinch ($\boldsymbol{j} \times \boldsymbol{B}$) force, which increases the pressure and drives flow, with flow speeds reaching several hundred meters per second. The strong radiative emission is used in arc lamps, e.g. in car headlights and data and movie projectors. In such lamps, a short, low-current arc is initiated between two electrodes in gas at several bars; the gas flow speed is much lower and is driven by buoyancy rather than the magnetic pinch force.

Non-transferred arc plasmas are also formed between two electrodes, but in this case, the electrodes are within a "plasma torch," which produces a jet of plasma through an aperture in one of the electrodes. While the temperature within the torch can reach 25,000 K or more, the jet temperature is typically lower than 15,000 K at the aperture and cools rapidly in the atmosphere. The high temperature within the torch leads to rapid thermal expansion, and the plasma jet reaches high speeds, hundreds to thousands of meters per second, and is often supersonic (even though the speed of sound increases with temperature because of the decreased mass density). The most important application of non-transferred arcs is plasma spraying, in which a ceramic powder is injected into the jet, which melts and accelerates the particles onto a substrate, forming a coating of thickness up to around 0.1 mm. Typical applications are thermal barrier coatings on jet engine turbine blades and wear-resistant coatings on paper rollers. Non-transferred plasmas are also used for plasma waste treatment and gasification, for gas heating (e.g. to simulate the heat fluxes that occur in space vehicle re-entry into the atmosphere), and for chemical production.

Inductively coupled RF plasmas are induced by a coil carrying RF current that is wound around a dielectric tube, with the plasma ignited inside the tube. The maximum temperature is significantly lower than in arc plasmas, typically around 12,000 K, and the temperature gradients are lower. Applications such as particle spheroidization and nanoparticle production take advantage of the relatively uniform plasma, which provides similar temperature histories for all precursors [8, 35]. In particle spheroidization, particles with sizes in the range of 10–100 μm are fed into the top of the plasma, where they melt and form spheres. A typical application is the production of metal powders for additive manufacturing. Nanoparticle production uses a solid, liquid, or gaseous precursor that is fed into the top of the plasma; a rapid quench of the flow at the bottom of the plasma leads to nucleation of particles in the 10–100 nm size range. Inductively coupled RF plasmas are also a critical component of analytical chemistry methods such as ICP-MS and ICP-OES (inductively coupled plasma – mass spectroscopy and optical emission spectroscopy); the chemical to be tested is dissolved in a liquid and injected into the plasma, where it is vaporized, and the elemental chemical composition is measured by MS, OES, or another analytical technique [36].

Thermal plasmas have been and continue to be successfully applied to large-scale chemical production [8, 16]. The Birkeland–Eyde process that was developed for $NO_x$ production will be discussed in Section 3.5. The Hüls process, initially developed in the 1940s, continues to be used for the production of acetylene from methane and hydrogen, with the methane introduced downstream of hydrogen in a non-transferred arc [8, 37]. The Tioxide process used a non-transferred arc to produce pigment-grade titanium dioxide from titanium tetrachloride vapor and oxygen [8]. Thermal plasmas are now being used to pyrolyze methane, producing carbon black and hydrogen. The largest such facility, run by Monolith in Ohio, USA, uses a transferred arc process [38]. The hydrogen is used as the input for producing ammonia by the Haber-Bosch process, while the carbon black is sold to the tire industry. The process uses less than 20% of the energy required to produce hydrogen by electrolysis of water. Non-transferred arcs are also being used for methane pyrolysis.

## 3.5 Role of Plasma in Ammonia and $NO_x$ Synthesis

The oxidation of nitrogen in a plasma to form $NO_x$ was first reported by Lord Rayleigh in 1897 [39]. He used a ~1 kW transferred arc plasma between two platinum electrodes in a 50 L glass vessel supplied with a mixture of 45% air and 55% oxygen. The gas produced was passed through an alkali solution to produce about 50 g of nitric acid per hour.

The production of $NO_x$ was one of the earliest industrial implementations of plasmas. The Birkeland–Eyde process [40, 41], developed in the early 1900s, used a thermal arc plasma between water-cooled copper electrodes to produce NO from air, with subsequent oxidation to $NO_2$ in a separate oxidation chamber, and reaction with water to form a nitric acid solution. The first plant was used in potassium nitrate manufacture in Notodden, Norway, with three 0.5 MW arc furnaces. A commercial plant was established in Niagara Falls, Canada, using hydroelectric power [42]. The power supplied increased from 2 kW in 1903 to 239 kW in 1928, with up to 38 kt of nitrogen fixed each year. The process produced 1–2 mol% NO and consumed 2.4–3.1 MJ $mol^{-1}$ N.

The invention of the Haber–Bosch process for ammonia production, whose energy requirements are much lower, eventually led to the demise of the Birkeland–Eyde process. Only recently, with the growing importance of renewable energy sources and carbon emission reduction, have electrically based processes such as those using plasmas become prominent again.

The production of $NO_x$ and $NH_3$ provides illustrative examples of the diverse chemical reactions that can occur in a plasma. Production of NO is an endothermic reaction:

$$N_2 + O_2 \rightarrow 2NO \quad \Delta H_0 = 90.3\ \text{kJ mol NO}^{-1} = 0.936\ \text{eV molecule NO}^{-1} \tag{3.21}$$

However, the standard entropy is positive (12.4 J $K^{-1}$ mol $NO^{-1}$), so the Gibbs free energy becomes negative as temperature increases. For a mixture of 80% nitrogen and 20% oxygen, the maximum NO production in chemical equilibrium is about 2% and occurs around 3,500 K. As the temperature increases further, the NO dissociates. The theoretical minimum energy requirement in a thermal plasma, based on the energy required to break the $N_2$ triple bond (945 kJ $mol^{-1}$ = 9.79 eV) and the $O_2$ double bond (498 kJ $mol^{-1}$ = 5.16 eV), is 0.72 MJ mol $NO^{-1}$.

The reaction to produce NO proceeds via the Zeldovich mechanism:

$$\begin{aligned} N_2 + O &\rightarrow NO + N \\ O_2 + N &\rightarrow NO + O \end{aligned} \tag{3.22}$$

The first of these reactions has an energy barrier of about 8 eV, which is reduced to about 3 eV for vibrationally excited nitrogen. This gives a theoretical minimum energy requirement of just 0.2 MJ mol $NO^{-1}$ in the case of optimum vibrational excitation [42]. Accordingly, warm plasmas such as spark, microwave, and gliding arc discharges, in which energy is coupled to vibrational modes, offer the best energy efficiencies, with approximately 2 MJ mol $NO^{-1}$ being obtained. This is still well above the theoretical maximum, partly because of VT coupling taking energy away for vibrational modes. Another problem is the reverse reaction, which is favored by chemical equilibrium as the plasma cools; rapid quenching is therefore beneficial.

In contrast to NO production, $NH_3$ production is exothermic at room temperature:

$$N_2 + 3H_2 \rightarrow 2NH_3 \quad \Delta H^0 = -45.9\,\text{kJ mol}\,{NH_3}^{-1} = -0.476\,\text{eV molecule}\,{NH_3}^{-1} \tag{3.23}$$

and the standard entropy is negative. Therefore, chemical equilibrium favors low temperatures. However, significant energy is required to overcome the energy barrier of the reaction. As discussed in Chapter 1, a catalyst is used to reduce the energy barrier; however, activation of the catalyst typically requires temperatures of around 500 °C, at which the equilibrium favors the reactants. The Haber–Bosch process used high pressures (over 100 bar) to shift the equilibrium toward the products and obtain reasonable $NH_3$ yields.

Plasmas can overcome these restrictions because the significant energy coupled to electrons allows the formation of N and H radicals, which can react with each other and $N_2$ and $H_2$ to form ammonia in the gas phase. Chemical equilibrium favors low-temperature plasma such as DBDs. Using such plasmas, ammonia can be formed at a gas temperature of 300 K and atmospheric pressure. However, the significant energy required to dissociate N and H means the energy efficiency is low; also, the $NH_3$ concentrations produced have been low. Consequently, researchers have focused on combining plasmas and catalysis. In this case, the lower energy pathways, such as vibrational excitation of $N_2$ in the plasma followed by dissociative adsorption on the catalyst, may be possible. Single-pass $NH_3$ yields of over 5% have been achieved, but while the energy efficiencies are much higher in the presence of a catalyst, they remain uncompetitive [15].

These are just two examples of the approaches used to fix nitrogen using plasma-based and plasma-assisted processes. Subsequent chapters, including Chapters 5, 6, 9, 10, 11 and 12, will provide a much more detailed assessment of the different approaches and their relative benefits and drawbacks.

## 3.6 Conclusions

Plasma technology harnesses the diverse chemical, physical, and electrical phenomena offered by electrical discharges to etch semiconductors, deposit thin and thick films, melt and process metals and minerals, treat surfaces, emit radiation, drive chemical reactions, and contribute to many other processes. Such applications are relied upon in industries including materials, metal, semiconductor, mineral and chemical processing, automotive and aerospace manufacturing, information technology, energy supply, environmental protection and remediation, and biomedical and are hence indispensable to the modern economy.

In this chapter, we have considered many of these applications from the perspective of the underlying plasma properties and processes. The huge range of electron densities and electron and heavy-particle temperatures that occur in different types of naturally occurring and technological plasmas can be classified according to Debye length, plasma frequencies, mean free paths and collision times, and other properties. These properties are helpful in understanding the different electrical breakdown processes and the wide variety of chemical reactions and physical phenomena that are taken advantage of in plasma applications. These chemical and physical phenomena, in turn, inform the choice of design parameters, including power supply frequency, voltage and current, background pressure, reactor geometry, and process gas composition and flow rate, that allow the underlying chemistry and physics to be tailored to a particular plasma application.

The renewed effort to electrify the chemical industry, motivated by the need to reduce carbon emissions, has highlighted the potential of plasmas for nitrogen fixation. The usefulness of plasmas driving the required reactions has been known for well over a century, and plasma technology was implemented for industrial $NO_x$ production for over 25 years in the early 1900s. The many advances in the understanding of plasma processes and their applications, some of which we have considered in this chapter, form the basis for the renewed effort to apply plasmas for this important goal, as will be seen in the remainder of this book.

### Nomenclature

| | |
|---|---|
| $A$, $B$ | constants [−] |
| $\mathbf{B}$ | magnetic field strength [T] |
| $D_{\text{ion}}$ | ionization length [m] |
| $e$ | electronic charge [C] |
| $E$ | electric field strength [V m$^{-1}$] |
| $E_a$ | activation energy [J] |

| | |
|---|---|
| $E_{af}$ | activation energy of forward reaction [J] |
| $E_{ar}$ | activation energy of reverse reaction [J] |
| $E_{\mathrm{ion}}$ | ionization energy [J] |
| $E_{\mathrm{kin}\,j}$ | kinetic energy of species $j$ [J] |
| $E_m$ | excitation energy of metastable atom [J] |
| $g_a$ | statistical weight of atoms [–] |
| $g_e$ | statistical weight of electrons [–] |
| $g_i$ | statistical weight of ions [–] |
| $g_j$ | statistical weight of state $j$ [–] |
| $g_0$ | statistical weight of ground state [–] |
| $h$ | Planck's constant [J s] |
| $\boldsymbol{j}$ | current density [A m$^{-2}$] |
| $k_B$ | Boltzmann's constant [J K$^{-1}$] |
| $k_R$ | reaction rate coefficient [m$^3$ mol$^{-1}$ s$^{-1}$] |
| $k_{R0}$ | pre-exponential term for ground-state molecule in $k_R$ [m$^3$ mol$^{-1}$ s$^{-1}$] |
| $l_e$ | electron mean free path [m] |
| $m_j$ | mass of species $j$ [kg] |
| $n$ | total number density [m$^{-3}$] |
| $n_a$ | number density of atoms [m$^{-3}$] |
| $n_0$ | number density of particles in ground state [m$^{-3}$] |
| $n_e$ | electron density [m$^{-3}$] |
| $n_j$ | number density of particles in state $j$ [m$^{-3}$] |
| $n_0$ | number density of particles in ground state [m$^{-3}$] |
| $Q_{ea}$ | momentum transfer cross-section for electron-atom collisions [m$^2$] |
| $P$ | pressure [Pa] |
| $T_e$ | electron temperature [eV or K] |
| $T_h$ | heavy-particle temperature [K] |
| $T_{\mathrm{vib}}$ | vibrational temperature [K] |
| $V_b$ | breakdown voltage [V] |
| $v_d$ | electron drift velocity [m s$^{-1}$] |
| $Z_j$ | charge number of species $j$ |
| $\boldsymbol{v}_j$ | velocity of species $j$ [m s$^{-1}$] |
| $v_j$ | average velocity of species $j$ [m s$^{-1}$] |
| $v_{de}$ | mean electron drift velocity [m s$^{-1}$] |
| $\tau_e$ | mean time between electron collisions [s] |
| $\alpha$ | ionization coefficient (first Townsend coefficient) [m$^{-1}$] |
| $\alpha_{\mathrm{FM}}$ | efficiency of excitation energy in overcoming reaction barrier [–] |
| $\beta$ | recombination coefficient (second Townsend coefficient) [m$^{-1}$] |
| $\Upsilon$ | secondary ionization coefficient (third Townsend coefficient) [m$^{-1}$] |
| $\Delta H^0$ | standard reaction enthalpy [J] |

$\eta$ recombination coefficient [$m^{-1}$]
$\varepsilon_0$ permittivity of free space [$F\,m^{-1}$]
$\lambda_D$ Debye length [m]
$\omega$ angular frequency of applied voltage [$rad\,s^{-1}$]
$\omega_{pj}$ plasma frequency of species $j$ [$rad\,s^{-1}$]

## References


1 Langmuir, I. (1928). Oscillations in ionized gases. *Proc. Natl. Acad. Sci.* 14 (8): 627–637. https://doi.org/10.1073/pnas.14.8.627.
2 Howatson, A.M. (1965). *An Introduction to Gas Discharges*. Oxford: Pergamon Press.
3 Von Keudell, A. and Schulz-von der Gathen, V. (2017). Foundations of low-temperature plasma physics—an introduction. *Plasma Sources Sci. Technol.* 26 (11): 113001. https://doi.org/10.1088/1361-6595/aa8d4c.
4 Devoto, R.S. (1973). Transport coefficients of ionized argon. *Phys. Fluids* 16 (5): 616–623. https://doi.org/10.1063/1.1694396.
5 Lelevkin, V.M., Ortobaev, D.K., and Schram, D.C. (1992). *Physics of Non-equilibrium Plasmas*. Amsterdam: North-Holland Publishing Company.
6 Itikawa, Y. (1974). Momentum-transfer cross sections for electron collisions with atoms and molecules. *At. Data Nucl. Data Tables* 14 (1): 1–10. https://doi.org/10.1016/S0092-640X(74)80026-4.
7 Itikawa, Y. (1978). Momentum-transfer cross sections for electron collisions with atoms and molecules. Revision and supplement, 1977. *At. Data Nucl. Data Tables* 21 (1): 69–75. https://doi.org/10.1016/0092-640X(78)90004-9.
8 Boulos, M.I., Fauchais, P.L., and Pfender, E. (2023). *Handbook of Thermal Plasmas*. Cham: Springer Nature https://doi.org/10.1007/978-3-030-84936-8.
9 Bruggeman, P.J., Iza, F., and Brandenburg, R. (2017). Foundations of atmospheric pressure nonequilibrium plasmas. *Plasma Sources Sci. Technol.* 26 (12): 123002. https://doi.org/10.1088/1361-6595/aa97af.
10 Pancheshnyi, S. (2005). Role of electronegative gas admixtures in streamer start, propagation and branching phenomena. *Plasma Sources Sci. Technol.* 14 (4): 645–653. https://doi.org/10.1088/0963-0252/14/4/002.
11 Raizer, Y.P. (1997). *Gas Discharge Physics*, 2e. Berlin: Springer.
12 Montijn, C. and Ebert, U. (2006). Diffusion correction to the Raether–Meek criterion for the avalanche-to-streamer transition. *J. Phys. D. Appl. Phys.* 39: 2979–2992. https://doi.org/10.1088/0022-3727/39/14/017.

13 Lowke, J.J. (1992). Theory of electrical breakdown in air-the role of metastable oxygen molecules. *J. Phys. D. Appl. Phys.* 25 (2): 202–210. https://doi.org/10.1088/0022-3727/25/2/012.

14 Naidis, G.V. (2005). Conditions for inception of positive corona discharges in air. *J. Phys. D. Appl. Phys.* 38 (13): 2211–2214. https://doi.org/10.1088/0022-3727/38/13/020.

15 Rouwenhorst, K.H.R., Engelmann, Y., Veer, K.v.'t. et al. (2020). Plasma-driven catalysis: green ammonia synthesis with intermittent electricity. *Green Chem.* 22 (19): 6258–6287. https://doi.org/10.1039/D0GC02058C.

16 Fridman, A. (1998). *Plasma Chemistry*. Cambridge: Cambridge University Press.

17 Snyder, S.C., Murphy, A.B., Hofeldt, D.L., and Reynolds, L.D. (1995). Diffusion of atomic hydrogen in an atmospheric-pressure free-burning arc discharge. *Phys. Rev. E* 52 (3): 2999–3009. https://doi.org/10.1103/PhysRevE.52.2999.

18 Oehrlein, G.S. and Hamaguchi, S. (2018). Foundations of low-temperature plasma enhanced materials synthesis and etching. *Plasma Sources Sci. Technol.* 27 (2): 023001. https://doi.org/10.1088/1361-6595/aaa86c.

19 Martinu, L. and Poitras, D. (2000). Plasma deposition of optical films and coatings: a review. *J. Vac. Sci. Technol. A* 18 (6): 2619–2645. https://doi.org/10.1116/1.1314395.

20 Gudmundsson, J.T., Anders, A., and von Keudell, A. (2022). Foundations of physical vapor deposition with plasma assistance. *Plasma Sources Sci. Technol.* 31 (8): 083001. https://doi.org/10.1088/1361-6595/ac7f53.

21 Zissis, G. and Kitsinelis, S. (2009). State of art on the science and technology of electrical light sources: from the past to the future. *J. Phys. D. Appl. Phys.* 42 (17): 173001. https://doi.org/10.1088/0022-3727/42/17/173001.

22 Kogelschatz, U. (2003). Dielectric-barrier discharges: their history, discharge physics, and industrial applications. *Plasma Chem. Plasma Process.* 23 (1): 1–46. https://doi.org/10.1023/A:1022470901385.

23 Schoenbach, K.H. and Becker, K. (2016). 20 years of microplasma research: a status report. *Eur. Phys. J. D - At. Mol. Opt. Phys.* 70: 29. https://doi.org/10.1140/epjd/e2015-60618-1.

24 Pietsch, G.J. and Gibalov, V.I. (2012). Dynamics of dielectric barrier discharges in different arrangements. *Plasma Sources Sci. Technol.* 21 (2): 024010. https://doi.org/10.1088/0963-0252/21/2/024010.

25 Massines, F., Gherardi, N., and Naudé, N.S. (2009). Recent advances in the understanding of homogeneous dielectric barrier discharges. *Eur. Phys. J. Appl. Phys.* 47 (2): 22805. https://doi.org/10.1051/epjap/2009064.

26 Bogaerts, A., Neyts, E.C., Guaitella, O., and Murphy, A.B. (2022). Foundations of plasma catalysis for environmental applications. *Plasma Sources Sci. Technol.* 31 (5): 053002. https://doi.org/10.1088/1361-6595/ac5f8e.

**27** Weltmann, K.-D., Kolb, J.F., Holub, M. et al. (2018). The future for plasma science and technology. *Plasma Process. Polym.* 16 (1): 1800118. https://doi.org/10.1002/ppap.201800118.

**28** Fridman, G., Peddinghaus, M., Balasubramanian, M. et al. (2006). Blood coagulation and living tissue sterilization by floating-electrode dielectric barrier discharge in air. *Plasma Chem. Plasma Process.* 26 (4): 425–442. https://doi.org/10.1007/s11090-006-9024-4.

**29** Bruggeman, P.J., Kushner, M.J., Locke, B.R. et al. (2016). Plasma–liquid interactions: a review and roadmap. *Plasma Sources Sci. Technol.* 25 (5): 053002. https://doi.org/10.1088/0963-0252/25/5/053002.

**30** Blotevogel, J., Thagard, S.M., and Mahendra, S. (2023). Scaling up water treatment technologies for PFAS destruction: current status and potential for fit-for-purpose application. *Curr. Opin. Chem. Eng.* 41: 100944. https://doi.org/10.1016/j.coche.2023.100944.

**31** Snoeckx, R. and Bogaerts, A. (2017). Plasma technology – a novel solution for $CO_2$ conversion? *Chem. Soc. Rev.* 46 (19): 5805–5863. https://doi.org/10.1039/C6CS00066E.

**32** Abdelaziz, A.A., Komuro, A.T., Schiorlin, Y.M. et al. (2024). Atmospheric-pressure plasmas for $NO_x$ production: short review on current status. *Curr. Opin. Green Sust. Chem.* 50: 100977. https://doi.org/10.1016/j.cogsc.2024.100977.

**33** Ullah, S., Gao, Y., Dou, L. et al. (2023). Recent trends in plasma-assisted $CO_2$ methanation: a critical review of recent studies. *Plasma Chem. Plasma Process.* 2023 (6): 1335–1383. https://doi.org/10.1007/s11090-023-10417-9.

**34** Murphy, A.B. and Uhrlandt, D. (2018). Foundations of high-pressure thermal plasmas. *Plasma Sources Sci. Technol.* 27 (6): 063001. https://doi.org/10.1088/1361-6595/aabdce.

**35** Boulos, M.I. (2016). The role of transport phenomena and modeling in the development of thermal plasma technology. *Plasma Chem. Plasma Process.* 36 (1): 3–28. https://doi.org/10.1007/s11090-015-9660-7.

**36** Broekaert, J.A.C. (2005). *Analytical Atomic Spectrometry with Flames and Plasmas*. Weinheim: Wiley.

**37** Gladisch, H. (1962). How Huels makes acetylene by DC arc. *Hydrocarbon Process. Petrol. Refiner* 41: 159–164.

**38** Dames, E., Rohani, V., and Fulcheri, L. (2023). Plasma Chemistry and Plasma Reactors for Turquoise Hydrogen and Carbon Nanomaterials Production. In: *Turquoise Hydrogen: An Effective Pathway to Decarbonization and Value Added Carbon Materials* (ed. M. Pelucchi and M. Maestri), 253–317. Cambridge, MA: Elsevier https://doi.org/10.1016/bs.ache.2023.05.001.

**39** Rayleigh, L. (1897). Observations on the oxidation of nitrogen gas. *J. Chem. Soc. Trans.* 71: 181–186. https://doi.org/10.1039/CT8977100181.

**40** Birkeland, K. (1906). On the oxidation of atmospheric nitrogen in electric arcs. *Trans. Faraday Soc.* 2 (2): 98–116. https://doi.org/10.1039/TF9060200098.

**41** Eyde, S. (1912). Oxidation of atmospheric nitrogen and development of resulting industries in Norway. *J. Ind. Eng. Chem.* 4 (10): 771–774. https://doi.org/10.1021/ie50046a025.

**42** Rouwenhorst, K.H.R., Jardali, F., Bogaerts, A., and Lefferts, L. (2021). From the Birkeland–Eyde process towards energy-efficient plasma-based $NO_X$ synthesis: a techno-economic analysis. *Energy Environ. Sci.* 14 (5): 2520–2534. https://doi.org/10.1039/d0ee03763j.

# 4

# Plasma Reactors

*Evgeny Rebrov* [1,2]

[1] *School of Engineering, University of Warwick, Coventry, UK*
[2] *Department of Chemical Engineering and Chemistry, Eindhoven University of Technology, Eindhoven, The Netherlands*

## 4.1 Introduction

The reactor, the "heart" of chemical processes, has received more attention than other unit operations in the context of process development. A large number of plasma reactors are either on the market already or under development, and this chapter introduces several of the most important reactors for $NO_x$ synthesis. $NO_x$ synthesis is performed in three different plasma reactors: thermal plasma, cold plasma, and "warm" plasma. Thermal plasma includes the electric arc, the spark discharge, or plasma torch. The electric arc plasma was used in the Birkeland-Eyde process, in which 1 vol.% of $NO_x$ was obtained, and subsequently used for the production of $HNO_3$ with an energy costs of 3.4–4.1 MJ $mol^{-1}$ [1]. The production of nitric oxide (NO) in thermal plasmas involves both thermal Zeldovich mechanism and electron-impact initiated reactions [2]. NO is the main reaction product in thermal plasmas, while the production of $NO_2$ and $O_3$ is suppressed by thermodynamics of the respective reactions.

Cold plasmas are nonthermal plasmas (NTPs) with a gas temperature close to room temperature. They have lower energy density than thermal plasmas, and ozone is normally produced, which can further oxidize NO to $NO_2$ [3]. Cold plasmas have lower energy efficiency; therefore, they are not employed for a large-scale NO production. NTP reactors are used to produce NO and reactive oxygen species for small-scale agricultural, biomedical, and therapeutic applications, where the equipment costs and convenience of operation are more important than the energy costs [4]. In these applications, both dielectric barrier discharge (DBD)

*Plasma-Assisted Nitrogen Fixation for Sustainable Process Industries*, First Edition. Edited by Volker Hessel, Annemie Bogaerts, Gabriele Centi, Evgeny Rebrov, and Nguyen Van Duc Long.

Published 2026 by John Wiley & Sons, Inc.

and plasma jet reactors were employed to obtained NO from the gas phase and a water solution.

Warm plasmas, which combine both thermal and nonthermal conditions, are used for nitrogen fixation, and a considerable progress to increase their energy efficiency and product selectivity was reported in the patent literature [5]. Among the different reactor configurations, gliding arc (GA) is the most efficient one for $NO_x$ synthesis. The arc, initially formed at the narrowest gap between electrodes, moves along the conical-shaped electrode by the gas flowing toward the wider end of the gap until it extinguishes. After that, a new cycle starts with the arc formation at the narrowest gap again. Other types of warm plasma reactors, such as the direct current (DC) corona, and microwave plasma are also reported and discussed in this chapter.

Due to the strong coupling between the electromagnetic, thermal, and chemistry transport phenomena that takes place in the plasma, the full three-dimensional (3D) modeling of large-scale plasma reactors remains a challenging task. Fluidized- and fixed-bed reactors are not compatible with the high-flow rate conditions needed for short residence time; therefore, reactors with low catalyst loadings or thin catalytic layers often give the best results. In these reactors, the catalysts and/or dielectric (or ferroelectric) materials are deposited onto the reactor walls to enhance the electric field near their surface without increasing hydrodynamic resistance to the gas flow. This, in turn, leads to more intense plasma in these areas. However, it should also be noted that a very high dielectric constant of the coatings leads to a nonuniform plasma field, where only the first few micrometer or millimeter near the coated surface are utilized for chemical reactions, lowering the overall productivity and often energy efficiency. Therefore, there exists an optimum in dielectric constant of materials for applications in plasma reactors.

For a better understanding of the underling phenomena, flow reactors with the diameter in the millimeter range (so-called millireactors) are often used, resulting in the formation of microplasma. Microplasma is a plasma of small dimensions, ranging from tens of micrometers to several millimeters. The narrow reaction channel allows to use relatively small breakdown voltage; therefore, the heat generation rate and the plasma can be made more uniform as compared to large-scale reactors that experience considerable thermal and concentration gradients. The larger reactors are more difficult to cool, since the heat generated by plasma is proportional to the reactor volume, whereas the surface available to dissipate a given heat output is decreased. However, due to their relatively low volume, the millireactors cannot handle high throughputs and are currently used mainly for research purposes. Furthermore, they are not very energy efficient due to high thermal losses to the environment.

Another important parameter that has a major impact on the conversion and energy efficiency is the hydrodynamics inside the reactor. Due to the small hydraulic diameter of the millireactors, low Reynolds numbers are observed, and these reactors operate in laminar flow regime. Transport of reactive species in such reactors is mainly controlled by diffusion, which is a slow process. The parabolic velocity profile observed under laminar flow conditions means that the gas velocity is the highest in the center of the channel and the velocity at the walls is equal to zero. This creates further a mismatch between the areas of the highest plasma intensity (near the reactor walls) and the areas of shortest residence time (in the middle of the channel). Such a large distribution in residence time and in the plasma intensity translates in large differences in local conversion. Therefore, it is clear that plasma intensity distribution and hydrodynamics inside the reactor have a significant impact on the conversion and energy efficiency of the reactor.

In a large-scale plasma reactors, a recycling gas flow driven by the motion of charged particles is observed in the radial direction between the two electrodes, in the course of the elementary electrical breakdown. These reactors could approach plug flow operation with minimum axial and radial dispersion, allowing very tight control of the residence time as each element of fluid spends exactly the same amount of time in each environment in the reactor. This leads to reduced byproducts formation, as elements of fluid are not exposed to the areas of temperature hot spots. Many industrial reactors operate under autothermal rather than isothermal conditions when the heat production rate is equal to the heat removal rate. This makes another major difference with "classical" chemical reactors as electron temperature and vibrational temperature in plasma reactors are more important than the gas temperature, except for the case of hot radiofrequency (RF) plasma reactors.

## 4.2 Microwave and RF Plasma

### 4.2.1 Microwave Plasma Torch

Nonequilibrium microwave-generated plasmas offer high ionization fraction (i.e. high electron density) at a relatively low mean electron energy of 1–3 eV resulting in a greater reactivity of molecules. In particular, self-sustained microwave discharges and their plasma jets could lead to effective dissociation of air molecules and the production of active species [6, 7]. The main advantages of such reactors are their high productivity, high concentrations of $NO_x$ after the quenching section, and a possibility of automation and advanced process control [8].

A microwave plasma torch (MPT) reactor consists of a magnetron; a waveguide; a cavity, in which microwave power is focused into a small volume by IR mirrors or by electromagnetic field; and a nozzle where the plasma is ignited. MPT reactors have many advantages, including high conversion rate from electrical to thermal energy, often above 75%; no electrode erosion by plasma species; high electron density and electron temperature; and ability to operate in a wide range of pressures below and slightly above the atmospheric pressure. Na et al. employed an MPT reactor for $NO_x$ generation [9]. In their design, an $N_2/O_2$ gas mixture enters the reactor through several inlets at a flow rate of 10 L min$^{-1}$ forming a vortex, where a plasma torch with a temperature of 6,000 K and a plasma density on the order of $10^{13}$ cm$^{-3}$ is generated at a 2.45-GHz microwave frequency. The gas is cooled via a heat-exchanger by a circulating water to reduce the gas temperature. The torch flame becomes longer as the electrical power increases; however, the authors did not quantify the exact concentration of $NO_x$ formed in their reactor. Increasing the microwave power in the submillimeter wavelength range allows to obtain inhomogeneous weakly collisional plasmas under near-atmospheric pressures [10]. The average electron concentration in such discharges exceeds the critical value for an electromagnetic field frequency (the highest frequency at which waves can penetrate plasma) and is equal to $10^{15}$–$10^{17}$ cm$^{-3}$ [11]. Their specific feature is the formation of plasma filaments with increased electron concentrations, which are attributed to nonlinear wave electrodynamics effects. Utilizing this principle, Sintsov et al. developed a hybrid plasmatron with a gas discharge at atmospheric pressure supported by microwave field with a frequency of 263 GHz (Figure 4.1) [12]. The ratio ($\upsilon_{ea}/\omega$) of the elastic electron–neutral

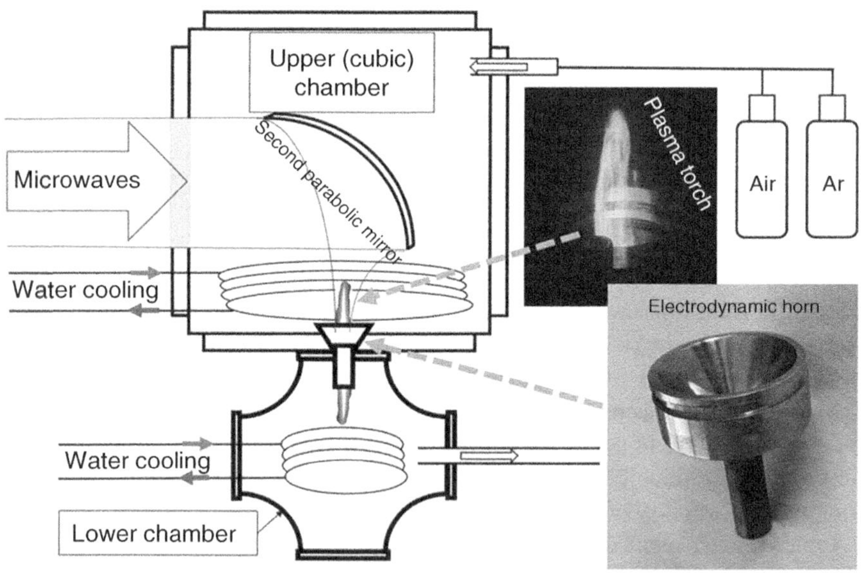


**Figure 4.1** Schematic diagram of the reactor module. The upper chamber has a plasma-forming gas mixture at atmospheric pressure. The parabolic mirror focused a submillimeter beam into a spot of 1.2 mm positioned exactly between the two chambers, where the gas flows (through a horn) in the lower chamber. The plasma torch in the gas flow was initiated inside the horn using a metal wire. The plasma temperature was 3,600–4,300 K. *Source:* Reproduced from Ref. [12] / with permission of Springer Nature.

collision frequency in plasma ($\upsilon_{ea}$) to the wave frequency ($\omega$) determines the interaction of microwaves with plasma [13]. This is an important parameter to obtain a high degree of absorption of the introduced MW irradiation power, which can exceed 80%. Using the same principle, Mansfeld et al. proposed a different configuration of a MW waveguide plasmatron, which created an electron density of $1\text{–}2 \times 10^{11}\ \text{cm}^{-3}$ at a frequency of 24 GHz in the 190–500 W power range [14]. The plasmatron is a continuation of a $TE_{11}$ waveguide with an inner diameter of 32.6 mm, which was connected via a boron nitride window. As in the previous example, the reactor design includes two sections: the waveguide with the plasma gas supply inlets that make an angle of 30° with the center axis of the waveguide and a heat-exchanger in the downstream section. The tangential gas inlets allow to form vortex flow, limiting the contact area between hot plasma and the waveguide walls. Such design allows to absorb up to 75% of MW power at a gas flow rate of $2\ \text{L min}^{-1}$. An $NO_x$ concentration of 1.35 vol.% was obtained at a flow rate of $10\ \text{L min}^{-1}$ at energy costs of 3.0–3.5 $\text{MJ mol}^{-1}$, which is considered to be a good starting point for further optimization of this reactor type [15]. The best results were obtained at the minimum input power, and a simultaneous increase in input power at the same specific energy input (SEI) reduces the energy efficiency. The authors suggested that the cavity dimensions and the number of inlet ports can further be optimized to decrease the energy costs. Swirling gas flow can stabilize the microwave plasmas in a surface wave sustained discharge mode [16]. Finally, the plasma can occupy a larger volume at reduced pressures, which may further improve the reactor performance.

In order to enlarge the plasma discharge volume, a circular waveguide was proposed by Lei et al. [17]. The 2.45-GHz microwave initially propagates through a standard rectangular tapered waveguide into a circular waveguide. To enhance the local electric field, the circular waveguide was loaded with double metal ridges. When the ends of the circular waveguide were closed, the electric field is a superposition of transverse electric $TE_{111}$ and transverse magnetic $TM_{010}$ modes [18]. In this propagation process, the electric field is strengthened by both the tapered waveguide and the metal ridges in the quartz reactor (i.d. 13 mm, o.d. 15 mm), creating the so-called local electric field enhancement (LEFE) effect. This reactor demonstrated a higher $NO_x$ concentration of 3.9 vol.% at a lower energy cost of $1.9\ \text{MJ mol}^{-1}$ compared to the rectangular waveguides.

Chepelev et al. presented a 1 kW, 2.45 GHz cylindrical microwave plasma jet reactor with six rod electrodes positioned each at 60° (Figure 4.2) [6]. The reactor housing was cooled with water. The electrodes were connected to a three-phase high-voltage (HV) power supply. The air passing through the discharge forms a plasma jet induced by interelectrode microwave discharges. The system allows to produce microwave discharges both in the waveguide path and in an external discharge chamber (plasma torch) attached to the main reactor body (data not

(a) (b)

**Figure 4.2** (a) Optical image of the plasma jet reactor. The diameter of the rod electrodes is 4 mm, and discharge gap is 4 mm. (b) The microwave discharge in the torch (view from the outlet side, gas flows toward an observer). *Source:* Ref. [7] / Springer Nature.

shown in Figure 4.2). The authors reported a gas temperature in the discharge channel of 1,200 K, an electron concentration of $10^{15}$–$10^{16}$ $cm^{-3}$ and an electron temperature of ~1 eV [19]. In the continuous regime, glow discharge in the form of a bright plasma filament between the tips of the electrodes and the inner reactor wall was observed. In the pulse regime, single-channel spark occurs randomly between the electrodes. While the initial experiments were performed with argon, it is considered that a glow-type microwave discharge holds promise for different applications, including $NO_x$ synthesis.

### 4.2.2 Surfaguide-Type Discharge

A microwave surfaguide-type discharge is another plasma source that uses electromagnetic waves to sustain plasma (Figure 4.3). The microwave irradiation is modulated by nearly square pulses with a repetition frequency ranging from 0.5 to 30 kHz. The reactor is often in a quartz tube with a diameter, shorter than the ¼ microwave wavelength, positioned in the waveguide [21]. If needed, the reactor can be cooled on a shell side by a microwave transparent oil flow. The degree of vibrational nonequilibrium (the ratio of vibrational to rotational temperature) in these reactors rapidly decreases as the pressure increases above 1–10 mbar, demonstrating that operation at higher pressures resulted in a significant loss in vibrational excitation pathway in the reactor. The residence time in this reactors should be similar to a characteristic vibrational-to-translational (VTT) energy transition time of about 0.1–2.0 ms. Longer residence times would reduce the positive effect of plasma, while shorter residence times would reduce the SEI and, therefore, the reactant conversion. Thus, this reactor type has limited flexibility, and the best operation can be achieved in a narrow range of flow rates. A pulsed microwave surfaguide-type discharge reactor was used for nitrogen fixation by Bahnamiri et al. [20]. The reactor was a quartz tube (o.d. 14 mm)

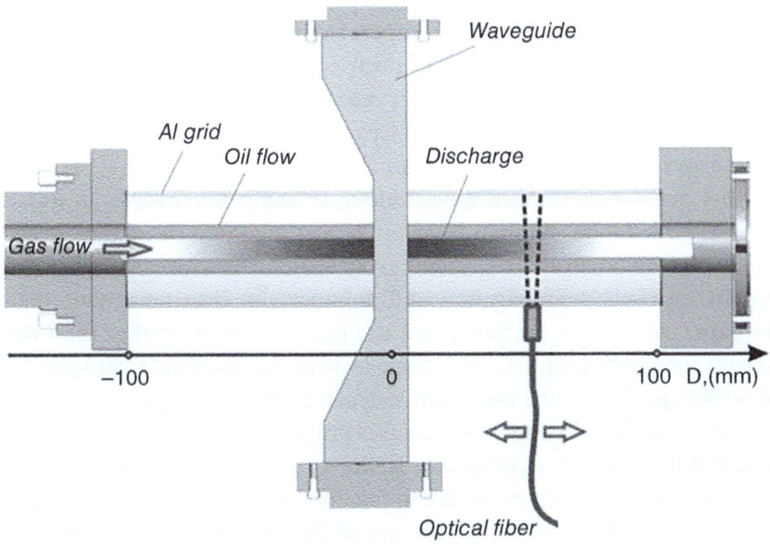


**Figure 4.3** Schematic view of the MW surfaguide-type plasma reactor. Mean applied power: 500–900 W, gas flow rate: 1 L $min^{-1}$, pressure: 1–12 Torr. *Source:* Reproduced from [20] / with permission of AIP Publishing.

(i.e. <¼ wavelength) with a length of 31 cm. Higher rotational and vibrational temperatures were observed in the active zone near the waveguide, decreasing along the reactor in both upstream and downstream directions. A higher SEI resulted in the enhanced production of N- and O-reactive species, primarily as a result of the electron-impact excitation and vibrational excitation of $N_2$ and $O_2$. The power modulation was reported to be beneficial to reduce the excessive gas heating and partially suppress unfavorable.

A rectangular tapered waveguide for a standard 2.45 GHz MW frequency with a cross section of 109.2 mm (width) × 54.6 mm (height) is the most widely used atmospheric microwave plasma source because of its simple structure, reliable performance, and low cost. However, due to the small generated plasma volume and short residence time, this plasma source has relatively low $NO_x$ production rate. Several approaches to reduce the energy costs and increase the $NO_x$ concentration were proposed, such as optimization of maximum microwave power, gas flow rate, gas ratio, and addition of pulse modulation to the MW power supply. In this way, Kelly and Bogaerts reported a 3.8 vol.% $NO_x$ concentration at an energy costs of 2 MJ $mol^{-1}$ in an atmospheric self-tuned microwave plasma reactor [16], demonstrating the importance of proper focus of microwave in a small reaction volume. They constantly measured the absorbed power by a Homer series auto-tuner using a six-port-reflectometer (SPR) method [22]. The maximum power in the discharge region was maintained by a system of auto-tuners in the waveguide. Increasing the electric field intensity increases the reduced electric field and the average energy of electrons and, therefore, the electron-impact reaction rate [23]. In an attempt to enhance local electric field in atmospheric microwave plasma, Wu et al. suggested to insert a tungsten rod array into the reactor (i.d. 26 mm, o.d. 30 mm) [24]. This array of parallel metal rods forms a waveguide to transmit microwaves in transverse electromagnetic mode. While it is generally assumed that microwaves cannot penetrate solid metal structures, metal array structures with small gaps are able to support microwave propagation and enhance the electromagnetic field intensity in gaps, even when the pitch and the gap are much smaller than the ¼ microwave wavelength [25]. Due to a huge impedance mismatch between air and two adjacent metal rods, the metal array structure forms an open-circuit resonator where the electric field intensity increases proportionally to the pitch to gap ratio, making the LEFE effect. In other words, the array of metal rods compresses the electric field component perpendicular to the direction of microwave propagation. The authors used an array with a length of 250 mm. The effect of several arrays including $3 \times 3$, $2 \times 2$, and $1 \times 2$ with different pitch/gap ratios was studied in the staggered and aligned configurations. In the staggered layout, the first and third rows were identical, but the second row was shifted. In such arrangement, the electric field intensity was enhanced by 385 times, and the energy costs were reduced by 20% compared to an empty reactor configuration. At the air flow rate of 17 L min$^{-1}$ and microwave power of 740 W, the energy costs reached the minimum of 1.78 MJ mol$^{-1}$ at an $NO_x$ concentration of 3.53 vol.%.

### 4.2.3 RF Plasma Torch

RF plasma is characterized by a low density of atoms and, therefore, a low frequency of collisions between electrons and heavy particles as compared to microwave plasmas of the same volume. To increase the energy density and generate air plasmas with an extremely high degree of temperature nonuniformity, the input power needs to be much higher. This makes the development of such reactors challenging as the temperatures may reach 6,000 K and a typical lifetime of 1,000 hours is required when using water as a coolant. These systems have a plasma generation efficiency, defined as the ratio of plume power to anode power, of about 50%, which is lower than in MPT. According to thermodynamics, a maximum NO concentration of 6.5 vol.% is achieved in an air at a temperature of 3,500 K. At the cooling rate of the synthesis products of $10^4$ K s$^{-1}$, the equilibrium concentration of NO is maintained by only 47%, while at a cooling rate of $10^7$, the product is preserved completely. Therefore, to obtain a high concentration of $NO_x$ in a thermal plasma process, rapid cooling with a rate of at least $10^6$ K s$^{-1}$ is required [26, 27]. The temperature inside the plasma plume needs to be reduced slowly to 3,500 K in the first quenching step and then rapid quenching should be applied to "freeze" the gas mixture in the second quenching step. To realize this two-step quenching concept, a flow modifier is typically attached to the plasma torch as an expanding water-cooled cone. The cone diameter increases four times from inlet to outlet position, and a typical length is 0.5 m. Using such design, Matveev et al. [8] obtained an $NO_x$ yield of 6.0 wt.%. The cone may have also several additional inlets for feeding additional gas flow to the precooled plasma plume, which may allow to reach the maximum NO concentration. Zinchenko et al. demonstrated a cooling rate of $2.9 \times 10^5$ K $s^{-1}$ with cooling air at a flow rate 5 g $s^{-1}$ in a 31-kW plasma torch operated with the same air flow rate [28]. This value is several times higher than the quenching rates of $5.22 \times 10^3 - 4.43 \times 10^4$, previously achieved using only water cooling [29]. The performance of classical heat-exchangers is limited by the inlet temperature of a cooling medium and the specific heat flux, and they cannot effectively "freeze" the desirable $NO_x$ concentration. Instead, the direct air quenching brings the expected outlet $NO_x$ concentration closer to the theoretical value determined by the Zeldovich mechanism. As a result, a very high $NO_x$ concentration of 7 wt.% was obtained in the product mixture.

The RF plasma torch reactors have a significant radial temperature nonuniformity near the torch nozzle, which affects the reaction kinetics. The magnitude of this temperature nonuniformity depends on the degree of spatial inhomogeneity of the plasma plume under the influence of electromagnetic forces as well as on hydrodynamics and heat transfer rate to the inner reactor walls. To reduce the radial temperature gradient, Matveev et al. proposed a multitorch RF system with simultaneous operation on one nozzle [30]. The authors suggested that the length of the mixing chamber should be 2–3 times larger than its internal diameter. Following this rule, the length of mixing chamber was fixed at 400 mm in their 225 kW RF reactor design. The three torches were positioned at a 30° angle to the mixing chamber center axis. In this design, the temperature nonuniformity for RF plasma torches was reduced by more than two times by blending three air plasma jets in a mixing chamber. The authors defined the temperature nonuniformity as a ratio of $(T_{max}-T_{av})/T_{max}$, where $T_{max}$ and $T_{av}$ are the maximum and average temperatures in the radial direction, respectively. They also investigated the effect of elevated pressure on the temperature nonuniformity in a single torch and three-torch plasma reactors [31]. A slightly elevated pressure of 1.1–1.5 bar in the plasma region was due to the pressure drop in the downstream chemical equipment. The authors concluded that both the single-torch system with a plasma power of 225 kW and a three-torch system with the same total power demonstrated similar thermal losses from the mixing chamber and as such both design demonstrated very similar energy performance. However, the three-torch reactor can be made more compact for transportation; therefore, this may be a decisive factor in the subsequent scale-up studies.

Increasing the pressure up to 4.0 MPa leads to a significant increase in the NO concentration [32]. With air, a maximum $NO_x$ concentration was 9.8 wt.% at 4,500 K. At an oxygen concentration of 37.5 wt.%, an $NO_x$ concentration of 13.3 wt.% was reported. To present their data, the authors used concentration in wt.% instead of vol.% (as in most other studies), their $NO_x$ concentration is a record-high among all other reactor types as the differences between molar concentration ($O_2$ = 20 vol.%, $N_2$ = 80 vol.%) and weight concentrations ($O_2$ = 24 wt.%, $N_2$ = 76 wt.%) in the air are relatively small. An increase in pressure in RF torch plasmas also leads to an increase in the process temperature. The latter has a positive effect on the $NO_x$ production rate without increasing the input power. The addition of water instead of oxygen has a negative effect on the reaction rate due to additional energy losses to hydrogen-containing species. The amount of $NO_x$ is also depends on the heat transfer rate during the quenching process, which was not optimized in this study.

Applied Plasma Technologies, Corp. (APT) has developed an APT-100-3 plasma system based on a high-pressure inductively coupled plasma (ICP) RF torch [32]. The second-generation ICP/RF plasma system with a regulated up to 50 kW power in plasma plume was demonstrated in air and $N_2/O_2$ mixtures of different composition in the pressure range from several mbar to 7 bar. Yet the development of more powerful RF sources above 225 kW range is limited by several factors including long development times (3–4 years), high costs ($5 million), and their less versatility for different applications [30].

## 4.3 Spark and High-frequency Pulsed Discharges

The transient spark discharge starts from a streamer phase, which is considered NTP and is subsequently transformed into short spark current pulses that generate thermal plasma [33]. Due to a short pulse time, thermalization of the plasma can be avoided. A relatively high electron density of $10^{17}$ cm$^{-3}$, which is an order of magnitude higher than in MPT, is achieved during the spark phase. Therefore, high-frequency spark and nanosecond pulsed spark discharges lead to enhance chemical activity as compared to MPT reactors. Spark discharge reactors are rather easy to construct; however, their productivity remains rather low due to the intermittent nature of discharge and relatively small plasma area inside the gas flow. The energy efficiency is somewhat similar to MPT and can be regulated by the heat transfer rate from the plasma zone. A long residence time is needed to increase the reactant conversion. Pei et al. reported an energy costs of 5.2 MJ mol$^{-1}$ (400 GJ tN$^{-1}$) in a pin-to-plane spark discharge reactor with a gap of 2 mm driven by a 20 kV ns pulsed power supply with a repetition rate of 10 kHz [34]. A tungsten pin with a diameter of 1.6 mm was used as the anode, and a copper plate

(10 mm × 100 mm) served as the cathode. Yet the ECs were higher than that reported in MPT reactors and similar to that observed in DC glow discharge, which is discussed in Section 4.6. The energy costs increased at large gap sizes as opposed to the case of high-frequency spark discharges.

Abdelaziz et al. employed a 20-kHz plate-to-plate spark discharge reactor with two perforated stainless-steel discs inserted a quartz tube (i.d. 15 mm, o.d. 18 mm, length: 170 mm) [35]. The lowest EC of 3 MJ $mol^{-1}$ with 1.2 vol.% of $NO_x$ was achieved at the largest discharge gap size of 10 mm and increased linearly with an increase in SEI, reaching 4.8 MJ $mol^{-1}$ at an $NO_x$ concentration of 2.1 vol.%. A similar trend in the rate of $NO_x$ formation with the electrode gap was observed by Rahman et al. in spark reactors for different peak currents and waveforms [36]. The addition of water vapors and reduction of the oxygen concentration to 5.0 vol.% decreased the EC to 2.1 MJ $mol^{-1}$; however, a relatively low $NO_x$ concentration of 0.36 vol.% was obtained [37]. Water vapor triggered extended Zeldovich mechanism, affecting formation and selectivity. The same electrodes were used for a 360 hour time on stream without experiencing significant damage. However, the quartz tube required regular cleaning to eliminate deposited metals, evaporated from the cathode, on its inner surface. The spark reactors allow to control the NO selectivity in a rather wide range between 20% and 95% [37, 38]. In general, lower power and shorter residence time lead to higher NO selectivity, while higher power and longer residence time favor the $NO_2$ formation [37]. Vervloessem et al. employed a pulsed plasma jet for direct $NO_x$ production from air [39]. While the pulsed regime allowed to get a record-low EC of 0.42 MJ $mol^{-1}$, the $NO_x$ concentration was limited to 0.02 vol.%, which is more than 18 times lower than that reported by Abdelaziz et al. in their pulsed spark discharge. Yet, these reactors demonstrated that it is possible to reach the theoretical minimum EC of plasma-based $NO_x$ formation and to meet the required EC target of 0.7 MJ $mol^{-1}$ as defined in a techno-economic analysis by Rouwenhorst et al. [40]. However due to the low $NO_x$ concentration, spark and jet-type pulsed plasma reactors are not suitable for their implementation in chemical industry. Yet, the concept can be utilized in other plasma applications.

## 4.4 Gliding Arc

The GA reactors work in nonequilibrium conditions at a maximum temperature of 1,500 K, thus achieving much higher yields of NO when compared with the thermal plasmas. At these temperature conditions, the NO selectivity is typically in the 80–85% range, decreasing at higher $O_2$ concentrations in the feed. Often, a supported heterogeneous catalyst, e.g. $MnO_x/Al_2O_3$, can be placed in the post-plasma zone to facilitate NO conversion to nitrogen dioxide utilizing the heat released by plasma [41]. This, however, does not change the total concentration of $NO_x$ in the process. There are two major types of GA reactors, which can be divided into 2D and 3D reactors (the latter is often called “rotating gliding arc” [RGA]) based on the discharge channel geometry: an arc (2D) and an helix (3D). 2D reactors have limited gas flow rate range and are mainly used in plasma research. They typically consist of two diverging knife-shaped electrodes often made from refractory metals, such as tungsten. Miniaturization of 2D GA reactors can improve the contact between reactive species and gas molecules, with a consequent increase in $NO_x$ production [42]. Often, the application of AC pulsed mode in 2D reactors can reduce the energy consumption (EC) as compared to continuous operation. Li et al. studied a 2D GA reactor for $NO_x$ synthesis with a duty cycle between 40% and 60% [43]. They measured the $NO_x$ concentration spatial profile, revealing higher EC near the inlet (9.59 MJ $mol^{-1}$) and the outlet (8.63 MJ $mol^{-1}$) and lower consumption in the middle section of the reactor (5.0–7.6 MJ $mol^{-1}$). An average EC of 6.95 MJ $mol^{-1}$ was reported, which is an improvement of 15% from the case of continuous operation.

An RGA reactor has a central cone electrode positioned inside a metal cylinder. The gas is fed via several (2–12) tangential inlets, facilitating a rotational gas motion that extends the gas residence time within the discharge zone. To reduce the breakdown voltage, a minimum gap between the two electrodes of 1 mm is often used. Denra et al. reported an $NO_x$ concentration of 0.23 vol.% in an AC powered RGA reactor consisted of an aluminum cone with a length of 45 mm (HV electrode) enclosed in a stainless-steel cylinder (i.d. 11 mm, ground electrode) [44]. A gas mixture with 20–40 vol.% $O_2$ in nitrogen was fed at 2 L $min^{-1}$ (SEI 1,800 J $L^{-1}$). The NO concentration decreased when the reactor was thermally insulated due to the increase in the gas temperature, which had a negative effect on the plasma-generated species [45]. The reactor was tested for 14 hours to confirm its long-term viability. Chen et al. studied the $NO_x$ synthesis in a 10 kV DC powered RGA reactor with a cone-shaped anode (base diameter: 36 mm, height: 55 mm) [46]. In the DC mode, the applied voltage was the least important parameter and the gas flow rate was the most important one affecting the reaction rate. The lowest EC of 4.2 MJ $mol^{-1}$ and $NO_x$ concentration of 0.8 vol.% was achieved at an air flow rate of 12 L $min^{-1}$. Angineni et al. employed an AC powered GA reactor with a stainless-steel cone-shaped HV electrode (base diameter: 10 mm, minimum gap: 2.5 mm) [41]. A voltage of 5 kV was required at a frequency of 14 kHz during breakdown, which was reduced to 1 kV in the postignition period. The authors obtained an $NO_x$ concentration of 0.6 vol.% at energy cost of 3.0 MJ $mol^{-1}$, with a total production of 9.7 g $h^{-1}$.

Muzammil et al. employed a very high-flow rate of 170 L $min^{-1}$ in an RGA to reduce the residence time of active species in the plasma region [47]. They used a specific nozzle to focus the arc into a small volume in order to increase its energy density and decrease heat losses [48]. A highly turbulent flow (Re number of 13,780) induced strong fluctuation of the current, resulting in a shorter arc shunt period and shorter arc length. An $NO_x$ concentration of 0.74 vol.% was obtained at an energy costs of 148 GJ per $t_N$.

In several studies, the shape of the ground electrode was changed from a cylinder to a helix attached to a nonconductive reactor housing. This allowed to observe arc behavior using axial and lateral high-speed imaging technics combined with current and voltage waveform analysis. Ideally, the helix trajectory should match the main flow path in the vortex. Such operation results in the existence of the optimal gas flow rate. Such reactor configuration was referred to as "rotating sliding arc" by several authors (Figure 4.4). Chen et al. studied the effect of flow regimes on the arc behavior in an AC powered RGA reactor at flow rates corresponding to Re numbers between 905 and 6,390 [49]. In the laminar flow regime ($Re < 2{,}300$), the energy provided to plasma was always sufficient to maintain the continuous existence of the arc due to the limited diffusion of ions in the discharge channel. The laminar regime provided the highest $NO_x$ concentration of 0.6 vol.%; however, the authors did not provide the respective energy costs. The longer afterglow region was observed at higher flow rates, opposite to the classical 2D GA discharge where the plasma contracts faster, confirming the role of turbulence. The concentrations of $NO_x$ drastically decreased in the turbulent flow regime, as the energy was not enough to maintain the arc channel in this electrode arrangement and the discharge mode was transformed into random spark discharge.

Three GA modes were observed by Sun et al. in a quartz RGA reactor with a helix electrode and two independent axial and tangential gas flow inlets to control the discharge modes by adjusting the respective flow rates [50]. A classical "breakdown-extinction-breakdown-extinction" mode between two successive current pulses was called U-B mode, where the arc is very bright at the breakdown moment with a large current pulse and large energy. When the input voltage was raised to 270 V, there was another mode in which the brightness changed periodically "off-dark-bright-dark-off" with the fluctuation of current; however, the average arc brightness was relatively weak. The authors called it S-G mode. The third mode possesses the

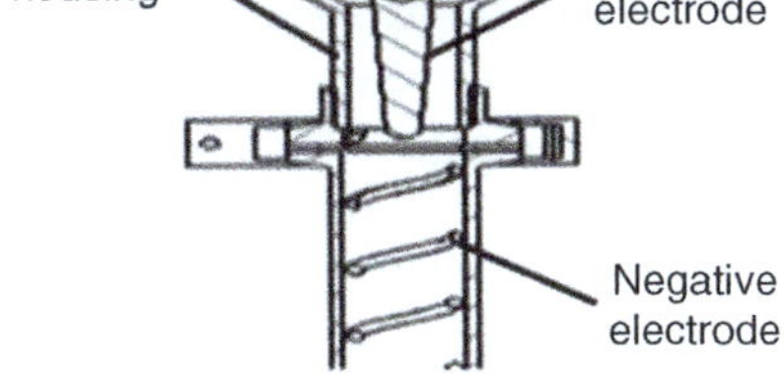


**Figure 4.4** A schematic view of the rotating sliding arc reactor. *Source:* Reproduced from [46] / with permission of IOP Publishing.

combined characteristics of U-B and S-G modes, and they called it transition mode. In the U-B mode, the main plasma parameters were higher than those in the S-G mode: reduced electric field (67 vs. 14 Td), electron density (3.00 vs. $1.14 \times 10^{20}$ $m^{-3}$), and electron temperature (1.90 vs. 0.64 eV), indicating a pulse discharge in the U-B mode. These results allow to select the appropriate discharge mode and achieve better plasma performance for $N_2$ fixation.

Jardali et al. performed computational flow dynamics (CFD) analysis to design an RGA reactor under reverse vortex flow (RVF) [51]. RVF RGA is a relatively new type of 3D plasma reactor. The external air flow in this reactor adopts a specific configuration of tangential intake, which generates a secondary reverse internal vortex flow to constrain the arc in the middle section of the reactor. This results in a good thermal insulation between the arc and the wall. At the same time, longer arc channels and residence times can better mix gases through the active plasma region, improving the energy efficiency of the reaction. The authors also observed two distinguished plasma modes in the reactor: an arc with variable length (mode I) and a steady arc (mode II). In mode I, the rotating arc treats the gas mixture several times with a characteristic time of about 4 ms, while the stable arc, located in the center of the reactor, increases the plasma–gas interaction time to 10 ms with a heating of the gas to 2,600 K. At this temperature, the VTT relaxations occur at a very high rate, and the gas translational temperature reaches the vibrational temperature. As a result, a high $NO_x$ concentration of 5.5 vol.% was achieved in the steady arc mode at an EC of 2.5 MJ $mol^{-1}$. The high reaction rate was explained by an essential additional contribution from the thermal (ground state $N_2$) pathway to the vibrationally promoted Zeldovich mechanism.

Van Alphen et al. improved the original design proposed by Jardali et al. [51] by the addition of an effusion nozzle, which acted as the reactor body [52]. The gas velocity decreased from 60 to 30 m $s^{-1}$ in the reactor body and further to 10 m $s^{-1}$ downstream of the discharge area near the gas outlet. The nozzle acts as a heat sink, causing a fast drop in temperature, which "freezes" the gas composition eliminating the backward reactions of recombination of NO with N and O atoms into nitrogen and oxygen. A small ceramic insert (length: 100 mm, diameter: 6 mm) was placed around the plasma zone to reduce the gas volume in the undesirable area so the plasma was formed only in a relatively small volume with a length of 1.4 mm and a diameter of 2.5 mm (Figure 4.5). The $NO_x$ concentration of 5.9 vol.% was obtained in an equimolar $N_2/O_2$ mixture with an energy costs of 2.1 MJ $mol^{-1}$. Overall, the enhanced flow turbulence increased the plasma–gas interaction. The authors suggested that further optimization of the nozzle shape could maximize the $NO_x$ concentration. In this design, the effect of elevated pressures (2–4 bar) was studied at different gas flow rates and gas compositions. In fact, reactor operation at slightly elevated pressures does not require major equipment modifications while improves the process efficiency. An increase in the $NO_x$ concentration to 6.3 vol.% was observed at a pressure of 4 bar in an equimolar $N_2/O_2$ mixture with an energy costs

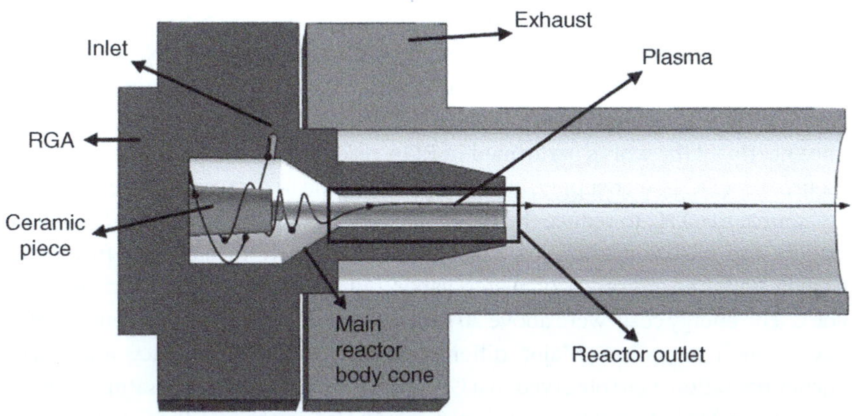


**Figure 4.5** Schematic view of the RGA plasma reactor showing gas velocity direction (black line) and the plasma in steady arc mode (purple zone). The cathode knob is indicated in light gray at the left, while the entire reactor body is at anode potential. *Source:* Ref. [52] / with permission of Elsevier.

of 1.8 MJ $mol^{-1}$ [53]. Simultaneously, operation at elevated pressures improved the selectivity to $NO_2$ due the shift of the equilibrium concentration in the oxidation of NO by O to $NO_2$. While an improvement at a higher pressure was observed, the best results gave only a minor improvement in $NO_x$ yield as compared to the operation at atmospheric pressure, which indicates that this reactor may have reached its performance improvement limits. Angineni et al. also employed an AC powered RGA reactor operated at a pressure of 3 bar [54]. The breakdown voltage was set at 5 kV, with the voltage stabilizing at around 1 kV after ignition following the protocol previously developed in Ref. [41]. They obtained an $NO_x$ production rate of 10.2 g $h^{-1}$ and an $HNO_3$ production rate of 8.83 g $h^{-1}$ in a subsequent gas/water mixing tank; however, their energy costs of 2.9 MJ $mol^{-1}$ were higher than those reported by Tsonev et al. [53] at the same elevated pressure.

Majeed et al. suggested to use a quench section downstream of the plasma zone to limit backward reactions and increase the $NO_x$ production rate. They used a RGA with a nozzle electrode connected to a 15-kW AC power supply [55]. Different quench gases ($N_2$, Ar, or $CO_2$) were fed to downstream jet of plasma reactor at three distances from reactor nozzle (25, 50, or 100 mm). By increasing the quench gas ratio from 0.25 to 0.5, the $NO_x$ concentration increases. The quench gas ratio here is the ratio between quench gas and total flow rate. Higher quench gas ratio leads to more temperature drop, resulting in higher $NO_x$ concentration. This increase was explained by change in rates of $NO_x$ formation and loss reactions in the post-plasma zone. The relative decrease in loss mechanism results in higher $NO_x$ concentration. Moreover, quenching with $CO_2$ led to higher concentration of $NO_x$ while argon resulted in the lowest effect, which was explained by the heat capacity of quench gases, which increases in the row Ar > $N_2$ > $CO_2$. The highest $NO_x$ concentration of 5.8 vol.% was observed with an equimolar $O_2/N_2$ gas mixture at an energy cost of 2.38 MJ $mol^{-1}$ at a rather large flow rate of 30 L $min^{-1}$ with the addition of $CO_2$ as quench gas.

Since nearly all the works performed in RGA reactors use either a high-frequency AC source or a DC power supply, an attempt was made to decrease the frequency of an AC source to 60 Hz to reduce the equipment costs [44]. Under the low-frequency operation, the $NO_x$ concentration decreased significantly to 0.25 and further to 0.10 vol.% with the presence of water vapor but increased at an $O_2$ content of 40 vol.%. The energy costs were above 20 MJ $mol^{-1}$, demonstrating a low energy efficiency of such operation. Major differences between low-frequency and high-frequency operation were observed in a GA reactor fed with an $N_2$ gas saturated with 39 vol.% water [56]. A low-frequency system (50 Hz) consumed three times more energy than a high-frequency system operating at 5 kHz at the same production rate of nitric/nitrate ions in water. The gas flow rate has a strong influence on the low-frequency system and little influence on the high-frequency system.

Böddeker et al. studied the formation of plasma under an axial magnetic field by the interaction of micro-vortices with the plasma channel [57]. They used an RGA reactor with a vortex flow in the transition flow regime with an Re number of 2,400, at a low DC current of 100 mA. At such conditions, the HV electrode surface experienced several melting and solidification cycles, which cannot be caused by a low-intensity DC current in the absence of air plasma. The flow instability in the transition flow regime created hot plasma spots near the gas inlets as a combination of magnetic constriction in the plasma together with the material fluxes from the electrode. Therefore, this regime should be avoided in upscaling RGA reactors to higher power densities and higher flow rates.

The very high energy efficiency at high-flow rates indicates the superiority of RGA reactors over other warm plasma reactors. While GA reactors are seen as the most promising type of warm plasma reactors, there are a number of problems with this type of reactors including vaporization of the electrode material, which leads to excessive wear of the electrodes and low residence contact of plasma with the surrounding gas.

## 4.5 Propeller Arc

In attempt to increase the gas–plasma interaction at very high gas flow rates, a propeller arc reactor design was studied by several groups. The propeller arc reactor typically consists of a rotating ground electrode, driven by a motor, with a fixed stainless-steel HV pin electrode. Two 35-mm-long propeller-like blades were made of stainless steel [31]. Using a DC power supply, the discharge can be extended up to a length of 82 mm as the ground electrode rotates. Lei et al. developed a DC

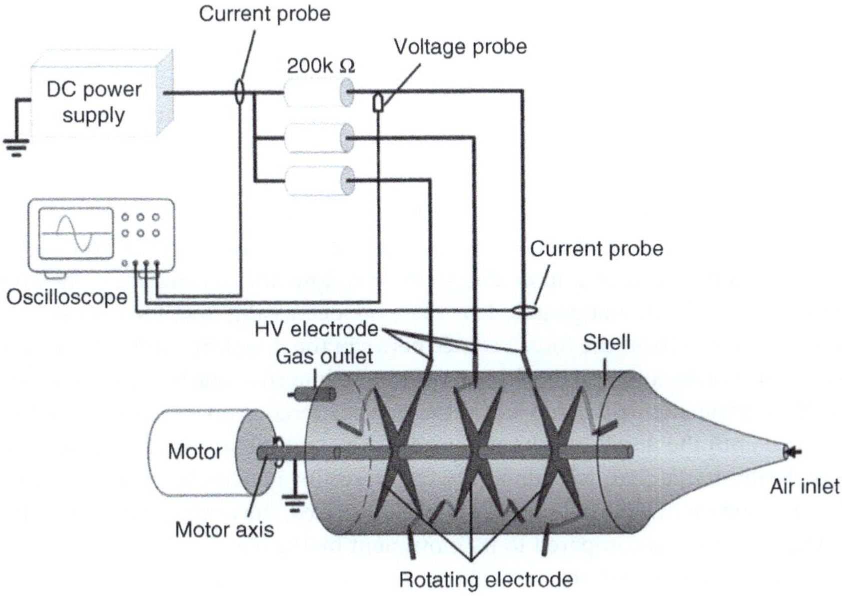


**Figure 4.6** Discharge configuration for $NO_x$ production using a three-level coupled rotating electrodes discharge. *Source:* Ref. [17] / Springer Nature.

powered three-level coupled rotating electrode plasma reactor (Figure 4.6) [17]. In this design, the frequency of the discharge was controlled by the motor rotational speed. Three HV electrodes were fixed near the reactor wall at an angular distance of 120° with a gap of 15 mm from the rotating ground electrode. The latter had three sets of four stainless-steel blades. The effect of blade numbers, rotating speed, gas flow rate, and $H_2O$ concentration on the $NO_x$ concentration and energy cost was investigated. The minimum energy costs of 2.26 MJ mol$^{-1}$ were reported at an air flow of 15 L min$^{-1}$ with a water concentration of 12 vol.%. This demonstrated a positive effect of water addition to the reaction mixture. A selectivity to NO above 90% was observed under all discharge conditions. The authors suggested that both the energy cost and the production rate can be improved by optimization of water content in the feed, gas flow rate, and the plasma discharge area. In a follow-up study, the reactor was loaded with $TiO_2/Al_2O_3$ rings (i.d. 12 mm, o.d. 38 mm, length: 5 mm) [58]. The packing was positioned downstream the rotating electrodes at a distance of 1 mm from them. The catalyst demonstrated a positive effect on the $NO_x$ formation rate with a maximum $NO_x$ concentration of 0.83 vol.% at the energy cost of 1.73 MJ mol$^{-1}$, which reduced by 22% as compared to the plasma-only discharge. It was assumed that high energy electrons and ions created oxygen defects onto the titania surface, resulting in additional surface pathways for $N_2$ fixation.

## 4.6 Glow Discharge

Another type of warm plasma, glow discharge, can be stabilized by the external DC magnetic field. Therefore, the arc length and the gas flow rate can be independently controlled by the applied voltage, the magnetic field strength, and the air flow rate, which is not the case in the classical GA reactors. The reactor configuration used is similar to a conventional 2D GA reactor, with the only exception that a DC magnetic field is applied. When the Lorentz force due to the DC magnetic field is opposite to the drag force due to the gas flow, the plasma arc is moving downstream in the discharge gap. However, the decreasing drag force along the discharge gap results in the force balance between the drag force and the Lorentz force at a certain location, so the discharge arc is stabilized inside the reactor. However, if the magnetic field is too weak, the Lorentz force is unable to balance the drag force until the discharge channel reaches the highest position in the reactor where the discharge channel is unable to stay, resulting in periodic operation. Also if the direction of the external magnetic field is reversed, the arc moves faster in the discharge channel as compared to its movement by the gas flow.

Li et al. [59] employed an atmospheric pressure DC powered glow discharge reactor stabilized by magnetic field (Figure 4.7). The energy cost of producing $NO_x$ was 2.65 MJ mol$^{-1}$ (189 GJ $t_N^{-1}$), which was 41% lower than the GA discharge in the same configuration without magnetic field. Subsequently, they further optimized their reactor, reducing the energy cost to 2.29 MJ mol$^{-1}$ at an $NO_x$ concentration of 1.59% [60]. Reducing both the electric field

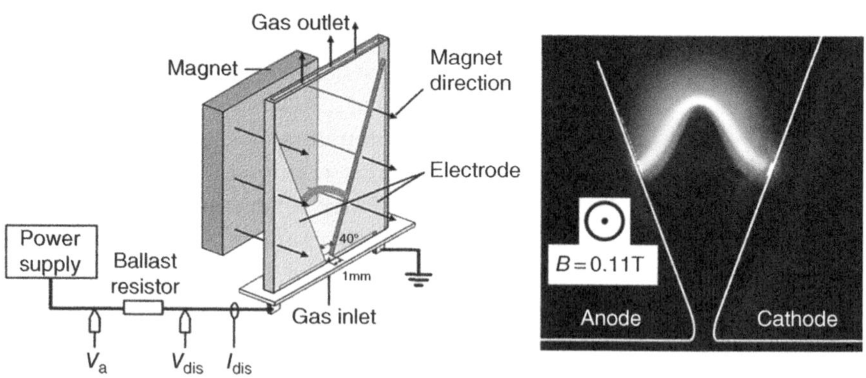


**Figure 4.7** (left) Schematic view of the experimental setup. (right) Photograph of the magnetic field stabilized glow discharge. The angle between the two electrodes is 40°. The applied voltage: 6.5 kV, and the air flow rate: 2 L min$^{-1}$. *Source:* Ref. [59] / John Wiley & Sons.

(from 44 to 31 Td) and the gas temperature increased the $NO_x$ production rate as it was formed primarily by the Zeldovich mechanism ($N_2 + O \rightarrow NO + N$). The optimal parameters were identified as a discharge current of 55 mA, a gas flow rate of 6 L $min^{-1}$, and an $O_2$ concentration of 40 vol.%. The rate of generation of both $NH_3$ (327 ppm) and $NO_x$ (0.397 vol.%) was greatly enhanced in an ethanol-saturated nitrogen gas. The lowest EC of 4.92 MJ $mol^{-1}$ was observed at an ethanol concentration of 50 vol.% and a discharge current of 40 mA. The authors observed that the decomposition of ethanol and water molecules in plasma produces a large number of H radicals. These radicals combine with N atoms produced ammonia precursors NH and $NH_2$ and finally $NH_3$ species, making an additional route for nitrogen fixation.

A DC pin-to-plane glow discharge reactor with a gap of 8 mm was studied by Pei et al. [34]. A temperature of 1,500–1,800 K in the discharge gap was reported. At these conditions, an EC of 500 GJ $t^{-1}$ were reported.

Tsonev et al. investigated the reactor scale-up from a flow rate of 5 to 300 L $min^{-1}$ in a DC powered pin-to-pin configuration [61]. The plasma in a small reactor was generated between two stainless-steel electrodes placed inside a quartz tube (i.d. 16 mm, o.d. 20 mm). The large reactor consists of two electrodes with a diameter of 30 mm, placed within a stainless-steel tube (i.d. 63 mm, o.d. 73 mm), so its volume was more than 10 times larger than that in the small reactor. A swirl flow was required at high-flow rates to maintain plasma in the large reactor operated in a torch configuration. The gas temperature in the torch configuration was significantly higher than in the pin-to-pin configuration. The flow pattern and the geometry facilitated fast mixing of gas mixture in the small reactor, demonstrating energy cost of 2.74 MJ $mol^{-1}$ for an $NO_x$ concentration of 1.72 vol.% corresponding to a production rate of 33 g $h^{-1}$. The maximum $NO_x$ concentration decreased to 0.45 vol.% in the large reactor. Yet, the torch configuration was recommended for reactor scale-up, due to better spatial distribution of plasma zone inside the reactor.

### 4.6.1 Triboelectric Nanogenerator

Following the implementation of a rotating ground electrode, the application of a triboelectric generator is another possible way to increase energy efficiency in plasma reactors. Triboelectric nanogenerator (TENG) is a mechanical energy acquisition device that can efficiently convert mechanical into electrical energy with pulsed DC current [62]. Usually, a TENG consists of three parts, viz. an electrode, a thin independent layer for power enhancement, typically a polytetrafluoroethylene (PTFE) film pasted onto the electrode, and a triboelectric layer. In TENG devices, charges are separated on the contact surfaces, and an electrical potential is generated between the surfaces. The TENG can be connected to a rectifier bridge to provide DC pulse current. The potential resulting from the rotational motion can be used for powering electric devices. The output voltage increases linearly with the increase in the rotor speed. To date, studies on TENG have focused primarily on optimizing the systems and circuit designs or exploring possible applications. The HV derived from TENG can be used to generate plasma; therefore, no additional plasma generator is required, reducing the total equipment costs. Therefore, this approach is suitable for a small-scale $N_2$ fixation with high efficiency.

This principle was adopted by Wang et al. in an atmospheric pressure plasma jet reactor consisted of a nozzle-shaped glass reactor (i.d. 18 mm, o.d. 26 mm) and two electrodes: negative tungsten needle and positive Pt sheet of 10 mm × 10 mm (Figure 4.8) [63]. The stator has four sectors made of a 60-μm-thick Cu film coupled to a disk (diameter: 25 mm) covered with a 4-mm-thick polymethylmethacrylate (PMMA) layer. The four copper electrodes were separated by a 10 mm gap. A flexible rabbit hair was used as the triboelectric layer, which was divided into two fan-shaped areas of equal size. The rabbit hair in these two areas was alternately pasted on the round PMMA substrate to form the rotor, and the stator was vertically fixed above the rotor. When rabbit hair contacts with the PMMA on the copper electrode, positive charges are generated on its surface and negative charges on the PTFE surface. The energy costs of 1.76 MJ $mol^{-1}$ at a flow rate of 2 L $min^{-1}$ were reported at a discharge distance of 2.0 mm. This value is the lowest record in the plasma-assisted $N_2$ fixation at room temperature and atmospheric pressure. The reactor was tested for 150 minutes, and an $NO_x$ production rate of 8 μmol $h^{-1}$ was reported.

In glow discharge-type reactors, the material of discharge electrodes is crucial as its oxidation and corrosion due to plasma's high temperatures and the presence of reactive oxygen species could alter the discharge mode. Under the influence of

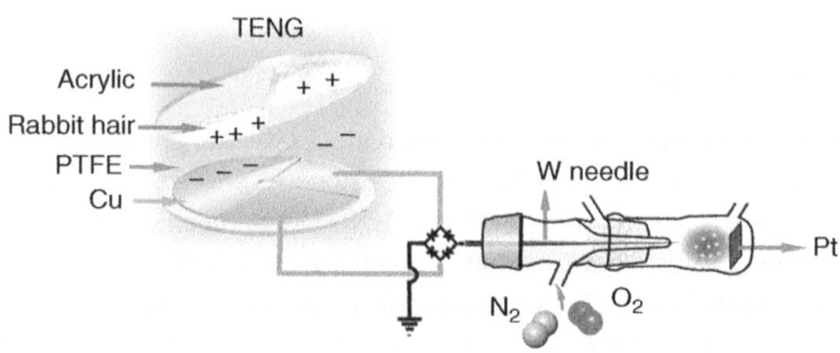


**Figure 4.8** Triboelectric plasma reactor. *Source:* Ref. [63] / with permission of Elsevier.

high temperatures and strong electric fields, different materials of metal electrodes can lead to various discharge modes, which then alters the chemical reactions within the plasma. The effect of different electrode materials, including iron, chromium, nickel, copper, and AISI304 stainless steel, on nitrogen fixation efficiency in glow discharges was evaluated [64]. Iron and 304 stainless-steel cathodes undergo a considerable oxidation at high currents but maintain their properties at low currents. With these two electrodes, the discharge mode transitions from the metal glow mode (mode 1) to the oxide cathode glow mode (mode 2) reduce the EC by 40% without decreasing the $NO_x$ production rate. The discharge mode 2 was stable at a minimum current of 35 mA. Other cathode materials (Cr, Ni, and Cu) did not exhibit any transition in discharge mode.

## 4.7 Dielectric Barrier Discharge

Among different reactor types, DBD reactors have a relatively simple design and can be operated by a nonspecialist. They also offer a possibility to load dielectric packings and/or heterogeneous catalysts directly in the plasma zone. Direct electron-impact dissociation (which often proceeds through electronic excitation) and ionization typically require more energy than it would be needed for bond dissociation and, therefore, DBD reactors exhibit a higher energy costs. The material of electrodes also influences the electron emission from the cathode [65]. There are two major mechanisms of electron emission in NTP: field emission and secondary electron emission, where the former mechanism often makes a significant contribution in discharge kinetics. Overall, the energy cost of $NO_x$ production is about an order of magnitude higher in DBD reactors than what is observed in spark/DC glow discharge reactors. The electrode surface microstructure and the geometry of the electrode system can improve the energy efficiency. Several nano- or micro-fabrication techniques to modify electrode surfaces to achieve enhanced energy utilization were investigated. Usually high aspect ratio materials, such as silicon nanowires, can create extremely strong electric field gradients, which, in turn, enable plasma inception at reduced total applied electric potential.

### 4.7.1 Micro-DBD Reactors

Patil et al. studied the direct synthesis of $NO_x$ by packing different catalysts in a DBD reactor [66]. Studying materials with different dielectric constants and particle shapes, the authors concluded that the noncatalytic pathways between the excited $N_2$ with $O_2$ species were dominant. Electronic excitations in DBD reactors

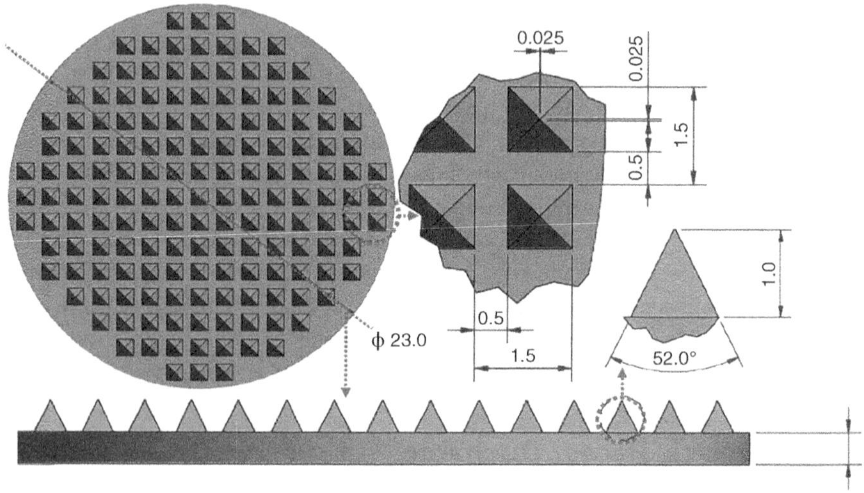


**Figure 4.9** A schematic view of the HV electrode with pyramid structure. *Source:* Ref. [68] / John Wiley & Sons / CC BY 4.0.

are responsible for $NO_x$ production while being limited by high energy costs and low efficiency [67]. Lamichhane et al. suggested a microstructured HV pyramid array electrode (Figure 4.9) in a plate-to-plate micro-DBD reactor with a gap size of 0.5 mm [68]. Both electrodes were enclosed in a cylindrical quartz window with a sealed peripheral region. The energy cost reached a minimum value of 3.72 MJ $mol^{-1}$ at an $O_2/N_2$ molar ratio of 2 when a thin titania layer was deposited onto the ground electrode. This value of energy costs is among the lowest observed in DBD reactors. However, the micro-DBD reactor has a relatively low productivity as the maximum flow rate is limited to 0.14 L $min^{-1}$.

Another example of a needle array microstructured HV electrode was presented by Li et al. [69] in a nanosecond pulse DBD reactor. The array consists of seven stainless-steel tubes (i.d. 0.9 mm, o.d. 1.26 mm) positioned at a short distance from an alumina covered stainless-steel disk served as ground electrode. The effect of pulse width in the range from 100 to 200 ns was studied at a constant peak voltage of 15 kV, pulse rising time of 100 ns, and pulse repetition rate of 1 kHz. The highest NO concentration of 1.11 vol.% was observed at a pulse width of 200 ns at a relatively low energy cost of 33 MJ $mol^{-1}$ at a gas flow rate of 200 mL $min^{-1}$. The authors explained increased $NO_x$ concentrations with pulse width by the increased SEI at longer discharge duration. Also they suggested that the number of discharge channel increased with the increased energy input [70]. A hybrid plasma system that combines a 20 kHz AC powered RGA followed by a commercial DBD ozone generator was proposed by Dinh et al. [48]. The ozone flow rate was controlled by tuning the electrical power. The RGA section has a conical HV copper electrode (o.d. 15 mm) surrounded by a helix ground electrode (helix diameter: 17 mm). The ECs were reduced to 8.0 MJ $mol^{-1}$ at an air flow rate of 15 L $min^{-1}$, while the exact production rate was not reported by the authors.

### 4.7.2 DBD Reactors for $N_2$/Water Plasma

Interest in exploring the interaction between plasma and water has grown due to its potential to enhance the efficiency of nitrogen fixation, producing an aqueous N-fertilizers. DBD reactors proposed for this purpose are only briefly discussed in this section while a more detailed information is provided in Chapter 12. A plasma jet reactor configuration induces reactions at the plasma/air, plasma/water interfaces, and also in the bulk volume of the liquid phase [71]. The ignition of plasma near a water interface is a challenging task due to the instability of the water surface, which should not touch the HV electrode. Roy et al. proposed a DBD reactor composed of two sections: a plasma discharge chamber and a series of bubblers, which were connected to the reactor outlet pipe to trap $NO_x$-containing gas into water [72]. They placed a glass tube (i.d. 23 mm, o.d. 26 mm) inside the water tank to withdraw excess of water to avoid its contact with the HV electrode. The ground electrode is placed directly in the water tank. The production rate of nitrate and nitrite ions was 12.9 μmol $min^{-1}$ at an energy costs of 20.7 MJ $mol^{-1}$.

The ammonia selectivity is typically 15–20% in the water produced by air jet with a certain air humidity. In general, the total nitrogen amount increases with increasing air humidity. For instance, in water saturated air at a flow rate of 0.2 L $min^{-1}$, the $NH_3$ concentration in the solution increases six times to 240 μmol $L^{-1}$, demonstrating an ammonia selectivity of 63% [73]. Gorbachev et al. observed three reaction regimes, depending on the $H_2O$ saturation of the nitrogen gas in a plasma jet reactor with a needle HV electrode inserted in a quartz tube (i.d. 2 mm, o.d. 5 mm) [73]. The plasma jet was positioned at a distance of 5 mm from the liquid surface. In the first regime, dry nitrogen plasma reacts with water, yielding almost equimolar amounts of $NH_3$ and $H_2O_2$ as previously reported by other authors [74, 75]. It was also suggested that plasma interacts with a vapor layer immediately above the liquid surface. In the second regime, observed at $H_2O$ concentrations between 2 and 10 vol. %, ammonia was formed with high selectivity. The third regime observed at an $H_2O$ concentration above 20 vol.%, with the formation of ammonia and nitrite/nitrate ions with a molar ratio of 2. Sakakura et al. proposed that such product distribution was observed due to the reactions of N atoms with $H_2O$ and/or H atoms leading to ammonia, and N atoms with •OH leading to nitrite/nitrate ions [76].

The $N_2$ conversion under all conditions studied remained rather low, with a maximum production rate of 0.4 μmol $min^{-1}$ at an energy costs of 15 MJ $mol^{-1}$. The jet flow rate has a major influence on the $NO_x$ absorption rate, while water mixing in the vessel and the water temperature have a little effect [77].

Without the supply of hydrogen, a low-temperature nitrogen plasma can generate hydrogen *in situ* and synthesize ammonia and other N-containing substances at ambient pressure. The synthesis rate increases with the addition of the metals in the aqueous solution [56]. Plasma-only operation yields an $NH_3$ concentration of 4.5 mg $L^{-1}$ (after 30 minutes), while in the presence of immersed Mg, Al, and Zn metals, the rate increased by 2.6, 2.3, and 2.2 times, respectively. These metals were oxidized in the aqueous solution and the released electrons reduced the proton into atomic hydrogen, which, in turn, reduced the plasma-excited nitrogen atoms into ammonia.

The falling film reactor (FFR) is the most popular type of gas–liquid contactor in gas–liquid reactors. In this reactor, the liquid flows down a structured reaction plate under the action of gravity and spreads to form an expanded thin film. The gas (or plasma) flow can be fed co-currently or counter-currently along the liquid film. A key feature of falling film microreactors is the comparably large specific surface area they afford compared to conventional reactors. The enhanced heat and mass transfer characteristics can be exploited for rapid gas–liquid reactions. A falling film DBD reactor for direct $N_2$–water interaction was proposed by Liu et al. [78]. A stainless-steel center electrode (o.d. 16 mm, length: 190 mm) was positioned into a quartz tube (i.d. 22 mm, o.d. 25 mm, length: 250 mm). The authors studied the relationship between the synthesis rate and the EC with different HV electrode dimensions, morphology (sheet, mesh), and their arrangements inside the reactor. It was found that five HV copper electrodes with a width of 15 mm positioned at a distance of 16 mm from each other provided the fastest synthesis rate at the highest energy efficiency. In this reactor configuration, a nitrite/nitrate production rate of 198 μmol $min^{-1}$ was reported at an energy costs of 39 MJ $mol^{-1}$, which was significantly higher than other DBD reactors. The synthesis rate was also considerably higher.

Ye developed a two-zone DBD reactor with a two-phase water–air flow via the discharge zones, intensifying the interfacial reactions for the distributed on-site production of N-fertilizers (Figure 4.10) [79]. A maximum production rate of 8,265 mg $min^{-1}$ was reported at an EC of 5.3 kWh $mol^{-1}$ (19.1 MJ $mol^{-1}$). While this EC value is an order of magnitude higher than those in RGA reactors, the author suggests there is still a lot of room to improve the performance of this DBD reactor by systematic optimization of several parameters. Such reactors can be positioned in remote farms with a direct connection to solar panels.

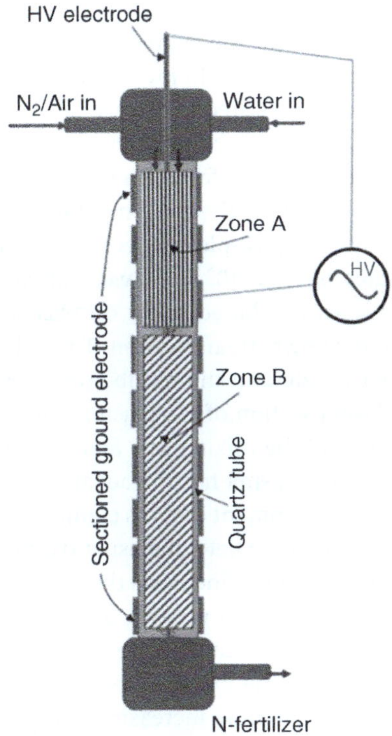


**Figure 4.10** The modular cNTP-$H_2O$ system consists of a two zones, configurable high-voltage (HV) electrode and ground electrode to generate dielectric barrier discharge (DBD) plasma. It is envisaged that the system can be scaled up by the parallel connection of a number of unit-cells and coupled to an irrigation line. *Source:* Ref. [79] / MDPI / CC BY 4.0.

## 4.8 Conclusions and Outlook

The plasma-assisted nitrogen fixation process is still quite far from the H-B process in terms of the EC and product yield. This chapter has systematically explored different plasma reactors employed in the $NO_x$ synthesis, the role of different flow conditions, heat transfer, and the influence of electrode geometry and gas quench section design. High-frequency and nanosecond pulsing in warm plasmas demonstrated the energy costs already within the industrial target range; however, the productivity of these reactors remains rather low.

The development in the area of thermal plasmas is targeted at microwave and ICP RF torches that have also a rather large equipment costs that are an order of magnitude larger as compared to other reactor types. In this category, a 200 kW RF plasma torch operated at 4 MPa produces the highest $NO_x$ fraction of 13 wt.% among all reactor types. These reactors have a large processing capacity, high product concentration, however, still rather low energy efficiency. The design of the quench section remains an important future target to further improve the product yield in hot plasma reactors as it allows effectively suppress backward reactions in the Zeldovich mechanism as their kinetics decreases much faster when cooling the product mixture. In thermal plasmas, cooling at a rate of up to one million $K\ s^{-1}$ is typically required to "freeze" the product mixture and get the highest concentration of $NO_x$ in the products. In most reactor designs, the quench section includes either a high surface area heat-exchanger or a cooled plate with small holes that is able to reduce the gas temperature at very high rates. Yet, the mixing with a cold gas in the post-plasma section is considered the most efficient method. A cooling rates of $2.9 \times 10^5\ K\ s^{-1}$ were already reported. Till now, the design of plasma torches was based solely on plasma physics with little attention to flow profiles in the downstream section. To further improve the cooling rate, the design of output nozzle and the number and the position of quench gas nozzles in the post-plasma section of the plasma torch should be performed to increase the gas flow rate and the mixing rate and, therefore, the convective heat transfer rate.

A GA plasma has the advantages of strong reactivity, high energy density, and lower equipment costs as compared to RF plasma torches, so it remains very competitive in the field of plasma-based $N_2$ fixation. The very high energy efficiency at high-flow rates indicates the superiority of RGA reactors over other warm plasma reactor types. In general, an operation in laminar flow regime gives higher $NO_x$ concentrations at higher energy costs. A highly turbulent flow induces strong fluctuation of the current, resulting in a shorter arc shunt period and shorter arc length, generally increasing energy efficiency but decreasing the $NO_x$ concentration in the products often below 0.5 vol.%. The operation in the transient flow regime (near Re number of 2,300) should be avoided as it results in a formation of local hot spots onto the HV electrode surface due to drastically fluctuating convective heat transfer rates. Among different RGA types, an RVF GA plasma reactor generates a secondary reverse internal vortex flow to constrain the arc in the middle section of the reactor. This results in a good thermal insulation between the arc and the wall at the best performance with an EC as low as $1.8\ MJ\ mol^{-1}$ with an $NO_x$ concentration of 6.3 vol.% obtained at a pressure of 4 bar in an equimolar $N_2/O_2$ mixture. However, based on different designs of this reactor type presented by several groups, it seems that this reactor has already reached the limit of its performance. Any further improvements would require computationally expensive CFD studies to better predict hydrodynamics in the plasma and post-plasma sections. While RGA reactors are seen as the most promising type of warm plasma reactors, there are a number of problems with this type of reactors including vaporization of the electrode material, which leads to excessive wear of the electrodes and low residence contact of plasma with the surrounding gas.

In glow discharge-type reactors, the material of electrodes is crucial as its oxidation and corrosion due to plasma's high temperatures and the presence of reactive oxygen species could alter the discharge mode. Under the influence of high temperatures and strong electric fields, different materials of metal electrodes can lead to various discharge modes, which then alters the chemical reactions within the plasma. Several stainless-steel types (AISI304, AISI316) seem to be a superior material for the cathode as their *in situ* oxidation can change the plasma to a low energy discharge mode and, therefore, increase the energy efficiency. Other cathode materials (nickel, copper) did not demonstrate such behavior.

The addition of $H_2O$ vapor to the feed has demonstrated a potential to improve $NO_x$ yields in warm plasmas through the formation of reactive NH and $NH_2$ species, while too high $H_2O$ concentration can change the discharge more and increase the energy costs.

## References


1 Birkeland, K.R. (1906). On the oxidation of atmospheric nitrogen in electric arcs. *Trans. Faraday Soc.* 2: 98–116.

2 Pavlovich, M.J., Ono, T., Galleher, C. et al. (2014). Air spark-like plasma source for antimicrobial $NO_x$ generation. *J. Phys. D. Appl. Phys.* 47: 505202. https://doi.org/10.1088/0022-3727/47/50/505202.

3 Rehbein, N. and Cooray, V. (2001). NO production in spark and corona discharges. *J. Electrost.* 51–52: 333–339. https://doi.org/10.1016/S0304-3886(01)00115-2.

4 F.J. Montgomery, C. Casper, D.P.L. Bathe, T. Gribb, Apparatus and method for generating nitric oxide in controlled and accurate amounts, US9573110B2, 2017.

5 S.A. Kuznetsov, S.V. Ardamakov, Plasmachemical chamber for making nitrogen oxide by direct oxidation, 2357922 C1, 2009.

6 Chepelev, V.M., Chistolinov, A.V., Antipov, S.N., and Gadzhiev, M.K. (2021). Chromatographic analysis of a cold plasma jet generated by an electrode microwave discharge in argon flow, interacting with atmospheric air. *J. Phys. Conf. Ser.* 2100: 012029. https://doi.org/10.1088/1742-6596/2100/1/012029.

7 Antipov, S.N., Gadzhiev, M.K., Il'ichev, M.V. et al. (2024). Analysis of gas composition of a cold plasma jet generated on the basis of atmospheric pressure microwave discharge. *Plasma Phys. Report.* 50: 653–658. https://doi.org/10.1134/S1063780X24600488.

8 Matveev, I.B., Serbin, S.I., and Zinchenko, A.E. (2022). A high-concentration $NO_x$ production system—Part 2: optimization of the nitrogen oxides yield. *IEEE Trans. Plasma Sci.* 50: 1635–1639. https://doi.org/10.1109/TPS.2021.3134678.

9 Na, Y.H., Kumar, N., Kang, M.-H. et al. (2015). Production of nitric oxide using a microwave plasma torch and its application to fungal cell differentiation. *J. Phys. D. Appl. Phys.* 48: 195401. https://doi.org/10.1088/0022-3727/48/19/195401.

**10** Gildenburg, V.B., Golubev, S.V., Gospodchikov, E.D. et al. (2024). Plasma-resonance-assisted filament in a high-pressure microwave discharge. *Phys. Plasmas* 31: 023507. https://doi.org/10.1063/5.0173489.

**11** Sidorov, A.V. (2022). Terahertz gas discharge: current progress and possible applications. *J. Phys. D. Appl. Phys.* 55: 293001. https://doi.org/10.1088/1361-6463/ac5556.

**12** Sintsov, S.V., Vodopyanov, A.V., Mansfeld, D.A. et al. (2024). Hybrid subterahertz atmospheric pressure plasmatron for plasma chemical applications. *J. Infrared Millim. Terahertz Waves* 45: 454–465. https://doi.org/10.1007/s10762-024-00987-w.

**13** Schluter, H. and Shivarova, A. (2007). Travelling-wave-sustained discharges. *Phys. Rep.* 443: 121–255. https://doi.org/10.1016/j.physrep.2006.12.006.

**14** Mansfeld, D.A., Vodopyanov, A.V., Sintsov, S.V. et al. (2023). Atmospheric pressure discharge sustained by millimeter radiation in a waveguide plasmatron. *Tech. Phys. Lett.* 49: 36. https://doi.org/10.21883/TPL.2023.01.55345.19384.

**15** Sintsov, S.V., Chekmarev, N.V., Mansfeld, D.A. et al. (2024). N2 fixation from air flow in an continous submillimeter microwave discharge. *Tech. Phys. Lett.* 50: 16–19.

**16** Kelly, S. and Bogaerts, A. (2021). Nitrogen fixation in an electrode-free microwave plasma. *Joule* 5: 3006–3030. https://doi.org/10.1016/j.joule.2021.09.009.

**17** Lei, X., Cheng, H., Nie, L., and Lu, X. (2022). Nitrogen fixation as NOx enabled by a three-level coupled rotating electrodes air plasma at atmospheric pressure. *Plasma Chem. Plasma Process.* 42: 211–227. https://doi.org/10.1007/s11090-021-10222-2.

**18** Collin, R.E. (2007). *Foundations for Microwave Engineering*. Wiley.

**19** Antipov, S.N., Gadzhiev, M.K., Il'ichev, M.V. et al. (2024). Operating regimes and structure of an atmospheric-pressure interelectrode microwave discharge in argon. *Tech. Phys.* 69: 1536–1540.

**20** Samadi Bahnamiri, O., Manaigo, F., Chatterjee, A. et al. (2023). Investigation of nitrogen fixation in low-pressure microwave plasma via rotational–vibrational NO and N2 kinetics. *J. Appl. Phys.* 133: 113303. https://doi.org/10.1063/5.0138298.

**21** N. Britun, G. Chen, T. Silva, T. Godfroid, M.P. Delplancke-Ogletree, R. Snyders, Enhancing the Greenhouse Gas Conversion Efficiency in Microwave Discharges by Power Modulation, in: *Green Chemical Processing and Synthesis*, INTECH, 2017. https://doi.org/10.5772/67875.

**22** V. Bilik, Six-port measurement technique: Principles, impact, applications, in: The International Conference RADIOELEKTRONIKA, 2002.

**23** Hong-Liang, D., Li-Ming, H., Yu-Dan, L., and Feng, W. (2011). Influence of reduced electric field on the evolvement characteristics of plasma under conditions of $N_2/O_2$ discharge. *Acta Phys. Sin.* 60: 115201. https://doi.org/10.7498/aps.60.115201.

**24** Wu, S., Liao, Y., Feng, K. et al. (2025). Nitrogen fixation by atmospheric microwave plasma with local electric field enhancement. *J. Phys. D. Appl. Phys.* 58: 015205. https://doi.org/10.1088/1361-6463/ad7d16.

**25** Liao, Y., Hong, T., Xiao, W., and Zhou, L. (2022). Local field enhancements of periodical metallic powders in microwaves. *IEEE Trans. Plasma Sci.* 50: 2049–2057. https://doi.org/10.1109/TPS.2022.3180043.

**26** Li, S., Medrano, J., Hessel, V., and Gallucci, F. (2018). Recent progress of plasma-assisted nitrogen fixation research: a review. *Processes* 6: 248. https://doi.org/10.3390/pr6120248.

**27** Ammann, P.R. and Timmins, R.S. (1966). Chemical reactions during rapid quenching of oxygen-nitrogen mixtures from very high temperatures. *AIChE J* 12: 956–963. https://doi.org/10.1002/aic.690120522.

**28** Zinchenko, A.E., Serbin, S.I., and Chernyak, V.Y. (2022). Investigation of the quenching of nitrogen oxide synthesis products in air plasma. *Ukrainian J. Phys.* 67: 44. https://doi.org/10.15407/ujpe67.1.44.

**29** Matveev, I.B., Serbin, S.I., and Zinchenko, A.E. (2021). A high-temperature quenching reactor. *IEEE Trans. Plasma Sci.* 49: 984–989. https://doi.org/10.1109/TPS.2021.3063906.

**30** Matveev, I.B. and Serbin, S.I. (2020). A multitorch RF plasma system as a way to improve temperature uniformity for high-power applications. *IEEE Trans. Plasma Sci.* 48: 332–337. https://doi.org/10.1109/TPS.2019.2950260.

**31** Matveev, I.B., Serbin, S.I., and Zinchenko, A.E. (2024). High-power RF plasma systems for the elevated pressure applications: single-torch versus multitorch design. *IEEE Trans. Plasma Sci.* 52: 1148–1152. https://doi.org/10.1109/TPS.2024.3367119.

**32** Matveev, I.B. and Serbin, S.I. (2019). Synthesis of nitrogen oxides in ICP/RF plasma. *IEEE Trans. Plasma Sci.* 47: 47–51. https://doi.org/10.1109/TPS.2018.2877453.

**33** Janda, M., Hoder, T., Sarani, A. et al. (2017). Cross-correlation spectroscopy study of the transient spark discharge in atmospheric pressure air. *Plasma Sources Sci. Technol.* 26: 055010. https://doi.org/10.1088/1361-6595/aa642a.

**34** Pei, X., Gidon, D., Yang, Y.-J. et al. (2019). Reducing energy cost of NO production in air plasmas. *Chem. Eng. J.* 362: 217–228. https://doi.org/10.1016/j.cej.2019.01.011.

**35** Abdelaziz, A.A., Teramoto, Y., Nozaki, T., and Kim, H.-H. (2023). Toward reducing the energy cost of $NO_x$ formation in a spark discharge reactor through pinpointing its mechanism. *ACS Sustain. Chem. Eng.* 11: 4106–4118. https://doi.org/10.1021/acssuschemeng.2c06535.

**36** Rahman, M., Cooray, V., Montaño, R. et al. (2011). $NO_X$ production by impulse sparks in air. *J. Electrost.* 69: 494–500. https://doi.org/10.1016/j.elstat.2011.06.008.

**37** Abdelaziz, A.A., Teramoto, Y., Nozaki, T., and Kim, H.-H. (2023). Performance of high-frequency spark discharge for efficient NO production with tunable selectivity. *Chem. Eng. J.* 470: 144182. https://doi.org/10.1016/j.cej.2023.144182.

**38** Zhang, S., Zong, L., Zeng, X. et al. (2022). Sustainable nitrogen fixation with nanosecond pulsed spark discharges: insights into free-radical-chain reactions. *Green Chem.* 24: 1534–1544. https://doi.org/10.1039/D1GC03859A.

**39** Vervloessem, E., Gorbanev, Y., Nikiforov, A. et al. (2022). Sustainable $NO_x$ production from air in pulsed plasma: elucidating the chemistry behind the low energy consumption. *Green Chem.* 24: 916–929. https://doi.org/10.1039/D1GC02762J.

**40** Rouwenhorst, K.H.R., Jardali, F., Bogaerts, A., and Lefferts, L. (2021). From the Birkeland–Eyde process towards energy-efficient plasma-based $NO_X$ synthesis: a techno-economic analysis. *Energy Environ. Sci.* 14: 2520–2534. https://doi.org/10.1039/D0EE03763J.

**41** Angineni, J., Reddy, P.M.K., Anga, S., and Somaiah, P.V. (2024). Sustainable nitrogen fixation by novel gliding arc plasma reactor for the production of nitrogen fertilizers. *Plasma Process. Polym.* 21: 2400059. https://doi.org/10.1002/ppap.202400059.

**42** Patil, B.S., Rovira Palau, J., Hessel, V. et al. (2016). Plasma nitrogen oxides synthesis in a milli-scale gliding arc reactor: investigating the electrical and process parameters. *Plasma Chem. Plasma Process.* 36: 241–257. https://doi.org/10.1007/s11090-015-9671-4.

**43** Li, S., van Raak, T., Kriek, R. et al. (2023). Gliding arc reactor under AC pulsed mode operation: spatial performance profile for $NO_x$ synthesis. *ACS Sustain. Chem. Eng.* 11: 12821–12832. https://doi.org/10.1021/acssuschemeng.3c03832.

**44** Denra, A., Saud, S., Nguyen, D.B. et al. (2024). Nitrogen fixation by rotational gliding arc plasma at surrounding conditions. *J. Clean. Prod.* 436: 140618. https://doi.org/10.1016/j.jclepro.2024.140618.

**45** Patil, B.S., Peeters, F.J.J., van Rooij, G.J. et al. (2018). Plasma assisted nitrogen oxide production from air: using pulsed powered gliding arc reactor for a containerized plant. *AIChE J* 64: 526–537. https://doi.org/10.1002/aic.15922.

**46** Chen, H., Wu, A., Mathieu, S. et al. (2021). Highly efficient nitrogen fixation enabled by an atmospheric pressure rotating gliding arc. *Plasma Process. Polym.* 18: 2000200. https://doi.org/10.1002/ppap.202000200.

**47** Muzammil, I., Lee, D.H., Dinh, D.K. et al. (2021). A novel energy efficient path for nitrogen fixation using a non-thermal arc. *RSC Adv.* 11: 12729–12738. https://doi.org/10.1039/D1RA01357B.

**48** Dinh, D.K., Muzammil, I., Kang, W.S. et al. (2021). Reducing energy cost of in situ nitrogen fixation in water using an arc-DBD combination. *Plasma Sources Sci. Technol.* 30: 055020. https://doi.org/10.1088/1361-6595/abff72.

**49** Chen, B., Liu, Q., Li, X. et al. (2024). Synthesis of NO by rotating sliding arc discharge reactor with conical-spiral electrodes. *Plasma Sci. Technol.* 26: 094010. https://doi.org/10.1088/2058-6272/ad6815.

**50** Sun, S.-R., Chen, F., Zheng, Y.-H. et al. (2024). Experimental study on the discharge characteristics of an air rotating gliding arc. *Plasma Sources Sci. Technol.* 33: 075013. https://doi.org/10.1088/1361-6595/ad5401.

**51** Jardali, F., Van Alphen, S., Creel, J. et al. (2021). $NO_x$ production in a rotating gliding arc plasma: potential avenue for sustainable nitrogen fixation. *Green Chem.* 23: 1748–1757. https://doi.org/10.1039/D0GC03521A.

**52** Van Alphen, S., Ahmadi Eshtehardi, H., O'Modhrain, C. et al. (2022). Effusion nozzle for energy-efficient $NO_x$ production in a rotating gliding arc plasma reactor. *Chem. Eng. J.* 443: 136529. https://doi.org/10.1016/j.cej.2022.136529.

**53** Tsonev, I., O'Modhrain, C., Bogaerts, A., and Gorbanev, Y. (2023). Nitrogen fixation by an arc plasma at elevated pressure to increase the energy efficiency and production rate of $NO_x$. *ACS Sustain. Chem. Eng.* 11: 1888–1897. https://doi.org/10.1021/acssuschemeng.2c06357.

**54** Angineni, J., Reddy, P.M.K., Anga, S., and Somaiah, P.V. (2024). Nitrogen fixation by simple gling arc plasma reactor at elevated pressure for synthesis of aqueous nitrogen fertilizer. *Plasma Process. Polym.* 22 (3): 2400209. https://doi.org/10.1002/ppap.202400209.

**55** Majeed, M., Iqbal, M., Altin, M. et al. (2024). Effect of thermal gas quenching on $NO_x$ production by atmospheric pressure rotating arc plasma: a pathway towards eco-friendly fertilizer. *Chem. Eng. J.* 485: 149727. https://doi.org/10.1016/j.cej.2024.149727.

**56** Lamichhane, P., Paneru, R., Nguyen, L.N. et al. (2020). Plasma-assisted nitrogen fixation in water with various metals. *React Chem Eng* 5: 2053–2057. https://doi.org/10.1039/D0RE00248H.

**57** Böddeker, S., Bracht, V., Hermanns, P. et al. (2020). Anode spots of low current gliding arc plasmatron. *Plasma Sources Sci. Technol.* 29: 08LT01. https://doi.org/10.1088/1361-6595/aba6a4.

**58** Lei, X., Cheng, H., Nie, L. et al. (2022). Plasma-catalytic $NO_x$ production in a three-level coupled rotating electrodes air plasma combined with nano-sized $TiO_2$. *J. Phys. D. Appl. Phys.* 55: 115201. https://doi.org/10.1088/1361-6463/ac3f0e.

**59** Li, Z., Nie, L., Liu, D., and Lu, X. (2022). An atmospheric pressure glow discharge in air stabilized by a magnetic field and its application on nitrogen fixation. *Plasma Process. Polym.* 19: 2200071. https://doi.org/10.1002/ppap.202200071.

**60** Li, Z., Wu, E., Nie, L. et al. (2023). Magnetic field stabilized atmospheric pressure plasma nitrogen fixation: effect of electric field and gas temperature. *Phys. Plasmas* 30: 083502. https://doi.org/10.1063/5.0155713.

**61** Tsonev, I., Ahmadi Eshtehardi, H., Delplancke, M.-P., and Bogaerts, A. (2024). Importance of geometric effects in scaling up energy-efficient plasma-based nitrogen fixation, Sustain. *Energy Fuel* 8: 2191–2209. https://doi.org/10.1039/D3SE01615C.

**62** Liu, L., Zhou, L., Liu, D. et al. (2021). Improved degradation efficiency of levofloxacin by a self-powered electrochemical system with pulsed direct-current. *ACS Nano* 15: 5478–5485. https://doi.org/10.1021/acsnano.1c00233.

**63** Wang, J., Zhang, B., Liu, Y. et al. (2023). Efficient N2 fixation in air enabled by mechanical-energy-driven triboelectric plasma jet. *Chem Catalysis* 3: 100647. https://doi.org/10.1016/j.checat.2023.100647.

**64** Li, Y., Luo, Y., Man, C. et al. (2024). Enhancing nitrogen fixation efficiency in glow-like discharge by reducing cathode-fall voltage. *Plasma Sources Sci. Technol.* 33: 105014. https://doi.org/10.1088/1361-6595/ad8307.

65 Miao, Y., Yokochi, A., Jovanovic, G. et al. (2023). Application-oriented non-thermal plasma in chemical reaction engineering: a review. *Green Energy Resource.* 1: 100004. https://doi.org/10.1016/j.gerr.2023.100004.
66 Patil, B.S., Cherkasov, N., Lang, J. et al. (2016). Low temperature plasma-catalytic $NO_x$ synthesis in a packed DBD reactor: effect of support materials and supported active metal oxides. *Appl Catal B* 194: 123–133. https://doi.org/10.1016/j.apcatb.2016.04.055.
67 Li, D., Zan, L., Chen, S. et al. (2022). Direct conversion of $N_2$ and $O_2$: status, challenge and perspective. *Natl. Sci. Rev.* 9: nwac042. https://doi.org/10.1093/nsr/nwac042.
68 Lamichhane, P., Pourali, N., Rebrov, E.V., and Hessel, V. (2024). Sustainable plasma-catalytic nitrogen fixation with pyramid shaped μ-electrode DBD and titanium dioxide. *ChemistrySelect* 9: e202401076. https://doi.org/10.1002/slct.202401076.
69 Li, Y., Qin, L., Wang, H.-L. et al. (2023). High efficiency $NO_x$ synthesis and regulation using dielectric barrier discharge in the needle array packed bed reactor. *Chem. Eng. J.* 461: 141922. https://doi.org/10.1016/j.cej.2023.141922.
70 Li, Y., Yang, D.-Z., Qiao, J.-J. et al. (2021). Discharge modes and characteristics optimization of nanosecond pulsed discharge in packed bed reactor. *J. Phys. D. Appl. Phys.* 54: 245206. https://doi.org/10.1088/1361-6463/abed6e.
71 Volkov, A.G., Bookal, A., Hairston, J.S. et al. (2021). Mechanisms of multielectron reactions at the plasma/water interface: interfacial catalysis, RONS, nitrogen fixation, and plasma activated water. *Electrochim. Acta* 385: 138441. https://doi.org/10.1016/j.electacta.2021.138441.
72 Roy, N.C., Maira, N., Pattyn, C. et al. (2023). Mechanisms of reducing energy costs for nitrogen fixation using air-based atmospheric DBD plasmas over water in contact with the electrode. *Chem. Eng. J.* 461: 141844. https://doi.org/10.1016/j.cej.2023.141844.
73 Gorbanev, Y., Vervloessem, E., Nikiforov, A., and Bogaerts, A. (2020). Nitrogen fixation with water vapor by nonequilibrium plasma: toward sustainable ammonia production. *ACS Sustain. Chem. Eng.* 8: 2996–3004. https://doi.org/10.1021/acssuschemeng.9b07849.
74 Haruyama, T., Namise, T., Shimoshimizu, N. et al. (2016). Non-catalyzed one-step synthesis of ammonia from atmospheric air and water. *Green Chem.* 18: 4536–4541. https://doi.org/10.1039/C6GC01560C.
75 Sakakura, T., Uemura, S., Hino, M. et al. (2018). Excitation of $H_2O$ at the plasma/water interface by UV irradiation for the elevation of ammonia production. *Green Chem.* 20: 627–633. https://doi.org/10.1039/C7GC03007J.
76 Sakakura, T., Murakami, N., Takatsuji, Y. et al. (2019). Contribution of discharge excited atomic N, $N_2^*$, and $N_2^+$ to a plasma/liquid interfacial reaction as suggested

by quantitative analysis. *ChemPhysChem* 20: 1467–1474. https://doi.org/10.1002/cphc.201900212.

**77** Rathore, V. and Nema, S.K. (2021). Optimization of process parameters to generate plasma activated water and study of physicochemical properties of plasma activated solutions at optimum condition. *J. Appl. Phys.* 129: 084901. https://doi.org/10.1063/5.0033848.

**78** Liu, Z., Tian, Y., Niu, G. et al. (2021). Direct oxidative nitrogen fixation from air and $H_2O$ by a water falling film dielectric barrier discharge reactor at ambient pressure and temperature. *ChemSusChem* 14: 1507–1511. https://doi.org/10.1002/cssc.202002794.

**79** Ye, X.P. (2024). Decarbonizing nitrogen fertilizer for agriculture with nonthermal plasma technology. *Eng* 5: 1823–1837. https://doi.org/10.3390/eng5030097.

# 5

# Plasma-Assisted Ammonia Synthesis

*Mateo Ruiz-Martín[1,2], Adrián Megías-Sánchez[1,2], Servando Marín-Meana[1,2], Manuel Oliva-Ramírez[1,2], Agustín R. González-Elipe[2], and Ana Gómez-Ramírez[1,2]*

[1] *Departamento de Física Atómica, Molecular y Nuclear, Universidad de Sevilla, Seville, Spain*
[2] *Laboratory of Nanotechnology on Surfaces and Plasma, Instituto de Ciencia de Materiales de Sevilla (CSIC-Universidad de Sevilla), Seville, Spain*

## 5.1 Introduction

Plasma processes driven by electricity stemming from renewable energy sources have emerged as a promising alternative to current chemical industry processes, although they still require to significantly increase yield and efficiency to make them competitive [1, 2]. Atmospheric plasmas have been used for gas-phase transformations in applications such as pollutant removal, ozone production, hydrocarbon reforming, ammonia synthesis, and $CO_2$ elimination and revalorization [3]. These processes are driven by various types of plasma sources, including those induced by microwaves (MWs), radiofrequency (RF), or direct current (DC) discharges. In all cases, the purpose is achieving an effective control over selectivity while improving energy efficiency, thus positioning plasma technology as a viable competitor to conventional thermally driven catalysis. Plasma reactors offer distinct advantages. First, they can operate at atmospheric pressure and room temperature and are well suited for harnessing a wide range of industrial gas synthesis and reforming processes. Second, they may proceed in a decentralized way by leveraging energy from renewable sources, such as photovoltaic plants and wind turbines, or surplus electricity from the grid during periods of low demand. Additionally, reactor scaling at different sizes is rather straightforward, while steady-state conditions can be reached after short induction times. All these

*Plasma-Assisted Nitrogen Fixation for Sustainable Process Industries*, First Edition. Edited by Volker Hessel, Annemie Bogaerts, Gabriele Centi, Evgeny Rebrov, and Nguyen Van Duc Long.

Published 2026 by John Wiley & Sons, Inc.

features make plasma technology a valuable option to advance in the electrification of chemical processes and the reduction of fossil fuels dependence [4].

The study of plasma-assisted ammonia synthesis began in the 1990s and since then an increasing number of publications have dealt with this topic. As discussed in Section 5.2.1, various types of plasma reactors have been utilized to produce ammonia. The most widely utilized rely on the so-called packed-bed plasma technology, which facilitates the incorporation of catalytically active materials in direct contact with the plasma phase. However, the efficiencies and chemical yields achieved to date with this technology are competitive neither for industrial applications nor for their implementation in small-scale decentralized facilities. Several bottlenecks hinder achieving competitive reaction yields and energy efficiencies. A major challenge is the occurrence of reverse reactions: once ammonia molecules are formed through the reaction $N_2 + 3H_2 \rightarrow 2NH_3$, interaction with the plasma can revert the process (i.e. $2NH_3 \rightarrow N_2 + 3H_2$) and drastically burden yield and efficiency. Therefore, current research extensively focuses on optimizing the chemical and energy yields and on decreasing the extent of back reactions through the adjustment of reactor operational parameters and geometric configurations, as well as through the use of suitable catalytic materials to improve reaction performance [1, 2, 5].

This chapter is organized as follows: it first provides an overview of the different reactor configurations used for plasma-assisted ammonia synthesis and their operating conditions. Next, the chapter introduces the fundamental principles governing *plasma-catalysis* processes and reviews recent studies on ammonia synthesis. Finally, it presents a critical evaluation of the current bottlenecks in plasma technology and reviews recent proposals to increase plasma reactor efficiency for ammonia synthesis.

## 5.2 Advanced Plasma Technologies for Ammonia Synthesis

As mentioned in Chapter 4, there are different types of plasma reactors that can be used to promote nitrogen fixation. For ammonia synthesis, the most popular ones are the so-called packed-bed reactors, although studies have also been carried out using other configurations and plasma types, such as classic dielectric barrier discharge (DBD), MW – ranging from 300 MHz to 30 GHz, RF – ranging from 40 kHz to 300 MHz, or gliding arc (GA) reactors [4, 6–9]. Particularly in RF and MW plasma processes, the operating pressure use to be low, in general smaller than atmospheric pressure. Depending on topologies and configurations, catalytic materials can be optionally incorporated in the reaction system. In packed-bed plasma reactors, though, catalysts are generally incorporated to promote the so-called *plasma-catalysis* processes (see Section 5.3) [5]. The purpose is to create synergies between surface interactions and processes involving excited plasma species, a combination expected to end up with improvements in reaction yield and energy efficiency.

During the early years of research on plasma-assisted ammonia synthesis, most common approach consisted of coupling a pure plasma reactor with a downstream catalytic reactor [10]. The temperature of the latter was typically lower than in classical catalytic processes, and it utilized the heat dissipated by the plasma reactor. More recent studies focused on the use of packed-bed reactors, with or without catalysts or a surface-active component deposited onto the dielectric/ferroelectric pellets acting as discharge moderator barrier. Results have been reported for both planar and cylindrical configurations, as well as for a wide variety of barrier architectures. The architecture of the packed-bed barrier and the dielectric properties of its constitutive materials are critical in determining the electric field distribution inside the reactor, and, consequently, specific plasma characteristics such as electron temperature and density. [2, 11–13]. Barium titanate ($BaTiO_3$), alumina ($Al_2O_3$), quartz (Q), quartz wool, and lead zirconate titanate (PZT) can be quoted among the most commonly used barrier materials. It is well known that materials with a high dielectric constant enhance the electric field in the gaps among pellets, increasing the number of microdischarges. Conversely, materials with a relatively lower dielectric constant promote surface discharges [14]. In addition, more complex structures, such as core-shell or mesoporous honeycomb-like designs, have been applied as barriers in packed-bed reactors, usually including catalysts [15].

Independently of the type of reactor, a critical step for $NH_3$ formation is the dissociation of the nitrogen molecule, an elementary reaction that may proceed through different mechanisms. In the case of packed-bed reactors and DBDs, usually working at atmospheric pressure, dominant mechanisms involve electron collisions, which result in ionizations and dissociations of the $N_2$ molecules (see Chapter 7). Conversely, in RF and MW plasmas, which usually operate at low pressures, vibrational and rotational excitations are more important, and ionization and molecular dissociation usually involve molecules in vibrationally excited states [16, 17]. Additionally, the gas temperature in these latter reactors is higher than in packed-bed and DBD reactors where, despite the occurrence of high electron temperatures, the gas temperature can be maintained at ambient levels. Meanwhile, GA reactors operate at higher electron temperatures, but their geometric configuration and operating mode make the incorporation of catalytic materials difficult or unsuitable.

### 5.2.1 Ammonia Synthesis Reactions and Plasma Types

We describe herein some key aspects of the $NH_3$ synthesis process in different plasma reactors. Since packed-bed reactors are the most widely used, they are addressed in a separate section (see Section 5.3).

#### 5.2.1.1 MW Reactions

In the late 1990s, Matsumoto studied the ammonia synthesis using MW plasmas at 650 Pa and discovered that ammonia molecules are primarily formed through the reaction between NH radicals and $H_2$ molecules, rather than with the H atoms formed in the nitrogen–hydrogen MW plasma [8]. In contrast to that, reaction with H atoms would lead to the dissociation of $NH_3$. Significantly, ammonia yield increased with plasma power (it was varied in the range 0–200 W) whenever a catalytic material based on iron was incorporated into the system. However, this came at the cost of energy efficiency. It turned out that the rate-limiting step for the ammonia synthesis under these conditions was the formation of $NH_x$ species at the catalytic surface.

#### 5.2.1.2 RF Discharge Reactions

Most recent studies carried out in RF plasma reactors make use of catalytic materials, including metals such as Ni, Cu, Pd, and Ag, apart from Fe. Shah et al. used and inductively coupled RF reactor operated at 0.3 Torr and reported that the energy efficiency achieved with a Ni-based metal–organic framework (MOF) catalyst (0.23 g-$NH_3$/kWh) surpasses that of pure Ni metal pieces and even those previously reported for other RF plasma systems [18]. The authors attributed this improvement to the porosity of the MOF, the presence of Ni metal sites available for reaction, and a reduction of hydrogen recombination probability at the surface. Unfortunately, high-power levels (power was varied from 100 to 300 W) adversely affected the porous framework, leading to amorphization and compromising the catalyst performance. Using a similar reactor, Shah et al. employed commercial metal meshes made of Cu, Pd, Ag, and Au as catalysts and reached energy efficiencies of 0.17–0.19 g-$NH_3$/kWh at 0.26 Torr [19]. These authors demonstrated that ammonia yield increases with the power (between 50 and 300 W) and that the obtained results could only be accurately justified with the participation of surface reactivity processes, not only on the catalyst surface but also on the quartz tube hosting the reaction. At higher pressures, Eley-Rideal (E-R) mechanisms (reactions between species adsorbed in the catalyst and plasma species) became also more important. These findings highlight the different role of a catalyst in traditional catalysis and in *plasma-catalysis* (see Section 5.3.3). In an atmospheric capacitive coupled RF plasma, Vervloedt and Von Keudell analyzed the ammonia synthesis varying the root mean square (rms) applied power between 100 and 500 V (which rendered applied plasma powers between 0 and 10 W) [9]. They showed that the incorporation of a catalytic material in a He diluted $N_2$ and $H_2$ plasma altered the discharge properties promoting the $N_2$ dissociation and enhancing the $NH_3$ formation. However, the plasma did not significantly alter or enhance surface reaction effects, which remained negligible for the global process.

#### 5.2.1.3 Plasma-Electrochemistry

RF plasma discharges can also be utilized to activate the ammonia synthesis in liquid electrolyzers. Sharma et al. studied the synthesis of ammonia from nitrogen and water using a plasma-activated proton-conducting solid oxide electrolyzer. The reactor was designed with two distinct compartments, enabling independent adjustment of experimental variables [20]. At the anode, proton species produced by water oxidation are transferred to the cathode through the proton-conducting membrane. The cathode is in contact with an RF plasma operated at low pressure and the diffused protons combine in this compartment with plasma-activated nitrogen to produce ammonia. To evaluate the role of electrochemical and plasma-assisted $NH_3$ synthesis, a transient experiment was conducted at 500 °C, in which the applied current (from 0 to 5 mA) and plasma power (from 0 to 80 W) were sequentially varied. Results showed a superior performance as compared with pure plasma experiments, with a Faradaic efficiency up to 88% and a $NH_3$ production rate of 26.8 nmol-$NH_3$/s·cm$^2$.

#### 5.2.1.4 Gliding Arc Reactions

Another important plasma method of ammonia synthesis relies on GA reactors. In these reactors, the geometry directly influences the chemical processes and determines the velocity of the gas flow and its regime, the size of the plasma region, and the extent by which the feed gas interacts with it. Xu et al. developed a nanosecond pulsed GA plasma that consists of a Laval nozzle and a stainless-steel bar used as high-voltage electrode, along with a cylindrical stainless steel hollow cone as ground electrode [21]. The authors operated the reactor with a pulse repetition frequency of 6 kHz and a peak pulse voltage of 16 kV and pulse width of 100 ns. The gas flow rate was fixed at 200 ml/min and the interelectrode distance at 1.4 mm. They also found that the vibrational excitation of nitrogen molecules plays a crucial role in the mechanism of plasma generation within the GA discharge, resulting in the production of a large number of vibrationally excited molecules. As the discharge gap distance decreases, the energy input per unit volume increases, leading to a higher collision frequency between electrons and nitrogen molecules. Increasing collisions raise the rotational temperature of nitrogen molecules. Simultaneously, the acceleration of the input gas flow rate contributes to enhance the generation of various types of free radicals. All these effects lead to a gradual linear increase in both the vibrational temperature of nitrogen molecules and the electron excitation temperature. Ultimately, energy efficiencies up to 0.28 g-$NH_3$/kWh were achieved.

Aiming the up-scaling of this technology, Long et al. studied the ammonia synthesis in a microreactor where one of the electrodes consisted of a series of micro pyramids capable of generating small arc discharges [22]. Additionally, they incorporated a catalytic material on the surface of the other electrode and analyze a wide range of operational parameters ($N_2$:$H_2$ ratio from 0.33 to 4, flow rates from 20 mL/min to 200 L/min, voltage from 1 to 20 kV, and frequency from 1 to 500 kHz). Special attention was paid to the flow distribution through the adjustment of three flow distributors at the reactor inlet that enable to control the gases mixing level before entering the reactor. Although plasma did not ignite homogeneously in all reactor pyramids, the pyramid structure further contributed to efficiently mixing the different gas molecules of reactants. Mixing was crucial for the maximization of the $NH_3$ outlet concentration, which might even double at low operational voltages whenever an efficient distributor ensured that the gas mixture of components was homogeneous when entering the reactor. The obtained results demonstrate the feasibility of carrying out the ammonia synthesis process in reactors with reduced dimensions and energy efficiencies of 1.76 g-$NH_3$/kWh. These results pave the way for future studies with reactor topologies ensuring short residence times and make the occurrence of reverse reactions unlikely.

### 5.2.2 Effects of Plasma Reactor Operational Conditions

Operational parameters are critical to control the efficiency of ammonia synthesis. Parameters include gas flow distribution within the reactor (this is particularly significant in GA reactors, where the contact time between the plasma arc and the reactive gases is key for controlling efficiency), the characteristics of electrical input signals (e.g. waveform – sinusoidal, square, or pulsed – frequency or peak voltage, in packed-bed reactors typically ranging from 1 to 100 kHz and 1 to 20 kV, respectively), the flow rate of the reactive gases (or, equivalently, the gas residence time within the reactor), the reactant gas mixture composition, the operating temperature or the working pressure. Most studies to date have been conducted at atmospheric pressure, with temperatures ranging from room to high temperatures, which in general are lower than those used in thermally driven catalytic processes (e.g. up to 400–500 °C) [3]. Figure 5.1 shows a sketch of some of the most relevant operational parameters for the plasma-catalysis synthesis of ammonia, while Table 5.1 compares the results obtained by different authors and

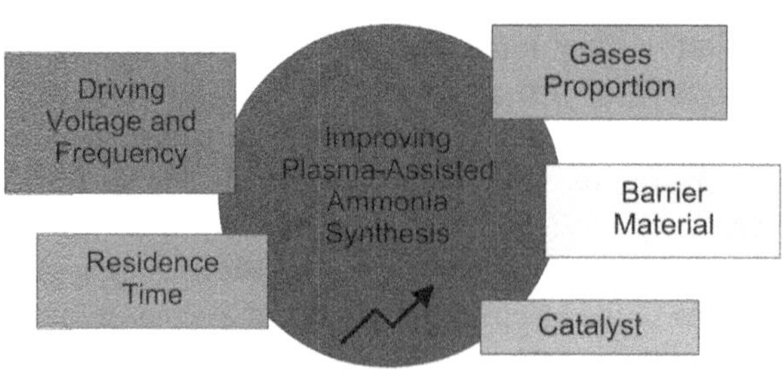


**Figure 5.1** Diagram of operational parameters in plasma packed-bed reactors that need to be optimized to achieve competitive yields for the plasma-assisted $NH_3$ synthesis.

**Table 5.1** Comparison of results of different authors for the plasma-assisted ammonia process.

| Plasma | Catalyst | $N_2/H_2$ ratio | Q gas (mL/min) | SEI (kJ/L) | Yield (%) | Energy efficiency (g-$NH_3$/kWh) | Ref. |
|---|---|---|---|---|---|---|---|
| DBD | Ru/γ-$Al_2O_3$ | 1:3 | 120 | 5 | 23.52 | 1.16 | [23] |
| DBD | Ru/γ-$Al_2O_3$ | 1:3 | 120 | 25 | 59.82 | 0.59 | [23] |
| DBD | Ni/MCM-out | 1:3 | 40 | 60 | 5.30 | 1.20 | [15] |
| DBD | Ni/MCM-out | 1:3 | 1,200 | 0.8 | 0.20 | 3.90 | [15] |
| DBD | VN | 2:3 | 100 | 60 | — | 1.50 | [24] |
| DBD | VN | 2:3 | 100 | 60 | — | 3.80 | [24] |
| DBD | VN | 2:3 | 100 | 60 | — | 4.50 | [24] |
| DBD | MgO | 1:1 | 20 | 11.4 | 12.43 | 0.54 | [17] |
| DBD | Ru/MgO | 1:1 | 20 | 11.4 | 32.52 | 1.41 | [17] |
| DBD | Co/MgO | 1:1 | 20 | 11.4 | 44.54 | 1.93 | [17] |
| DBD | Pt/MgO | 1:1 | 20 | 11.4 | 35.61 | 1.54 | [17] |
| DBD | Cu/MgO | 1:1 | 20 | 11.4 | 41.79 | 1.81 | [17] |
| DBD | Pd/MgO | 1:2 | 20 | 11.4 | 33.82 | 1.46 | [17] |
| Arc plasma | — | 3:1 | 200 | 0.21 | 2.15 | 7.56 | [22] |
| Arc plasma | — | 3:1 | 200 | 2.91 | 3.55 | 0.90 | [22] |
| DBD | SBA-15 | 1:1 | 20 | 32.3 | 45.79 | 0.70 | [25] |
| DBD | Ni/SBA-15 | 1:1 | 20 | 32.3 | 62.14 | 0.95 | [25] |
| DBD | $MgTiO_3$ | 1:1 | 20 | 65.7 | 31.69 | 1.19 | [26] |
| DBD | Ru/$MgTiO_3$–6-500 | 1:1 | 20 | 65.7 | 39.94 | 1.50 | [26] |
| DBD | Ru-Co/$MgTiO_3$–6-500 | 1:1 | 20 | 65.7 | 41.01 | 1.54 | [26] |
| DBD | Co/$MgTiO_3$—6-500 | 1:1 | 20 | 65.7 | 38.61 | 1.45 | [26] |
| DBD | PZT | 1:3 | 23 | 13.8 | 1.12 | 1.02 | [27] |
| DBD | $Al_2O_3$/PZT | 1:3 | 23 | 7.0 | 0.49 | 0.89 | [27] |
| DBD | Ru-$Al_2O_3$/PZT | 1:3 | 23 | 26.1 | 1.38 | 0.68 | [27] |
| DBD | Acidic γ-$Al_2O_3$ | 1:3 | 100 | 14.5 | — | 5.80 | [28] |
| DBD | Alkaline γ-$Al_2O_3$ | 1:3 | 100 | 14.5 | — | 6.60 | [28] |

(*Continued*)

**Table 5.1** (Continued)

| Plasma | Catalyst | $N_2/H_2$ ratio | Q gas (mL/min) | SEI (kJ/L) | Yield (%) | Energy efficiency ($g\text{-}NH_3/kWh$) | Ref. |
|---|---|---|---|---|---|---|---|
| DBD | Neutral $\gamma\text{-}Al_2O_3$ | 1:3 | 100 | 14.5 | — | 6.20 | [28] |
| DBD | Ru(2) Mg(5)/ $\gamma\text{-}Al_2O_3$ | 1:4 | 2,000 | 0.12 | — | 35.70 | [23] |
| RF plasma | Ni-MOF-74 | 1:4 | 20 | 150 | — | 0.23 | [18] |
| RF plasma | Cu | 1:4 | 20 | 900 | 16.7 | 0.17 | [19] |
| RF plasma | Au | 1:4 | 20 | 900 | 19.1 | 0.19 | [19] |
| RF plasma and electrolysis | Pt/BCZY | — | — | — | — | 0.10 | [20] |
| Gliding arc plasma | — | 1:1 | 200 | 11.7 | 19 | 0.20 | [21] |
| Gliding arc plasma | — | 1:1 | 200 | 7.1 | 15 | 0.26 | [21] |

laboratories working on this topic. According to the data reported in this table, the highest energy efficiency so far achieved is 37.5 $g\text{-}NH_3/kWh$ [23]. This value is significantly lower than the efficiency of the Haber-Bosch process (600 $g\text{-}NH_3/kWh$, excluding hydrogen production cost) and approximately four times lower than the targeted cost required for plasma technology (values as high as 150 $g\text{-}NH_3/kWh$ has been recently claimed for this operational mode) [3, 29]. It is discussed in the final section of this chapter that various variables can be optimized to enable a competitive usage of energy for the ammonia synthesis reaction driven through *plasma-catalysis* [1].

Next, we discuss how a proper selection of operational parameters can contribute to optimize the efficiency of the ammonia synthesis process. However, it must be borned in mind that many recent papers published on the topic lack the conditions required to enable a systematic comparison of results from different laboratories, thereby making the establishment of rigorous and standardized criteria for process optimization difficult. Therefore, rather than in a quantitative evaluation of specific results, discussion will focus on the general tendencies found when varying each parameter, highlighting the most relevant findings in each case.

#### 5.2.2.1 Carrier Gas

Ammonia synthesis is carried out using $H_2$ and $N_2$ as reactant gases. Ideally, no carrier gas should be used since any extra gas would increase the industrial costs of the process. However, at laboratory level, carrier gases, mainly Ar and He, have been employed to increase the plasma electron temperature, a key factor to enhance the plasma dissociation and excitation probabilities of nitrogen molecules and expectedly the reaction yield. Vervloedt and Keudell used an RF discharge to study the $NH_3$ synthesis in a $N_2$ and $H_2$ discharge diluted in He [9]. Liu et al. looked for the effect of adding Ar to a packed-bed plasma reactor [30]. They varied the Ar proportion between 0 and 30% vol. and demonstrated that Ar increased the nitrogen conversion by 15% and the $NH_3$ production by 18.6%. The enhancement was attributed to electron impact and Ar-collision reactions. However, the practical implementation of this Ar dilution approach has proved unsuitable due to the elevated costs it involves.

#### 5.2.2.2 Gases Proportion

Early *plasma-catalysis* studies focused on determining the optimal gas composition for this process, since the $N_2$:$H_2$ ratio does not necessarily have to match the stoichiometric 1:3 N/H proportion in the ammonia molecule. The said ratio has been widely varied in the works dedicated to the topic, with most studies agreeing that ratios between 1:1 and 1:3 maximize the chemical yield. An exception is that of Ref. [23] showing good results for a 4:1 ratio. Recent studies utilizing an Artificial Neural Network model have demonstrated that this parameter ($N_2$:$H_2$ ratio) has the greatest impact on the synthesis process, with a quantified yield of approximately 42% [23]. In a very recent approach, Hoang et al. have checked the use of more sustainable and cheap reactants, using seawater vapor as hydrogen source, and achieving yields of 3.2 g-$NH_3$/kWh for applied powers as low as 2 W [31].

#### 5.2.2.3 Residence Time and Gas Flow Regime

Gases flow rate, or equivalently, the residence time of gases within the reactor has also a high effect on process efficiency. Around this topic, there is consensus in the scientific community in the sense that low-flow rates tend to decrease the reaction yields [3, 7, 32]. This has been attributed to an increase in the likelihood that ammonia molecules interact with plasma-excited species and become destroyed (i.e. promoting the back reaction $2NH_3 \rightarrow N_2 + 3H_2$). Conversely, high-flow rates minimize this effect because the formed ammonia molecules spend less time inside the reactor and have less chances to interact with excited plasma species and be destroyed. In this line, it has been also shown that long residence times, aside reverse reactions, may favor the occurrence of other inefficient reactions. For instance, hydrogen exchange between ammonia and hydrogen molecules (i.e. $\mathbf{NH_3} + H_2 \rightarrow NH_3 + \mathbf{H_2}$), a process that does not generate new reaction products but contributes to decrease energy efficiency [33]. To mitigate these effects, various laboratories have proposed the use of plasma microreactors to limit the impact of reverse reactions on the overall process efficiency. In this line, Long et al. stressed the importance of properly mixing the gases at the microreactor inlet to ensure an optimal process performance [22]. Another often-overlooked aspect is the gas flow regime. While this is irrelevant in low-pressure discharges such as MW and RF, it becomes important in packed-bed and GA reactors. In packed-bed reactors, the barrier induces pressure variations that affect gas flow, ultimately influencing the chemical yield and energy efficiency. In GA reactors, turbulent flow is typically employed to maximize the residence time of gases within the plasma arc. This type of flow has proven highly beneficial for other processes, such as $NO_x$ synthesis. However, its impact on $NH_3$ synthesis remains unexplored [21].

#### 5.2.2.4 Driving Voltage and Frequency

Electrical input signal features are also critical to control efficiency and power consumption. MW reactors usually operate at 2.45 GHz and power consumptions range around hundreds of watts (100–300 W) [8]. RF reactors typically operate at 13.56 MHz, and power consumption depends on the coupling method. Power consumption for $NH_3$ synthesis using RF inductive coupling is within the same order of magnitude than for MW discharges, whereas reported values are lower for capacitive coupling (around 10 W) [9, 18, 19]. GA reactors can operate with either DC or AC (alternating current), but AC is usually preferred, using frequencies in the kHz range and peak voltage amplitudes between 5 and 20 kV [21, 22]. Packed-bed reactors operate with AC, usually at frequencies ranging from 1 to 100 kHz and a sinusoidal voltage variation [7, 34]. However, square waves and even pulsed AC signals have been also tested aiming at increasing process performance. Operating voltages generally range between 1 and 10 kV, with currents in the range of 10–100 mA, depending on the size of electrodes (these parameters render consumed powers between 5 and 100 W) [22, 32]. These values, however, vary greatly depending on reactor geometric configuration, the type of barrier material, and the presence or absence of catalysts. Overall, the prevailing trend indicates that operating at high voltages and frequencies enhances nitrogen conversion, though at the expense of energy efficiency [4].

#### 5.2.2.5 Barrier Materials and Catalysts

Both the influence of the barrier materials and the catalysts are described in detail in Section 5.3, focused on packed-bed reactors.

## 5.3 Plasma-Catalysis of Ammonia: Seeking Synergies to Improving Energy Efficiency

### 5.3.1 Plasma-Catalysis: A Brief Introduction

The role of a heterogeneous catalyst in the presence of plasma (herein we will designate the materials interacting with the plasma as "catalyst," although its actual role in the reactions is not exactly the same than that of a conventional heterogeneous catalyst) is to promote the adsorption of molecules or fragmented species onto its surface to induce surface reactions, which presenting reduced activation energies, lead to the synthesis of intermediate reactants or final products [35]. It is important to highlight that plasma activation is not necessarily linked to a reduction in activation energy; but to procure distinct physicochemical activation mechanisms. The differential effect in a plasma is that energy is supplied to the molecules via electromagnetic fields, which, unlike thermal activation, can effectively facilitate certain molecular and atom processes even at relatively low temperatures. We should remark in this regard that reactive molecular moieties eventually adsorbed on the catalysts encompass not only neutral ground state molecules but also charged and other kinds of plasma activated species.

The study of *plasma-catalysis* processes is rather complex due to the variety of physicochemical interactions involved [14, 16, 36–38]. Catalysts formed by an active phase (typically metal nanoparticles) and an inert substrate in powder form are often incorporated into the reactors by supporting them on the pellets that form the barrier, usually oxides or other inert dielectric materials. Therefore, the influence of barrier materials on process efficiency is also crucial. Figure 5.2 summarizes the mechanisms and interactions that may take place during a *plasma-catalysis* reaction, in addition to the L-H (Langmuir-Hinshelwood) and

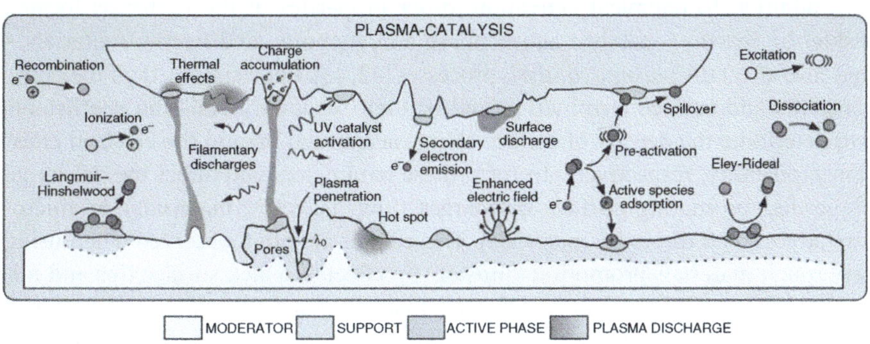


**Figure 5.2** Schematic representation of the different physicochemical processes that can occur during a *plasma-catalysis* process.

E-R mechanisms. In a typical L-H mechanism, the reactants are all adsorbed onto the catalyst surface active sites, while in the E-R mechanism, the synthesis reaction occurs between a species adsorbed on the catalyst surface and another species from the gas phase. An important difference with respect to thermal catalysis is that in *plasma-catalysis,* the L-H or E-R processes may take place on the catalyst and other exposed surfaces (e.g. from the support or the barrier materials) and that other series of physicochemical processes may also take place in the plasma phase (e.g. ionization, electronic and vibrational excitation, dissociation, formation of new molecules and reaction intermediates) or via three-body reactions with reactor walls. In addition, other activation processes of the reactor materials such as ultraviolet (UV) light photocatalytic activation, secondary electron emission, and surface discharge or tip effects might also take place. A factor adding complexity to the system is that surface active sites for L-H or E-R processes may exist not only on the catalyst surfaces (as in classical catalysis) but also at the surface of barrier materials, which have been shown to adsorb and induce the reaction of certain plasma species [37, 39]. Regardless of the location of the active sites, the key issue is that radicals must have a sufficiently long lifetime to reach the involved surfaces. The average lifetime of plasma species is typically less than 100 μs, although this depends on the type of gas, radicals density (as bulk plasma reactions between radicals accelerate their depletion), and the characteristic dimensions of the discharge in each reactor [40]. Within this short timescale, radicals must travel from the plasma bulk to the surface, either crossing the plasma sheath or diffusing into the pores in the case of porous materials [38]. In the specific case of $NH_3$ synthesis, Liu et al. developed a microkinetic model demonstrating that N∗ and H∗ radicals govern the process [41]. Their findings indicate that, even if only a minor fraction of generated radicals reaches the catalyst surface, this may be an averaging factor ensuring that different catalysts may exhibit a similar activity.

In addition to chemical activation, other properties of the materials located inside the reactor – catalyst active phase and supports, and barrier materials – may also affect the *plasma-catalysis* processes [42, 43]. For example, their dielectric constants and surface termination and/or their secondary emission coefficients will determine the density of surface charge accumulation and the electron emission probability, respectively. In turn, these parameters will affect the discharge properties, promoting surface discharges (low dielectric materials) or micro-discharges (high dielectric materials). These latter are often observed when using very rough materials promoting kinds of "tip effect." In fact, surface tips and hillocks produce variations in the electric field distribution, which contribute to increase the density and intensity of microdischarges and the population of excited species at such surface prominences. In the same line, porosity of supports and barrier materials may also play a key role: depending on pore size, plasma may or may not penetrate the pores (it will penetrate pores larger than the Debye length but not smaller ones), where active phase particles may be also located. Additionally, the thermal properties of the barrier and catalytic materials may condition local heating effects, which are crucial for catalytic processes (local temperatures in plasma reactors nominally working at ambient conditions are known to reach values as high as 120 °C [44]). Figure 5.2 also shows how UV radiation resulting from the de-excitation of plasma species can activate the catalyst, particularly if they are made of photoactive materials, and facilitate light-induced surface reactions [37, 39].

At this point, it is also important to remark that plasma interactions may render traditional ammonia synthesis catalysts ineffective for plasma-assisted reactions. As described in Section 5.3.2 and 5.3.3, there is a vivid debate about the precise role of catalysts trying to differentiate between purely catalytic processes and other mechanisms that may also contribute to enhance process efficiency.

### 5.3.2 Barrier Materials and Catalysts in Packed-Bed Plasma Reactors for $NH_3$ Synthesis

In the following, we describe the influence of barrier materials and catalysts on the energy efficiency and chemical yield of the ammonia synthesis process in packed-bed reactors. To this end, it is worth recalling some key aspects of packed-bed reactors: they are typically operated with sinusoidal signals and frequencies ranging from 0 to 30 kHz, with operating voltages and currents generally below 10 kV and 100 mA, respectively. Although little attention has been given to the gas flow regime within the barrier, there is a consensus on the influence of residence time, which should be minimized to prevent the occurrence of backward reactions.

#### 5.3.2.1 Barrier Materials

The chemical characteristics of the barrier material used as moderator in packed-bed plasma reactors have deserved less attention than those of catalysts. However, the barrier material is a critical component of any reactor design because it controls the inner distribution of electric field, while its dielectric constant and morphological and surface properties are key controlling factors of the plasma properties [37, 43]. Most used barrier materials include dielectrics, such as alumina and quartz, with dielectric constants of approximately 8 and 3, respectively, and ferroelectrics, such as $BaTiO_3$, which has a dielectric constant ranging from 1,200 to 10,000 at temperatures between 25 and 120 °C. Recent studies have shown that materials with a high dielectric constant promote the formation of localized microdischarges, while those with low dielectric constants favor the distributed formation of surface discharges [38, 45]. However, the advantages for the ammonia synthesis reaction of each one of these two types of discharge still remain unclear and require a clear understanding of their influence on reaction mechanisms. In the work of Gómez-Ramírez *et al.*, the authors utilize PZT as barrier material, which has both a high dielectric constant and an elevated Curie temperature (332 °C) [32]. Comparing the results obtained with PZT and $BaTiO_3$ under identical experimental conditions, they demonstrated that reaction efficiency was higher with PZT. Furthermore, using isotopic labeling techniques, they showed that employing high dielectric constant materials as barrier and no catalysts, the rate-limiting step for the $NH_3$ synthesis is the plasma nitrogen dissociation, followed by the formation of $NH^*$ or $NH_2{}^*$ species, also in the gas phase [33]. These findings agree with those reported by Hong *et al.*, but contrast with other investigations proposing that the dissociative surface adsorption of dinitrogen molecules, whether excited, is the rate-limiting step of the process [11, 46–48]. The importance of surface reactions was also stressed by Tsung-Wey *et al.*, who proposed that reactions occurring on the surface of catalysts involve plasma-generated N and H radicals [41]. Interestingly, the results of this work revealed that two different metals acting as catalytic active phase (Fe and Ag) behaved similarly in the presence of plasma despite that they depict very different catalytic activity toward nitrogen adsorption and reaction. Very likely, both plasma and surface-driven mechanisms are compatible, and their relative importance will depend on factors such as operating temperature or the presence of materials acting as effective catalysts [49].

It is also noteworthy that while the gas temperature in packed-bed reactors can be maintained close to room temperature, localized heating effects above 120 °C – the Curie temperature of $BaTiO_3$ – may also occur. This means that selecting ferroelectrics as discharge moderators with a high Curie temperature is essential to prevent the loss of ferroelectric properties [44]. In addition to the dielectric properties of the barrier materials, its architecture also plays a significant role for the control of the discharge. In this regard, features such as the shape, surface termination and size of the pellets, or the coating of electrodes with one or more dielectric layers should also be taken into account because they can greatly influence the ammonia synthesis efficiency.

#### 5.3.2.2 Catalyst: Active Phase and Support

In *plasma-catalysis,* features such as the chemical composition of the catalyst, its shape, porosity, crystalline or amorphous structure, characteristics of active sites, or acid–base properties of the support determine both the catalyst selectivity and the effectiveness of the plasma–catalyst interaction. This interaction refers to the way the plasma phase becomes modified by the catalyst and vice versa. Recent studies have shown that plasma properties are modified by the catalytic materials, significantly influencing the chemical and energy efficiency of *plasma-catalysis* processes [14, 16, 36–38].

At an industrial scale, the availability and low cost of iron-based catalysts have made them the catalyst of choice for the Haber-Bosch ammonia synthesis. In academic oriented studies, other active phases, such as ruthenium, are also used, as they depict a high activity for the dissociation of nitrogen molecules, the rate-limiting step in the ammonia synthesis. Although the selection of catalysts in *plasma-catalysis* processes has been one of the most thoroughly studied issues for optimizing the ammonia synthesis in packed-bed reactors, there are not yet clear criteria for the selection of catalyst. Significantly, some studies demonstrate that certain catalysts effectively enhance process efficiency, while others fail to induce any improvement. In particular, the role of the active phase in the ammonia synthesis process remains a highly relevant and debated topic. Beyond their potential catalytic effect, metallic particles may significantly influence the electrical properties of the discharge. This means that process efficiency may be affected not only by specific catalytic effects but also by the electrical properties and morphology of the ensemble formed by the catalyst (including active phase and support) and barrier materials within the reactor [2, 5, 14, 16, 36, 38]. In this regard, new concepts such as that of an effective medium material – encompassing in this definition the overall effect of catalyst active phase and support, as well as the moderator – with specific electrical properties might contribute to define standardized criteria for comparison of results from different laboratories [50].

At present, the series of published *plasma-catalysis* works dealing with the ammonia synthesis is quite large and elusive for systematization because not always a clear distinction is made between catalytic effects and other electrical and plasma processes. Zhu et al. investigated the influence of Ru particles deposited on various supports held by glass wool in a packed-bed plasma reactor and observed a significant increase in $NH_3$ production (ranging from 163% to 387% in comparison with the yield without catalyst) that followed the order: Ru/activated carbon > Ru/ZSM-5 > Ru/$\alpha$-$Al_2O_3$ > Ru/$SiO_2$ [42]. Additional studies in the same laboratory found that the surface basicity of packing materials (i.e. its ability to accept protons) positively affects the plasma-assisted $NH_3$ synthesis process, with pronounced improvements for alkaline alumina with respect to neutral or acidic alumina [28]. Similarly, Peng et al. reported an increase in energy yield (2.3 g-$NH_3$/kWh) when using Ru particles supported on carbon nanotubes as barrier and Cs as promotor [34]. Wang et al. used various transition metal particles (Fe, Ni, Cu) supported on $Al_2O_3$ and found a higher yield ammonia production in comparison with that obtained with a clean $Al_2O_3$ support [51]. An important finding in this work was that the plasma electrical properties were equivalent irrespective of the metal, suggesting that the variations in yield were due to the different surface chemistry of the active phase. Shah et al. achieved an enhanced reaction performance using zeolite 5A as moderator and found a remarkably high energy yield of 15.5 g-$NH_3$/kWh [52]. This achievement was attributed to the modifications induced by the zeolite 5A on plasma properties, particularly its ability to promote microdischarges, enhancing the catalytic activity of the zeolite. Notably, without zeolite, the energy yield was 50 times lower. Zhang et al. also reported an improved reactor performance (i.e. an energy efficiency of 1.54 g-$NH_3$/kWh) when employing a barrier formed by a catalyst including bimetallic RuCo particles supported on $MgTiO_3$ [26]. Rouwenhorst and Lefferts used Ru, Co, Pt, Pd, Cu, and Ag metal particles as active phases supported on MgO, with quartz wool as barrier material. Experiments were carried out at temperatures ranging from room temperature to 500 °C [17]. Low $NH_3$ concentrations were used to prevent reverse reactions and properly analyze the reaction mechanisms. They found that the rate-limiting step involves an E-R reaction between hydrogen atoms adsorbed on the surface and N radicals in the plasma. An exception was the case of silver, where deviations toward an L-H type mechanisms were observed. Zhang et al. have shown that catalyst supports with oxygen vacancies can enhance the energy efficiency for $NH_3$ synthesis [53]. They also found a linear correlation between these two parameters, achieving a maximum yield of 1.7 g-$NH_3$/kWh. However, a critical comparison of all these investigations does not reveal any clear trend as a function of the experimental conditions, although they present in common the use of low-dielectric-constant materials as support or barriers to induce the plasma discharge.

Recently, Luo et al. have explored the use of novel catalysts based on metal nitrides, reaching an energetic yield of 1.43 g-$NH_3$/kWh [24]. This chemical formulation has recently deserved much interest for other electrochemical and thermocatalytic processes. The authors justified this choice with the argument that lattice nitrogen atoms can directly interact with hydrogen species from the plasma, forming $NH_3$ and creating nitrogen vacancies that can be replenished by N species from the plasma in a cyclic process.

The pore size and porosity distribution within the support has been also specifically investigated for the *plasma-catalysis* of ammonia. In a recent work, Wang et al. used Ni particles supported on a mesoporous channeled silica (MCM-41) [15]. Figure 5.3 shows an image of the catalyst structure incorporating the metallic particles. The work reports that ammonia, once formed, diffuses within the mesopores, where it is protected from the plasma interaction, avoiding the occurrence of reverse reactions. The authors claim that the catalyst architecture offers a kind of "shielding effect" contributing to increase the reaction yield. As a result, authors achieved an efficiency of 1.5 g-$NH_3$/kWh (with an ammonia yield of about 5% at 60 kJ/L).

The catalytic influence of metal particle size in packed-bed reactors was first analyzed by Gorky et al. using a well-ordered porous support. Their results indicate that using smaller particle sizes (5.6 nm) can even double the ammonia

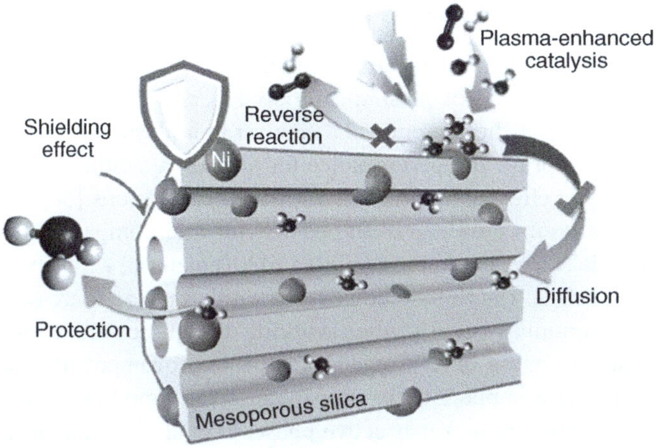


**Figure 5.3** Schematic of the Ni-loaded mesoporous MCM41 catalyst used as a barrier and representation of shielding effect. *Source:* Ref. [15] / American Chemical Society / CC BY 4.0.

synthesis rate in comparison with that obtained with larger nanoparticles (13.5 nm) [54]. Similarly, Gorky et al. examined the use as barrier of microporous silico-alumina-phosphates (SAPOs) with varying porosities and Si/Al ratios [55]. The authors demonstrated that a SAPO with a low Si/Al ratio and a high porosity contributes to increase the $NH_3$ synthesis, likely due to the diffusion of $N_2$ and other excited species into the pores where a more efficient reaction would take place.

In summary, it seems that both the use of disruptive barrier architectures facilitating the selective adsorption of specific species, and the incorporation of novel catalysts – distinct from those traditionally employed in classical catalysis – are promising routes for the optimization of the ammonia synthesis process in the presence of plasma. However, identifying commonalities across different studies to establish a comprehensive framework of results remains challenging. Reactor geometries, barrier materials, and experimental conditions exhibit significant differences, and there are not yet clear criteria for homogenization. Notably, when the inclusion of catalysts leads to increased yields, barriers with low dielectric constants are typically employed. In contrast, the use of materials with high dielectric constants does not result in similar increases. Moreover, in some cases, better results are achieved in the absence of metal catalysts. These observations underscore the importance of the electrical properties of barrier materials in plasma behavior and, consequently, in the overall efficiency of the process. A discussion about new approaches and concepts is presented in Section 5.3.3.

### 5.3.3 New Paradigms in Plasma-Catalysis for Ammonia Synthesis

Recently, new concepts have been proposed in the quest for breakthrough concepts in the current understanding of *plasma-catalysis* [14, 36, 38]. In 2022, Gershman et al. highlighted that the presence of a catalyst in a plasma environment does not necessarily improve the performance for ammonia synthesis and discarded a classical catalytic effect as responsible for their observations [56]. In these investigations, the authors used a packed-bed reactor and compared the effect of two barrier materials: bare mesoporous silica and this material with silver particles deposited on its surface. The results showed that without metal, there was an improvement in the ammonia synthesis when working at high powers, while the addition of Ag as active phase only provided a little improvement at low powers. Using the same silica support as barrier material but modified with MgO and with the incorporation of nickel as active phase, Li et al. demonstrated that the presence of Ni particles can be detrimental for the $NH_3$ synthesis, limiting the energy efficiency to a maximum of around 1.05 g-$NH_3$/kWh [25]. Navascués et al. have also shed some doubts about the involvement of classical catalytic processes by the plasma-catalysis synthesis of ammonia when using high dielectric constant materials [27]. These authors used a packed-bed ferroelectric reactor with planar geometry and PZT pellets acting as moderator either uncoated or coated with a Ru-$Al_2O_3$ catalysts or just the $Al_2O_3$ support powder. They conducted experiments at ambient or at elevated temperatures at which catalytic effects can be promoted by the metal. Results showed that at ambient temperature, the incorporation of Ru/$Al_2O_3$ catalyst did increase the plasma current, while the $Al_2O_3$ support cooled down the plasma, reducing the current through the reactor electrodes and the electric field in the interpellet space. This effect mirrored the evolution of nitrogen conversion. Differences disappeared at high temperatures, when the current curves obtained for the different barrier configurations and the nitrogen conversion curves overlap. Significantly, the energy efficiency for the barrier configuration incorporating the Ru-$Al_2O_3$ catalyst was lower than that obtained with the PZT and PZT-$Al_2O_3$ configurations. This result pointed out that the involvement of some surface catalytic process when using the Ru-$Al_2O_3$-coated pellets is negligible when working in packed-bed reactors incorporating moderator materials with a high dielectric constant. Furthermore, results emphasized the importance of thoroughly characterizing barrier materials (dielectric constants and Curie temperatures), as well as the electrical behavior of reactors incorporating catalytic materials that may affect microdischarge density and overall plasma current. In the same line, Liu *et al.* have analyzed the role of dielectric constant and thermal conductivity for various materials typically used as moderator in packed-bed reactors (BeO, AlN, $TiO_2$, $BaTiO_3$, and $SiO_2$) [43]. These authors demonstrated that materials with high thermal conductivity and dielectric constants enhance the yields in plasma-assisted ammonia synthesis. Moreover, their results suggest that applying a high specific energy (defined as the ratio of the plasma absorbed power to the gas flow rate) can be detrimental for the synthesis of ammonia, because a considerable part of energy may be used to increase the $NH_3$ decomposition. A similar effect was found in the work of Navascues et al. discussed above where, using PZT as barrier material and beyond a certain voltage/frequency, efficiency begins to drop due to an increase in inefficient processes, contributing to the destruction of $NH_3$ [27].

The role of catalysts in the $NH_3$ synthesis has also been critically examined by Ndayirinde et al. [12]. These authors used an alumina-packed-bed barrier with several types of cobalt-based metal catalysts deposited by means of a wet impregnation method ($Co/Al_2O_3$, $CoCeAl_2O_3$, $CoLa/Al_2O_3$, $CoCeLa/Al_2O_3$, and $CoCeMg/Al_2O_3$). Experiments were conducted with a 1:1 $N_2$:$H_2$ mixture at atmospheric pressure and temperatures below 120 °C, i.e., conditions at which no conventional catalytic effects are expected, but local heating from microdischarges could not be discarded. The results showed that the presence of the catalysts significantly enhanced nitrogen conversion and reduced energy consumption. However, there were noticeable differences in the plasma electrical properties, and, for certain conditions, local heating effects associated to the microdischarges. The authors concluded that the chemical composition of the catalyst did not affect the plasma properties but induced changes in plasma power and $I(t)$ characteristic curves. Based on this behavior, they designate the catalysts as "plasma modifiers," emphasizing their active role in modifying discharge properties. Furthermore, they suggest that materials favoring surface discharges instead of microdischarges would be more favorable for the synthesis of ammonia. Additional experimental and simulation studies by the same authors confirmed these results, reinforcing the view that the role of catalytic materials in plasma environments should not be understood in the same way as in classical catalysis [1, 39].

These new approaches and considerations have been highlighted in a recent review article by Ruiz-Martín et al. on the impact of catalysts on plasma behavior in various *plasma-catalysis* processes, concluding that incorporating catalysts into packed-bed reactors significantly influences discharge properties [37]. This effect can sometimes outweigh the importance of classical catalytic effects, indicating that catalysts in plasma environments do not necessarily need to meet the traditional requirements of classical catalysis. The authors suggested the term Plasma-Catalysis Promoters (PCPs), previously used by Akay to designate high dielectric constant materials that promote plasma generation and catalytic effects [57–59]. That is, PCPs would refer to any kind of material incorporated into plasma reactors that enhances the efficiency of conventional moderators. This review article also underscores the need of a holistic approach to properly characterize *plasma-catalysis* processes and the involvement of specific reaction mechanisms in the presence of these PCPs. In this scheme both catalyst-related activation and discharge-related properties (e.g. electrical characteristics) would be determining factors of reactivity. However, to determine the relative importance of each type of processes, it is deemed necessary to proceed to a thorough catalyst characterization before and after the exposure to the plasma discharge and to unravel the surface reaction mechanisms through *in situ* or *in operando* experiments. For the latter, the characterization of electrical properties of the discharge and the determination of gas-phase reactions, electron temperatures and densities, electric field distribution, and discharge regimes would be requirements for a proper assessment of reactivity. Moreover, the authors highlight in this review the need of developing advanced theoretical and simulation models and to exploit the possibilities offered by the new computational procedures based on artificial intelligence techniques to precisely identify the dominant reaction mechanisms in plasma-assisted chemical reactions involving PCPs.

## 5.4 Conclusions

Plasma-assisted ammonia synthesis is carried out at significantly lower temperatures and pressures than those dictated by thermodynamics. However, published results to date indicate that the efficiencies achieved are substantially below the levels required for industrial viability. To make this process competitive in small plants powered by renewable energy, an efficiency of approximately 150 g-$NH_3$/kWh is needed, which contrasts with the highest efficiency achieved so far of only 35.7 g-$NH_3$/kWh [3, 23, 29]. The scenario differs when regarding chemical yields, with values comparable to those of the Haber-Bosch process or even higher (see Table 5.1; the highest efficiency is 62% [25]). Unfortunately, at such high yields, energy efficiency is significantly low. This means that a significant increase in energy efficiency is a key issue to make the synthesis of ammonia competitive with plasma-assisted processes. This substantial limitation calls for new experimental and theoretical approaches to identify and optimize the reaction mechanisms occurring in plasma reactors and for the engineering upgrade of plasma reactors.

In *plasma-catalysis* to synthesize ammonia, it is important to differentiate catalytic from pure plasma-driven processes. It is equally important to determine the precise impact of catalytic materials on the plasma properties, i.e., the way they affect the electrical characteristics of the discharge (e.g. density and intensity of microdischarges, promotion of surface discharges, localized increases in electron density), and conversely, how plasma influences the catalyst activity. Addressing these issues requires additional experimental, theoretical, and simulation studies where the sole variable is the presence or the absence of a catalyst in the plasma reactor. To date, only a limited number of studies have thoroughly compared ammonia synthesis with and without catalysts while keeping fixed all other variables. Since the operation of plasma reactors requires control of many operating parameters and magnitudes – voltage, frequency, reactant flow rate, temperature, pressure, etc. – the benchmarking of results from different laboratories is a rather complicated but necessary task. Cross-investigations will be essential not only to understand the role of catalysts in the plasma processes, but also to identify the most suitable catalytic material for the plasma-assisted synthesis of ammonia [38]. Current research suggests that conventional catalysts may not necessarily be the optimal choice for *plasma-catalysis* processes. Instead, they could be replaced with less expensive materials that exhibit longer lifetime and avoid issues like poisoning or deactivation when exposed to plasmas. The term *Plasma-Catalysis Promoter* (PCP) has been introduced to describe materials that enhance reaction yield and efficiency even if they do not behave as classical catalysts [37]. However, a deep understanding of PCP behavior requires their thorough characterization both "*in situ*" and "*in operando*" conditions using, for example, vibrational techniques such as Fourier transform infrared spectroscopy (FTIR), Raman spectroscopy, or DRIFTS [16]. Undoubtedly, the presence of plasma adds complexity to these measurements in comparison to classical catalysis experiments and demands the design of *ad hoc* reactors for this type of analysis. Additionally, characterization of PCPs before and after the plasma driven reaction, using methods such as scanning electron microscopy (SEM) and x-ray photoelectron spectroscopy (XPS), will provide detailed information about the physical and chemical characteristics of surfaces under reaction conditions. It is expected that such studies would help defining optimal PCPs to enhance the energy efficiency of reactors. The promotion of surface mechanisms, which involve interactions between species adsorbed on the PCP surfaces and plasma species to form $NH_3$, is a possible route for yield improvement. At the same time, PCPs should prevent to occurrence of inefficient processes, such as $NH_3$ decomposition or hydrogen atom exchange between $NH_3$ molecules and hydrogen species in the plasma [33]. It is also expected that PCPs will contribute to optimizing the electrical properties of the plasma discharge, facilitating reactor operation along a wide range of temperatures and pressures with minimal degradation costs over multiple operational cycles.

It has been also identified that improving chemical and energy yields of *plasma-catalysis* processes requires the optimization of the discharge electrical properties. Achieving this goal requires not only the appropriate selection of ferroelectric/dielectric barrier materials and PCPs (in *plasma-catalysis* processes) but also a precise control over the discharge characteristics and its spatial distribution. This optimization demands a detailed analysis of $I(t)$, $Q(t)$, and $V(t)$ curves, power consumption, and the characterization of density and intensity of microdischarges. Since localized heating caused by microdischarges may favor surface catalytic processes, determining the dominant reaction mechanisms in the vicinity of microdischarges and in more homogeneous plasmas (e.g. surface or diffuse discharges) is also a key issue to improve operation conditions and reactor designs [13, 60, 61]. For a proper assessment of reaction mechanisms, isotopic labeling has demonstrated to be a valuable tool for unveiling the mechanisms involved in the ammonia synthesis under plasma conditions and differentiating between competing pathways of reactions [33, 62].

Minimizing reverse reactions is currently a great challenge to improve efficiency. Landmarks in this regard can be achieved by integrating porous materials (as *ad hoc* designed zeolites) into reactors to trap $NH_3$ molecules once they are formed. Some studies have already demonstrated the effectiveness of using porous materials for this purpose [15]. However, for industrial implementation, it would be also necessary to propose alternative functioning modes, including, for example, two-step processes, where desorption of $NH_3$ by heating should follow the plasma-assisted reaction, or the incorporation of selective membranes for gas separation during or downstream the process. Another possibility would be to work in parallel with many small size reactors, as this would decrease the residence time of the gases, minimizing the occurrence of inefficient reactions [22].

## References

1 Gorbanev, Y., Fedirchyk, I., and Bogaerts, A. (2024). Plasma catalysis in ammonia production and decomposition: use it, or lose it? *Curr Opin Green Sustain Chem* 47: 100916. https://doi.org/10.1016/j.cogsc.2024.100916.

2 Rouwenhorst, K.H.R., Engelmann, Y., van 't Veer, K. et al. (2020). Plasma-driven catalysis: green ammonia synthesis with intermittent electricity. *Green Chem.* 22: 6258–6287. https://doi.org/10.1039/D0GC02058C.

3 Bogaerts, A., Neyts, E.C., Guaitella, O., and Murphy, A.B. (2022). Foundations of plasma catalysis for environmental applications. *Plasma Sources Sci. Technol.* 31: 053002. https://doi.org/10.1088/1361-6595/AC5F8E.

4 Adamovich, I., Agarwal, S., Ahedo, E. et al. (2022). The plasma roadmap: low temperature plasma science and technology. *J. Phys. D. Appl. Phys.* 55 (2022): 373001. https://doi.org/10.1088/1361-6463/ac5e1c.

5 Zhang, Y., Niu, J., Chen, S. et al. (2024). Ammonia synthesis by nonthermal plasma catalysis: a review on recent research progress. *J. Phys. D. Appl. Phys.* 57: 323001. https://doi.org/10.1088/1361-6463/ad4717.

6 Meng, X., Lu, N., Shang, K., and Jiang, N. (2024). Dielectric-boosted gliding arc discharge for N2 fixation into $NO_x$. *Plasma Chem. Plasma Process.* 44: 1513–1528. https://doi.org/10.1007/s11090-024-10474-8.

**7** Gómez-Ramírez, A., Montoro-Damas, A.M., Cotrino, J. et al. (2017). About the enhancement of chemical yield during the atmospheric plasma synthesis of ammonia in a ferroelectric packed bed reactor. *Plasma Process. Polym.* 14: 1600081. https://doi.org/10.1002/ppap.201600081.

**8** Matsumoto, O. (1998). Plasma catalytic reaction in ammonia synthesis in the microwave discharge. *J. Phys. IV: JP* 8: Pr7-411. https://doi.org/10.1051/jp4:1998736.

**9** Vervloedt, S.C.L. and von Keudell, A. (2024). Ammonia synthesis by plasma catalysis in an atmospheric RF helium plasma. *Plasma Sources Sci. Technol.* 33: 045005. https://doi.org/10.1088/1361-6595/ad38d6.

**10** Hosseini, H. (2023). Dielectric barrier discharge plasma catalysis as an alternative approach for the synthesis of ammonia: a review. *RSC Adv.* 13: 28211–28223. https://doi.org/10.1039/D3RA05580A.

**11** Mehta, P., Barboun, P.M., Engelmann, Y. et al. (2020). Plasma-catalytic ammonia synthesis beyond the equilibrium limit. *ACS Catal.* 10: 6726–6734. https://doi.org/10.1021/acscatal.0c00684.

**12** Ndayirinde, C., Gorbanev, Y., Ciocarlan, R.-G. et al. (2023). Plasma-catalytic ammonia synthesis: packed catalysts act as plasma modifiers. *Catal. Today* 419: 114156. https://doi.org/10.1016/j.cattod.2023.114156.

**13** Bogaerts, A., Centi, G., Hessel, V., and Rebrov, E. (2023). Challenges in unconventional catalysis. *Catal. Today* 420: 114180. https://doi.org/10.1016/j.cattod.2023.114180.

**14** De Meyer, R., Gorbanev, Y., Ciocarlan, R.G. et al. (2024). Importance of plasma discharge characteristics in plasma catalysis: dry reforming of methane vs. ammonia synthesis. *Chem. Eng. J.* 488: 150838. https://doi.org/10.1016/j.cej.2024.150838.

**15** Wang, Y., Yang, W., Xu, S. et al. (2022). Shielding protection by mesoporous catalysts for improving plasma-catalytic ambient ammonia synthesis. *J. Am. Chem. Soc.* 144: 12020–12031. https://doi.org/10.1021/jacs.2c01950.

**16** Lefferts, L. (2024). Leveraging expertise in thermal catalysis to understand plasma catalysis. *Angew. Chem. Int. Ed.* 63: e202305322. https://doi.org/10.1002/anie.202305322.

**17** Rouwenhorstm, K.H.R. and Lefferts, L. (2023). Plasma-catalytic ammonia synthesis via Eley-Rideal reactions: a kinetic analysis. *ChemCatChem* 15: e202300078. https://doi.org/10.1002/cctc.202300078.

**18** Shah, J., Wu, T., Lucero, J. et al. (2019). Nonthermal plasma synthesis of ammonia over Ni-MOF-74. *ACS Sustain. Chem. Eng.* 7: 377–383. https://doi.org/10.1021/acssuschemeng.8b03705.

**19** Shah, J., Wang, W., Bogaerts, A., and Carreon, M.L. (2018). Ammonia synthesis by radio frequency plasma catalysis: revealing the underlying mechanisms. *ACS Appl Energy Mater* 1: 4824–4839. https://doi.org/10.1021/acsaem.8b00898.

**20** Sharma, R.K., Patel, H., Mushtaq, U. et al. (2021). Plasma activated electrochemical ammonia synthesis from nitrogen and water. *ACS Energy Lett.* 6: 313–319. https://doi.org/10.1021/acsenergylett.0c02349.

**21** Xu, X., Sun, M., Song, Q. et al. (2024). Nanosecond pulsed gliding arc plasma for ammonia synthesis: better insight from discharge mode and vibrational temperature. *J. Phys. D. Appl. Phys.* 57: 415206. https://doi.org/10.1088/1361-6463/ad5f3d.

**22** Van Duc Long, N., Pourali, N., Lamichhane, P. et al. (2024). Catalytic ammonia formation in a microreaction chamber with electrically intensified arc plasma. *ChemCatChem* 16: e202400005. https://doi.org/10.1002/cctc.202400005.

**23** Wang, X., Du, X., Chen, K. et al. (2023). Predicting the ammonia synthesis performance of plasma catalysis using an artificial neural network model. *ACS Sustain. Chem. Eng.* 11: 4543–4554. https://doi.org/10.1021/acssuschemeng.2c04715.

**24** Luo, S., Liu, Y., Song, Y. et al. (2024). Plasma-induced nitrogen vacancy-mediated ammonia synthesis over a VN catalyst. *Chem. Commun.* 60: 3295–3298. https://doi.org/10.1039/d4cc00042k.

**25** Li, S., Shao, Y., Chen, H., and Fan, X. (2022). Nonthermal plasma catalytic ammonia synthesis over a Ni catalyst supported on MgO/SBA-15. *Ind. Eng. Chem. Res.* 61: 3292–3302. https://doi.org/10.1021/acs.iecr.1c04968.

**26** Zhang, Y., Li, S., Yuan, Z. et al. (2022). Mechanochemical synthesis of RuCo/$MgTiO_3$ catalysts for nonthermal plasma-assisted ammonia synthesis. *Ind. Eng. Chem. Res.* 61: 14199–14210. https://doi.org/10.1021/acs.iecr.2c02216.

**27** Navascués, P., Garrido-García, J., Cotrino, J. et al. (2023). Incorporation of a metal catalyst for the ammonia synthesis in a ferroelectric packed-bed plasma reactor: does it really matter? *ACS Sustain. Chem. Eng.* 11: 3621–3632. https://doi.org/10.1021/acssuschemeng.2c5877.

**28** Zhu, X., Hu, X., Wu, X. et al. (2020). Ammonia synthesis over $\gamma$-$Al_2O_3$ pellets in a packed-bed dielectric barrier discharge reactor. *J. Phys. D. Appl. Phys.* 53: 164002. https://doi.org/10.1088/1361-6463/ab6cd1.

**29** Gorky, F., Lucero, J.M., Crawford, J.M. et al. (2021). Insights on cold plasma ammonia synthesis and decomposition using alkaline earth metal-based perovskites. *Cat. Sci. Technol.* 11: 5109–5118. https://doi.org/10.1039/D1CY00729G.

**30** Liu, J., Zhu, X., Hu, X. et al. (2022). Plasma-assisted ammonia synthesis in a packed-bed dielectric barrier discharge reactor: effect of argon addition. *Vacuum* 197: 110786. https://doi.org/10.1016/j.vacuum.2021.110786.

**31** Nguyen, H.M., Gorky, F., Guthrie, S., and Carreon, M.L. (2023). Sustainable ammonia synthesis from nitrogen wet with sea water by single-step plasma catalysis. *Catal. Today* 418: 114141. https://doi.org/10.1016/j.cattod.2023.114141.

32 Gómez-Ramírez, A., Cotrino, J., Lambert, R.M., and González-Elipe, A.R. (2015). Efficient synthesis of ammonia from $N_2$ and $H_2$ alone in a ferroelectric packed-bed DBD reactor. *Plasma Sources Sci. Technol.* 24: 065011. https://doi.org/10.1088/0963-0252/24/6/065011.

33 Navascués, P., Obrero-Pérez, J.M., Cotrino, J. et al. (2020). Unraveling discharge and surface mechanisms in plasma-assisted ammonia reactions. *ACS Sustain. Chem. Eng.* 8: 14855–14866. https://doi.org/10.1021/acssuschemeng.0c04461.

34 Peng, P., Li, Y., Cheng, Y. et al. (2016). Atmospheric pressure ammonia synthesis using non-thermal plasma assisted catalysis. *Plasma Chem. Plasma Process.* 36: 1201–1210. https://doi.org/10.1007/s11090-016-9713-6.

35 Xin Tu, J.C.W.T.N. (2019). *Plasma Catalysis Fundamentals and Applications.* Springer.

36 Maslova, V., Nastase, R., Veryasov, G. et al. (2024). Current status and challenges of plasma and plasma-catalysis for methane coupling: a review. *Prog. Energy Combust. Sci.* 101: 101096. https://doi.org/10.1016/j.pecs.2023.101096.

37 Ruiz-Martín, M., Oliva-Ramírez, M., González-Elipe, A.R., and Gómez-Ramírez, A. (2024). Plasma catalysis for gas conversion – impact of catalyst on the plasma behavior. *Curr Opin Green Sustain Chem* 51: 100990. https://doi.org/10.1016/j.cogsc.2024.100990.

38 Loenders, B., Michiels, R., and Bogaerts, A. (2023). Is a catalyst always beneficial in plasma catalysis? Insights from the many physical and chemical interactions. *J. Energy Chem.* 85: 501–533. https://doi.org/10.1016/J.JECHEM.2023.06.016.

39 Engelmann, Y., van 't Veer, K., Gorbanev, Y. et al. (2021). Plasma catalysis for ammonia synthesis: a microkinetic modeling study on the contributions of Eley–Rideal reactions. *ACS Sustain. Chem. Eng.* 9: 13151–13163. https://doi.org/10.1021/acssuschemeng.1c02713.

40 Kim, H.H., Teramoto, Y., Negishi, N., and Ogata, A. (2015). A multidisciplinary approach to understand the interactions of nonthermal plasma and catalyst: a review. *Catal. Today* 256: 13–22. https://doi.org/10.1016/J.CATTOD.2015.04.009.

41 Liu, T.W., Gorky, F., Carreon, M.L., and Gomez-Gualdron, D.A. (2023). Plasma radicals as kinetics-controlling species during plasma-assisted catalytic $NH_3$ formation: support from microkinetic modeling. *ACS Sustain. Chem. Eng.* 11: 16749–16763. https://doi.org/10.1021/acssuschemeng.3c03813.

42 Zhu, X., Liu, J., Hu, X. et al. (2022). Plasma-catalytic synthesis of ammonia over Ru-based catalysts: insights into the support effect. *J. Energy Inst.* 102: 240–246. https://doi.org/10.1016/j.joei.2022.02.014.

43 Liu, J., Zhu, X., Hu, X., and Tu, X. (2022). Plasma-assisted ammonia synthesis in a packed-bed dielectric barrier discharge reactor: roles of dielectric constant and thermal conductivity of packing materials. *Plasma Sci. Technol.* 24: 025503. https://doi.org/10.1088/2058-6272/ac39fb.

44 Gómez-Ramírez, A., Álvarez, R., Navascués, P. et al. (2021). Electrical and reaction performances of packed-bed plasma reactors moderated with ferroelectric or dielectric materials. *Plasma Process. Polym.* 18: 2000193. https://doi.org/10.1002/ppap.202000193.

45 Butterworth, T. and Allen, R.W.K. (2017). Plasma-catalyst interaction studied in a single pellet DBD reactor: dielectric constant effect on plasma dynamics. *Plasma Sources Sci. Technol.* 26: 065008. https://doi.org/10.1088/1361-6595/aa6c35.

46 Mehta, P., Barboun, P., Herrera, F.A. et al. (2018). Overcoming ammonia synthesis scaling relations with plasma-enabled catalysis. *Nat. Catal.* 1: 269–275. https://doi.org/10.1038/s41929-018-0045-1.

47 Mehta, P., Barboun, P., Go, D.B. et al. (2019). Catalysis enabled by plasma activation of strong chemical bonds: a review. *ACS Energy Lett.* 4: 1115–1133. https://doi.org/10.1021/acsenergylett.9b00263.

48 Hong, J., Prawer, S., and Murphy, A.B. (2018). Plasma catalysis as an alternative route for ammonia production: status, mechanisms, and prospects for progress. *ACS Sustain. Chem. Eng.* 6: 15–31. https://doi.org/10.1021/acssuschemeng.7b02381.

49 Akay, G. and Zhang, K. (2017). Process intensification in ammonia synthesis using novel coassembled supported microporous catalysts promoted by nonthermal plasma. *Ind. Eng. Chem. Res.* 56: 457–468. https://doi.org/10.1021/acs.iecr.6b02053.

50 Paula Navascués, Mateo Ruiz-Martín, Guillermo Regodón, Alberto Palmero, J. Cotrino, Agustín Rodríguez González-Elipe, Ana Gómez-Ramírez. 2025. Ammonia to hydrogen conversion in a ferroelectric packed-bed plasma reactor: The effect of a high-permittivity effective medium material. *Plasma Processes and Polymers*, 22 (10). e70067. Doi:https://doi.org/10.1002/ppap.70067

51 Wang, Y., Craven, M., Yu, X. et al. (2019). Plasma-enhanced catalytic synthesis of ammonia over a $Ni/Al_2O_3$ catalyst at near-room temperature: insights into the importance of the catalyst surface on the reaction mechanism. *ACS Catal.* 9: 10780–10793. https://doi.org/10.1021/acscatal.9b02538.

52 Shah, J.R., Gorky, F., Lucero, J. et al. (2020). Ammonia synthesis via atmospheric plasma catalysis: zeolite 5A, a case of study. *Ind. Eng. Chem. Res.* 59: 5167–5176. https://doi.org/10.1021/acs.iecr.9b05220.

53 Zhang, Y., Li, S., Qiu, B. et al. (2024). MgAl layered double hydroxide (LDH) for promoting ammonia synthesis in non-thermal plasma: role of surface oxygen vacancy. *Chem. Eng. Process. Process Intensif.* 195: 109608. https://doi.org/10.1016/j.cep.2023.109608.

54 Gorky, F., Best, A., Jasinski, J. et al. (2021). Plasma catalytic ammonia synthesis on Ni nanoparticles: the size effect. *J. Catal.* 393: 369–380. https://doi.org/10.1016/j.jcat.2020.11.030.

55 Gorky, F., Nambo, A., Carreon, M.A., and Carreon, M.L. (2024). Plasma catalytic conversion of nitrogen and hydrogen to ammonia over silico alumino phosphate

(SAPO) zeolites. *Plasma Chem. Plasma Process.* 44: 1357–1368. https://doi.org/10.1007/s11090-023-10397-w.

**56** Gershman, S., Fetsch, H., Gorky, F., and Carreon, M.L. (2022). Identifying regimes during plasma catalytic ammonia synthesis. *Plasma Chem. Plasma Process.* 42: 731–757. https://doi.org/10.1007/s11090-022-10258-y.

**57** Akay, G. (2020). Plasma generating—chemical looping catalyst synthesis by microwave plasma shock for nitrogen fixation from air and hydrogen production from water for agriculture and energy technologies in global warming prevention. *Catalysts* 10: 152. https://doi.org/10.3390/catal10020152.

**58** Akay, G. (2023). Hydrogen, ammonia and symbiotic/smart fertilizer production using renewable feedstock and $CO_2$ utilization through catalytic processes and nonthermal plasma with novel catalysts and in situ reactive separation: a roadmap for sustainable and innovation-based technology. *Catalysts* 13: 1287. https://doi.org/10.3390/catal13091287.

**59** Akay, G. (2025). Green ammonia, nitric acid, advanced fertilizer and electricity production with in situ $CO_2$ capture and utilization by integrated intensified nonthermal plasma catalytic processes: a technology transfer review for distributed biorefineries. *Catalysts* 15: 105. https://doi.org/10.3390/catal15020105.

**60** Andersen, J.A., Holm, M.C., van't Veer, K. et al. (2023). Plasma-catalytic ammonia synthesis in a dielectric barrier discharge reactor: a combined experimental study and kinetic modeling. *Chem. Eng. J.* 457: 141294. https://doi.org/10.1016/j.cej.2023.141294.

**61** Liu, T.-W., Gorky, F., Carreon, M.L., and Gómez-Gualdrón, D.A. (2022). Energetics of reaction pathways enabled by N and H radicals during catalytic, plasma-assisted $NH_3$ synthesis. *ACS Sustain. Chem. Eng.* 10: 2034–2051. https://doi.org/10.1021/acssuschemeng.1c5660.

**62** Navascués, P., Obrero-Pérez, J.M., Cotrino, J. et al. (2019). Isotope labelling for reaction mechanism analysis in DBD plasma processes. *Catalysts* 9: 9–11. https://doi.org/10.3390/catal9010045.

# 6

# Plasma-Assisted $NO_x$ Synthesis

*Tianyu Li, Haoxuan Jiang, Rusen Zhou, Jing Sun, and Renwu Zhou*

*State Key Laboratory of Electrical Insulation and Power Equipment, Center for Plasma Biomedicine, Xi'an Jiaotong University, Xi'an, PR China*

## 6.1 Introduction

Oxidized nitrogen plays a crucial role in the nitrogen cycle [1], not only serving as a fundamental element for life activities but also finding wide applications in industrial production, while exerting significant impacts on the environment. For instance, it can be used to produce nitric acid, nitrates, and other nitrogen-containing compounds, which have extensive applications in agriculture, the chemical industry, and the pharmaceutical field [2]. We collectively define the neutral and anionic forms of nitrogen oxides as $NO_x$. The neutral nitrogen oxides, including nitric oxide (NO) and nitrogen dioxide ($NO_2$), are characterized by low synthesis costs and a wide range of raw material sources, capable of being directly extracted from the air. Among non-biological sources, a significant contributor to $NO_x$ on Earth is lightning generated by electrical discharges between clouds, where conditions are optimum for $NO_x$ formation [3]. The anionic forms of nitrogen oxides mainly include nitrate ($NO_3^-$) and nitrite ($NO_2^-$). In the early Earth, when organisms died, ammonium ($NH_4^+$) would return to the environment. In the presence of oxygen, these processes involve $NH_4^+$ being oxidized to $NO_3^-$ by specific nitrifying bacteria and fungi, denitrifying bacteria reducing $NO_3^-$ to $NO_2^-$, and ultimately reducing it to $N_2$ to return to the atmosphere, thereby closing the nitrogen cycle. However, biological and geochemical nitrogen fixation account for only a small portion of nitrogen fixation. The annual total production of non-biological nitrogen oxides is 171 million tons [1, 4]. The ever-increasing global population demands more fixed nitrogen than what the Earth can provide.

*Plasma-Assisted Nitrogen Fixation for Sustainable Process Industries*, First Edition. Edited by Volker Hessel, Annemie Bogaerts, Gabriele Centi, Evgeny Rebrov, and Nguyen Van Duc Long.

Published 2026 by John Wiley & Sons, Inc.

The chemists have known for many years how to convert atmospheric inert free nitrogen into compounds of nitrogen in the laboratory. The history of artificial nitrogen fixation through nitrogen oxidation can be traced back to 1781 when Cavendish observed the formation of nitric acid during the combustion of $H_2$ in air [5]. In 1800, Sir Humphrey Davy produced NO by passing air over an electrically heated wire [5]. However, the products obtained in this early work were in pitifully small quantities. It was not until 1902 that two Americans, C. S. Bradley and R. Lovejoy, established the first commercial plant for NO production at Niagara Falls, New York [5]. Before the Niagara plant ceased operations, in 1903, Professors Kristian Birkeland and Samuel Eyde in Norway developed and tested the first plasma-based process for synthesizing $NO_x$ (the process of nitrogen oxidation using plasma was realized industrially earlier than the Haber-Bosch process), known as the Birkeland-Eyde (B-E) process, which had proven to be both mechanically effective and economically viable [5, 6]. In the B-E process (Figure 6.1a),

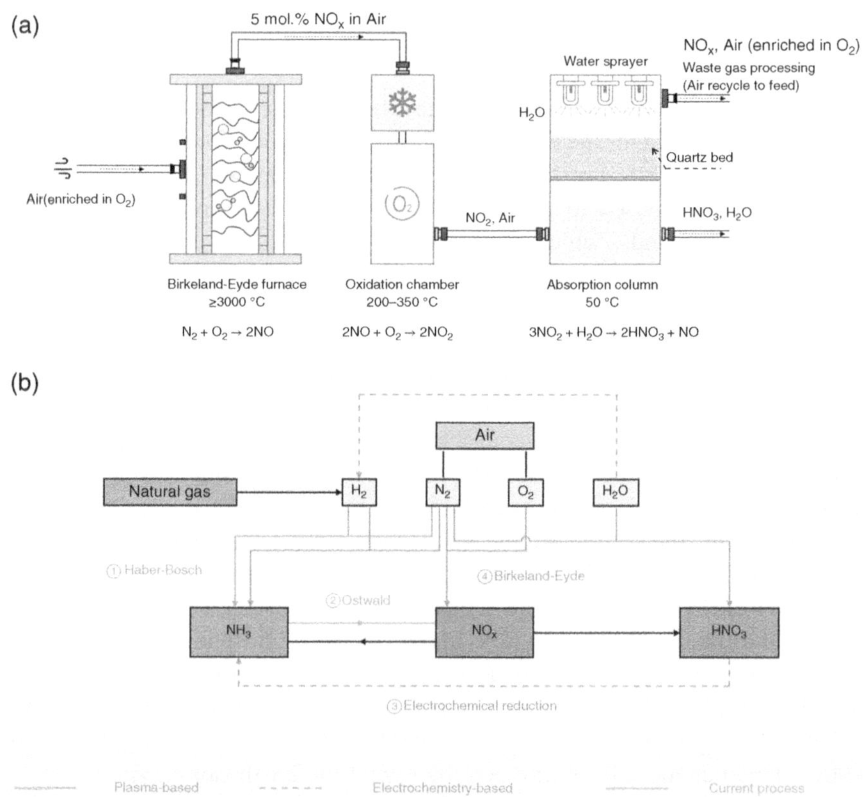


**Figure 6.1** (a) Process scheme for the Birkeland-Eyde industrial nitrogen fixation process. (b) Schematic diagram of major nitrogen fixation routes. *Source:* Adapted from Patil et al. [7] and adapted from Rouwenhorst et al. [8].

air passes through an electric arc, which is a thermal plasma, resulting in the production of NO (Eq. 6.1) and $NO_2$ (Eq. 6.2). Subsequently, $NO_2$ is concentrated and absorbed into water, forming $HNO_3$ (Eq. 6.3) [9].

To address the escalating demands of the growing population, a process for fixing atmospheric nitrogen, known as the "Haber-Bosch ammonia synthesis process"(H-B), was developed in 1908 and is regarded as one of the most important discoveries of the twentieth century. This process directly synthesizes ammonia ($NH_3$) from $N_2$ and $H_2$ (Eq. 6.4). Presently, nitrogen is nearly exclusively fixed through the H-B process. This process now sustains about 40% of the world's population [10, 11], yet it also consumes nearly 1–2% of global energy production [12] and results in substantial carbon dioxide emissions (approximately 420 million tons per year) [13].

Despite these inherent limitations, efforts have been made over the decades to progressively improve and develop the H-B process. After several decades of refinement, the current H-B process consumes approximately 0.25 MJ $mol_N^{-1}$ of energy, which is considered to have reached the theoretical limit in terms of ammonia energy efficiency [2]. With the advent of the H-B process and lower cost fossil fuels such as coal and natural gas, the B-E process became economically disadvantageous (including higher investment and equipment maintenance costs). Ultimately, in 1928, the H-B process surpassed the B-E process, and some B-E process plants were permanently shut down.

$$N_2 + O_2 \leftrightharpoons 2NO \text{ with } \Delta H_r^o = 180\,kJ \cdot mol^{-1} \tag{6.1}$$

$$2NO + O_2 \rightarrow 2NO_2 \text{ with } \Delta H_r^o = -144\,kJ \cdot mol^{-1} \tag{6.2}$$

$$3NO_2 + H_2O \rightarrow 2HNO_3 + NO \text{ with } \Delta H_r^o = -117\,kJ \cdot mol^{-1} \tag{6.3}$$

$$3H_2 + N_2 \rightarrow 2NH_3 + NO \text{ with } \Delta H_r^o = -92\,kJ \cdot mol^{-1} \tag{6.4}$$

$$4NH_3 + 5O_2 \rightarrow 6H_2O + 4NO \text{ with } \Delta H_r^o = -905\,kJ \cdot mol^{-1} \tag{6.5}$$

Until now, the demand for nitrogen oxides is primarily met through the production of nitric acid ($HNO_3$) via the Ostwald process, which involves the catalytic oxidation of $NH_3$ produced by the H-B process at high temperatures [14]. This oxidation process was developed by Wilhelm Ostwald, who patented the ammonia oxidation process in 1906 [15]. Currently, nearly all $HNO_3$ is produced through the oxidation of $NH_3$ via the Ostwald process. In this process, $NH_3$ is oxidized at 600–800 °C and 4–10 atm in the presence of rhodium–platinum gauze to generate NO and $H_2O$ (Eqs. 6.2 and 6.5). Then, the NO is cooled to around 50 °C, subsequently oxidized to $NO_2$, and absorbed by $H_2O$ to produce dilute $HNO_3$. The untreated NO is recovered, and the $HNO_3$ is distilled and concentrated. The overall yield of the Ostwald process is typically 98%.

Given the increasing demand for nitrogen-fixed products, the scalability of the main industrial nitrogen fixation process (i.e. the H-B process) in decentralized production is poor. Furthermore, since the H-B process operates at high temperatures and pressures (350–500 °C and 100–300 bar), it is challenging to operate under varying loads of intermittent renewable energy [16]. The resulting high-energy intensity and environmental issues have made it imperative to develop and integrate more sustainable processes [17, 18]. Several alternative (nonconventional) technologies are currently being researched to enable nitrogen fixation under mild conditions, such as biological [19, 20], plasma-based [21, 22], and electrochemical nitrogen fixation [23]. Among the various technologies mentioned, plasma-based $N_2$ fixation is particularly attractive due to its flexibility and synergy with sustainable energy sources [6, 24].

As mentioned earlier, natural lightning contributes significantly to global nitrogen fixation, which is also called the thermal plasma. However, similar to the B-E process outlined above, thermal plasma operates at very high temperatures (on the order of $10^4$ K), which can easily lead to the decomposition of newly formed NO back into $N_2$ and $O_2$. Therefore, rapid quenching of the gas at rates of millions of Kelvin per second is necessary [25]. The energy cost of the electric arc to generate the thermal plasma for NO synthesis is 2.4–3.1 MJ $mol_N^{-1}$ [7, 24, 26]. Even with further optimization of the reactor, the minimum theoretical energy efficiency is only 0.72 MJ $mol_N^{-1}$ [8], meaning that the energy efficiency of thermal plasma cannot compete with that of the H-B process based on electrolysis for producing $HNO_3$ (approximately 0.6 MJ $mol_N^{-1}$). A significant portion of the energy converted to the plasma is not used for $NO_x$ production but is dissipated as heat, light radiation, and other forms of energy. Therefore, one of the main challenges in plasma oxidation for nitrogen fixation is the high-energy cost at high production rates.

In comparison, nonthermal plasma (NTP) holds great promise. The electrons in NTP have temperatures much higher than those of heavier particles [27, 28]. In most cases, the electron temperature is on the order of 1 eV, while the gas temperature remains close to room temperature. This high electron temperature arises from the small mass of electrons, which allows them to be easily accelerated by an applied electromagnetic field, whereas heavier particles – even ions – are not readily accelerated [29]. Electrons can be considered initiators for highly reactive and complex chemical reactions. This is one of the key advantages of NTP: it may allow the activation of molecules with strong chemical bonds, including $N_2$ (with a bond energy of about 9.79 eV) [30], under mild conditions. There is no need to heat the entire reactor or gas, as the discharge and related reactions can be easily initiated by applying an electric field. Furthermore, it is a highly flexible process, as it can be easily turned on and off instantaneously. Another advantage of NTP is the low investment cost for reactors and its wide range of application scales, spanning from watts to megawatts, as evidenced by the successful development of ozone generators due to their excellent scalability [31]. Figure 6.1b illustrates the scheme of the current and plasma-based alternative routes for nitrogen fixation.

In summary, atmospheric-pressure NTP offers a unique perspective, as they are capable of inducing chemical reactions in gases at ambient pressure and temperature with limited energy costs, while also enabling distributed and on-demand production. This topic has also been thoroughly investigated within the framework of a large EU project called "Microwave, Acoustic and Plasma SYNtheses" (MAPSYN) [18, 32]. After decades of intensive research, significant progress has been made recently in plasma-assisted nitrogen oxide synthesis technology. Next, this chapter summarizes the formation pathways of nitrogen oxides, discusses the latest technologies and potential improvement approaches for plasma-based nitrogen oxide synthesis, and provides references for efficient nitrogen oxidation through plasma technology.

## 6.2 The Mechanism of Plasma-Assisted Nitrogen Oxidation

The main pathway for nitrogen oxidation via NTP is shown in Figure 6.2. One of the key steps is the formation of NO, which, as indicated by Eq. (6.1), is a highly endothermic reaction [34]. Given that the bond energy of the O=O (about 5.12 eV) is weaker compared to that of the N≡N, the dissociation of $O_2$ is more readily achieved. Therefore, the most challenging and crucial step in achieving nitrogen fixation through oxidation to form NO via NTP is the breaking of the N≡N bond.

The Zeldovich mechanism is considered one of the lowest energy cost pathways for NO formation (Eqs. 6.7 and 6.8), which is a chemical reaction mechanism that describes the formation of NO from N and O. It is named after Yakov Borisovich Zel'dovich, who proposed the theory in 1946 [35]. Plasma can activate $N_2$ molecules through three mechanisms: electron collision excitation, dissociation, and ionization, thereby lowering the reaction energy barrier and enhancing the production of NO (Eq. 6.6). The activated $N_2$ then reacts with O atoms to produce NO and N (Eq. 6.7). Subsequently, N atoms react with $O_2$ molecules to generate a second NO molecule and a new O atom (Eq. 6.8). The reaction chain closes when the new O atom reacts with the next excited $N_2$ molecule. By combining Eqs. (6.7) and (6.8), the net energy consumption for $NO_x$ synthesis is calculated to be 0.2 MJ $mol_N^{-1}$, which is lower than that of $NO_x$ synthesis based on the electrolysis-driven H-B process combined with the Ostwald process [24].

However, in practice, the energy cost is higher due to vibrational–translational (V–T) relaxation and the decomposition of $NO_x$ after plasma treatment if the temperature does not drop quickly enough. The radicals excited by the plasma may also recombine to form $O_2$ and $N_2$, meaning that all the energy is lost as heat. As the concentration of NO in the discharge region increases, it can terminate

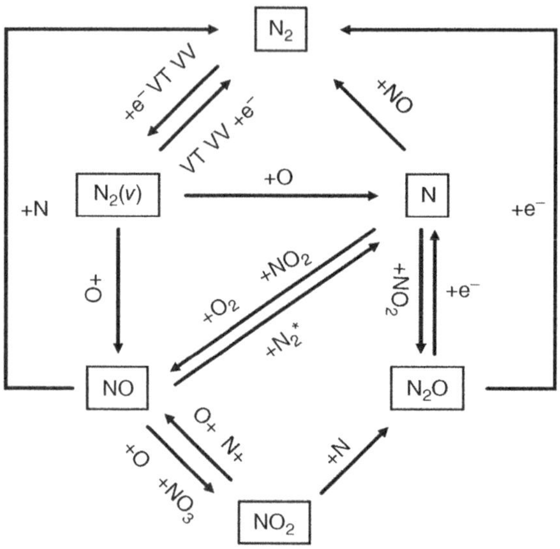


**Figure 6.2** The reaction scheme illustrates the primary pathways in nitrogen oxide chemistry via NTP. In this context, $N_2^*$ indicates all electronically excited states of the $N_2$ molecule, e indicates electrons, VT indicates vibrational–translational relaxation, and VV represents vibrational–vibrational relaxation. *Source:* Reproduced from [33] / with permission of John Wiley & Sons.

the Zeldovich chain. Therefore, further promotion of NO synthesis is possible if its reverse reaction (Eq. 6.9) can be reduced. Additionally, the second important process for NO formation is the reaction of N atoms with $O_2$ molecules (in either the ground or excited state) (Eq. 6.8). Thus, efforts should be made to direct the N atoms generated in Eq. (6.7) to react with $O_2$ molecules in Eq. (6.8) rather than with NO molecules in the reverse Eq. (6.9), to optimize NO synthesis and improve energy efficiency. For example, at a fixed specific energy input (SEI), by increasing the gas flow rate, more feed gas will be exposed to the plasma. The local NO concentration within the discharge region will decrease, but the overall NO production will increase, thereby enhancing energy efficiency. $NO_2$ is primarily formed through the oxidation of NO by reacting with O atoms (Eq. 6.10). Under high or low oxygen conditions, $NO_2$ primarily reacts back to NO by reacting with N or O atoms (Eqs. 6.11 and 6.12).

$$e^- + N_2 \rightarrow e^- + N_2(v)/N_2^* \tag{6.6}$$

$$N_2 + O \rightarrow NO + N \tag{6.7}$$

$$N + O_2 \rightarrow NO + O \tag{6.8}$$

$$N + NO \rightarrow O + N_2 \tag{6.9}$$

$$O + NO \rightarrow NO_2 \tag{6.10}$$

$$O + NO_2 \rightarrow NO + O_2 \tag{6.11}$$

$$N + NO_2 \rightarrow 2NO \tag{6.12}$$

$N_2$ can be either electronically excited $N_2^*$ or vibrationally excited $N_2(v)$, and the same applies to $O_2$. During the discharge process in an $N_2/O_2$ mixture, electrons convert electrical energy into rotational, vibrational, and electronic excited molecules, leading to ionization and dissociation of the molecules. Then, through processes such as rotation–translation (R–T), V–T relaxation, and de-excitation, the energy is ultimately converted into heat for the gas and chemical energy for the newly formed gas components [36–44]. Therefore, utilizing the nonequilibrium phenomena in plasma offers a promising approach to enhance the energy efficiency of the plasma nitrogen fixation process.

Depending on the actual electron temperature, electrons can excite molecules to different vibrational and electronic excited states. According to previous literature reports, the vibrational constant of $N_2(X, v)$ is approximately 0.29 eV [2]. Therefore, when $N_2$ is vibrationally excited to $N_2(X, v > 12)$, Eq. (6.7) can be initiated. In addition, electronically excited $N_2^*$ such as $N_2(A)$, $N_2(B)$, and $N_2(C)$ with excitation energies of 6.23, 7.39, and 11.05 eV, respectively, can also trigger Eq. (6.7) [2]. However, there is currently a significant lack of experimental techniques for observing short-lived excited state species, including $N_2^*$ and $N_2(X, v)$, which makes it unclear which of these species plays a dominant role in NO formation. Different literature sources even provide completely contrasting conclusions.

Some early studies suggested that electronically excited $N_2^*$ plays a dominant role. Burnette et al. employed positive polarity pulses (with a voltage of 15 kV and a pulse duration of ~200 ns) to generate diffuse filament electric discharges in dry air and $N_2$ between a pair of spherical copper electrodes spaced 10 mm apart and with a diameter of 7.5 mm, to investigate the dynamics of vibrational energy transfer and plasma-chemical reactions at 100 Torr by a suite of laser diagnostics [45]. The results indicated that NO reached a peak concentration of approximately $2 \times 10^{15}/cm^3$ within a few microseconds [45]. They proposed that the population of $N_2(v)$ at high vibrational levels, which is measured by picosecond coherent anti-Stokes Raman spectroscopy, seemed too low to influence the NO formation rate, even if vibrational excitation effectively overcame the activation energy for Eq. (6.7) [45]. Therefore, the impact of vibrationally excited $N_2(X, v)$ molecules and excited N atoms on the kinetics of NO formation is estimated to be negligible. In nanosecond pulse discharges, the formation of NO primarily occurs through reactions involving electronically excited $N_2^*$ on the microsecond timescale because of the high electron temperature.

By integrating kinetic models of these processes and comparing their predictions with current experimental results, Shkurenkov and Burnette et al. further elucidated the kinetic mechanisms of formation and decay for ground-state N, O, and NO during discharge and afterglow and identified excited states generated in the discharge but not directly measured as the main "precursors" for these species [46]. Through kinetic modeling, they found that the population density of

$N_2(X, v > 12)$ was only $10^{-3}$, which is two orders of magnitude lower than the measured NO density in the experiments [46]. When only the three electronically excited states $N_2(A)$, $N_2(B)$, and $N_2(C)$ were considered in the model, the resulting NO concentration was about five times lower and the decay rate was two orders of magnitude higher than those observed in the experiments. A good agreement between the simulation results and experimental data was only achieved when the model included $N_2(W)$, $N_2(B')$, $N_2(E)$, $N_2(a')$, $N_2(a)$, $N_2(W)$, and $N_2(a'')$. This indicates that the quenching of other electronically excited $N_2^*$ contributes to the NO formation rate in the afterglow, suggesting that the quenching of $N_2^*$ is the most likely dominant mechanism for NO formation in the afterglow [46].

Other scholars believe that vibrationally excited $N_2(X, v)$ plays a crucial role in efficient NO formation [47, 48], which is currently a more widely accepted view in the field of plasma nitrogen oxidation. Zhang et al. utilized nanosecond pulse discharges to drive plate-to-plate spark discharges with a fixed gas gap of 3 mm, achieving a maximum $NO_x$ concentration of 10,900 ppm and an energy efficiency of ~11 g/kW·h. They combined experiments with chemical kinetic models to explore the underlying mechanisms of $NO_x$ formation. The results indicated that over 50% of NO was generated through chain reactions involving O and N radicals with vibrationally excited $N_2$ and $O_2$ molecules (Eqs. 6.7 and 6.8), while 72.7% of $NO_2$ was formed by further oxidation of NO species (Eq. 6.11). The presence of O and N spectral lines after the discharge further confirmed the significant role of radical chain reactions [49]. Kinetic simulation results showed that vibrationally excited $N_2(X, v)$ and $O_2(X, v)$ could lower the energy barrier for the Zeldovich reaction, providing an energy-efficient pathway for NO production. Additionally, they proposed that most of the electron energy is initially stored in vibrationally excited $N_2(X, v)$ and electronically excited $O_2^*$. The energy is then transferred to $N(^4S)$ atoms, NO molecules, and vibrationally excited $O_2(X, v)$ [49].

Wang et al. for the first time employed a chemical kinetic model to investigate the synthesis process of $NO_x$ in a pulse-driven gliding arc (GA) reactor and validated the model's effectiveness through experiments [33]. The results revealed that, compared with thermal processes, the vibrationally excited $N_2(X, v)$ in the GA activates $N_2$ molecules and leads to an energy-efficient $NO_x$ generation pathway [33]. At the same time, they compared their results with those of thermal $NO_x$ synthesis. Due to the nonequilibrium nature of the plasma, the $NO_x$ yield and energy efficiency obtained in their GA were much higher than thermal values, as the chemical reactions of the conversion process were initiated by high-energy electrons [33]. Finally, they proposed solutions for improving the $NO_x$ synthesis process based on a GA. For example, by optimizing the reactor's electrical operating parameters, actively adjusting the reduced electric field (i.e. the E/N ratio), promoting vibrational excitation, and selectively delivering energy to the Zeldovich chemical reaction for NO synthesis in an energy-efficient manner.

Vervloessem et al. developed a low-power pulsed plasma jet for the direct production of $NO_x$ from air, achieving a low-energy consumption of 0.42 MJ $mol_N^{-1}$ [50]. The model revealed that on average, >99% of NO in the jet is formed through vibrationally excited $N_2(X, v)$, even though only a small portion of the gas passes through the arc, which is sufficient to generate a sufficient vibrational population for $NO_x$ formation. The exact contribution of each energy level depends on the energy level itself (i.e. the higher the energy level, the more the activation energy of the Zeldovich mechanism is reduced) and the population density of that level. $N_2(X, v_{12})$ is considered to contribute the most between and during pulses [50]. Meanwhile, the temperature drop between pulses prevents vibrational energy loss caused by V–T relaxation and inhibits V–V exchange between $N_2$ and $O_2$, which primarily connects the vibrational energy of $N_2$ to the V–T relaxation of $O_2$–$O_2$ [51]. It remains unclear which pathway predominates, which may be related to the type of plasma reactor and the power supply driving the plasma. Future work should focus on understanding which pathway of nitrogen fixation plays a major role and which pathway has the lowest energy consumption.

Further research by Bogaerts and colleagues has recently revealed new insights, suggesting that V–T nonequilibrium does not actually exist in typical warm plasmas, such as GA and microwave (MW). Due to the high gas temperature, vibrationally excited levels rapidly relax to the ground state, leading to further gas heating [52]. Thermal mechanisms predominantly govern nitrogen fixation. They investigated the energy consumption and $NO_x$ yield of both small-scale and large-scale reactors operating in a pin-to-pin configuration, powered by the same direct current (DC) power supply unit. The results showed that their energy consumption nearly approaches the thermodynamic minimum; however, the $NO_x$ concentration was significantly lower than the values predicted by thermodynamic equilibrium calculations [52]. This discrepancy was attributed to the dilution of unreacted gases. This illustrates the need to improve the reactor's geometry and optimize flow conditions to expose the maximum amount of feed gas to the active plasma (arc) [33].

Besides optimizing the plasma, the introduction of appropriate catalysts also represents an effective pathway to further reduce the energy costs of plasma-assisted $NO_x$ production. The results from multiple research groups have demonstrated that the combination of air plasma with heterogeneous catalysis can significantly enhance the energy efficiency of nitrogen fixation under certain conditions [8, 53, 54]. Pei et al. introduced catalysts into a DC glow discharge system, thereby reducing the energy costs of the production process [55]. They positioned the catalysts downstream of the plasma jet plume or injected floating catalyst powder into the air plasma jet, achieving a minimum energy cost of 2.9 MJ $mol^{-1}$ [55]. They propose that the reactive species within the discharge plasma channel diffuse onto the surface of the activated alumina and are subsequently adsorbed, significantly enhancing the yield of NO [55]. Lei et al. developed a novel three-stage coupled rotating electrode air plasma combined with nano-$TiO_2$ photocatalyst for $NO_x$ production, reducing the energy cost from 2.91 to 1.73 MJ mol$^{-1}$, which is attributed to the generation of high-energy electrons and ions on the $TiO_2$ surface with oxygen defects activated by the plasma [56].

Although plasma catalysis is increasingly gaining attention in various applications, the underlying mechanisms remain unclear due to the intricate interactions between the plasma and the catalyst. Currently, research in plasma catalysis often relies on a trial-and-error approach. The catalyst may primarily influence the electrical behavior of the plasma rather than exerting a limited surface chemical catalytic effect. Many catalysts simply quench the reactive species generated by the plasma rather than interacting with them, resulting in a higher yield for the pure plasma process compared to the plasma-catalyzed process. Developing more comprehensive kinetic models and more advanced characterization based on existing technologies may aid in the design of catalysts suitable for plasma environments.

## 6.3 Nitrogen Oxidation Achieved by Different Types of Plasma

In typical NTP, such as dielectric barrier discharge (DBD), the electron temperature is usually a few electronvolt, which primarily induces electronic excitation [6], with most of the electron energy going into electronic excitation, dissociation, and ionization of $N_2$ (and $O_2$) molecules [57]. The N atoms generated by direct electron-impact dissociation of $N_2$ molecules can react with $O_2$ molecules to produce NO. The synthesis mechanism of $NO_x$ in DBD involves charged and electronically excited species, and the energy cost for forming these species is high. Regardless of the catalyst packing, it is difficult for NTP to achieve a $NO_x$ yield of 1% and an energy cost below 10 MJ mol$^{-1}$. Between thermal plasma and NTP, we can define so-called warm plasmas, such as GA, atmospheric pressure glow discharges (APGD), and MW plasmas, where the electron temperature is still higher than the gas temperature, but the latter can reach 2,000–3,000 K [6], with electron temperatures typically around 1–2 eV. Therefore, these warm plasmas can better create nonequilibrium conditions, and due to this higher gas temperature, they can influence chemical kinetics.

Following early research on thermal plasmas (i.e. the B–E process) [58–60], various plasma types and reactors have been utilized for $NO_x$ production. These include spark discharges [61–64], radio frequency discharges [65], corona discharges [61, 66], glow discharges [54, 62], DBD [53, 62], GA [33, 60, 62, 67, 68], MW [69–71], and so on. The reported energy costs and product concentrations for various plasma reactors are listed in Table 6.1. It is worth noting that the reported energy consumption pertains only to the reactor, not the entire process.

**Table 6.1** Comparison of energy consumption for $NO_x$ production in various plasma reactors.

| Plasma type | Product (concentration) | Energy cost ($MJ\ mol_N^{-1}$) | Ref. |
|---|---|---|---|
| Electric arc (Birkeland-Eyde) | NO (1–2%) | 2.4–3.1 | [58, 59, 72] |
| Spark discharge | NO and $NO_2$ | 20.27, 40 | [61, 64] |
| Transient spark discharge | NO and $NO_2$ | 8.6 | [73] |
| Pin-to-plane ns-pulsed spark discharge | NO and $NO_2$ | 5.0–7.7 | [62] |
| Pin-to-plate ns-pulsed spark discharge | NO and $NO_2$ | 6.5 | [67] |
| Radio frequency-crossed discharge | $HNO_3$ | 24–108 | [65] |
| (Positive/negative) DC corona discharge | NO and $NO_2$ | 1,057/1,673 | [61] |
| Pulsed corona discharge | $HNO_3$ | 186 | [74] |
| Pin-to-plane DC glow discharge | NO and $NO_2$ | 7 | [62] |
| Pin-to-pin DC glow discharge | NO and $NO_2$ (0.7%) | 2.8 | [54] |
| DBD | NO and $NO_2$ (0.6%) | 56–140 | [62] |
| DBD with γ-$Al_2O_3$ catalyst | NO and $NO_2$ (0.5%) | 18 | [53] |
| DBD with $BaTiO_3$ catalyst | NO and $NO_2$ | 15.9 | [75] |
| DC plasma arc jet | NO (6.5%) | 3.6 | [68] |
| Propeller arc | NO and $NO_2$ (0.4%) | 4.2 | [62] |
| Pulsed milli-scale gliding arc | NO and $NO_2$ (1–2%) | 2.8–4.8 | [33, 76] |
| Gliding arc | NO and $NO_2$ (1.5%) | 3.6 | [67] |
| Pulsed arc discharge | NO and $NO_2$ | 10.6 | [77] |
| Rotating gliding arc | NO and $NO_2$ (5.4%) | 2.5 | [60] |
| Rotating gliding arc (nozzle) | NO and $NO_2$ (5.9%) | 2.1 | [78] |
| Rotating gliding arc | NO and $NO_2$ (0.28%) | 0.67 | [79] |
| 2D-GAR dgap 6.20 mm | NO and $NO_2$ (1.43%) | 6.95 | [80] |

(*Continued*)

**Table 6.1** (Continued)

| Plasma type | Product (concentration) | Energy cost ($MJ\ mol_N^{-1}$) | Ref. |
|---|---|---|---|
| Single 2D-GAR dgap 2.0 mm | NO and $NO_2$ (0.91%) | 2.89 | [81] |
| Series 2D-GARs | NO and $NO_2$ (1.56%) | 2.29 | [81] |
| Rotating gliding arc (elevated pressure) | NO and $NO_2$ (2.4%) | 1.8 | [82] |
| Microwave plasma | NO and $NO_2$ (0.6%) | 3.76 | [69] |
| Microwave plasma with catalyst | NO (6%) | 0.84 | [70] |
| Pulsed microwave discharge | NO (6%) | 0.6 | [83] |
| Microwave plasma with magnetic field | NO (14%) | 0.28 | [71] |
| Atmospheric pressure microwave plasma | NO (3.8%) | 2.0 | [84] |
| Pulsed plasma jet | NO and $NO_2$ (0.02%) | 0.42 | [50] |

The results reported in the literature vary significantly among different plasma types, with NO yields typically in the % and energy cost ranging from 0.3 to 1,600 $MJ\ mol^{-1}$. Recent research has shown that warm plasmas have made significant progress in reducing energy costs and increasing yields for $NO_x$ production, as vibrationally excited $N_2(X, v)$ is considered important in the energy-efficient production of $NO_x$ based on plasma technology. As mentioned above, Vervloessem et al. achieved a record-low energy consumption of 0.42 $MJ\ mol_N^{-1}$ for direct $NO_x$ production from air using a pulsed-driven jet plasma; however, only a 200 ppm $NO_x$ was obtained [50]. Therefore, in terms of both $NO_x$ concentration and energy cost, pulsed MW plasma operating at low pressure (0.01–0.07 atm) performs best among all reported works, achieving a low-energy cost of 0.6 $MJ\ mol_N^{-1}$ and a 6% NO content simultaneously [83]. In MW with catalysts, the same NO yield was obtained at an energy cost of 0.84 $MJ\ mol^{-1}$ [70]. Due to the lower number density of gas molecules, lower gas pressure will result in less V–T relaxation, thus leading to a more pronounced excess of higher vibrational levels, which will contribute more significantly to the vibrational dissociation pathway [85].

However, the aforementioned energy cost only includes the portion for exciting the plasma. If the energy required for maintaining low pressure and reactor cooling is also considered, the energy cost of low-pressure MW would be much higher. Furthermore, the low-pressure operation of these MW requires vacuum equipment, which makes their application in industrial-scale processes more challenging, and the energy demand of such vacuum equipment should be included in the calculation of energy cost for fairness. Operating MW reactors at higher pressures is also possible, but thermal losses increase due to the increased collision frequency [86]. GA is considered one of the most effective and promising plasmas for plasma gas conversion at present [6, 33, 67, 76, 87–92]. This is because they can operate within the ambient pressure range at various power levels and flow rates, achieving promising results in terms of energy cost and yield, especially when compared to the literature based on cold plasmas [29, 93, 94]. GA is characterized by an E/N below 100 Td, resulting in electron energies of approximately 1 eV. Such electron energy is most favorable for the vibrationally excited $N_2$ molecules, which is regarded as the most effective way to assist in conversion or synthesis (Figure 6.3) [95].

The rapid development of $NO_x$ production using GA in adjusting various parameters to find effective conditions for $NO_x$ production. Among these, the $N_2/O_2$ ratio is the most important parameter and has been widely studied. Van Alphen et al. investigated the effect of feed gas ratio on $NO_x$ generation in a rotating GA reactor and determined that a 40–60% $N_2$–$O_2$ mixture can produce the maximum $NO_x$ concentration of 3.35% with the lowest energy consumption of 2.4 MJ $mol_N^{-1}$ [96]. Although other gas mixtures have potential efficiency, air remains the primary feed gas for plasma-assisted $NO_x$ synthesis. The ease of air acquisition is a major motivation for its use, both in the laboratory and in foreseeable applications [97].

The reactor geometry and the connection between reactors also have a significant impact on $NO_x$ production performance. Van Raak et al. utilized two series GA reactors with a discharge gap of 2 mm, achieving energy consumptions ranging from 2.29 to 2.42 MJ $mol_N^{-1}$ and yields of 124.6–158.3 mmol·N/h [81]. By extending the post-plasma oxidation time, consuming excess $O_2$ in the feed, and utilizing the low-temperature conditions at the reactor outlet, the selectivity of $NO_2$ can be improved. Compared to a single reactor, $NO_x$ production can be doubled while energy cost is reduced by 20.9% [81]. Angineni et al. employed a GA reactor with a conical electrode to ensure all feed gas passes through the plasma, achieving a $NO_x$ content of 0.6% vol. and an energy consumption of 3.0 MJ $mol_N^{-1}$ [98].

Increasing pressure is also key to enhancing $NO_x$ production. For instance, Tsonev et al. conducted nitrogen fixation using an enhanced-pressure rotating GA plasma [82]. They demonstrated that by increasing the pressure from 1 to 4 atm, the $NO_x$ production rate and the oxidation rate of NO can be improved while reducing energy consumption. This effect is more pronounced at higher

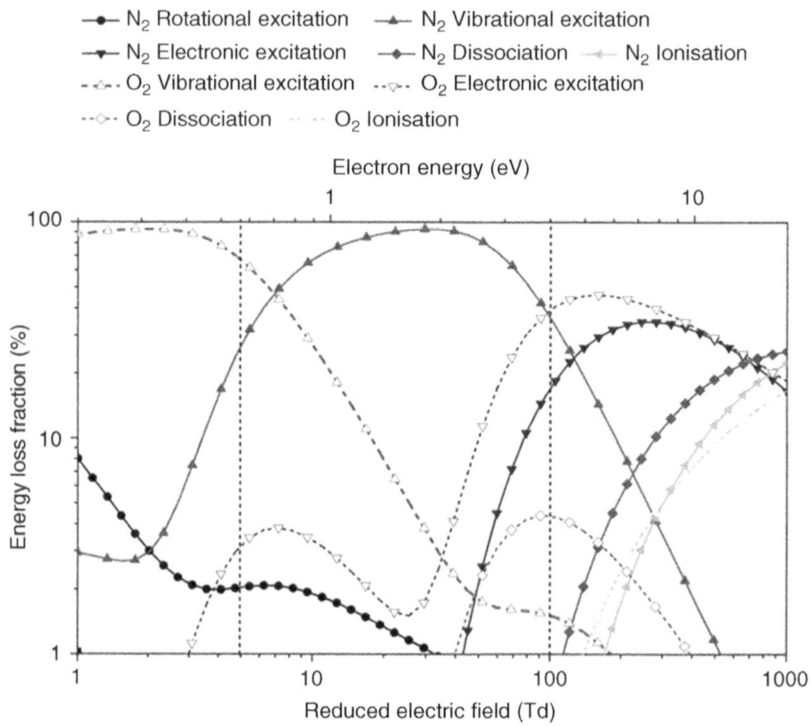


**Figure 6.3** The dominant plasma-activation channels in 50:50 $N_2$:$O_2$ stream. *Source:* Reproduced from Ref. [33] / with permission of John Wiley & Sons. Reduced electric fields of 5–100 Td correspond to GA and MW plasmas, while the region above 100 Td corresponds to a DBD reactor [6].

gas flow rates (12 L min$^{-1}$). Under these conditions, the maximum $NO_x$ production rate can reach 69 g h$^{-1}$, with a $NO_2$ selectivity as high as 94% [82].

To achieve higher $NO_x$ yields, a higher SEI is required, which results in a high-temperature environment within the plasma. At high temperatures (>3,000 K), the V–T relaxation rate coefficient increases with gas temperature, further accelerating the vibrational depopulation process. The generated $NO_x$ can return to $N_2$ through reactions with N or radical NO losses [76]. However, high temperatures can reduce the E/N and electron temperature, favoring vibrational excitation over electronic excitation, ionization, or dissociation [11, 99]. Meanwhile, the formation of NO, being an endothermic reaction, is considered chemically kinetically favorable at elevated temperatures. This necessitates finding a balance. It should be recognized that the gas temperature in the plasma generally varies with the applied power and various chemical reactions and is not easily controlled [100].

Quenching by removing heat from the reaction flow can be an effective method to prevent NO losses during the reaction, as these recombination reactions accelerate at higher temperatures. Mixing with cold gas is one of the common techniques for cooling hot gas streams [76]. Muhammad et al. enhanced the maximum attainable $NO_x$ concentration from 5.16% to 5.8% through quenching [101]. This improvement was achieved by mitigating reverse reactions via quenching, and the conclusion was supported by both experimental results and a zero-dimensional chemical kinetic model [101]. However, gas cooling also increases the overall energy cost.

Ultimately, regardless of how other parameters are adjusted, the overall $N_2$ conversion is limited by the proportion of gas processed by the GA. For example, it is estimated that only 15% of the gas passes through the plasma arc in the GA plasmatron [88, 102], while the rest bypasses the reactor without contacting the plasma, thereby limiting the amount of $NO_x$ produced.

Typical NTP, such as packed-bed DBD reactors, are also used for $NO_x$ production. They hold great potential for enhancing chemical reactions due to their easy integration with catalysts, which can improve product selectivity and energy efficiency [103]. DBD can operate at atmospheric pressure and is easily scalable. Currently, most research focuses on investigating the impact of packing materials on the coupled electrical and chemical processes in plasma-assisted $NO_x$ synthesis. For example, Patil et al. conducted a systematic study for the first time on the plasma synthesis of $NO_x$ in a DBD reactor using supported catalysts packed with $\gamma$-$Al_2O_3$, $\alpha$-$Al_2O_3$, $TiO_2$, MgO, $BaTiO_3$, and Quartz Wool [53]. A maximum $NO_x$ concentration of only about 5,700 ppm and an energy cost of 18 MJ $mol_N^{-1}$ were achieved with 5% $\gamma$-$Al_2O_3$ particles sized between 250 and 160 μm. Although the $WO_3/\gamma$-$Al_2O_3$ catalyst further increased the $NO_x$ concentration by approximately 10%, this was mainly attributed to micro-discharges generated by the plasma at the sharp edges of the catalyst particles [53]. Ma et al. investigated the influence of packing with $Al_2O_3$ and $BaTiO_3$ microbeads in a packed-bed DBD reactor on plasma-enhanced $NO_x$ synthesis from air [104]. The results indicated that when $BaTiO_3$ beads were placed in the DBD reactor, the energy consumption for $NO_x$ generation was the lowest, reaching 15.9 MJ $mol_N^{-1}$. This can be attributed to the increased average electron energy due to the packing, which favored the formation of more excited $N_2$ species in the reaction [104]. Therefore, selecting an appropriate packing material can effectively reduce the energy cost of $NO_x$ synthesis in DBD plasma.

In summary, the presence of packing materials can generate more high-energy electrons, promoting the formation of chemical reaction species, including excited $N_2$ molecules, thereby effectively accelerating the synthesis of nitrogen oxides. However, compared to other atmospheric-pressure plasma reactors, DBD does not perform as well. These unfavorable results obtained in DBD can be explained by the high reduced electric field, which is mostly in the range of 100–200 Td, generating high-energy electrons that primarily lead to most of the electron energy (70–80%) being used for electronic excitation rather than vibrational excitation.

As a result, the most energy-efficient pathway for $NO_x$ formation through vibrationally induced Zeldovich mechanism is not utilized [6].

Due to differences in discharge characteristics (such as structure, discharge mode, and current distribution), it is challenging to compare different discharge forms. Different discharges may also utilize distinct chemical reaction pathways to produce nitrogen oxides. Pei et al. studied four types of discharges, namely DBD, glow, spark, and arc, and identified a key dimensionless parameter $\chi$ (Eq. 6.13) [62] that influences the energy cost of $NO_x$ production. The study revealed that the efficiency of $NO_x$ production is primarily controlled by the average electric field and average gas temperature.

$$\chi = \frac{\overline{E} \times \overline{T}}{E_r \times T_r} \tag{6.13}$$

$\overline{E}$ and $\overline{T}$ represent the average electric field and average gas temperature of the discharge plasma being studied, respectively. $E_r$ and $T_r$ are taken as 1.43 kV cm$^{-1}$ and 1,800 K, respectively, as baseline conditions for normalization. There is an almost linear relationship between $\chi$ and the energy cost of $NO_x$ production.

However, the underlying mechanisms have not been well explained. The authors suggest that the electric field is related to electron temperature, which, in turn, affects the energy transfer efficiency from electrons to vibrational and electronically excited $N_2$ [62]. Meanwhile, a lower gas temperature may favor the reverse reaction of effective NO formation. By reducing the $\chi$ factor, e.g. by decreasing the electric field and/or the average gas temperature during discharge, the energy cost can be reduced. The important mechanisms controlling the energy efficiency of $NO_x$ production in any type of discharge can be summarized as two points [105]: (i) efficient electron-impact activation of $N_2$ to promote the formation of $NO_x$ and (ii) rapid thermal quenching of NO to prevent its conversion back to $N_2$ and $O_2$ molecules when the gas temperature drops slowly [33]. However, reducing the $\chi$ factor is not easily achieved in practice. In any specific discharge, $\overline{E}$ and $\overline{T}$ are usually correlated and cannot be manipulated independently and directly. For example, a higher current corresponds to a lower $\overline{E}$, which typically results in an increase in $\overline{T}$. The authors also suggest several methods in the paper to reduce $\chi$, such as lowering the plasma gas temperature through cooling or using short, high-current pulses, or extending the discharge length [62].

Adjusting the reactor geometry, gas flow rate, discharge form, and excitation source can further enhance the production of $NO_x$. However, it is also noteworthy that accurate measurement of the discharge power significantly affects performance evaluation, regardless of the plasma reactor used [106]. The $NO_x$ concentration (>1,000 ppm) and energy cost of plasma-assisted $NO_x$ synthesis from some latest reports are presented in Figure 6.4. The current research results are still far from our goals, particularly as there is often a trade-off between $NO_x$ concentration

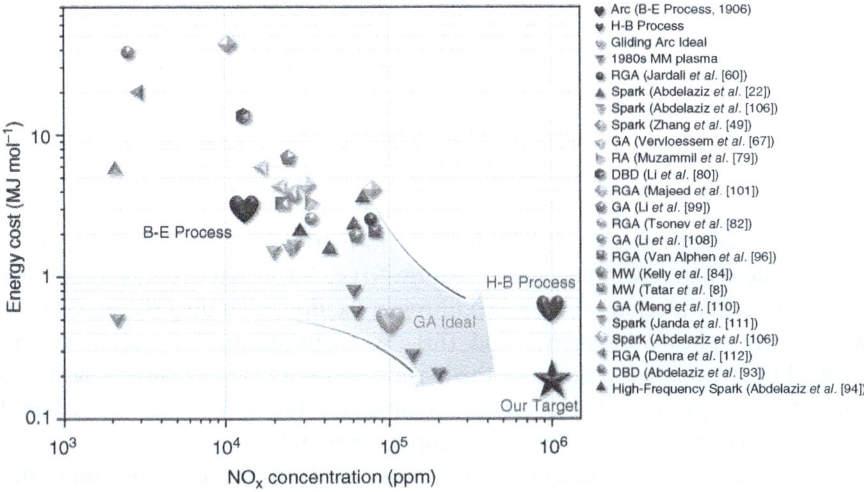


**Figure 6.4** Comparison of energy consumption for $NO_x$ production in various plasma reactors. The data points were obtained from the literature [8, 49, 60, 67, 70, 78, 79, 82, 84, 93, 94, 101, 106–112]. The $H\text{-}B_{Gross}$ is 0.51 MJ/mol-$NH_3$, encompassing hydrocarbon reforming and catalytic $NH_3$ production. When only considering the net energy of catalytic $NH_3$ production, the $H\text{-}B_{net}$ is 0.1 MJ/mol-$NH_3$ [69].

and energy cost. It has been reported that MW with a magnetic field achieved the highest NO concentration of 14% and the lowest energy cost of 0.28 MJ $mol_N^{-1}$. However, these values were reported in the 1980s and have not been reproduced in recent years [71]. A similar situation exists for plasma-based carbon dioxide splitting, where results from the 1980s have not been reproducible with similar reactors in recent years [113]. Therefore, the energy cost and yield calculations for plasma-based $NO_x$ synthesis in MW plasma reported in the 1980s should be rigorously evaluated. Additionally, when nanosecond repetitive pulsed plasma is used for precise control of chemical reactions, the voltage and current signals contain rapid rises or falls on the nanosecond time scale, making this measurement issue particularly critical [22].

## 6.4 Plasma–Water-Based Nitrogen Fixation

Another approach involves direct nitrogen fixation through the reaction of $H_2O$ with $N_2$. This process is sustainable and offers many benefits under environmentally conditions similar to biological organisms [114]. NTP generates reactive nitrogen species (RNS) and reactive oxygen species (ROS) at the gas–liquid interface, which then diffuse into the bulk water and subsequently produce useful products, such as nitric acid and $H_2O_2$. These products are beneficial for seed germination rates, plant growth and development, antibacterial effects, and crop yields. The introduction of $H_2O$ into the reaction complicates the process further, as $H_2O$ is decomposed to produce ·H and ·OH radicals (Eq. 6.14).

$$e + H_2O \rightarrow OH + H + e \tag{6.14}$$

The ·OH radicals play a crucial role in the formation of $NO_x$ ($NO_2^-$ and $NO_3^-$) [115], while H can recombine with N to form $NH_4^+$. Various types of plasmas have been utilized for plasma–water-based nitrogen fixation, including (i) "plasma-on-water," (ii) "plasma-in-water," and (iii) "water-in-plasma," primarily distinguished by the relative position of the plasma and $H_2O$ [116]. Figure 6.5a shows a schematic diagram of plasma generation on water (type *i*, continuous gas–liquid phase), which is currently the most common approach.

For "plasma-on-water" reactors, $H_2O$ often participates in the reaction only after the discharge. However, stagnant water can lead to saturation of active nitrogen and oxygen species generated by the plasma, thus limiting further diffusion. Due to the high gas temperature produced by the plasma and the intense evaporation at the plasma/liquid interface, only a small amount of $H_2O$ is decomposed, and the water primarily serves as a receptor pool for $NO_x$. Additionally, the evaporation process itself can reduce efficiency as it acts as an energy sink.

Figure 6.5b illustrates the schematic of a plasma source that generates plasma within water (type ii, dispersed gas phase with continuous liquid phase). In this

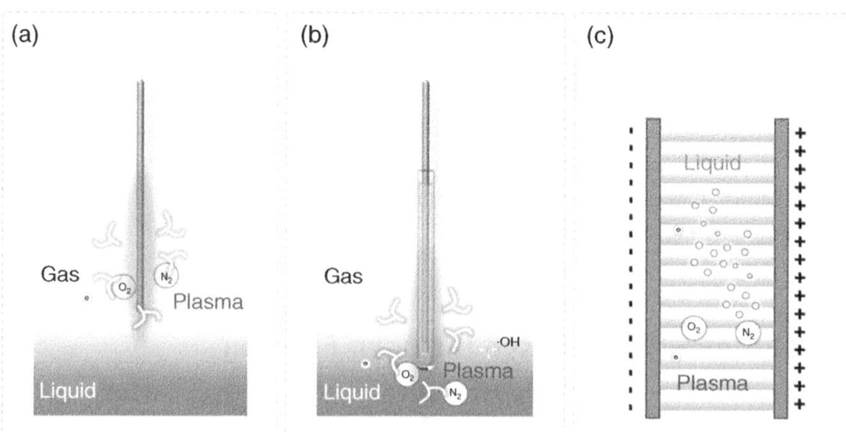


**Figure 6.5** The schematics of plasma–water-based nitrogen fixation sources. (a) "Plasma-on-water" type. (b) "Plasma-in-water" type. (c) "Water-in-plasma" type.

configuration, the presence of more $H_2O$ facilitates the collection of nitrogen fixation products. But this approach introduces some additional complexities in reactor design. Various submerged pin-to-pin or wire-to-pin geometries have been attempted [117]; however, the properties of the solution become critical as the liquid phase acts as an electrode in the discharge system. These typically require solutions with low electrical conductivity, such as deionized water (DI), to prevent the plasma discharge from becoming uncontrollable while controlling energy consumption.

Figure 6.5c displays the schematic of a plasma source where water enters the plasma volume in the form of molecules or small droplets (aerosols) (type iii, continuous gas phase with dispersed liquid phase). More direct contact between water and plasma generates more ·OH radicals. For "water-in-plasma" reactors, $H_2O$ vapor or $H_2O$ aerosols (small droplets) are mixed into the $N_2$ or air gas flow, and the energetic electrons generated by the plasma efficiently dissociate $H_2O$ into ·H and ·OH radicals. In the presence of humid air, ·OH produced by the discharge can generate NO through the extended Zeldovich mechanism (Eq. 6.15), which is beneficial due to its fast reaction rate. However, small droplets can quickly reach saturation with nitrogen and oxygen reactive species, reducing the efficiency of their entry into the liquid. On the other hand, the introduction of excess water vapor can lead to V–T relaxation, quenching a significant amount of vibrationally excited $N_2(X, v)$, which is crucial for the formation of $NO_x$ species [118, 119] and is significantly easier to dissociate.

More droplets can suppress the plasma, requiring more power and reducing energy efficiency. For direct liquid treatment, this is largely unavoidable. When the air contains 3% $H_2O$, the concentrations of some highly reactive species, such as $O(^1D)$, $O(^1S)$, $N_2(A)$, $N(^2D)$, $N(^2P)$, drop to zero within $10^{-6}$ seconds [120, 121]. These reactive species typically convert into NO, NH, or OH. NO is further oxidized by O, $O_3$, and ·OH to form $NO_2$, $N_2O_3$, $N_2O_4$, $N_2O_5$, $HNO_2$, and $HNO_3$ (Eqs. 6.15–6.20). Most of these gaseous $NO_x$ species diffuse into the water layer and gradually convert into $HNO_{x(aq)}$. Therefore, if the plasma discharge fails to form $O_3$ due to high gas temperatures, the oxidation of $N_2$ in the extended Zeldovich process will be limited (Eqs. 6.16 and 6.17) [122–125].

The Henry's law constant is considered a key parameter for comparing the solubility of different gases in a solution. NO and $NO_2$, as the main products of plasma gas-phase oxidative nitrogen fixation, have Henry's law constants of $1.9 \times 10^{-4}$ and $1.2 \times 10^{-4}$ mol/m$^3$·P [126, 127], respectively, which means that most $NO_x$ species are difficult to absorb by the solution (typically $H_2O$). Although NO and $NO_2$ can directly react with $H_2O$ in water to form $NO_2^-$ and $NO_3^-$, this mechanism is not considered the primary pathway due to the existence of Henry's law constants. Therefore, NO and $NO_2$ continue to react with ·OH radicals in the gas phase to form $HNO_2$ and $HNO_3$ (Eqs. 6.19 and 6.20), which then dissolve in water. Since

$HNO_2$ (Henry's law constant = $4.8 \times 10^{-1}$ mol/m$^3$·P) and $HNO_3$ (Henry's law constant = $2.6 \times 10^4$ mol/m$^3$·P) have high solubility, this mechanism is considered the main pathway for the formation of $NO_x$ in water.

$$N + OH \rightarrow NO + H, \tag{6.15}$$

$$NO + O_3 \rightarrow NO_2 + O_2 \tag{6.16}$$

$$NO_2 + O_3 \rightarrow NO_3 + 2O_2 \tag{6.17}$$

$$NO + NO_2 + M \rightarrow N_2O_5 + M \tag{6.18}$$

$$NO + OH \rightarrow HNO_2 \tag{6.19}$$

$$NO_2 + OH \rightarrow HNO_3 \tag{6.20}$$

Simultaneously, ·OH radicals recombine to form $H_2O_2$. Since $H_2O_2$ (Henry's law constant = $9.1 \times 10^2$ mol/m$^3$·P) is highly soluble in water, it facilitates the rapid conversion of $NO_2^-$ to $NO_3^-$ through reactions during the plasma treatment process. The oxidation of $N_2$ in air plasma is much higher than that in pure $N_2$ plasma due to the presence of $O_2$, which generates more $NO_x$ through reactions.

Table 6.2 lists some of the main products and energy efficiencies of plasma reactors used for water-based oxidative nitrogen fixation. Although these figures have shown progress in recent years, they still need improvement compared to the energy costs of direct H-B and B-E processes [6, 50, 59]. It is worth noting that the most efficient single reactor currently achieves an energy consumption of around 20 MJ mol$^{-1}$ [128, 139], which is significantly higher (by 1–2 orders of magnitude) than the energy cost achieved by most reactors for direct gas-phase $NO_x$ production listed in Table 6.1 above. In fact, the energy cost in Table 6.2 is partly dependent on the gas-phase generation of $NO_x$, which can be attributed to mass transfer losses when attempting to deliver gas-phase $NO_x$ to the liquid phase [139]. This implies that the efficient transport of nitrogen–oxygen reactive species generated by the plasma into the liquid is of greater importance [142, 143].

Increasing the exposure time of nitrogen–oxygen reactive particles generated by the plasma to the liquid and expanding the plasma–liquid interface area are effective ways to promote the transport of nitrogen–oxygen reactive species into the treated water and improve efficiency. Mass transfer efficiency may be improved by generating plasma in bubbles within the target liquid [144] or by bubbling the plasma exhaust into the liquid [145]. Man et al. achieved low-energy cost (21.08 MJ mol$^{-1}$) and considerable yield (10.7 mg min$^{-1}$) using a DBD plasma driven by nanosecond repetitive pulses in microbubbles [128]. The higher gas–liquid surface area generated by microbubble discharge and the bubble fragments produced after bubble collapse can persist in water for a long time, facilitating the diffusion of nitrogen–oxygen reactive species [128]. However, current exploration of this form is relatively limited, especially for underwater bubble discharge.

**Table 6.2** Comparative summary of recent studies on plasma–water-based $NO_x$ production.

| Plasma type | Main product | Energy cost ($MJ\ mol^{-1}$)[a] | Ref. |
|---|---|---|---|
| Microbubble-enabled DBD | $NO_2^-$, $NO_3^-$, and $H_2O_2$ | 21.08 | [128] |
| DBD | $NO_3^-$ and $NO_2^-$ | 1,217.039 | [129] |
| Cyl-DBD | $NO_3^-$ and $H_2O_2$ | 77.538 | [130] |
| Indirect air DBD | $NO_2^-$, $NO_3^-$, and $H_2O_2$ | 250 | [131] |
| Helical coil DBD reactor | $NO_2^-$, $NO_3^-$, and $H_2O_2$ | 45 | [132] |
| Corona | $NO_2^-$, $NO_3^-$, and $H_2O_2$ | 4,144.37 | [133] |
| Microplasma bubbles gas–liquid DBD discharge | $NO_2^-$, $NO_3^-$, and $H_2O_2$ | 105.88 | [134] |
| In-plane DBD | $NO_2^-$, $NO_3^-$, and $H_2O_2$ | 2,000 | [135] |
| Surface-wave microwave discharge | $NO_2^-$, $NO_3^-$, and $H_2O_2$ | 367,624.8 | [136] |
| Falling film dielectric barrier discharge | $NO_3^-$ | 39 | [75] |
| Pulse discharge | $NO_2^-$, $NO_3^-$, and $H_2O_2$ | 403 | [66] |
| Plasma–water droplet reactor | $NO_2^-$, $NO_3^-$, and $H_2O_2$ | 1,900 | [137] |
| Arc-DBD combination | $NO_2^-$ and $NO_3^-$ | 8.7 | [138] |
| DBD | $NO_2^-$, $NO_3^-$, and $NH_4^+$ | 20.7 | [139] |
| Plasma jet | $NO_2^-$ and $NO_3^-$ | 1,016.9 | [140] |
| Bubble discharge reactor | $NO_2^-$, $NO_3^-$, and $NH_4^+$ | 696.4 | [141] |

[a] The energy consumption here refers to the energy consumption for $NO_x$ production.

Underwater discharge primarily occurs in the form of spark discharge at the gas–liquid interface, which is a complex discharge process coupling electric, fluid, and thermal fields. The processes of bubble growth, rupture, and phase transition at the gas–liquid interface remain to be explored, which are crucial for achieving efficient transport of nitrogen–oxygen reactive species.

Furthermore, the commercialization of the B-E process provides us with valuable insights [59]. By oxidizing NO to higher valence states such as $NO_2$, $NO_3$ (Henry's law constant = $3.8 \times 10^{-4}$ mol/m$^3$·P), these products can subsequently be efficiently converted to $NO_3^-$ in water. Dinh et al. combined DBD (as an efficient $O_3$ source) with GA (as an efficient $NO_x$ source), utilizing the strong oxidizing power of $O_3$ to oxidize NO to higher valence, water-soluble nitrogen oxides like $NO_2$, achieving nearly zero emissions of $NO_x$ and $O_3$, and thus obtaining a low-energy consumption of 8 MJ mol$^{-1}$ [138]. Li et al. used a series connection of DBD and three-stage spark discharge to convert gaseous $NO_x$ from NO and $NO_2$ to water-soluble $N_2O_5$, enhancing the yield of plasma-based water nitrogen fixation and reducing energy cost to 25.66 MJ mol$^{-1}$ [99]. By adding $O_3$ to the $N_2$ plasma gas flow, $NO_2$ with a higher Henry's law constant is generated, along with $H_2O_2$ in the water; these strongly oxidizing substances further oxidize $NO_2^-$ to $NO_3^-$. Therefore, 99% of the nitrogen fixation products are $NO_3^-$ [138].

It is evident that plasma–water-based nitrogen fixation is cleaner and more intense, making it more suitable for generating $NO_x$. Simultaneously, their potential for high-speed nitrogen fixation offers opportunities for sustainable ammonia synthesis under mild conditions by combining them with complementary technologies (such as electrochemical catalysis for converting $NO_x$ to ammonia) [146].

## 6.5 Conclusion and Outlook

With the growing global demand for a low-carbon society, plasma-assisted processes have attracted significant attention due to their ecological friendliness, mild reaction conditions, and potential for low-energy cost, especially when combined with renewable energy sources. In recent years, researchers have conducted in-depth studies on the reaction mechanisms, kinetic models, and related applications of plasma-assisted $NO_x$ synthesis. Furthermore, experiments and numerical calculations have been performed on various types of discharge reactors and their designs. After decades of continuous research, plasma-assisted nitrogen oxide synthesis technology has re-emerged into the spotlight after a century. This chapter analyzes the latest research status of different types of atmospheric pressure low-temperature plasma-assisted $NO_x$ synthesis, focusing on plasma-assisted reactions, including $N_2/O_2$ (air) and $N_2$(air)/$H_2O$. We have evaluated the most advanced plasma-assisted $NO_x$ synthesis technologies, elucidated the reaction pathways and mechanisms, and proposed potential improvements to enhance nitrogen oxidation efficiency. Despite the considerable attention and rapid development of plasma-assisted $NO_x$ synthesis technology, it still faces several challenges:

i) Compared to NTPs, warm plasma is considered more promising for plasma-assisted $NO_x$ synthesis as thermal mechanisms play a dominant role. The most significant progress has been made using warm plasma in the past five years. These developments have led to $NO_x$ yields exceeding 1% and energy costs reduced to below 2 MJ $mol^{-1}$, even surpassing the B-E process. However, research data indicates that the energy cost of plasma-assisted $NO_x$ synthesis process remains too high compared to the H-B process based on electrolysis combined with the Ostwald process. Most plasma chemistry exhibits a trade-off between $NO_x$ concentration and energy cost, particularly evident when $NO_x$ yields exceed 2%. The process strongly depends on the effective SEI delivered to the plasma rather than the overall (macroscopic, measured) SEI supplied to the reactor. Enhancing energy deposition in the limited fraction of gas flow passing through the plasma is considered crucial. If the energy consumption of most studies can be further reduced to below 1–1.5 MJ $mol^{-1}$ through intelligent reactor design [8], adjustment of reaction kinetics, selection of appropriate excitation sources, and avoidance of reverse reactions, the plasma-assisted $NO_x$ synthesis process will become a highly competitive alternative to the combined H-B and Ostwald processes.

ii) In plasma–water-based nitrogen fixation system, the diversity of interactions between the plasma reactor and water complicates comparative research. New plasma reactors continue to be developed, and due to their uniqueness, they cannot fully leverage the extensive pre-existing body of work. Micro- and nano-bubble discharges are considered potential pathways to achieve a breakthrough in plasma-based nitrogen fixation in water, but they require more detailed experimental and modeling approaches to explore, particularly regarding multiphase flow fields. Despite its potential, the high-energy consumption of the $N_2$(air) and $H_2O$ reaction system remains a challenge. Considering that gas-phase reactive species can undergo both formation and dissolution processes, potentially leading to reactions in both phases, should we focus more on the synthetic process of $NO_x$ in the gas phase or the transfer process of reactive species?

iii) The gap between laboratory-scale and real-world applications of plasma-assisted $NO_x$ synthesis remains vast. Laboratory-scale equipment is not designed to meet the production demands of real-world applications. It should be noted that, in this regard, the B-E process is highly industrialized. It involves reaction temperatures exceeding 3,000 °C, a 3-m-wide plasma chamber, plasma exhaust flowing through multiple 4-story-high towers, a 3-MW reactor, and dozens of workers [59]. This is far more intensive than any experimental reactor currently being developed. Currently, no research has achieved a level of technological maturity comparable to the B-E process. In particular, the energy cost provided in the current literature only accounts for the energy required to excite the plasma. The efficiency of power transmission from the plug to the plasma power supply, as well as capital and depreciation costs, are also crucial and need to be carefully considered in practical applications. However, before investing effort in researching and developing specific industrial-scale designs, it is prudent to have a clear idea of how to scale up for its application.

Lastly, plasma-assisted $NO_x$ synthesis represents an emerging technology for sustainable energy conversion, opening up new and exciting opportunities for cleaner industries and a greener future. It is anticipated that more disciplines will become involved, leading to increased interdisciplinary collaboration in the field of plasma-assisted sustainable $NO_x$ synthesis. With the efforts of researchers worldwide, significant discoveries and technological advancements are expected in the near future. Although the energy efficiency of plasma processes currently cannot match that of established methods, plasma processes aim to utilize inexpensive surplus renewable energy, which is a prerequisite for their advantage. Understanding the plasma-assisted $NO_x$ synthesis process is essential, and developing it from an industrial perspective is even more challenging.

## References


1 Canfield, D.E., Glazer, A.N., and Falkowski, P.G. (2010). The evolution and future of Earth's nitrogen cycle. *Science* 330 (6001): 192–196.

2 Liu, J., Nie, L., Liu, D., and Lu, X. (2024). Plasma for nitrogen fixation by using $N_2/O_2$ mixture: reaction pathway, energy flow, and plasma reactor. *Plasma Process. Polym.* 21 (2): 2300153.

3 Miller, J.A. and Bowman, C.T. (1989). Mechanism and modeling of nitrogen chemistry in combustion. *Prog. Energy Combust. Sci.* 15 (4): 287–338.

4 Gerba, C. (2009). Environmentally Transmitted Pathogens. In: *Environmental Microbiology* (ed. I.L. Pepper, C.P. Gerba, and T.J. Gentry), 445–484. Elsevier.

5 Ernst, F.A. (1928). *Fixation of Atmospheric Nitrogen*. D. Van Nostrand Company.

6 Bogaerts, A. and Neyts, E.C. (2018). Plasma technology: an emerging technology for energy storage. *ACS Energy Lett.* 3 (4): 1013–1027.

7 Patil, B., Wang, Q., Hessel, V., and Lang, J. (2015). Plasma $N_2$-fixation: 1900–2014. *Catal. Today* 256: 49–66.

8 Rouwenhorst, K.H., Jardali, F., Bogaerts, A., and Lefferts, L. (2021). From the Birkeland–Eyde process towards energy-efficient plasma-based $NO_x$ synthesis: a techno-economic analysis. *Energy Environ. Sci.* 14 (5): 2520–2534.

9 Egeland, A. and Burke, W.J. (2005). *Kristian Birkeland*. Springer.

10 Cowling, M. (2004). The growth–profit nexus. *Small Bus. Econ.* 22 (1): 1–9.

**11** Erisman, J.W., Sutton, M.A., Galloway, J. et al. (2008). How a century of ammonia synthesis changed the world. *Nature Geosci.* 1 (10): 636–639.

**12** Tanabe, Y. and Nishibayashi, Y. (2013). Developing more sustainable processes for ammonia synthesis. *Coord. Chem. Rev.* 257 (17–18): 2551–2564.

**13** MacFarlane, D.R., Cherepanov, P.V., Choi, J. et al. (2020). A roadmap to the ammonia economy. *Joule* 4 (6): 1186–1205.

**14** Raymond, J., Siefert, J.L., Staples, C.R., and Blankenship, R.E. (2004). The natural history of nitrogen fixation. *Mol. Biol. Evol.* 21 (3): 541–554.

**15** Thiemann, M., Scheibler, E., and Wiegand, K.W. (2000). Nitric Acid, Nitrous Acid, and Nitrogen Oxides. In: *Ullmann's Encyclopedia of Industrial Chemistry* (ed. F. Ullmann), 1–43. Wiley.

**16** Armijo, J. and Philibert, C. (2020). Flexible production of green hydrogen and ammonia from variable solar and wind energy: case study of Chile and Argentina. *Int. J. Hydrog. Energy* 45 (3): 1541–1558.

**17** Anastasopoulou, A., Wang, Q., Hessel, V., and Lang, J. (2014). Energy considerations for plasma-assisted N-fixation reactions. *Processes* 2 (4): 694–710.

**18** Hessel, V., Cravotto, G., Fitzpatrick, P. et al. (2013). Industrial applications of plasma, microwave and ultrasound techniques: nitrogen-fixation and hydrogenation reactions. *Chem. Eng. Process. Process Intensif.* 71: 19–30.

**19** Burris, R.H. and Roberts, G.P. (1993). Biological nitrogen fixation. *Ann. Rev. Nutrition* 13 (1): 317–335.

**20** Bazhenova, C.T. and Shilov, A. (1995). Nitrogen fixation in solution. *Coord. Chem. Rev.* 144: 69–145.

**21** Li, J., Yao, S., and Wu, Z. (2020). $NO_x$ production in plasma reactors by pulsed spark discharges. *J. Phys. D. Appl. Phys.* 53 (38): 385201.

**22** Abdelaziz, A.A., Komuro, A., Teramoto, Y. et al. (2024). Atmospheric-pressure plasmas for $NO_x$ production; short review on current status. *Curr. Opin. Green Sustain. Chem.* 50: 100977.

**23** Smith, C., Hill, A.K., and Torrente-Murciano, L. (2020). Current and future role of Haber–Bosch ammonia in a carbon-free energy landscape. *Energy Environ. Sci.* 13 (2): 331–344.

**24** Cherkasov, N., Ibhadon, A., and Fitzpatrick, P. (2015). A review of the existing and alternative methods for greener nitrogen fixation. *Chem. Eng. Process. Process Intensif.* 90: 24–33.

**25** Leigh, G.J. (2002). *Nitrogen Fixation at the Millennium*. Elsevier.

**26** Rouwenhorst, K.H., Krzywda, P.M., Benes, N.E. et al. (2020). Ammonia, 4. Green Ammonia Production. In: *Ullmann's Encyclopedia of Industrial Chemistry* (ed. F. Ullmann), 1–20. Wiley.

**27** Snoeckx, R. and Bogaerts, A. (2017). Plasma technology–a novel solution for $CO_2$ conversion? *Chem. Soc. Rev.* 46 (19): 5805–5863.

**28** Li, T., Gao, Y., Zhou, R. et al. (2024). Outlook for improving energy efficiency, conversion rates, and selectivity of plasma-assisted $CO_2$ conversion. *Curr. Opin. Green Sustain. Chem.* 47: 100915.

**29** Fridman, A. (2008). *Plasma Chemistry*. Cambridge University Press.

**30** Mehta, P., Barboun, P., Go, D.B. et al. (2019). Catalysis enabled by plasma activation of strong chemical bonds: a review. *ACS Energy Lett.* 4 (5): 1115–1133.

**31** Kogelschatz, U. (2003). Dielectric-barrier discharges: their history, discharge physics, and industrial applications. *Plasma Chem. Plasma Process.* 23 (1): 1–46.

**32** Hessel, V., Anastasopoulou, A., Wang, Q. et al. (2013). Energy, catalyst and reactor considerations for (near)-industrial plasma processing and learning for nitrogen-fixation reactions. *Catal. Today* 211: 9–28.

**33** Wang, W., Patil, B., Heijkers, S. et al. (2017). Nitrogen fixation by gliding arc plasma: better insight by chemical kinetics modelling. *ChemSusChem* 10 (10): 2145–2157.

**34** Laue, W., Thiemann, M., Scheibler, E., and Wiegand, K.W. (2000). Nitrates and Nitrites. In: *Ullmann's Encyclopedia of Industrial Chemistry* (ed. F. Ullmann), 153–157. Wiley.

**35** Zeldvich, Y.B. (1946). The oxidation of nitrogen in combustion and explosions. *J. Acta Physicochimica* 21: 577.

**36** Popov, N. (2001). Investigation of the mechanism for rapid heating of nitrogen and air in gas discharges. *Plasma Phys. Report* 27: 886–896.

**37** Popov, N. (2011). Fast gas heating in a nitrogen–oxygen discharge plasma: I. Kinetic mechanism. *J. Phys. D. Appl. Phys.* 44 (28): 285201.

**38** Aleksandrov, N., Kindysheva, S., Nudnova, M., and Starikovskiy, A.Y. (2010). Mechanism of ultra-fast heating in a non-equilibrium weakly ionized air discharge plasma in high electric fields. *J. Phys. D. Appl. Phys.* 43 (25): 255201.

**39** Flitti, A. and Pancheshnyi, S. (2009). Gas heating in fast pulsed discharges in $N_2$–$O_2$ mixtures. *Eur. Phys. J. Appl. Phys.* 45 (2): 21001.

**40** Komuro, A. and Ono, R. (2014). Two-dimensional simulation of fast gas heating in an atmospheric pressure streamer discharge and humidity effects. *J. Phys. D. Appl. Phys.* 47 (15): 155202.

**41** Zhu, Y. and Starikovskaia, S. (2018). Fast gas heating of nanosecond pulsed surface dielectric barrier discharge: spatial distribution and fractional contribution from kinetics. *Plasma Sources Sci. Technol.* 27 (12): 124007.

**42** Pintassilgo, C.D. and Guerra, V. (2015). On the different regimes of gas heating in air plasmas. *Plasma Sources Sci. Technol.* 24 (5): 055009.

**43** Pintassilgo, C., Guerra, V., Guaitella, O., and Rousseau, A. (2014). Study of gas heating mechanisms in millisecond pulsed discharges and afterglows in air at low pressures. *Plasma Sources Sci. Technol.* 23 (2): 025006.

**44** Piper, L., Caledonia, G., and Kennealy, J. (1981). Rate constants for deactivation of $N_2(A_3\Sigma^+$ u, v′= 0, 1) by O. *J. Chem. Phys.* 75 (6): 2847–2852.

**45** Burnette, D., Montello, A., Adamovich, I., and Lempert, W. (2014). Nitric oxide kinetics in the afterglow of a diffuse plasma filament. *Plasma Sources Sci. Technol.* 23 (4): 045007.

**46** Shkurenkov, I., Burnette, D., Lempert, W.R., and Adamovich, I.V. (2014). Kinetics of excited states and radicals in a nanosecond pulse discharge and afterglow in nitrogen and air. *Plasma Sources Sci. Technol.* 23 (6): 065003.

**47** Rusanov, V.D., Fridman, A., and Sholin, G.V. (1981). The physics of a chemically active plasma with nonequilibrium vibrational excitation of molecules. *Soviet Phys. Uspekhi* 24 (6): 447.

**48** Amouroux, J., Cavvadias, S., and Rapakoulias, D. (1979). Réacteur de synthèse et de trempe dans un plasma hors d'équilibre: application à la synthèse des oxydes d'azote. *Rev. Phys. Appl.* 14 (12): 969–976.

**49** Zhang, S., Zong, L., Zeng, X. et al. (2022). Sustainable nitrogen fixation with nanosecond pulsed spark discharges: insights into free-radical-chain reactions. *Green Chem.* 24 (4): 1534–1544.

**50** Vervloessem, E., Gorbanev, Y., Nikiforov, A. et al. (2022). Sustainable $NO_x$ production from air in pulsed plasma: elucidating the chemistry behind the low energy consumption. *Green Chem.* 24 (2): 916–929.

**51** Adamovich, I.V., Macheret, S.O., Rich, J.W., and Treanor, C.E. (1998). Vibrational energy transfer rates using a forced harmonic oscillator model. *J. Thermophys. Heat Transf.* 12 (1): 57–65.

**52** Tsonev, I., Eshtehardi, H.A., Delplancke, M.-P., and Bogaerts, A. (2024). Importance of geometric effects in scaling up energy-efficient plasma-based nitrogen fixation. *Sustain. Energy Fuels* 8 (10): 2191–2209.

**53** Patil, B., Cherkasov, N., Lang, J. et al. (2016). Low temperature plasma-catalytic $NO_x$ synthesis in a packed DBD reactor: effect of support materials and supported active metal oxides. *Appl. Catal. B Environ.* 194: 123–133.

**54** Pei, X., Gidon, D., and Graves, D.B. (2019). Specific energy cost for nitrogen fixation as $NO_x$ using DC glow discharge in air. *J. Phys. D. Appl. Phys.* 53 (4): 044002.

**55** Pei, X., Li, Y., Luo, Y. et al. (2024). Nitrogen fixation as $NO_x$ using air plasma coupled with heterogeneous catalysis at atmospheric pressure. *Plasma Process. Polym.* 21 (1): 2300135.

**56** Lei, X., Cheng, H., Nie, L. et al. (2021). Plasma-catalytic $NO_x$ production in a three-level coupled rotating electrodes air plasma combined with nano-sized $TiO_2$. *J. Phys. D. Appl. Phys.* 55 (11): 115201.

**57** Snoeckx, R., Heijkers, S., Van Wesenbeeck, K. et al. (2016). $CO_2$ conversion in a dielectric barrier discharge plasma: $N_2$ in the mix as a helping hand or problematic impurity? *Energy Environ. Sci.* 9 (3): 999–1011.

**58** Birkeland, K. (1906). On the oxidation of atmospheric nitrogen in electric arcs. *Trans. Faraday Soc.* 2 (December): 98–116.

59 Eyde, S. (1912). Oxidation of atmospheric nitrogen and development of resulting industries in Norway. *Industr. Eng. Chem.* 4 (10): 771–774.

60 Jardali, F., Van Alphen, S., Creel, J. et al. (2021). $NO_x$ production in a rotating gliding arc plasma: potential avenue for sustainable nitrogen fixation. *Green Chem.* 23 (4): 1748–1757.

61 Rehbein, N. and Cooray, V. (2001). $NO_x$ production in spark and corona discharges. *J. Electrost.* 51: 333–339.

62 Pei, X., Gidon, D., Yang, Y.-J. et al. (2019). Reducing energy cost of $NO_x$ production in air plasmas. *Chem. Eng. J.* 362: 217–228.

63 Janda, M., Martišovitš, V., Hensel, K., and Machala, Z. (2016). Generation of antimicrobial $NO_x$ by atmospheric air transient spark discharge. *Plasma Chem. Plasma Process.* 36: 767–781.

64 Pavlovich, M., Ono, T., Galleher, C. et al. (2014). Air spark-like plasma source for antimicrobial $NO_x$ generation. *J. Phys. D. Appl. Phys.* 47 (50): 505202.

65 Partridge, W.S., Parlin, R.B., and Zwolinski, B.J. (1954). Fixation of nitrogen in a crossed discharge. *Industr. Eng. Chem.* 46 (7): 1468–1471.

66 Bian, W., Shi, J., and Yin, X. (2008). Nitrogen fixation into water by pulsed high-voltage discharge. *IEEE Trans. Plasma Sci.* 37 (1): 211–218.

67 Vervloessem, E., Aghaei, M., Jardali, F. et al. (2020). Plasma-based $N_2$ fixation into $NO_x$: insights from modeling toward optimum yields and energy costs in a gliding arc plasmatron. *ACS Sustain. Chem. Eng.* 8 (26): 9711–9720.

68 J. Coudert, J. Baronnet, J. Rakowitz, P. Fauchais, Synthesis of nitrogen oxides in a plasma produced by a jet arc generator, Symp. Int. Chim. Plasmas, 1977.

69 Kim, T., Song, S., Kim, J., and Iwasaki, R. (2010). Formation of $NO_x$ from air and $N_2/O_2$ mixtures using a nonthermal microwave plasma system. *Jpn. J. Appl. Phys.* 49 (12R): 126201.

70 Mutel, B., Dessaux, O., and Goudmand, P. (1984). Energy cost improvement of the nitrogen oxides synthesis in a low pressure plasma. *Rev. Phys. Appl.* 19 (6): 461–464.

71 Asisov, R., Givotov, V., Rusanov, V., and Fridman, A. (1980). High energy chemistry (Khimia Vysokikh energij). *Sov. Phys* 14: 366.

72 T. Taugbøl, E.M. Andersen, U. Grønn, B.F. Moen, Rjukan-Notodden industriarv. Nominasjon til Unescos verdensarvliste, 2014.

73 Janda, M., Martišovitš, V., Hensel, K., and Machala, Z. (2016). Generation of antimicrobial $NO_x$ by atmospheric air transient spark discharge. *Plasma Chem. Plasma Process.* 36: 767–781.

74 Bian, W., Song, X., Shi, J., and Yin, X. (2012). Nitrogen fixed into $HNO_3$ by pulsed high voltage discharge. *J. Electrost.* 70 (3): 317–326.

75 Liu, Z., Tian, Y., Niu, G. et al. (2021). Direct oxidative nitrogen fixation from air and $H_2O$ by a water falling film dielectric barrier discharge reactor at ambient pressure and temperature. *ChemSusChem* 14 (6): 1507–1511.

**76** Patil, B.S., Peeters, F., van Rooij, G.J. et al. (2018). Plasma assisted nitrogen oxide production from air: using pulsed powered gliding arc reactor for a containerized plant. *AIChE J* 64 (2): 526–537.

**77** Namihira, T., Katsuki, S., Hackam, R. et al. (2002). Production of nitric oxide using a pulsed arc discharge. *IEEE Trans. Plasma Sci.* 30 (5): 1993–1998.

**78** Van Alphen, S., Eshtehardi, H.A., O'Modhrain, C. et al. (2022). Effusion nozzle for energy-efficient $NO_x$ production in a rotating gliding arc plasma reactor. *Chem. Eng. J.* 443: 136529.

**79** Muzammil, I., Lee, D.H., Dinh, D.K. et al. (2021). A novel energy efficient path for nitrogen fixation using a non-thermal arc. *RSC Adv.* 11 (21): 12729–12738.

**80** Li, S., van Raak, T., Kriek, R. et al. (2023). Gliding arc reactor under AC pulsed mode operation: spatial performance profile for $NO_x$ synthesis. *ACS Sustain. Chem. Eng.* 11 (34): 12821–12832.

**81** van Raak, T., van den Bogaard, H., De Felice, G. et al. (2024). Numbering up and sizing up gliding arc reactors to enhance the plasma-based synthesis of $NO_x$. *Cat. Sci. Technol.* 14 (18): 5405–5421.

**82** Tsonev, I., O'Modhrain, C., Bogaerts, A., and Gorbanev, Y. (2023). Nitrogen fixation by an arc plasma at elevated pressure to increase the energy efficiency and production rate of $NO_x$. *ACS Sustain. Chem. Eng.* 11 (5): 1888–1897.

**83** Polak, L., Ovsyannikov, A., Slovetskii, D., and Vurzel, F. (1975). Theoretical and applied plasma chemistry. *M.: Nauka* 304.

**84** Kelly, S. and Bogaerts, A. (2021). Nitrogen fixation in an electrode-free microwave plasma. *Joule* 5 (11): 3006–3030.

**85** Berthelot, A. and Bogaerts, A. (2017). Modeling of $CO_2$ splitting in a microwave plasma: how to improve the conversion and energy efficiency. *J. Phys. Chem. C* 121 (15): 8236–8251.

**86** Bruggeman, P.J., Iza, F., and Brandenburg, R. (2017). Foundations of atmospheric pressure non-equilibrium plasmas. *Plasma Sources Sci. Technol.* 26 (12): 123002.

**87** Czernichowski, A. (1994). Gliding arc: applications to engineering and environment control. *Pure Appl. Chem.* 66 (6): 1301–1310.

**88** Ramakers, M., Trenchev, G., Heijkers, S. et al. (2017). Gliding arc plasmatron: providing an alternative method for carbon dioxide conversion. *ChemSusChem* 10 (12): 2642–2652.

**89** Indarto, A., Yang, D.R., Choi, J.-W. et al. (2007). Gliding arc plasma processing of $CO_2$ conversion. *J. Hazard. Mater.* 146 (1–2): 309–315.

**90** Nunnally, T., Gutsol, K., Rabinovich, A. et al. (2011). Dissociation of $CO_2$ in a low current gliding arc plasmatron. *J. Phys. D. Appl. Phys.* 44 (27): 274009.

**91** Tao, X., Bai, M., Li, X. et al. (2011). $CH_4$-$CO_2$ reforming by plasma–challenges and opportunities. *Prog. Energy Combust. Sci.* 37 (2): 113–124.

**92** Petitpas, G., Rollier, J.-D., Darmon, A. et al. (2007). A comparative study of non-thermal plasma assisted reforming technologies. *Int. J. Hydrog. Energy* 32 (14): 2848–2867.

93 Abdelaziz, A.A., Teramoto, Y., Nozaki, T., and Kim, H.-H. (2023). Toward reducing the energy cost of $NO_x$ formation in a spark discharge reactor through pinpointing its mechanism. *ACS Sustain. Chem. Eng.* 11 (10): 4106–4118.

94 Abdelaziz, A.A., Teramoto, Y., Nozaki, T., and Kim, H.-H. (2023). Performance of high-frequency spark discharge for efficient $NO_x$ production with tunable selectivity. *Chem. Eng. J.* 470: 144182.

95 Winter, L.R. and Chen, J.G. (2021). $N_2$ fixation by plasma-activated processes. *Joule* 5 (2): 300–315.

96 Van Alphen, S., Jardali, F., Creel, J. et al. (2021). Sustainable gas conversion by gliding arc plasmas: a new modelling approach for reactor design improvement. *Sustain. Energy Fuels* 5 (6): 1786–1800.

97 Robinson, C. and Stapelmann, K. (2024). Plasma treating water for nitrate based nitrogen fertilizer-a review of recent device designs. *Curr. Opin. Green Sustain. Chem.* 50: 100978.

98 Angineni, J., Reddy, P.M.K., Anga, S., and Somaiah, P.V. Sustainable nitrogen fixation by novel gliding arc plasma reactor for the production of nitrogen fertilizers. *Plasma Process. Polym.* 21 (9): e2400059.

99 Li, J., Lan, C., Nie, L. et al. (2023). Distributed plasma-water-based nitrogen fixation system based on cascade discharge: generation, regulation, and application. *Chem. Eng. J.* 478: 147483.

100 Bogaerts, A. and Centi, G. (2020). Plasma technology for $CO_2$ conversion: a personal perspective on prospects and gaps. *Front Energy Res.* 8: 111.

101 Majeed, M., Iqbal, M., Altin, M. et al. (2024). Effect of thermal gas quenching on $NO_x$ production by atmospheric pressure rotating arc plasma: a pathway towards eco-friendly fertilizer. *Chem. Eng. J.* 485: 149727.

102 Cleiren, E., Heijkers, S., Ramakers, M., and Bogaerts, A. (2017). Dry reforming of methane in a gliding arc plasmatron: towards a better understanding of the plasma chemistry. *ChemSusChem* 10 (20): 4025–4036.

103 Mei, D., Zhu, X., He, Y.-L. et al. (2014). Plasma-assisted conversion of $CO_2$ in a dielectric barrier discharge reactor: understanding the effect of packing materials. *Plasma Sources Sci. Technol.* 24 (1): 015011.

104 Ma, Y., Wang, Y., Harding, J., and Tu, X. (2021). Plasma-enhanced $N_2$ fixation in a dielectric barrier discharge reactor: effect of packing materials. *Plasma Sources Sci. Technol.* 30 (10): 105002.

105 Ingels, R. and Graves, D.B. (2015). Improving the efficiency of organic fertilizer and nitrogen use via air plasma and distributed renewable energy. *Plasma Med.* 5 (2–4): 257–270.

106 Abdelaziz, A.A., Teramoto, Y., Kim, D.-Y. et al. (2024). Critical considerations in power measurements for the precise estimation of energy costs in plasma $NO_x$ synthesis. *Plasma Chem. Plasma Process.* 44 (4): 1493–1512.

**107** Shmelev, V.M., Saveliev, A.V., and Kennedy, L.A. (2009). Plasma chemical reactor with exploding water jet. *Plasma Chem. Plasma Process.* 29: 275–290.

**108** Li, Y., Qin, L., Wang, H.-L. et al. (2023). High efficiency $NO_x$ synthesis and regulation using dielectric barrier discharge in the needle array packed bed reactor. *Chem. Eng. J.* 461: 141922.

**109** Li, Z., Nie, L., Liu, D., and Lu, X. (2022). An atmospheric pressure glow discharge in air stabilized by a magnetic field and its application on nitrogen fixation. *Plasma Process. Polym.* 19 (12): 2200071.

**110** Meng, X., Lu, N., Shang, K., and Jiang, N. (2024). Dielectric-boosted gliding arc discharge for $N_2$ fixation into $NO_x$. *Plasma Chem. Plasma Process.* 44 (4): 1–16.

**111** Janda, M., Hensel, K., Machala, Z., and Field, T.A. (2023). The influence of electric circuit parameters on $NO_x$ generation by transient spark discharge. *J. Phys. D. Appl. Phys.* 56 (48): 485202.

**112** Denra, A., Saud, S., Nguyen, D.B. et al. (2024). Nitrogen fixation by rotational gliding arc plasma at surrounding conditions. *J. Clean. Prod.* 436: 140618.

**113** Azizov, R.I., Vakar, A.K., Zhivotov, V.K. et al. (1983). Nonequilibrium plasmachemical process of $CO_2$ decomposition in a supersonic microwave discharge. *Akad. Nauk SSSR Doklady* 271: 94–98.

**114** R.D. Milton, R. Cai, S. Abdellaoui, D. Leech, A.L. De Lacey, M. Pita, S.D. Minteer, Bioelectrochemical Haber–Bosch process: an ammonia-producing H2/N2 fuel cell, *Angew. Chem. Int. Ed.* 56(10) (2017) 2680–2683.

**115** Zhou, D., Zhou, R., Zhou, R. et al. (2021). Sustainable ammonia production by non-thermal plasmas: status, mechanisms, and opportunities. *Chem. Eng. J.* 421: 129544.

**116** Huang, Z., Xiao, A., Liu, D. et al. (2022). Plasma-water-based nitrogen fixation: status, mechanisms, and opportunities. *Plasma Process. Polym.* 19 (4): 2100198.

**117** Mizoi, K., Rodríguez-González, V., Sasaki, M. et al. (2022). Interactions between pH, reactive species, and cells in plasma-activated water can remove algae. *RSC Adv.* 12 (13): 7626–7634.

**118** Sremački, I., Gromov, M., Leys, C. et al. (2020). An atmospheric pressure non-self-sustained glow discharge in between metal/metal and metal/liquid electrodes. *Plasma Process. Polym.* 17 (6): 1900191.

**119** Gromov, M., Kamarinopoulou, N., De Geyter, N. et al. (2022). Plasma-assisted nitrogen fixation: the effect of water presence. *Green Chem.* 24 (24): 9677–9689.

**120** Sieck, L.W., Heron, J.T., and Green, D.S. (2000). Chemical kinetics database and predictive schemes for humid air plasma chemistry. Part I: positive ion–molecule reactions. *Plasma Chem. Plasma Process.* 20: 235–258.

**121** Herron, J.T. and Green, D.S. (2001). Chemical kinetics database and predictive schemes for nonthermal humid air plasma chemistry. Part II. Neutral species reactions. *Plasma Chem. Plasma Process.* 21: 459–481.

**122** Eisenberg, G. (1943). Colorimetric determination of hydrogen peroxide. *Industr Eng Chem Anal. Ed.* 15 (5): 327–328.

**123** Zhou, L. and Boyd, C.E. (2016). Comparison of Nessler, phenate, salicylate and ion selective electrode procedures for determination of total ammonia nitrogen in aquaculture. *Aquaculture* 450: 187–193.

**124** Deng, X., Nikiforov, A.Y., Vanraes, P., and Leys, C. (2013). Direct current plasma jet at atmospheric pressure operating in nitrogen and air. *J. Appl. Phys.* 113 (2): 023305.

**125** Xiao, D., Cheng, C., Shen, J. et al. (2014). Characteristics of atmospheric-pressure non-thermal $N_2$ and $N_2/O_2$ gas mixture plasma jet. *J. Appl. Phys.* 115 (3): 033303.

**126** Hong, J., Prawer, S., and Murphy, A.B. (2018). Plasma catalysis as an alternative route for ammonia production: status, mechanisms, and prospects for progress. *ACS Sustain. Chem. Eng.* 6 (1): 15–31.

**127** Sander, R. (2015). Compilation of Henry's law constants (version 4.0) for water as solvent. *Atmos. Chem. Phys.* 15 (8): 4399–4981.

**128** Man, C., Zhang, C., Fang, H. et al. (2022). Nanosecond-pulsed microbubble plasma reactor for plasma-activated water generation and bacterial inactivation. *Plasma Process. Polym.* 19 (6): 2200004.

**129** Šimečková, J., Krčma, F., Klofáč, D. et al. (2020). Influence of plasma-activated water on physical and physical–chemical soil properties. *Water* 12 (9): 2357.

**130** Sivachandiran, L. and Khacef, A. (2017). Enhanced seed germination and plant growth by atmospheric pressure cold air plasma: combined effect of seed and water treatment. *RSC Adv.* 7 (4): 1822–1832.

**131** Traylor, M.J., Pavlovich, M.J., Karim, S. et al. (2011). Long-term antibacterial efficacy of air plasma-activated water. *J. Phys. D. Appl. Phys.* 44 (47): 472001.

**132** Subramanian, P., Rao, H., Shivapuji, A.M. et al. (2021). Plasma-activated water from DBD as a source of nitrogen for agriculture: specific energy and stability studies. *J. Appl. Phys.* 129 (9): 093303.

**133** Xu, Z., Cheng, C., Shen, J. et al. (2018). In vitro antimicrobial effects and mechanisms of direct current air-liquid discharge plasma on planktonic *Staphylococcus aureus* and *Escherichia coli* in liquids. *Bioelectrochemistry* 121: 125–134.

**134** Zhou, R., Zhou, R., Wang, P. et al. (2019). Microplasma bubbles: reactive vehicles for biofilm dispersal. *ACS Appl. Mater. Interfaces* 11 (23): 20660–20669.

**135** Park, J.Y., Park, S., Choe, W. et al. (2017). Plasma-functionalized solution: A potent antimicrobial agent for biomedical applications from antibacterial therapeutics to biomaterial surface engineering. *ACS Appl. Mater. Interfaces* 9 (50): 43470–43477.

**136** Kutasi, K., Popović, D., Krstulović, N., and Milošević, S. (2019). Tuning the composition of plasma-activated water by a surface-wave microwave discharge and a kHz plasma jet. *Plasma Sources Sci. Technol.* 28 (9): 095010.

**137** Toth, J.R., Abuyazid, N.H., Lacks, D.J. et al. (2020). A plasma-water droplet reactor for process-intensified, continuous nitrogen fixation at atmospheric pressure. *ACS Sustain. Chem. Eng.* 8 (39): 14845–14854.

**138** Dinh, D.K., Muzammil, I., Kang, W.S. et al. (2021). Reducing energy cost of in situ nitrogen fixation in water using an arc-DBD combination. *Plasma Sources Sci. Technol.* 30 (5): 055020.

**139** Roy, N.C., Maira, N., Pattyn, C. et al. (2023). Mechanisms of reducing energy costs for nitrogen fixation using air-based atmospheric DBD plasmas over water in contact with the electrode. *Chem. Eng. J.* 461: 141844.

**140** Kubota, Y., Koga, K., Ohno, M., and Hara, T. (2010). Synthesis of ammonia through direct chemical reactions between an atmospheric nitrogen plasma jet and a liquid. *Plasma Fusion Res.* 5: 042–042.

**141** Peng, P., Schiappacasse, C., Zhou, N. et al. (2019). Plasma in situ gas–liquid nitrogen fixation using concentrated high-intensity electric field. *J. Phys. D. Appl. Phys.* 52 (49): 494001.

**142** Rathore, V. and Nema, S.K. (2022). A comparative study of dielectric barrier discharge plasma device and plasma jet to generate plasma activated water and post-discharge trapping of reactive species. *Phys. Plasmas* 29 (3): 033510.

**143** Shih, K.Y. and Locke, B.R. (2009). Effects of electrode protrusion length, pre-existing bubbles, solution conductivity and temperature, on liquid phase pulsed electrical discharge. *Plasma Process. Polym.* 6 (11): 729–740.

**144** Sun, J., Alam, D., Daiyan, R. et al. (2021). A hybrid plasma electrocatalytic process for sustainable ammonia production. *Energy Environ. Sci.* 14 (2): 865–872.

**145** Rathore, V., Jamnapara, N.I., and Nema, S.K. (2023). Enhancing the physicochemical properties and reactive species concentration of plasma activated water using an air bubble diffuser. *Phys. Lett. A* 482: 129035.

**146** Sun, J., Zhou, R., Hong, J. et al. (2024). Sustainable ammonia production via nanosecond-pulsed plasma oxidation and electrocatalytic reduction. *Appl. Catal. B Environ.* 342: 123426.

# Part III

# Mechanisms of Nitrogen Fixation

# 7

# Ammonia Synthesis with Plasma Catalysis: Mechanisms

*Kevin Rouwenhorst*[1,2] *and Leon Lefferts*[1]

[1] *Catalytic Processes & Materials, Department of Chemical Engineering, Faculty of Science and Technology, University of Twente, Enschede, The Netherlands*
[2] *Ammonia Energy Association, Ashburn, VA, USA*

## 7.1 Introduction

The combination of plasma and catalysis can be attractive, from the perspective of either thermodynamics or kinetics. Reactions can be pushed beyond thermodynamic equilibrium compared to molecules in the ground state, via excitation of reactant molecules. Typical examples are dissociation of $CO_2$ and $H_2O$, as well as $NO_X$ formation from $N_2$ and $O_2$. On the other hand, kinetics can be enhanced via excitation of the reactant molecules to make reactions possible at mild temperatures by decreasing the activation barrier to be passed thermally. Examples include ammonia synthesis and oxidation reactions of organic compounds for, e.g. air cleaning. In fact, the advantage for ammonia synthesis could be twofold, because lower operation temperature also induces a favorable shift of the equilibrium, which, in turn, would enable operation at lower pressures compared to Haber-Bosch (HB).

## 7.2 Methods to Study Mechanisms in Catalysis

Information on the mechanism of catalytic reactions can be obtained, based on experimental kinetic information, by varying concentrations, pressures, temperature, and residence time. This is usually combined with microkinetic modeling, typically based on Langmuir-Hinshelwood mechanistic models. The interested

*Plasma-Assisted Nitrogen Fixation for Sustainable Process Industries*, First Edition. Edited by Volker Hessel, Annemie Bogaerts, Gabriele Centi, Evgeny Rebrov, and Nguyen Van Duc Long.

Published 2026 by John Wiley & Sons, Inc.

reader is referred to Ref. [1] and references therein for more details. For more information about mechanisms in the plasma phase without catalyst, the reader is referred to a recent review of Qu et al. [2]

In short, mechanistic models consist of elementary reactions, which may involve reactant molecules, active sites, adsorbed intermediates, and product molecules. One of these steps is usually rate-determining step (RDS), while all other steps are assumed to be in quasi-equilibrium. This allows for the derivation of rate equations expressing the activity per active site (TOF, turnover frequency) as a function of temperature and pressures of reactants and products. Such a rate equation contains the rate constant of the RDS according to the Arrhenius equation ($k = k_0 e^{-\frac{E_{act}}{RT}}$), as well as the equilibrium constants of all other reaction steps.

In case the resulting rate equation cannot describe the observations correctly (after optimization of unknown rate constants, activation barriers, or equilibrium constants of elementary reactions), the mechanistic model can be excluded. If a good fit can be obtained with realistic values for the fitting parameters, the model remains a valid hypothesis.

As an example, dissociation of $N_2$ on the catalyst surface is generally the RDS in ammonia synthesis over Fe and Ru catalysts [3], although reaction conditions can cause changes in the RDS [4]. Fe and Ru are suitable catalysts because the interaction with $N_2$ is strong enough to cause dissociation of $N_2$ on the surface, while the interaction is weak enough to enable further hydrogenation of adsorbed N-atoms to ammonia, according to the Sabatier principle [5], as illustrated in Figure 7.1.

A second main source of information is obtained by detecting reaction intermediates in operando experiments, i.e. detection of surface intermediates with spectroscopic techniques, while the catalyst is active under reaction conditions. Preferably, these conditions are similar to practical industrial conditions. However, this is not possible in many cases, because high gas pressure interferes with

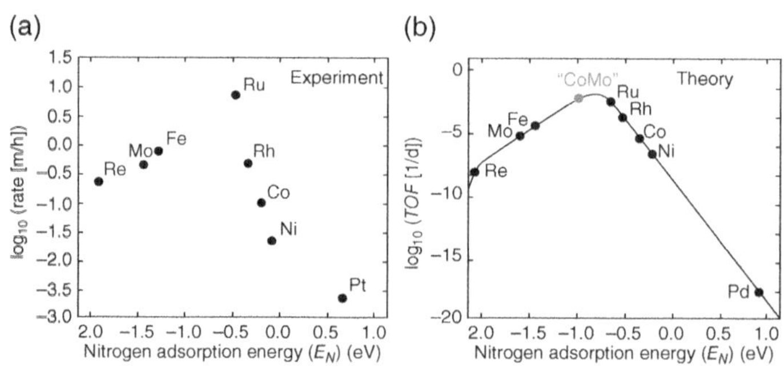


**Figure 7.1** Volcano plots according experiments (a) and theory (b) illustrating that the activity of catalyst for ammonia synthesis is maximal at intermediate bonding strength of nitrogen. *Source:* Ref. [5] / with permission of Elsevier.

the spectroscopic techniques. Therefore, many studies are performed under reactive conditions, but at much lower pressures [4], and by measuring model catalysts like single crystals [6]. Still, these kind of experiments give insight into what types of surface species are involved in the catalytic reaction.

Finally, theoretical density functional theory (DFT)-based models for chemical reactions are increasingly used to gain understanding by calculating the energy of hypothesized surface intermediate species, as well as the activated complexes of elementary reactions. This can reveal the influence of detailed characteristics (e.g. coordination numbers of surface metal atoms) or description of surface intermediate species that cannot be observed experimentally. Unfortunately, testing of these models with experimental observations is often not possible.

## 7.3 Experimental Kinetics: From Catalysis to Plasma Catalysis

Kinetic data contains information on the mechanism when considering the following conditions regarding experimental procedures and data interpretation. These are based on best practices in catalysis research but need special attention in the case of plasma catalysis [1].

- **Control over reaction conditions**: Precise control over the concentration or pressure of reactants and temperature is critical. Determination of experimental activation barriers requires experiments at different, well-controlled temperatures. Temperature gradients can develop in catalyst beds in the case of highly exothermic or highly endothermic reactions, which would strongly influence the observed reaction rates. In the case of catalytic reactions, such an effect can be minimized, e.g. by decreasing the activity of the catalyst via a low loading of the active phase, by choosing lower reactant concentrations, or by mixing the catalyst with inert material [7]. In the case of plasma catalysis, there is always uncertainty about temperature, even in the case of a dielectric barrier discharge (DBD) plasma reactor, because of the heating effect of the plasma itself. Any enhancement of reaction rate versus thermal catalysis might be simply caused by local plasma heating. Complexity is further increased by the fact that the heating effect of the plasma is heterogeneous in space as well as dynamic in time, because of local short-lived streamers.
- **Blank experiments**: Catalytic reactors can contribute to conversion in the absence of a catalyst, via gas-phase reactions, especially at high temperatures, or catalytic contributions of reactor wall and internals. This is especially important in the case of plasma catalysis, because the reactor without a catalyst converts via plasma-chemical reactions. Experiments with an empty plasma reactor

are generally not a solution because, first, the presence of a solid influences the plasma via its dielectric properties, and second, the residence time decreases when the reactor is filled with catalyst. The last effect may be compensated by decreasing the flow rate, but that would result in a change in specific energy input (SEI) by the plasma. To the best of our knowledge, the best solution is using a plasma reactor filled with bare catalyst support as a blank. The actual measurement with a catalyst is best performed with a relatively low loading of catalytically active material on the support, to minimize the effect on the plasma properties and thereby keeping the conversion in the plasma-phase constant. Thus, the additional conversion caused by the active phase can be considered the result of catalyst activity.

- **Differential or integral experiments**: Differential experiments are experiments with conversion less than typically 5% so that the concentration of the reactants can be considered quasi-constant. In this case, it is possible to calculate the catalyst activity at that specific reactant concentration directly based on the conversion, flow rate, and amount of catalyst. This would result in an activity per gram catalyst, or per active site (TOF), if the number of active sites is determined with appropriate methods like chemisorption or electron microscopy. Integral experiments imply higher conversion so that the concentration of the reactant varies significantly with the position in the reactor. Therefore, the catalytic rate also varies inside the reactor, and a direct calculation would result in an incorrect average catalyst activity. A correct calculation of catalyst activity would require integration over the reactor, requiring a reactor model on top of the kinetic model. On the other hand, experiments at high conversion are more relevant for practical application.
- **Reverse reactions**: Usually, one considers exclusively the forward reaction from reactant(s) to product(s). However, when equilibrium is approached, e.g. by increasing the residence time, the reverse reaction will contribute more and more. If this is neglected, one would erroneously conclude that the activity of the catalyst decreases with residence time. Thus, the reverse reaction needs to be included in the kinetic analysis. This is even more important for plasma catalysis, because accelerated electrons in the plasma will excite not only reactant molecules but also product molecules, such as ammonia. Thus, the reverse reaction must be considered, both for thermal-catalytic conversion and for plasma-catalytic conversion. If this is neglected, the resulting apparent catalytic activity is lower than the real value. The consequence is that for obtaining correct kinetic data, the product concentration should be limited to even lower values than needed for doing differential experiments as discussed above. The concentration that is still acceptable can be determined by adding product molecules to the feed stream, such as ammonia to the $N_2/H_2$ feed mixture, checking what maximal ammonia concentration is acceptable, and preventing significant ammonia decomposition [8].

- **Mass transfer**: Heterogeneous catalysts usually consist of nanoparticles of an active phase, e.g. a metal, supported on a microporous material. Reactants and products diffuse through the pores to reach the supported nanoparticles and to diffuse out of the catalyst particle, respectively. This often leads to significant concentration gradients in the support particles, depending on the activity of the nanoparticles, the loading of nanoparticles, the size of the support particles, the porosity and tortuosity of the pore system, and the diffusion coefficients of reactants. In case internal concentration gradients develop, the apparent reaction rate is lower than the intrinsic reaction rate, i.e. the rate that would be achieved in the absence of any mass transfer limitation.

  Pores in catalyst supports are usually in the range of a few nanometers to 100 nm. In such a limited space, electrons cannot be sufficiently accelerated in an electrical field to generate a plasma inside the pores. Furthermore, activated species are short-lived and are not able to diffuse in the pores before reacting or quenching. Thus, a thin outermost layer of the porous support particles is likely to exclusively contribute to plasma-catalytic conversion. Obviously, the calculation of TOFs based on the total number of active sites will result in the underestimation of the actual TOF. Metal particles at the outer surface will have the largest effect on the plasma-catalytic performance. Dense support particles with metal particles on the external surface would be preferred, with the disadvantage of a relatively low number of active sites per unit of reactor volume.
- **Catalyst stability**: Finally, catalyst stability should be considered. Most plasma-catalytic experiments are performed in continuously operated reactors so that any change in performance with time on stream is easily observed. Plasma heating can, e.g. cause accelerated aggregation of nanoparticles.

Numerous papers in plasma catalysis, including ammonia synthesis, do not consider these points, with the consequence that correct kinetic data is not obtained. Therefore, many papers report exclusively on the overall performance (conversion, yield, etc.) of a certain combination of plasma reactor and catalyst. It is important to note that kinetic information of the catalyst considered is not obtained in that case (reaction rate per gram catalyst, or TOF). There are only a few experimental studies that consider these issues, which are discussed in more detail in Section 7.6.

## 7.4 Beyond Equilibrium and Reverse Reactions

Thermodynamics and kinetics are coupled in thermal catalysis, as the equilibrium constant, determined by the free enthalpy of the reaction, is equal to the ratio of the forward- and backward-rate constant. In the case of ammonia synthesis at low temperature and atmospheric pressure (such as in a DBD plasma), the equilibrium

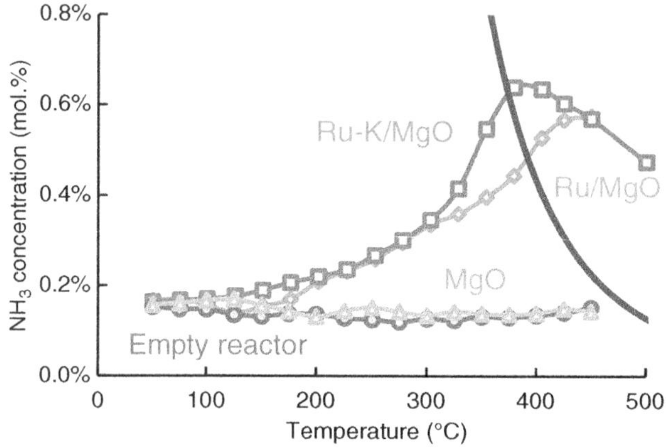


**Figure 7.2** Activity for plasma-catalytic $NH_3$ synthesis (and decomposition) for an empty reactor (only the spacer and quartz wool, circles), MgO (triangles), Ru/MgO (diamonds), and Ru–K/MgO (squares). Total flow rate 20 mL min$^{-1}$, $H_2$:$N_2$ = 1:1 (no $NH_3$ co-feed), catalyst loading 130 mg (250–300 μm), plasma power 3.8 W (SEI = 11.4 kJ L$^{-1}$). The solid line presents the thermodynamic equilibrium. *Source:* Ref. [8] / Royal Society of Chemistry / CC BY 4.0.

conversion to ammonia becomes very low (Figure 7.2) when the temperature is increased to typical temperatures used in thermal-catalytic ammonia synthesis.

Although plasma catalysis aims at operation at low temperature and mild pressure, knowledge about the effect of temperature is important. From a practical point of view, the temperature should be high enough to enable effective cooling of the heat caused by both the exothermic reaction and the heat formation by the plasma. On the other hand, variation of temperature in experiments is necessary to determine activation barriers. Furthermore, temperatures need to be sufficient to allow for desorption of ammonia from the catalyst, as discussed in Section 7.5.

Conversion beyond thermodynamic equilibrium is reported by Mehta et al. [9] in well-controlled experiments, as well as microkinetic modeling on plasma-catalytic ammonia synthesis with supported Ni and Pt catalysts at elevated temperatures. Rouwenhorst et al. [8] performed similar experiments on Ru supported on MgO, with and without a K promoter, including experiments with co-feeding of ammonia. This demonstrated that ammonia decomposition is significantly contributing during plasma-catalytic ammonia synthesis, becoming significant when approaching thermodynamic equilibrium by increasing the temperature. This is in good agreement with the results of the study by Patil et al. [10], who reported that ammonia decomposition via thermal catalysis over Ru and Fe, the most active catalysts for thermal ammonia synthesis, sets in at temperatures above 250 °C. Iwamoto et al. [11] reported ammonia decomposition assigned to plasma-irradiated gas-phase reactions. These findings imply that catalysts with thermal activity are not particularly useful when aiming at conversion beyond equilibrium and that apparent forward reaction rates underestimate the true forward reaction rates.

Ammonia decomposition and $NH_3$-$D_2$ exchange reactions are also reported in DBD reactors filled with dielectric materials instead of any catalyst, where surface reactions may still contribute [12]. The same group also reported that the addition of Ru decreases ammonia yield under their conditions, caused by ammonia decomposition [13].

The importance of ammonia decomposition reactions is also illustrated by experiments that prevent ammonia decomposition by integrating plasma-catalytic synthesis with the separation of ammonia. Protective adsorption of ammonia in a DBD reactor was demonstrated by Rouwenhorst et al. [14] using zeolite 4A as an adsorbent for *in situ* ammonia removal from the plasma phase. The nanopores in zeolites do not allow any significant formation of plasma, nor can activated species diffuse into the nanoporous system, so that adsorbed ammonia is protected against decomposition. Upon ammonia desorption via increasing the temperature in the absence of a plasma, the ammonia yield increased by a factor of 2 compared to steady-state plasma-chemical ammonia synthesis. A similar suggestion has been made by Wang et al. [15], unfortunately without experimental evidence; protected ammonia in the porous support was not desorbed in the absence of plasma so favorable results in terms of energy yield cannot be ascribed to protective adsorption. Clearly, protective adsorption can be achieved only via chemical looping, desorbing ammonia in the absence of plasma. Nevertheless, several authors [15–17] also reported improved ammonia yields in steady-state operation with zeolites but note that the reason cannot be protective adsorption. Peng et al. [18] studied Mg salts, forming $Mg_3N_2$ and $Mg(NH_3)_6Cl_2$ in a $N_2/H_2$ plasma, which was converted to ammonia by dissolving in water in a separate step, thus preventing plasma-phase ammonia decomposition. Gorky et al. [19] demonstrated a membrane reactor with molecular sieve membranes as an alternative approach for the integration of conversion and separation, protecting ammonia from plasma decomposition.

## 7.5 Effect of Catalyst on Plasma

The concept of plasma catalysis relies on exciting reactant molecules in the plasma, with subsequent adsorption on the catalyst, followed by surface reactions. However, the presence of a catalyst, or any solid, in a plasma locally influences the electrical field, plasma composition, and the chemical conversion in the plasma phase [20]. The complexity of the mutual influence of plasma and catalyst is discussed elsewhere, e.g. in ref. [21], and is schematically presented in Figure 7.3.

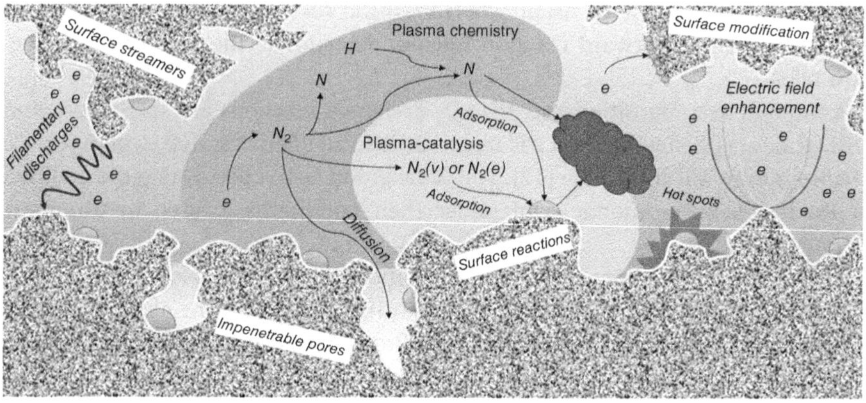


**Figure 7.3** Complex mutual influence of plasma and catalyst. *Source:* Ref. [21] / Royal Society of Chemistry / CC BY 4.0.

This chapter focuses on conversion via the catalyst. Therefore, any experimental observations of catalytic reactions should be corrected for conversion in the plasma phase via blank experiments, as described in Section 7.3. This approach was applied by Rouwenhorst et al. for ammonia synthesis (Figure 7.2), which shows that a temperature of at least 200 °C is needed to observe a significant contribution of plasma-catalytic formation of ammonia, in addition to plasma chemistry as observed in the absence of any Ru. At lower temperatures, hydrogenation of surface N species and/or ammonia desorption is too slow.

Wang et al. [22] reported very similar discharge characteristics for alumina and alumina-supported Ni, Cu, and Fe catalysts in a packed-bed DBD reactor, using a transparent water electrode. Similarly, Herrera et al. [23] report that the macroscopic properties of plasma with a packed-bed $Al_2O_3$ is not influenced by the presence of ~5 wt.% Fe, Co, or Ni. On the other hand, Patil et al. [10] reported that this is not always true for high metal loading, as 10 wt.% of Mn or Fe on alumina suppresses plasma formation. In short, it cannot be ruled out that local properties of the plasma are influenced by the presence of a relatively low amount of supported metal nanoparticles.

## 7.6 Kinetics of Plasma-Catalytic Ammonia Synthesis

Plasma-catalytic synthesis of ammonia can be regarded as an excellent model reaction to study the synergy between plasma and catalysis. The reaction is relatively simple with only one product so that selectivity issues are not relevant. Catalytic ammonia synthesis is extremely well studied and is the best described and understood industrial-catalytic reaction. As explained in Section 7.3, it is rather complicated to extract valid kinetic information from plasma-catalytic experiments. Therefore, only a few studies are considered here.

Barboun et al. [24] reported on a systematic study on ammonia synthesis on Ru, Co, and Ni supported on γ-alumina-supported transition metal catalysts in a DBD reactor, considering several of the issues in Section 7.3. The overall reaction and plasma-phase reaction are first order in both $N_2$ and $H_2$. In contrast, the rate contributions due to plasma-catalyst interactions are first order in $N_2$ but zeroth order in $H_2$. Mehta et al. [9] reported on Pt and Ni supported on γ-alumina, with similar experiments and microkinetic modeling.

Rouwenhorst et al. [25] studied ammonia synthesis on MgO-supported Ru, Co, Pt, Pd, Cu, and Ag catalysts in a DBD plasma at temperatures between room temperature and 500 °C and plasma power between 3.8 and 6.4 W. Underestimation of TOFs, caused by ammonia decomposition, was minimized using exclusively observations at low ammonia concentration, far away from equilibrium. This resulted in very similar TOFs ($10^{-3}$–$10^{-2}$ second$^{-1}$ at 250 °C) for plasma-catalytic ammonia synthesis on Ru/MgO, Co/MgO, Pt/MgO, Pd/MgO, Cu/MgO, and Ag/MgO. Also very similar apparent activation barriers in the range 18–24 kJ mol$^{-1}$ were obtained, suggesting the same RDS for the different metals under these conditions. In contrast, the apparent activation barrier on Ag/MgO was significantly higher at 30 kJ mol$^{-1}$, which may indicate a different RDS.

## 7.7 Mechanism of Plasma-Catalytic Ammonia Synthesis

A key question is the dominant pathway of $N_2$ dissociation. Either $N_2$ molecules are plasma excited with subsequent enhanced dissociative adsorption on the catalyst surface, or $N_2$ is dissociated in the plasma and N-radicals interact with the catalyst surface or adsorbed species on the catalyst surface. These two options are presented in Figure 7.4 and discussed in Sections 7.7.1 and 7.7.2, whereas Section 7.7.3 discusses possible reaction surface intermediates in the pathway to ammonia via hydrogenation of N-atoms on the metal surface.

### 7.7.1 Dominant Pathway: Catalytic Dissociation of Excited $N_2$

Mehta et al. [26] claimed a reaction pathway via vibrational excitation followed by dissociative adsorption based on modeling, supported by experimental data for a supported metal catalyst. Later, it was realized, though, that the differences in the experimental data were actually too small to support this convincingly. In

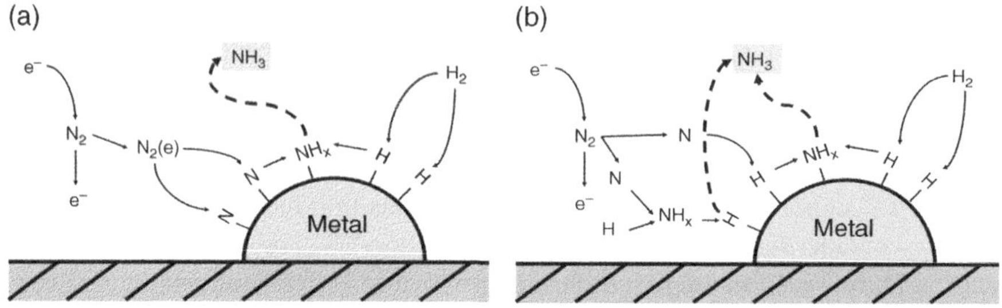


**Figure 7.4** Schematic representation of (a) reaction pathway via dissociative adsorption of plasma-activated $N_2$ and (b) via $N_2$ dissociation in plasma.

addition, nitrogen activation may also occur via electronic excitation as a dominant pathway, instead of via vibrational excitation.

Rouwenhorst et al. [27] obtained experimental support for this hypothesis when operating in very weak plasma, with SEI in the order of 200 J $L^{-1}$, using unpromoted and promoted alumina-supported ruthenium catalysts at temperatures between 200 and 500 °C. Note that this resulted in very low ammonia concentrations, well below 1,000 ppm. Small reduced electrical fields are needed to achieve dominant plasma excitation of molecular $N_2$. [21] The apparent activation energy for thermal-catalytic ammonia synthesis, ranging between ~60 and ~115 kJ $mol^{-1}$ depending on the promoters, is much higher than the apparent activation energy for plasma-catalytic ammonia synthesis (~20 to ~40 kJ $mol^{-1}$), consistent with the hypothesis that ammonia synthesis is enhanced via plasma-induced vibrational or electronic excitations of $N_2$. Furthermore, it is observed that the effect of K and Mg promotion on activity is similar for thermal catalysis and plasma catalysis. It is well known that these electronic promoters weaken the N—N bond in the adsorbed $N_2$ molecules, enhancing the $N_2$ dissociation and RDS in thermal catalysis. The observation of the same effect in plasma catalysis suggests that breaking the N—N bond is still the RDS in plasma-catalytic ammonia synthesis under these conditions.

Plasma activation of $N_2$ results in a longer N—N bond distance. Rouwenhorst et al. [28] suggested that this can allow for a direct dissociation pathway, with both N-atoms in the excited $N_2$ molecule interacting with adjacent metal atoms in the surface, resulting in a lower apparent activation barrier compared to thermal catalysis.

Iwamoto et al. [11] suggested dissociation of plasma-excited $N_2$ as the dominant pathway using metal wool as both catalyst and electrode, based on the experiments and DFT calculations.

Modeling studies by Engelman et al. [29] showed that TOFs of plasma-catalytic $NH_3$ synthesis, via vibrational or electronic excitation of $N_2$ on noble metal

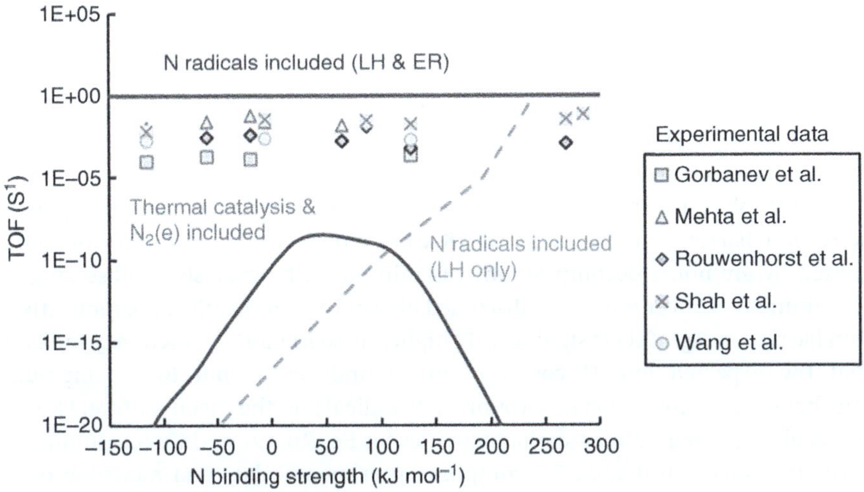


**Figure 7.5** Modeled as well as experimental TOFs as a function of the N binding strength on several metals. Modeling data originally published in [29], containing experimental data. *Source:* Adapted from Gorbanev [30], Mehta [26], Shah [31], and Wang [22], revised with additional data from Rouwenhorst [25].

catalysts (Rh, Ni, Pt, and Pd), depend on the population of the higher energy levels of excitation. The population of higher energy levels seems more important than the vibrational temperature.

In short, excitation of $N_2$ molecules might be relatively more important than plasma dissociation at low SEIs. In that case, the $NH_3$ TOF follows the volcano behavior, rather than the horizontal trends predicted for reactions with nitrogen radicals (see Figure 7.5).

### 7.7.2 Dominant Pathway: $N_2$ Dissociation in Plasma

In a DFT modeling study, Engelmann et al. [29] considered micro-discharges in a DBD plasma, existing locally in the reactor for a short time, as well as the following afterglow, considering dynamic operation locally in a DBD plasma. NH formation proceeds dominantly on the catalyst surface in both regimes, namely the micro-discharges and the afterglow. The results show that volcano curves are obtained when it is assumed that the route via vibrationally excited molecules dominates, for both micro-discharges and afterglows. When radicals are present, the radical pathway dominates, resulting in hardly any influence of the catalyst on TOFs. This results in a horizontal line rather than a volcano curve, as shown in Figure 7.5.

Experimental data suggests that the reaction pathway via plasma-generated radicals dominates for plasmas with a high SEI. The typical suggested pathway is the Eley-Rideal (ER) reaction between surface-adsorbed H-atoms with N-radicals in the plasma [25, 29]. Rouwenhorst et al. [25] reported similar TOFs for MgO-supported Ru, Co, Pt, Pd, Cu, and Ag catalysts in a DBD plasma at temperature between room temperature and 500 °C, and for plasma powers between 3.8 and 6.4 W, implying an SEI larger than 10 kJ $L^{-1}$, with very similar apparent activation barriers in the range of 18–24 kJ $mol^{-1}$. Underestimation of TOFs caused by ammonia decomposition was minimized by exclusively observing at low ammonia concentration, as discussed above. In contrast, the apparent activation barrier on Ag/MgO is significantly higher at 30 kJ $mol^{-1}$, which may indicate that the expected low H coverage on Ag induces a shift to a Langmuir-Hinshelwood pathway, via adsorption of N-radicals on the metal surface. Gorbanev et al. also reported very similar activities for Fe, Ru, Co, and Cu on alumina in a DBD plasma [30]. Figure 7.5 compiles experimental data and modeling data, demonstrating that pathways via N-radicals generated in the plasma are dominant in both streamers and afterglow.

Liu and coworkers [32] suggested that dissociation in the plasma of both $N_2$ and $H_2$ controls the kinetics of ammonia formation. Plasma $N_2$ dissociation is more important on metals to bond N only weakly, e.g. Au, while $H_2$ plasma dissociation is suggested to be important on catalyst bonding N strongly, like Fe.

Hong et al. [33] modeled ammonia synthesis on $Fe/Al_2O_3$ catalyst in a DBD plasma, assuming a homogeneous plasma to describe the experimental data. They suggested that the $NH_3$ synthesis on non-noble catalysts follows a pathway where $N_2$ is dissociated in the plasma and $H_2$ dissociates both in plasma and via dissociative adsorption on the catalyst surface. It is proposed that NH forms in the plasma, which subsequently reacts via an ER reaction with the adsorbed H to form ammonia.

Shah et al. [34] studied Fe, Cu, Pd, and Ag metal mesh catalysts in low-pressure radio frequency (RF) plasma experimentally and via modeling. It is suggested that plasma dissociation of both $H_2$ and $N_2$ is dominant under these conditions, while ammonia is mostly formed on the catalyst surface after adsorption of N- and H-radicals. The authors suggested formation of surface-NH via interaction of surface-N with H in the plasma, probably due to the low concentration of radicals at low pressure.

Sun et al. [35] modeled and measured ammonia synthesis on Fe/alumina in a nanosecond pulsed discharge plasma and concluded that the reaction pathway via N-radicals dominates, more specifically $N(^2D)$ radicals.

### 7.7.3 Surface Intermediate Species

Spectroscopic techniques can be applied to catalysts under controlled conditions or even reaction conditions to identify potential surface intermediate species. It should be noted that such measurements are complex for plasma catalysis, because the presence of an electrical field or electromagnetic field easily interferes with spectroscopic techniques like infrared (IR) or UV–Vis spectroscopy. Therefore, experimental information on intermediate species is rare.

Zhao et al. [36] reported the detection of gas-phase species NNH and $N_2H_2$ during plasma-assisted ammonia synthesis at 533 K and atmospheric pressure in a coaxial DBD plasma reactor with mass spectrometry, and these species were present in the plasma discharge at levels comparable to that of $NH_3$. Adding supported metal particles to alumina caused a decrease in the concentration of $N_2H_Y$ while the ammonia concentration increased, suggesting that $N_2H_Y$ is an intermediate in the formation of ammonia. A potential explanation is that catalyst surfaces are typically covered with H-atoms under plasma catalysis conditions, allowing $N_2H_Y$ species to be hydrogenated to form ammonia.

Yamijala et al. [37] studied the performance of Fe, Pt, and Cu metal foils in low-pressure RF plasmas, based on *ab initio* molecular dynamics calculations supported by optical emission spectroscopy (OES). It is concluded that mass transfer of N-radicals is rate limiting. N-radicals were found to interact directly with surface H on Cu, while the Pt surface is mostly covered with NH under these conditions. This result is consistent with the fact that Pt is much more active than Cu in dissociative adsorption of $H_2$.

Modeling and experiments by Liu et al. [32] suggest $N_2H_Y$ as an intermediate species and $N_2H_4$ as a product in their DFT models for DBD plasma, suggesting that this might boost the rate of ammonia formation. $N_2Hy$ has been detected in the plasma phase with electron-impact-molecular-beam mass spectrometry. In any case, the hydrazine formation is in the orders of magnitude smaller than ammonia formation, which is in good agreement with the experimental results.

Winter et al. [38] reported that $N_2H_Y$ species can be detected on Fe catalysts in a DBD plasma with FTIR (Fourier Transform infrared), whereas $NH_X$ species dominate on Ni. The formation of $N_2H_Y$ species is proposed to decrease the rate of ammonia formation. The role of charged species was discarded because these are expected to be less abundant than radicals.

In contrast, Lele et al. [39] suggest that surface charging breaks the traditional scaling relationships. The effect of surface charging on adsorption strengths of H, $H_2$, NH, $NH_2$, and $NH_3$ is different for the different metals studied with DFT calculations, namely Ni, Co, Ru, Pt, and Fe. As a result, the $NH_3$ linear scaling relations might not hold in plasma environment. Hence, surface charging could provide a new explanation for the enhancement of catalytic $NH_3$ synthesis by plasma. However, the authors mention that additional simulations would be required to quantify these effects properly. Also, electron densities vary over time in plasmas, and the effects of surface charging may be very different during micro-discharges and afterglows.

### 7.7.4 Other Mechanisms

Luo et al. [40] suggested that a Mars-Van Krevelen mechanism is dominant when exposing unsupported VN to an $N_2/H_2$ DBD plasma. Vacancies in VN were detected with electron paramagnetic resonance (EPR) and x-ray photoelectron spectroscopy (XPS), while the formation of ammonia is reported during exposure of VN to $H_2$ plasma.

Van Raak et al. [41] proposed that Ru catalysts supported on titania-ceria outperform ceria-supported catalyst because of the higher capacity of the support to the adsorbed H, offering an alternative pathway to ammonia formation via diffusion to Ru via spill-over.

## 7.8 Energy Efficiency

Under most plasma conditions, $N_2$ dissociation in the plasma seems dominant, rather than vibrational or electronic excitation of $N_2$ molecules, in agreement with the conclusions of Kang et al. [42] in a recent review. Molecular excitation seems relevant only in very mild plasma, resulting in very low conversion and low ammonia concentrations. $N_2$ dissociation in the plasma requires significant energy, such that the overall energy input for plasma-catalytic ammonia synthesis is typically high. As breaking the N—N bond requires 945 kJ $mol^{-1}$, the resulting theoretical minimum energy cost would be 131 g-$NH_3$ per kWh. Figure 7.6 presents the reported energy costs for plasma-catalytic ammonia synthesis, showing that DBD plasma is superior in terms of energy yield. The best reported energy yields for plasma-catalytic ammonia synthesis are in the range of 30 g-$NH_3$ $kWh^{-1}$ as reported by Kim et al. [44]. Obviously, this will never be competitive with the HB process on a large scale, compressing $H_2$ and $N_2$ to 100–450 bar to produce liquid ammonia at an energy cost of about 1,800 g-$NH_3$ per kWh (2 MJ/kg-$NH_3$) without considering heat integration [45]. Production of $H_2$ and $N_2$ is not included for both cases, i.e. plasma catalysis and HB. Note that the production of $H_2$ is highly energy intensive and the main cause of $CO_2$ emission (in the case of steam reforming) or power consumption (in the case of water electrolysis).

Upon scale-down of the HB process, e.g. in the case of local production with intermittent green electricity, the efficiency of the process decreases due to heat losses. This results in an energy cost of up to 76 g-$NH_3$ per kWh for a plant producing typically 100 kg-$NH_3$ per day [46]. The consequence of the apparent dominance of the route via N radicals is that it needs to be operated very close to the theoretical minimum energy cost (131 g-$NH_3$/kWh). That would imply that

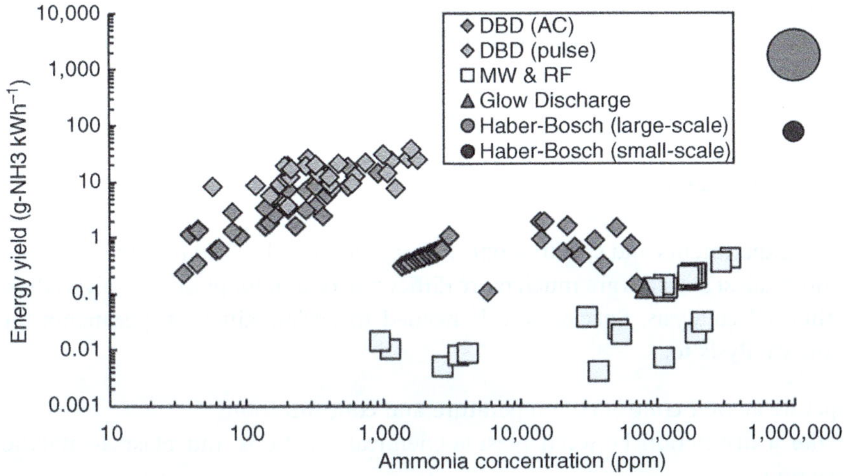


**Figure 7.6** Reported energy yield *vs.* ammonia concentration for DBD reactors (AC and pulse). Constructed from [21] and extended with entries for DBD results [10, 15, 43], as well as entries for ammonia synthesis with HB at large scale (typically 400 ktons/year) as well as small scale (10–50 tons/year).

non-productive reactions like radical recombination and, even more importantly, ammonia decomposition need to be almost completely prevented.

Even if that were achieved, a second main hurdle remains. The entries for the HB process in Figure 7.6 include the energy cost of the separation of ammonia from the synthesis loop with an ammonia concentration of typically 15%. That is the reason why both HB entries report production of pure ammonia. As can be seen in Figure 7.6, the concentration achieved with plasma catalysis is much lower, especially with DBD plasma. The energy cost of separation is not included in Figure 7.6 for all plasma entries.

On the one hand, low ammonia concentration is favorable because ammonia decomposition is suppressed, decreasing the energy cost. On the other hand, low ammonia concentration is highly unfavorable because the separation of ammonia and the cost of recycling of unconverted synthesis gas will be extremely high, both in terms of energy consumption and capital cost.

The most appealing approach to improve the energy yield, based on the mechanistic insight, would be optimizing the plasma conditions to achieve selective vibrational and/or electronic excitation of $N_2$, with minimal N-radical generation. Also, it is clear that the integration of ammonia synthesis and ammonia separation in a multifunctional reactor will be required to minimize ammonia decomposition by the plasma. This also implies the development of an integrated separation technology that can handle low concentration of ammonia. Catalysts active for thermal ammonia synthesis are not preferred as these will also be active for thermal ammonia decomposition.

## 7.9 Conclusions

Kinetic experiments and spectroscopic information of adsorbed intermediate species on catalyst surfaces are much more difficult to obtain for plasma catalysis than for thermal catalysis. Special care is needed to design kinetic experiments for plasma catalysis to:

- operate at well-controlled temperature and concentrations
- make a distinction between plasma-chemical reactions and plasma-catalytic reactions
- minimize the influence of the catalytically active phase on (local) plasma
- prevent that integral, high conversion data is interpreted as they were differential, low conversion data
- consider reverse reactions, via plasma chemistry, thermal catalysis, as well as plasma catalysis. This is always important but especially when pushing the conversion beyond thermodynamic equilibrium.
- prevent internal diffusion limitation, e.g. using dense support particles

Unfortunately, many studies have not considered most of these limitations, resulting in reporting performance data without achieving kinetic data, that would allow testing of reaction mechanisms. Also, detection of adsorbed species during plasma-catalytic experiments with spectroscopy techniques is rare. To the best of our knowledge, only Winter et al. [38] observed $N_2H_Y$ and $NH_X$ surface species with FTIR on Fe and Ni catalysts, respectively, in a DBD plasma. All other claims on adsorbed species rely on logical speculation and/or theoretical modeling.

In short, the industrial feasibility of plasma-catalytic ammonia synthesis is questionable, because of the too low energy efficiency and low ammonia concentration. In addition, ammonia synthesis is an exothermic reaction, implying that additional heat is generated by the reaction itself so that the amount of heat produced is further increased. This heat would be lost unless some kind of heat integration is possible, which depends highly on local heat requirements at the production site. Therefore, production of nitrate-based fertilizers via plasma-based $NO_X$ synthesis for nitric acid ($HNO_3$) production with green electricity is likely more attractive, as this reaction is both endothermic and endergonic.

Ammonia production via plasma catalysis as a hydrogen carrier would clearly need significant improvement of energy efficiency in order to compete with other options for storage and transportation of renewable energy.

## References


1 Lefferts, L. (2024). Leveraging expertise in thermal catalysis to understand plasma catalysis. *Angew. Chem. Int. Ed.* 63 (10): e202305322. https://doi.org/10.1002/anie.202305322.

2 Qu, Z., Zhou, R., Sun, J. et al. (2024). Plasma-assisted sustainable nitrogen-to-ammonia fixation: mixed-phase, synergistic processes and mechanisms. *ChemSusChem* 17 (6): e202300783. https://doi.org/10.1002/cssc.202300783.

3 Aika, K.-i. (2017). Role of alkali promoter in ammonia synthesis over ruthenium catalysts—effect on reaction mechanism. *Catal. Today* 286: 14–20. https://doi.org/10.1016/j.cattod.2016.08.012.

4 Goodwin, C.M., Lömker, P., Degerman, D. et al. (2024). Operando probing of the surface chemistry during the Haber–Bosch process. *Nature* 625 (7994): 282–286. https://doi.org/10.1038/s41586-023-06844-5.

5 Medford, A.J., Vojvodic, A., Hummelshøj, J.S. et al. (2015). From the Sabatier principle to a predictive theory of transition-metal heterogeneous catalysis. *J. Catal.* 328: 36–42. https://doi.org/10.1016/j.jcat.2014.12.033.

6 Ertl, G. (1980). Surface science and catalysis—studies on the mechanism of ammonia synthesis: the P. H. Emmett Award Address. *Catal. Rev.* 21 (2): 201–223. https://doi.org/10.1080/03602458008067533.

7 Berger, R.J., Pérez-Ramírez, J., Kapteijn, F., and Moulijn, J.A. (2002). Catalyst performance testing: bed dilution revisited. *Chem. Eng. Sci.* 57 (22): 4921–4932. https://doi.org/10.1016/S0009-2509(02)00273-7.

8 Rouwenhorst, K.H.R., Burbach, H.G.B., Vogel, D.W. et al. (2021). Plasma-catalytic ammonia synthesis beyond thermal equilibrium on Ru-based catalysts in non-thermal plasma. *Cat. Sci. Technol.* 11 (8): 2834–2843. https://doi.org/10.1039/D0CY02189J.

9 Mehta, P., Barboun, P.M., Engelmann, Y. et al. (2020). Plasma-catalytic ammonia synthesis beyond the equilibrium limit. *ACS Catal.* 10 (12): 6726–6734. https://doi.org/10.1021/acscatal.0c00684.

10 Patil, B.S., Cherkasov, N., Srinath, N.V. et al. (2021). The role of heterogeneous catalysts in the plasma-catalytic ammonia synthesis. *Catal. Today* 362: 2–10. https://doi.org/10.1016/j.cattod.2020.06.074.

11 Iwamoto, M., Akiyama, M., Aihara, K., and Deguchi, T. (2017). Ammonia synthesis on wool-like Au, Pt, Pd, Ag, or Cu electrode catalysts in nonthermal

atmospheric-pressure plasma of N2 and H2. *ACS Catal.* 7 (10): 6924–6929. https://doi.org/10.1021/acscatal.7b01624.

**12** Navascués, P., Obrero-Pérez, J.M., Cotrino, J. et al. (2020). Unraveling discharge and surface mechanisms in plasma-assisted ammonia reactions. *ACS Sustain. Chem. Eng.* 8 (39): 14855–14866. https://doi.org/10.1021/acssuschemeng.0c04461.

**13** Navascués, P., Garrido-García, J., Cotrino, J. et al. (2023). Incorporation of a metal catalyst for the ammonia synthesis in a ferroelectric packed-bed plasma reactor: does it really matter? *ACS Sustain. Chem. Eng.* 11 (9): 3621–3632. https://doi.org/10.1021/acssuschemeng.2c05877.

**14** Rouwenhorst, K.H.R., Mani, S., and Lefferts, L. (2022). Improving the energy yield of plasma-based ammonia synthesis with in situ adsorption. *ACS Sustain. Chem. Eng.* 10 (6): 1994–2000. https://doi.org/10.1021/acssuschemeng.1c08467.

**15** Wang, Y., Yang, W., Xu, S. et al. (2022). Shielding protection by mesoporous catalysts for improving plasma-catalytic ambient ammonia synthesis. *J. Am. Chem. Soc.* 144 (27): 12020–12031. https://doi.org/10.1021/jacs.2c01950.

**16** Gorky, F., Carreon, M.A., and Carreon, M.L. (2020). Experimental strategies to increase ammonia yield in plasma catalysis over LTA and BEA zeolites. *IOP SciNotes* 1 (2): 024801. https://doi.org/10.1088/2633-1357/aba1f8.

**17** Shah, J.R., Gorky, F., Lucero, J. et al. (2020). Ammonia synthesis via atmospheric plasma catalysis: zeolite 5A, a case of study. *Ind. Eng. Chem. Res.* 59 (11): 5167–5176. https://doi.org/10.1021/acs.iecr.9b05220.

**18** Peng, P., Chen, P., Addy, M. et al. (2019). Atmospheric plasma-assisted ammonia synthesis enhanced via synergistic catalytic absorption. *ACS Sustain. Chem. Eng.* 7 (1): 100–104. https://doi.org/10.1021/acssuschemeng.8b03887.

**19** Gorky, F., Nguyen, H.M., Lucero, J.M. et al. (2022). CC3 porous organic cage crystals and membranes for the non-thermal plasma catalytic ammonia synthesis. *Chem. Eng. J. Adv.* 11: 100340. https://doi.org/10.1016/j.ceja.2022.100340.

**20** Ndayirinde, C., Gorbanev, Y., Ciocarlan, R.-G. et al. (2023). Plasma-catalytic ammonia synthesis: packed catalysts act as plasma modifiers. *Catal. Today* 419: 114156. https://doi.org/10.1016/j.cattod.2023.114156.

**21** Rouwenhorst, K.H.R., Engelmann, Y., Van' T Veer, K. et al. (2020). Plasma-driven catalysis: green ammonia synthesis with intermittent electricity. *Green Chem.* 22 (19): 6258–6287. https://doi.org/10.1039/d0gc02058c.

**22** Wang, Y., Craven, M., Yu, X. et al. (2019). Plasma-enhanced catalytic synthesis of ammonia over a $Ni/Al_2O_3$ catalyst at near-room temperature: insights into the importance of the catalyst surface on the reaction mechanism. *ACS Catal.* 9 (12): 10780–10793. https://doi.org/10.1021/acscatal.9b02538.

**23** Herrera, F.A., Brown, G.H., Barboun, P. et al. (2019). The impact of transition metal catalysts on macroscopic dielectric barrier discharge (DBD) characteristics in an ammonia synthesis plasma catalysis reactor. *J. Phys. D. Appl. Phys.* 52 (22): 224002. https://doi.org/10.1088/1361-6463/ab0c58.

**24** Barboun, P., Mehta, P., Herrera, F.A. et al. (2019). Distinguishing plasma contributions to catalyst performance in plasma-assisted ammonia synthesis. *ACS Sustain. Chem. Eng.* 7 (9): 8621–8630. https://doi.org/10.1021/acssuschemeng.9b00406.

**25** Rouwenhorstm, K.H.R. and Lefferts, L. (2023). Plasma-catalytic ammonia synthesis via Eley-Rideal reactions: a kinetic analysis. *ChemCatChem* 15 (12): e202300078. https://doi.org/10.1002/cctc.202300078.

**26** Mehta, P., Barboun, P., Herrera, F.A. et al. (2018). Overcoming ammonia synthesis scaling relations with plasma-enabled catalysis. *Nature Catalys* 1 (4): 269–275. https://doi.org/10.1038/s41929-018-0045-1.

**27** Rouwenhorst, K.H.R., Kim, H.H., and Lefferts, L. (2019). Vibrationally excited activation of N2 in plasma-enhanced catalytic ammonia synthesis: a kinetic analysis. *ACS Sustain. Chem. Eng.* 7 (20): 17515–17522. https://doi.org/10.1021/acssuschemeng.9b04997.

**28** Rouwenhorst, K.H.R. and Lefferts, L. (2021). On the mechanism for the plasma-activated N2 dissociation on Ru surfaces. *J. Phys. D. Appl. Phys.* 54 (39): 393002. https://doi.org/10.1088/1361-6463/ac1226.

**29** Engelmann, Y., van' t Veer, K., Gorbanev, Y. et al. (2021). Plasma catalysis for ammonia synthesis: a microkinetic modeling study on the contributions of Eley–Rideal reactions. *ACS Sustain. Chem. Eng.* 9 (39): 13151–13163. https://doi.org/10.1021/acssuschemeng.1c02713.

**30** Gorbanev, Y., Engelmann, Y., van' t Veer, K. et al. (2021). Al2O3-supported transition metals for plasma-catalytic $NH_3$ synthesis in a DBD plasma: metal activity and insights into mechanisms. *Catalysts* 11 (10): 1230.

**31** Shah, J., Gorky, F., Psarras, P. et al. (2020). Enhancement of the yield of ammonia by hydrogen-sink effect during plasma catalysis. *ChemCatChem* 12 (4): 1200–1211. https://doi.org/10.1002/cctc.201901769.

**32** Liu, T.W., Gorky, F., Carreon, M.L., and Gomez-Gualdron, D.A. (2023). Plasma radicals as kinetics-controlling species during plasma-assisted catalytic $NH_3$ formation: support from microkinetic modeling. *ACS Sustain. Chem. Eng.* 11 (47): 16749–16763. https://doi.org/10.1021/acssuschemeng.3c03813.

**33** Hong, J., Pancheshnyi, S., Tam, E. et al. (2017). Kinetic modelling of $NH_3$ production in $N_2$–$H_2$ non-equilibrium atmospheric-pressure plasma catalysis. *J. Phys. D. Appl. Phys.* 50 (15): 154005. https://doi.org/10.1088/1361-6463/aa6229.

**34** Shah, J., Wang, W., Bogaerts, A., and Carreon, M.L. (2018). Ammonia synthesis by radio frequency plasma catalysis: revealing the underlying mechanisms. *ACS Appl Energy Mater* 1 (9): 4824–4839. https://doi.org/10.1021/acsaem.8b00898.

**35** Sun, J., Chen, Q., Zhao, X. et al. (2022). Kinetic investigation of plasma catalytic synthesis of ammonia: insights into the role of excited states and plasma-enhanced surface chemistry. *Plasma Sources Sci. Technol.* 31 (9): 094009. https://doi.org/10.1088/1361-6595/ac8e2c.

**36** Zhao, H., Song, G., Chen, Z. et al. (2022). In situ identification of NNH and $N_2H_2$ by using molecular-beam mass spectrometry in plasma-assisted catalysis for $NH_3$ synthesis. *ACS Energy Lett.* 7 (1): 53–58. https://doi.org/10.1021/acsenergylett.1c02207.

**37** Yamijala, S.S.R.K.C., Nava, G., Ali, Z.A. et al. (2020). Harnessing plasma environments for ammonia catalysis: mechanistic insights from experiments and large-scale ab initio molecular dynamics. *J. Phys. Chem. Lett.* 11 (24): 10469–10475. https://doi.org/10.1021/acs.jpclett.0c03021.

**38** Winter, L.R., Ashford, B., Hong, J. et al. (2020). Identifying surface reaction intermediates in plasma catalytic ammonia synthesis. *ACS Catal.* 10 (24): 14763–14774. https://doi.org/10.1021/acscatal.0c03166.

**39** Lele, A.D., Xu, Y., and Ju, Y. (2024). Modelling the effect of surface charging on plasma synthesis of ammonia using DFT. *Phys. Chem. Chem. Phys.* 26 (12): 9453–9461. https://doi.org/10.1039/D3CP06050K.

**40** Luo, S., Liu, Y., Song, Y. et al. (2024). Plasma-induced nitrogen vacancy-mediated ammonia synthesis over a VN catalyst. *Chem. Commun.* 60 (24): 3295–3298. https://doi.org/10.1039/d4cc00042k.

**41** van Raak, T.A.B.J., Li, S., and Gallucci, F. (2023). Prevailing surface reactions in the plasma-catalytic ammonia synthesis with Ru/$CeO_2$ and Ru/Ti-$CeO_2$. *Chem. Eng. J.* 455: 140691. https://doi.org/10.1016/j.cej.2022.140691.

**42** Lim, K.H., Yue, Y., Bella, X. et al. (2023). Sustainable hydrogen and ammonia technologies with nonthermal plasma catalysis: mechanistic insights and technoeconomic analysis. *ACS Sustain. Chem. Eng.* 11 (13): 4903–4933. https://doi.org/10.1021/acssuschemeng.2c06515.

**43** Ma, Y., Tian, Y., Zeng, Y., and Tu, X. (2021). Plasma synthesis of ammonia in a tangled wire dielectric barrier discharge reactor: effect of electrode materials. *J. Energy Inst.* 99: 137–144. https://doi.org/10.1016/j.joei.2021.09.002.

**44** Kim, H.-H., Teramoto, Y., Ogata, A. et al. (2017). Atmospheric-pressure nonthermal plasma synthesis of ammonia over ruthenium catalysts. *Plasma Process. Polym.* 14 (6): 1600157. https://doi.org/10.1002/ppap.201600157.

**45** Jennings, J.R. (1991). *Catalytic Ammonia Synthesis*. New York, NY: Springer.

**46** Rouwenhorst, K.H. and Lefferts, L. (2020). Feasibility study of plasma-catalytic ammonia synthesis for energy storage applications. *Catalysts* 10 (9): 1–19.

# 8

# Mechanisms of Plasma-Driven $NO_x$ Synthesis

*Weitao Wang and Xin Tu*

*Department of Electrical Engineering and Electronics, University of Liverpool, Liverpool, UK*

## 8.1 Introduction

The synthesis of nitrogen oxides ($NO_x$) through plasma-driven processes has attracted significant attention as a sustainable alternative to traditional nitrogen fixation methods. The inherent flexibility of plasma technology makes it promising for mitigating the environmental and energy challenges associated with conventional processes like the Haber-Bosch method [1]. In particular, plasma has demonstrated the potential to drive $NO_x$ synthesis under ambient conditions, owing to their unique ability to produce highly reactive species within a nonequilibrium environment. This chapter delves into the mechanisms that govern plasma-driven $NO_x$ synthesis, offering insights into the key processes, intermediates, and factors influencing efficiency and yield. Nonthermal plasma (NTP), such as dielectric barrier discharge (DBD) and corona discharge, as well as warm plasmas, such as microwave (MW) discharge and gliding arc (GA), generate electrons that excite, ionize, and dissociate molecular species. This selective energy input enables efficient activation of nitrogen ($N_2$) and oxygen ($O_2$), essential precursors for $NO_x$ formation.

Central to the plasma-driven $NO_x$ synthesis mechanism is the Zeldovich mechanism [2], which involves the interaction of nitrogen and oxygen molecules with dissociated atomic species. Vibrational excitation, facilitated by electron collisions, reduces the energy barrier for bond dissociation, enabling the formation of reactive intermediates. These intermediates drive the synthesis of nitric oxide (NO) and nitrogen dioxide ($NO_2$) through a series of reactions [3]. However, the efficiency

*Plasma-Assisted Nitrogen Fixation for Sustainable Process Industries*, First Edition. Edited by Volker Hessel, Annemie Bogaerts, Gabriele Centi, Evgeny Rebrov, and Nguyen Van Duc Long.

Published 2026 by John Wiley & Sons, Inc.

of these processes depends on various factors, including plasma type, feed gas composition, pressure, and gas temperature. Indeed, the Zeldovich mechanism is not always enhanced by vibrational excitation; it often also occurs with ground-state molecules (thermal Zeldovich mechanism). Understanding the interaction of these factors is crucial for optimizing $NO_x$ yields and minimizing energy consumption.

In addition to plasma-only conditions, integrating catalysts within plasma systems has emerged as a strategy to tune reaction pathways and reduce energy consumption. Catalysts can modulate discharge characteristics and introduce reactive sites, possibly creating synergistic effects that improve $NO_x$ production [4, 5]. These plasma–catalyst interactions underscore the importance of integrated system design in achieving sustainable $NO_x$ synthesis.

This chapter provides a comprehensive analysis of the mechanisms underlying plasma-driven $NO_x$ synthesis, drawing on insights from chemical kinetic modeling, experimental studies, and advanced diagnostics. It begins with a focus on $NO_x$ synthesis without catalysts, highlighting the role of plasma properties, critical intermediates, and reaction dynamics. Subsequently, the possibility of synergistic effects of plasma–catalytic systems is explored, emphasizing their potential to advance nitrogen fixation processes. By unraveling these mechanisms, this work has the potential to guide the development of next-generation plasma reactors and operational strategies, paving the way for sustainable and energy-efficient $NO_x$ production.

## 8.2 $NO_x$ Synthesis Without a Catalyst

### 8.2.1 Plasma Physics Relevant to $NO_x$ Formation

To date, various plasma-based methods and reactors have been employed for $NO_x$ production. The results exhibit differences in energy efficiency and $NO_x$ concentration across different plasma types. These differences can largely be attributed to variations in plasma properties. Common plasma reactors used for $NO_x$ production include DBD, corona discharge, plasma jets, and GA. Understanding these plasma properties is crucial for elucidating the mechanisms underlying $NO_x$ generation across different plasma systems.

The classification presented in Figure 8.1 provides an overview of plasma types, ranging from NTP to thermal plasmas, each with unique properties and application potential [7]. All thermal, warm, and NTPs have been explored for $NO_x$ production. One of the earliest attempts, the "Birkeland-Eyde" (B-E) process, utilized thermal plasma generated by an electric arc for industrial nitrogen fixation [8, 9]. However, thermal plasmas typically suffer from limited energy efficiency, as a significant portion of the energy is expended on heating the gas. Studies have shown

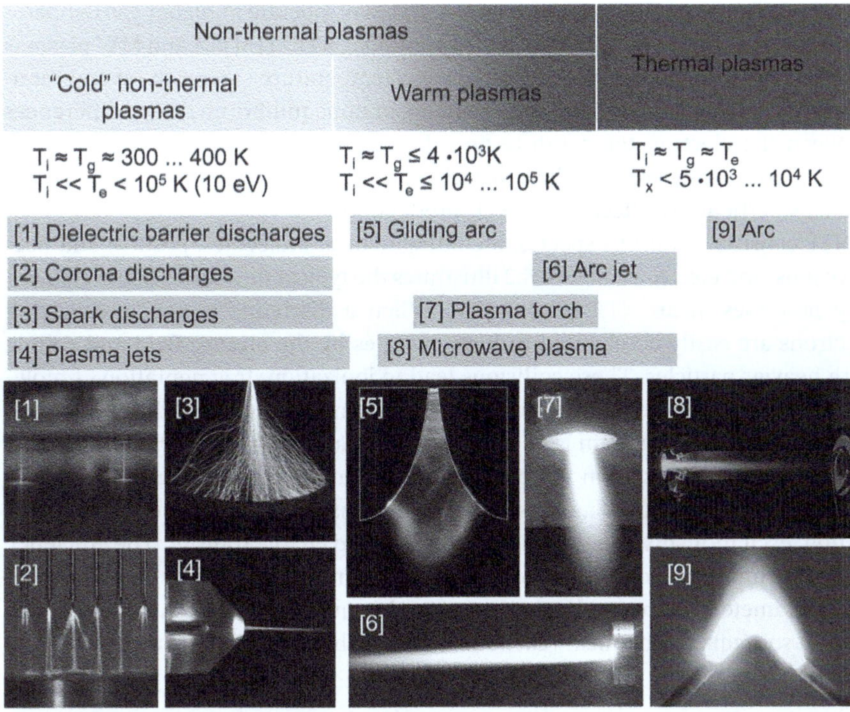


**Figure 8.1** Classification of atmospheric pressure plasma techniques based on thermal regimes. *Source:* Ref. [6] / Springer Nature.

that at temperatures exceeding 1,200 K, thermal dissociation of NO occurs [10], leading to reduced $NO_x$ concentrations.

In contrast, warm plasmas have demonstrated better performance on $NO_x$ production, both theoretically [3] and experimentally [11], due to the combination of thermal effects and nonequilibrium nature between electrons and heavy particles, although the chemistry is mainly thermal (vibrational–translational equilibrium) [12]. This characteristic allows for targeted energy transfer to electrons rather than bulk gas heating, making such plasmas a more viable option for efficient $NO_x$ production.

NTP is characterized by high electron temperatures ($T_e$) and low gas temperatures ($T_g$). The gas temperature of certain NTPs can be as low as room temperature and, therefore, are referred to as "cold" plasmas. Examples include DBD, corona discharges, spark discharges, and plasma jets, as shown in Figure 8.1. Among these, DBD is particularly characterized by its low gas temperature, which facilitates the efficient generation of ozone. The operational mode of DBD is highly adaptable, highlighting the flexibility of cold plasmas for various applications, particularly where control over chemical pathways is required.

In contrast, warm plasmas, such as GA and MW, operate at higher gas temperatures and are, therefore, regarded as "translational NTP" [13] GA and MW plasmas exhibit a transitional behavior where the gas temperature can reach several thousand Kelvin [14, 15], yet the plasma remains in nonequilibrium due to differences between the electron temperature and the gas temperature (although in some warm plasmas they are very close to each other). These characteristics make these plasmas particularly effective for $NO_x$ production.

NTP consists of various species, including neutral molecules, positive and negative ions, and electrons. Figure 8.2 illustrates the typical timescales of the elementary processes in an NTP system [16]. When a discharge occurs, pre-existing electrons are easily accelerated to high energies by the electric field and collide with heavier particles. These collisions lead to ionization, fragmentation of molecules, and the excitation (electronic, vibrational, and rotational) of the neutral gas. The elementary processes in NTP can be broadly categorized into primary and secondary processes, based on the timescale of streamer propagation. The primary processes, which occur on a timescale of approximately $10^{-8}$ seconds, include ionization, excitation, dissociation, light emission, and charge transfer. The efficiency of these primary processes depends significantly on the energization methods and their parameters, such as voltage rise time and frequency. The secondary processes involve subsequent chemical reactions between the products of the primary processes (radicals, ions, and excited molecules).

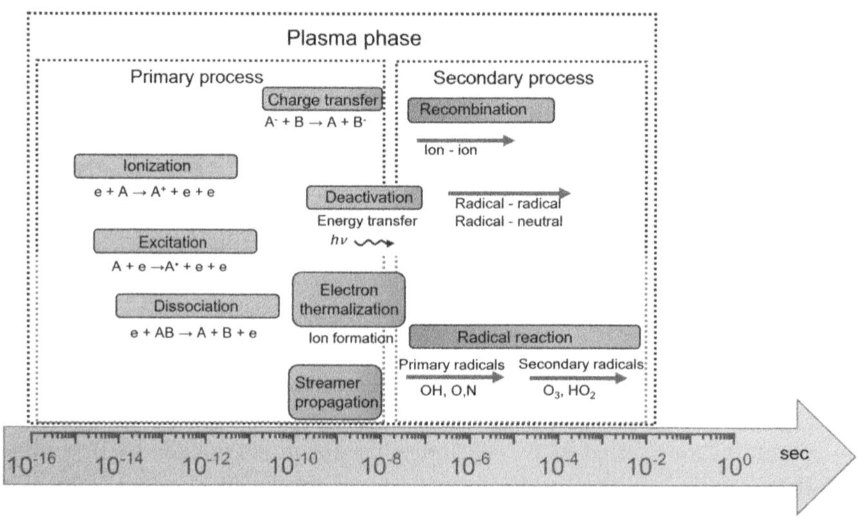


**Figure 8.2** Timescale of different reaction types within a plasma regime. *Source:* Ref. [16] / John Wiley & Sons.

Electron-impact reactions with $N_2$ and $O_2$ molecules drive the formation of $NO_x$ in NTP. Figure 8.3 illustrates how electron energy is transferred to different channels (excitation, ionization, and dissociation) of both $N_2$ and $O_2$ molecules in $N_2/O_2$ (80/20) gas mixtures at atmospheric pressure, as a function of the reduced electric field ($E$/N, where $E$ is the electric field and N is the concentration of neutral particles) in the discharge. The reduced electric field is a key parameter for distinguishing different types of plasma, as it determines the average electron energy and, consequently, the rate of various electron-impact reactions. The electron energy values corresponding to the reduced electric field are also shown in Figure 8.3. MW and GA typically operate at reduced electric field between 5 and 50 Td [17–20], while a DBD, commonly used in gas conversion applications, typically operates at values above 100 Td [21, 22]. As shown in Figure 8.3, above approximately 100 Td, most of the electron energy is directed toward electronic excitation, dissociation, and ionization of $N_2$ (and $O_2$) molecules. The N atoms produced by direct electron-impact dissociation of $N_2$ molecules can then react with $O_2$ molecules to form NO.

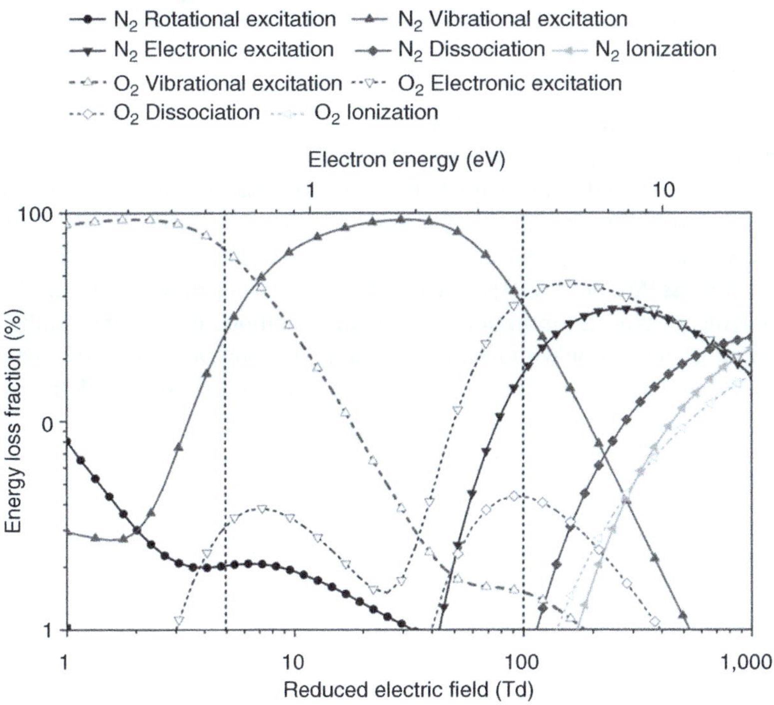


**Figure 8.3** Energy loss fraction of different channels of $N_2$ and $O_2$ as a function of $E$/N. The results are calculated under $N_2/O_2$ = 80/20 at atmospheric pressure. *Source:* Ref. [3] / John Wiley & Sons.

### 8.2.2 Plasma Chemistry and Key Reaction Mechanisms

#### 8.2.2.1 Electrons Induced Reactions

The electron-impact dissociation of $N_2$ and $O_2$ molecules, driven by the electric field, is the initial step in $NO_x$ production within plasma systems, as confirmed by several studies [16, 17]. For instance, a modeling study demonstrated that in a GA containing 80% $N_2$ and 20% $O_2$, the densities of both N and O atoms peak within 0.2 ms before decreasing to steady-state levels. Notably, the density of O atoms is 2–3 orders of magnitude higher than that of N atoms due to the significantly lower bond dissociation energy of $O_2$ (5.15 eV) compared to $N_2$ (9.79 eV) [23], making $O_2$ much easier to dissociate.

The direct dissociation for $N_2$ molecules is

$$e^- + N_2 \rightarrow N_2^* + e^- \rightarrow e^- + N\cdot + N\cdot \tag{8.1}$$

$$e^- + O_2 \rightarrow O_2^* + e^- \rightarrow e^- + O\cdot + O\cdot \tag{8.2}$$

The above two reactions require high-energy threshold; therefore, $NO_x$ production through direct electron-impact dissociation is energy demanding.

The possible alternatives will be electron-vibrational excitation as suggested:

$$e^- + N_2 \rightarrow e^- + N_2\,(v) \tag{8.3}$$

The electron-vibrational excitation pathway plays a significant role in warm plasmas, where the electron temperatures (~1 eV) closely match the threshold energies for vibrational excitation. Under these conditions, electron energy is efficiently transferred through vibrational excitation channels, facilitating further molecular dissociation. The presence of metastable species further enhances the dissociation process. Studies [3, 24] have observed a stepwise dissociation mechanism involving electronic excitation sequences. Additionally, the metastable states of gas molecules, along with nitrogen and oxygen atoms, actively participate in chemical reactions, forming NO or $NO_2$. This process is discussed in detail in Section 8.2.2.2.

Nitrogen vibrational-state species act as energy reservoirs in plasma, playing a key role in many heavy-particle reactions. Vibrational translational (V–T) relaxation processes, which contribute to gas heating, significantly impact the gas temperature and the overall reaction dynamics [25]. Vibrationally excited $N_2$ molecules can either collide directly with free electrons, driving further excitation or de-excitation processes, or interact with other neutral molecules, facilitating energy redistribution [26].

#### 8.2.2.2 Formation and Loss Processes of NO and $NO_2$

Understanding the processes involved in $NO_x$ formation and loss is essential for understanding $NO_x$ production in NTP systems. Table 8.1 summarizes the possible

**Table 8.1** Overview of the possible formation and loss reactions for NO and $NO_2$.

| NO formation processes | | NO loss processes | |
|---|---|---|---|
| **F1** | $NO_2 + M \rightarrow NO + O\cdot + M$ | **L1** | $NO + O\cdot\ (+M) \rightarrow NO_2\ (+M)$ |
| **F2** | $O\cdot + NO_2 \rightarrow NO + O_2$ | **L2** | $NO + O_2(v) \rightarrow O\cdot + NO_2$ |
| **F3** | $O\cdot + N_2(g) \rightarrow NO + N\cdot$ | **L3** | $NO + O_2(g) \rightarrow O\cdot + NO_2$ |
| **F4** | $O\cdot + N_2(v) \rightarrow NO + N\cdot$ | **L4** | $O\cdot + NO \rightarrow N\cdot + O_2$ |
| **F5** | $N\cdot + O_2(g) \rightarrow NO + O\cdot$ | **L5** | $N\cdot + NO \rightarrow O\cdot + N_2$ |
| **F6** | $N\cdot + O_2(v) \rightarrow NO + O\cdot$ | **L6** | $NO + NO + O_2(g) \rightarrow NO_2 + NO_2^*$ |
| **F7** | $N_2O_3 + M \rightarrow NO + NO_2 + M$ | **L7** | $NO + NO + O_2(v) \rightarrow NO_2 + NO_2^*$ |
| **F8** | $NO_2 + NO_2 \rightarrow NO_3 + NO$ | **L8** | $NO_3 + NO \rightarrow NO_2 + NO_2$ |
| **F9** | $NO_2 + NO_2 \rightarrow NO + NO + O_2^*$ | **L9** | $NO + NO_2 + M \rightarrow N_2O_3 + M$ |
| **F10** | $N\cdot + NO_2 \rightarrow NO + NO$ | **L10** | $NO + NO_2 \rightarrow NO^* + NO_2$ |
| **$NO_2$ formation processes** | | **$NO_2$ loss processes** | |
| **f1 = L1** | $NO + O\cdot\ (+M) \rightarrow NO_2\ (+M)$ | **l1 = F1** | $NO_2 + M \rightarrow NO + O\cdot + M$ |
| **f2 = L2** | $NO + O_2(v) \rightarrow NO_2 + O\cdot$ | **l2 = F2** | $O\cdot + NO_2 \rightarrow NO + O_2$ |
| **f3 = L3** | $NO + O_2(g) \rightarrow NO_2 + O\cdot$ | **l3 = F8** | $NO_2 + NO_2 \rightarrow NO_3 + NO$ |
| **f4 = L8** | $NO + NO_3 \rightarrow NO_2 + NO_2$ | **l4 = F9** | $NO_2 + NO_2 \rightarrow NO + NO + O_2^*$ |
| **f5 = L6** | $NO + NO + O_2(g) \rightarrow NO_2 + NO_2^*$ | **l5** | $NO_2 + NO_2 + M \rightarrow N_2O_4 + M$ |
| **f6 = L7** | $NO + NO + O_2(v) \rightarrow NO_2 + NO_2^*$ | **l6 = L9** | $NO + NO_2 + M \rightarrow N_2O_3 + M$ |
| **f7** | $N_2O_4 + M \rightarrow NO_2 + NO_2 + M$ | **l7 = F10** | $N\cdot + NO_2 \rightarrow NO + NO$ |
| **f8 = F7** | $N_2O_3 + M \rightarrow NO + NO_2 + M$ | **l8** | $N\cdot + NO_2 \rightarrow O\cdot + N_2O$ |

*Source:* [27] / American Chemical Society.

reactions responsible for the formation and loss of NO and $NO_2$. The time- and space-averaged reaction rates for these processes depend on the type of plasma used and the operating conditions, such as pressure, flow rate, gas composition, power supply, and discharge power. Reactions **F1** (**f1**) and **L1** (**l1**), as well as **F2** (**f2**) and **L2** (**l2**), play a major role in both the formation and loss of NO and $NO_2$. However, these reactions facilitate the interconversion between NO and $NO_2$, rather than contributing to their net production from $N_2$ and $O_2$. For instance, a modeling study revealed that over 90% of NO (and $NO_2$) formation and loss is attributed to **F1**(**f1**) and **L1**(**l1**), as illustrated in Figure 8.4 [27]. This indicates that the main formation pathways for NO serve as the primary loss pathways for $NO_2$, and vice versa.

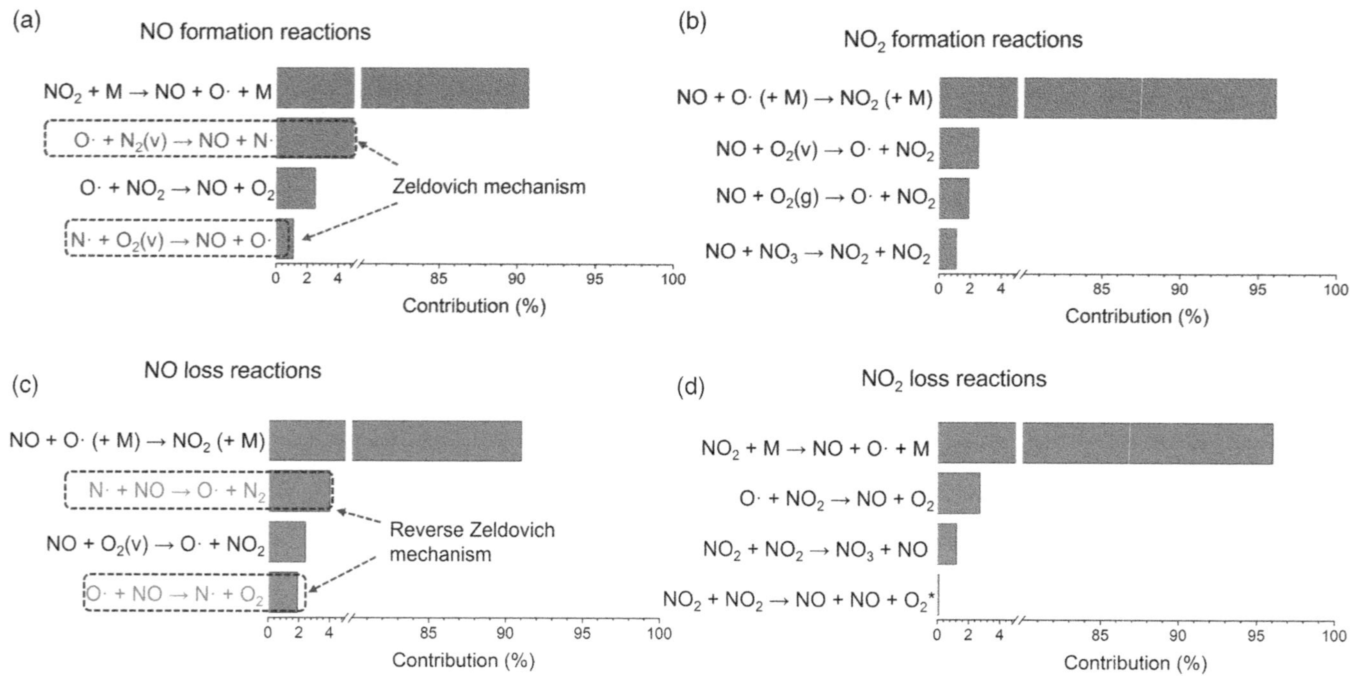


**Figure 8.4** Main reactions for the formation and loss of NO and $NO_2$ and their relative contributions at an 80/20 $N_2/O_2$ ratio in a GA reactor. (a) NO formation reaction, (b) $NO_2$ formation reaction, (c) NO loss reaction and (d) $NO_2$ loss reaction. *Source:* Reproduced using data from ref. [27] with permission from American Chemical Society.

Another modeling study [3] showed that reactions involving $N_2O_4$ (**f7** and **l5**) dominate the formation and loss mechanisms of $NO_2$ at $N_2/O_2$ feed ratios ranging from 0.2 to 10. However, the absolute reaction rates for these pathways are nearly balanced, resulting in a negligible net contribution to $NO_2$ formation. Similarly, the reactions involving $N_2O_3$ (**f8** and **l6**), collectively referred to as **N6** in Table 8.2, also have an insignificant net effect on $NO_2$ production.

To gain a deeper understanding of the overall reaction mechanisms in $NO_x$ chemistry, it is essential to examine the net reaction rates (i.e. the forward reaction rate minus the reverse reaction rate). Table 8.2 summarizes the main net formation and loss processes of NO and $NO_2$ in plasma-enabled $NO_x$ production.

Vervloessem et al. [11] summarized the reaction pathways (Figure 8.5) for NO and $NO_2$ formation and loss in an 80/20 $N_2/O_2$ feed gas mixture in a GA reactor. Their modeling study revealed that the primary mechanism for NO formation involves the combination of vibrationally excited $N_2$ and $O_2$ molecules with N· and O· atoms generated by electron impact-induced dissociation, although recent modeling insights from the same group, with improved rate coefficient data, reveal that these reactions mainly occur with ground-state molecules. The reaction NO + O· → $NO_2$ was identified as the main consumption pathway for NO and simultaneously the main formation route for $NO_2$.

These findings align with those of Zhang et al. [28], who used a chemical kinetics model to identify the key intermediate species in $NO_x$ reaction pathways under spark discharge conditions. Their results show that, while N and O atoms are initially produced via electron impact-induced dissociation, their generation is primarily sustained by vibrationally excited $N_2$ and $O_2$ molecules. Notably, unlike GA and spark discharges, in DBD plasma, the low temperature allows $O_2$ to react with atomic oxygen to form ozone ($O_3$). It is believed that ozone formation is the

**Table 8.2** Main net NO and $NO_2$ formation or loss processes.

| Main net NO and $NO_2$ formation or loss processes | | |
|---|---|---|
| **N1** | = F2 − (L2 + L3) | $NO_2$ + O· → NO + $O_2$ |
| **N2** | = (F3 + F4) − L5 | O· + $N_2$(g/v) → NO + N· |
| **N3** | = (F5 + F6) − L4 | N· + $O_2$(g/v) → NO + O· |
| **N4** | = L1 − F1 | NO + O· (+ M) → $NO_2$ (+ M) |
| **N5** | = l3 − f4 | $NO_2$ + $NO_2$ → NO + $NO_3$ |
| **N6** | = L9 − F7 | NO + $NO_2$ + M → $N_2O_3$ + M |
| **N7** | = l5 − f7 | $NO_2$ + $NO_2$ + M → $N_2O_4$ + M |

*Source:* [27] / American Chemical Society.

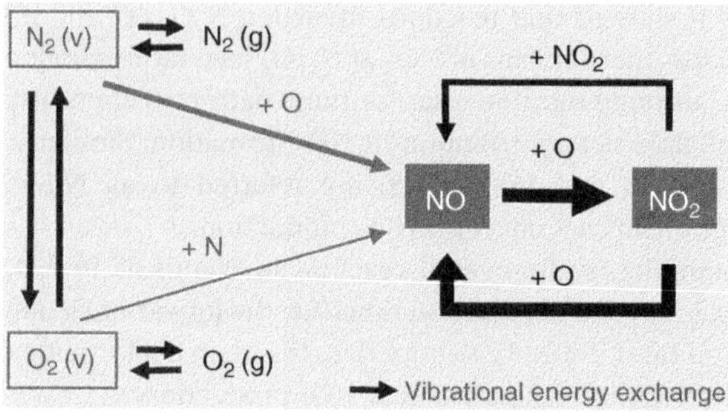


**Figure 8.5** Reaction scheme of the main reaction paths for $NO_x$ synthesis in a GA reactor for an 80/20 $N_2/O_2$ mixture. The thickness of the arrow lines corresponds to the importance of the net reactions. *Source:* Ref. [11] / American Chemical Society.

most significant competing process against NO generation in air plasmas [29]. The stability of ozone strongly depends on temperature, even in the absence of other reactive species. At higher temperatures, more NO is produced due to the interactions between nitrogen species and ozone [30].

In addition to the $NO_x$ loss pathways discussed earlier, vibrational kinetics can also limit $NO_x$ production. Vibrational-induced dissociation is the most energy-efficient pathway, as low-energy electrons first populate the lowest vibrational levels of molecules through electron impact. These levels are then progressively filled through vibrational–vibrational (V–V) exchanges.

However, as illustrated in Figure 8.5, V–V exchanges occur not only among $N_2$ molecules and $O_2$ molecules individually but also during collisions between $N_2$ and $O_2$ [11]. Since the vibrational levels of $O_2$ are more easily depopulated, these exchanges lead to a net loss of $N_2$ vibrational energy and an increase in $O_2$ vibrational energy. A modeling study showed that excluding $N_2$-$O_2$ V–V exchanges in the model significantly enhanced the calculated $NO_x$ yield. However, minimizing these exchanges is challenging. One potential approach is to preheat the feed gas, which could facilitate the thermal dissociation of $O_2$ and boost its vibrational population. While promising, further research is needed to evaluate its practicality and effectiveness.

#### 8.2.2.3 The Key Role of the Zeldovich Mechanism

As discussed earlier, vibrationally excited- or ground-state $N_2$ molecules are the primary source of NO production in both spark and GA discharges [3, 27, 28]. Zhang et al. [28] employed a chemical kinetics model to investigate the $NO_x$ formation mechanism in pulsed spark discharges, demonstrating that more than half of the NO is produced from vibrationally excited $N_2$ molecules. These vibrational states of $N_2$ and $O_2$ are generated through electronic excitation and further elevated to higher vibrational levels via rapid V–V relaxation processes.

Another modeling study [27] revealed that in GA, over 98% of NO is formed through reactions involving vibrationally excited $N_2$ molecules. In this process, the efficient dissociation of the N≡N bond occurs as follows:

$$O\cdot + N_2(v) \rightarrow NO + N\cdot \tag{8.4}$$

This is followed by

$$N\cdot + O_2(g, v) \rightarrow NO + O\cdot \tag{8.5}$$

O· atoms can react with vibrationally excited $N_2$ molecules to complete the cycle described in reaction (8.4), where both N and O atoms are produced and consumed. These reactions (8.4) and (8.5) are collectively referred to as the Zeldovich mechanism and represent the most energy-efficient pathway for $NO_x$ production, also when they proceed with ground-state $N_2$ molecules. The mechanism is driven by N and O atoms, which are initially formed through dissociation processes. Among these reactions, (8.4) is considered the rate-limiting step. Further oxidation of NO to $NO_2$ follows with a strong correlation with oxygen concentration and temperature.

Minimizing the reverse Zeldovich mechanism is critical for improving energy efficiency in $NO_x$ production. Vervloessem et al. [11] conducted calculations to evaluate the impact of various $NO_x$ loss reactions in a GA. Their findings indicate that the primary contributor to NO loss is the reaction of NO with either O atoms or $O_2$ molecules to form $NO_2$. Conversely, $NO_2$ loss reactions ultimately generate NO, but they do not constitute a net loss of $NO_x$. Notably, reactions involving O or N atoms reacting with NO to form $N + O_2$ or $O + N_2$, collectively known as the reverse Zeldovich mechanism, are significant contributors to NO loss, despite occurring at a lower rate. These reverse reactions compete with the NO production pathway and effectively terminate the reaction chain. Model simulations suggest that excluding these reverse reactions from the chemical kinetics set could boost $NO_x$ yield by a factor of 10. As a result, reducing the reverse Zeldovich reactions is a promising strategy for enhancing NO production efficiency. This can be achieved through rapid quenching of the plasma and *in situ* removal of NO.

### 8.2.3 Factors Influencing Reaction Pathways

#### 8.2.3.1 Impact of Plasma Types

The energy efficiency of $NO_x$ formation is primarily determined by the method used to break the strong (~10 eV) bond of the $N_2$ molecule. In thermal plasmas, where heavy species (gas molecules, radicals, etc.) and electrons are in thermal equilibrium, a significant portion of the input energy is lost to gas heating. This results in high gas temperatures, increasing energy consumption and hindering effective gas conversion by decomposing the formed products.

In "warm" or quasi-thermal plasmas, such as GA or MW plasmas, the electron temperature (~1 eV) exceeds that of the heavy species, though the latter can still reach several thousand Kelvin. This moderate electron temperature is ideal for efficiently exciting $N_2$ molecules vibrationally, which is the most energy-efficient pathway for $N_2$ dissociation and $NO_x$ formation via the vibrationally enhanced Zeldovich mechanism. However, the elevated gas temperatures in these plasmas promote V–T relaxation, where vibrational energy dissipates into heat through collisions with ground-state molecules, increasing gas temperature, so that the thermal Zeldovich mechanism is also efficient.

In contrast, "cold" plasmas, such as DBD plasmas, maintain near-room-temperature gas conditions while achieving electron temperatures of around 10 eV. This feature directs energy toward gas conversion rather than heating. However, the electron temperatures in cold plasmas are often too high for efficient vibrational excitation of $N_2$ molecules. As shown in Figure 8.3, at reduced electric fields exceeding 100 Td, most electron energy is diverted toward electronic excitation, dissociation, and ionization of $N_2$ and $O_2$ molecules. While dissociation of $N_2$ via direct electron impact can produce N atoms that react with $O_2$ to form NO, this process is inherently limited by the high dissociation energy threshold of $N_2$, resulting in low energy efficiency (~3%) [3].

In addition to variations in electron energy levels caused by differences in the reduced electric field, various plasma types also exhibit distinct gas temperatures. However, isolating the effect of temperature on plasma-based nitrogen fixation or determining the optimal temperature for this process is challenging. Contradictory trends regarding the impact of temperature on $NO_x$ production performance have been reported in different studies [23, 31].

To evaluate $NO_x$ production across different discharges, Pei et al. [29] introduced a dimensionless parameter, $\chi$, which incorporates both the electric field and gas temperature. The parameter is defined as follows:

$$\chi = \frac{\bar{E} \times \bar{T}}{E_r \times T_r} \tag{8.6}$$

where $\bar{E}$ is the average electric field with units of kV cm$^{-1}$; $\bar{T}$ is the average temperature mentioned with units of K; and $E_r$ and $T_r$ are chosen to normalize the parameter at a reference condition.

The study revealed that the $\chi$ factor is closely correlated with the energy cost of $NO_x$ production across vastly different types of discharge [29]. This suggests that $\chi$ encapsulates certain implicit time-and-space-dependent phenomena that fundamentally influence $NO_x$ production. The researchers proposed that $\chi$ relates to two key mechanisms governing the energy efficiency of $NO_x$ production in NTP. The first mechanism involves the efficient electron-impact activation of $N_2$, which facilitates $NO_x$ formation. The second mechanism is the rapid thermal quenching of NO to prevent its reconversion into $N_2$ and $O_2$ as the gas temperature decreases. Based on the trend, reducing the $\chi$ factor could serve as a guiding principle for lowering the energy cost of $NO_x$ production. However, achieving a reduction in $\chi$ is not straightforward. The average electric field ($\bar{E}$) and gas temperature ($\bar{T}$) are often interdependent within a given discharge, and direct manipulation of these parameters is generally not feasible.

#### 8.2.3.2 Impact of Gas Composition

The feed gas composition significantly affects $NO_x$ concentration, the $NO/NO_2$ ratio, and energy consumption by influencing reaction pathways. The optimal feed ratio for $NO_x$ production, typically expressed as the percentage of $O_2$, varies between 30% and 50%, depending on the type of plasma and operating conditions [3, 32]. The rates of $NO_x$ formation and loss reactions in NTP vary with the fraction of $N_2$ in the feed gas. Figure 8.6 provides a detailed analysis of the net formation

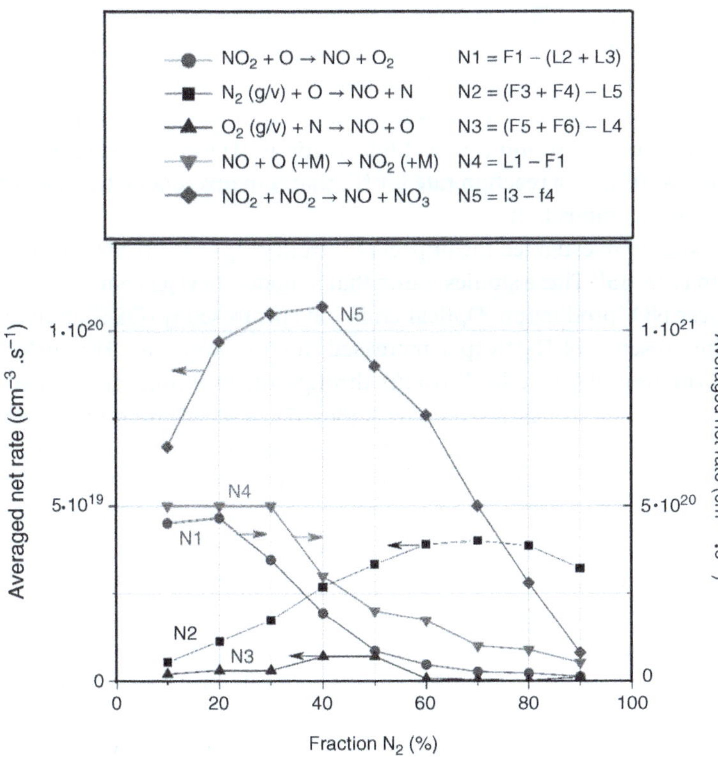


**Figure 8.6** Net rates of the most important formation and loss process of NO (and $NO_2$) as a function of the $N_2$ fraction in the feed gas in a GA reactor. *Source:* Ref. [11] / American Chemical Society.

and loss rates for key reactions at different $N_2/O_2$ ratios in the mixture in a GA reactor. As shown in Figure 8.6, low $N_2$ fractions enhance the interconversion between NO and $NO_2$ but do not contribute to their net formation. The primary $NO_x$ production mechanisms, collectively referred to as **N2** and **N3** (the Zeldovich mechanism), are driven by N and O atoms. The optimal $N_2$ fraction for $NO_x$ production is around 70%, as it maximizes the combined rate of **N2** and **N3** reactions.

The Zeldovich mechanism begins with the generation of N and O atoms. Vervloessem et al. [11] demonstrated that N atoms are predominantly formed by electron-impact dissociation of $N_2$ (in ground or vibrational states), making their production proportional to the $N_2$ fraction. Meanwhile, O atoms, generated through electron impact on $O_2$ or collisions with excited $N_2$, are largely unaffected by feed composition. The net production rate of **N2** reactions increases with rising $N_2$ fractions, peaking at 70–80%. This peak occurs below the expected value at equal $N_2$ and $O_2$ fractions due to V–V exchanges between excited $N_2$ and $O_2$, which reduce $N_2(v)$ populations and slow the **N2** reaction rate. These exchanges transfer vibrational energy from $N_2$ to $O_2$, depopulating $N_2$ vibrational levels since $O_2$ vibrational levels are more easily depopulated.

In contrast, the rate of the **N3** process is nearly independent of gas composition. At low $N_2$ fractions, N atom production becomes the limiting factor, while at high $N_2$ fractions, $O_2$ availability is reduced. Since both $O_2(v)$ (derived from $O_2$ molecules) and N atoms are required for the **N3**, N atoms are limiting at low $N_2$ fractions, while $O_2$ molecules are limiting at high fractions. These opposing effects counterbalance, resulting in a reaction rate for **N3** that remains relatively constant regardless of the gas composition.

Several studies have investigated the impact of introducing $H_2O$ on $NO_x$ production in air plasma [33–35]. These studies found that a moderate $H_2O$ content in the feed gas enhances $NO_x$ production. Optical emission spectroscopy (OES) analysis showed that the presence of $H_2O$ vapor increased the formation of OH and NH radicals. The formation of OH radicals occurs through the electron impact dissociation of $H_2O$ molecules, as described by (8.7) [36, 37]. N atoms then react with $H_2O$ molecules or OH radicals to generate NH· radicals, as shown in reactions (8.8)–(8.9) [38]. Both OH and NH radicals contribute to NO production through (8.10)–(8.13). Additionally, OH radicals promote the oxidation of NO to $NO_2$ (8.14), which increases the overall $NO_x$ production. However, excessively high $H_2O$ concentrations can reduce the energy efficiency of $NO_x$ production, as much of the plasma energy is consumed in heating and dissociating $H_2O$. This results in decreased overall energy efficiency. The optimal $H_2O$ content for $NO_x$ energy efficiency in a rotating gliding arc (RGA) reactor was found to be 12% [39]. Further increasing the $H_2O$ concentration reduces the mean electron energy and the intensity of $N_2$, $N_2^+$, and NH due to the quenching effect of $H_2O$ molecules in the plasma [40–42].

$$e^- + H_2O \rightarrow e^- + H\cdot + \cdot OH \tag{8.7}$$

$$N\cdot + H_2O \rightarrow \cdot OH + NH\cdot \tag{8.8}$$

$$N\cdot + \cdot OH \rightarrow O\cdot + NH\cdot \tag{8.9}$$

$$N\cdot + \cdot OH \rightarrow NO + H\cdot \tag{8.10}$$

$$NH\cdot + O\cdot \rightarrow NO + H\cdot \tag{8.11}$$

$$NH\cdot + \cdot OH \rightarrow NO + H_2 \tag{8.12}$$

$$NH\cdot + O_2 \rightarrow NO + \cdot OH \tag{8.13}$$

$$NO + \cdot OH \rightarrow NO_2 + H\cdot \tag{8.14}$$

A modeling study by Vervloessem et al. [27] proposed that ozone ($O_3$) could enhance $NO_x$ production by suppressing the reverse reactions of the Zeldovich mechanism. These reverse reactions, where O or N atoms react with NO to form $N\cdot + O_2$ or $O\cdot + N_2$, effectively terminate the reaction chain. Theoretically, removing NO from the mixture would prevent these reverse reactions. Although current technology cannot directly achieve this, adding $O_3$ to the feed gas offers a potential alternative. $O_3$ reacts with NO to convert it into $NO_2$, effectively reducing NO levels and promoting $NO_x$ production. Simulations indicated that introducing $O_3$ could increase $NO_x$ yield while lowering energy costs. However, further research is required to validate this approach and determine whether the benefits of $O_3$ addition outweigh its associated costs.

#### 8.2.3.3 Impact of Pressure

Unlike the high pressures (>100 bar) required for the industrial Haber-Bosch nitrogen fixation process [43], plasma processes can operate efficiently at atmospheric pressure, with elevated pressures of a few bars also being relatively easy to achieve. The recently published low-temperature plasma roadmap underscores the significance of plasmas operating at pressures above atmospheric levels [44]. High-pressure plasmas in various gases, including arc discharges, have been extensively investigated through experimental and theoretical studies [45–47]. Elevated-pressure plasmas have been applied to processes such as $CO_2$ splitting [48], hydrocarbon synthesis [49, 50], and dry reforming [51], using techniques like MW discharge [52], arc [49], spark discharge [51], and DBD plasmas [53]. However, their potential in nitrogen fixation remains largely unexplored.

A recent study utilizing the RGA plasma reactor revealed that increasing pressure generally enhances $NO_x$ concentration, production rates, and energy efficiency, as long as the experiments remain within the thermal equilibrium peak for optimal NO production [54]. This highlights the promising role of elevated-pressure plasmas in advancing nitrogen fixation technology. Figure 8.7 illustrates that NO concentration declines once the temperature surpasses its optimal value

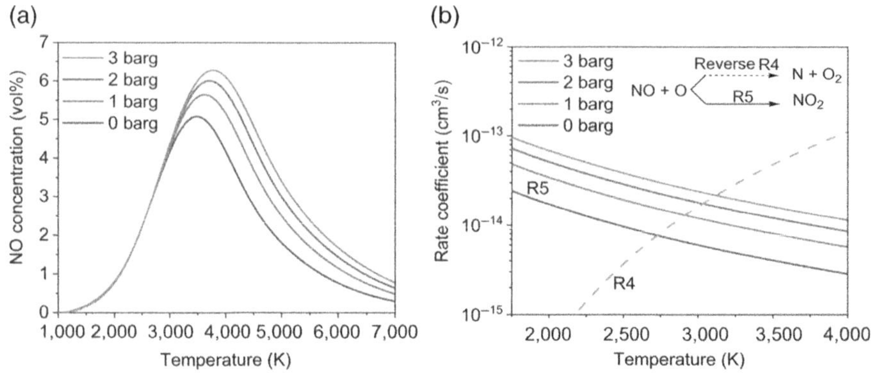


**Figure 8.7** (a) Equilibrium concentration of NO as a function of temperature at different pressures (0–3 bar), and (b) rate coefficients of the pressure-independent back reaction of R4 and the pressure-dependent forward reaction of R5 as a function of temperature. *Source:* Ref. [54] / American Chemical Society.

at a given pressure. However, at higher pressures, the NO concentration remains elevated beyond the peak across all temperatures. The authors attribute this enhanced performance to two key factors: an increase in the equilibrium NO concentration and the accelerated oxidation of NO by O radicals to form $NO_2$ (R5 in Figure 8.7b). This oxidation process minimizes the recombination of NO with O radicals (R4 in Figure 8.7b), thereby sustaining higher $NO_x$ concentrations.

#### 8.2.3.4 Impact of Pulsed Discharge

The positive impact of pulsed plasma on $N_2$ fixation has been widely documented in the literature [28, 29, 55, 56]. The lowest energy cost for atmospheric-pressure plasma reported to date was achieved using a pulsed jet [55]. The authors attributed this exceptional energy efficiency to the pulsed power and corresponding pulsed gas temperature profile, which established a strong vibrational–translational nonequilibrium, thereby enhancing the nonthermal Zeldovich mechanism.

Figure 8.8a illustrates the power and gas temperature profiles in their pulsed plasma system, with the gas temperature derived from OES measurements. During the plasma pulses, the gas temperature was calculated to be $1{,}750 \pm 150$ K, decreasing exponentially to 330 K during the interpulse period. Additionally, the use of pulsed power resulted in periodic variations in electron temperature, electron density, and the reduced electric field, as depicted in Figure 8.8b.

Electrons with temperatures around 1 eV are optimal for efficiently exciting $N_2$ molecules vibrationally, thereby promoting energy-efficient $NO_x$ formation through the vibrationally enhanced Zeldovich mechanism, although recent

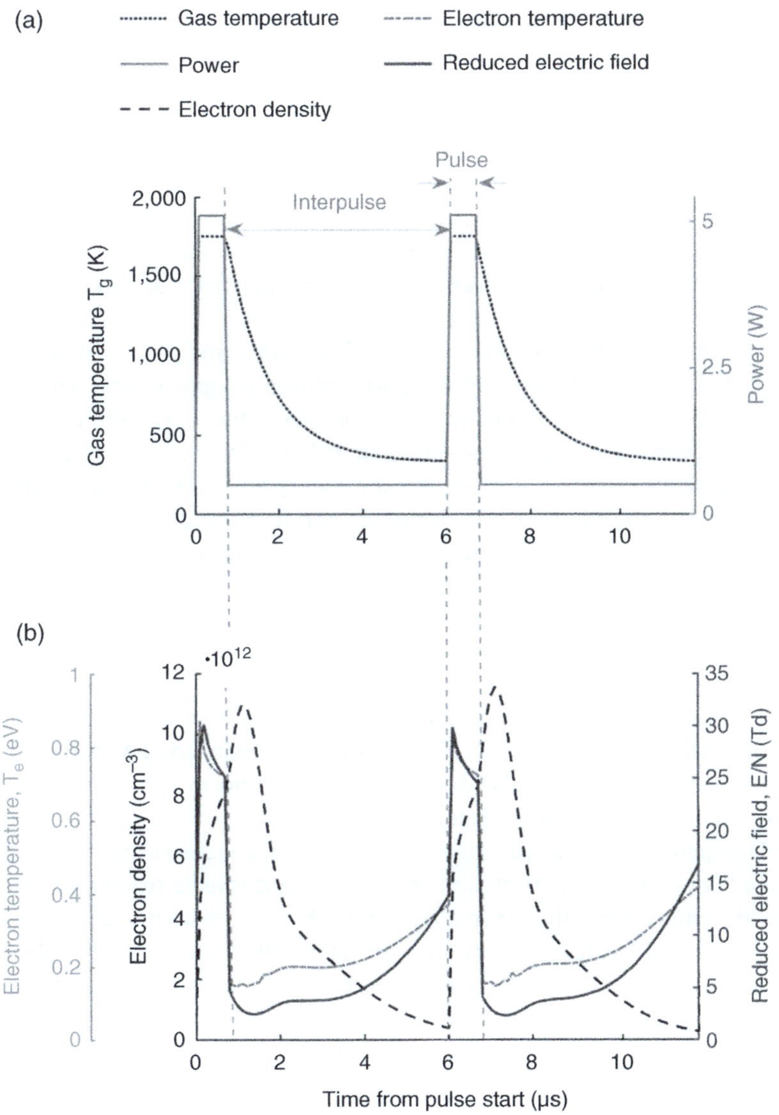


**Figure 8.8** (a) Input plasma parameters for the model, including gas temperature and power for two successive pulses. (b) Simulated plasma characteristics, including electron temperature, electron density, and reduced electric field, calculated for a feed gas composition of 80% $N_2$ and 20% $O_2$ in a spark discharge. *Source:* Ref. [55] / The Royal Society of Chemistry.

insights reveal that the thermal Zeldovich mechanism is also efficient in warm plasmas, due to their high temperature. Arc-based plasmas, such as GA and arc jet, are well-suited for producing electrons within this temperature range due to their relatively low reduced electric fields (<50 Td). However, the relatively high gas temperature in these plasmas diminishes the vibrational population through V–T relaxation, where vibrational energy dissipates via collisions with ground-state molecules, leading to further gas heating. One approach to sustain a high vibrational population of $N_2$ molecules is operating plasma at low pressures, where collisions are less frequent. However, low-pressure systems can significantly increase process costs.

Another promising method involves the use of pulsed plasma power. In pulsed plasma systems, the short duration of the pulses minimizes gas heating. During these pulses, $N_2(v)$ populations build up through electron-impact vibrational excitation and V–V energy exchanges between excited $N_2$ and $O_2$ molecules. Between pulses, the temperature drop mitigates two key energy losses: (i) It limits vibrational energy loss caused by V–T relaxation. (ii) It suppresses $N_2$–$O_2$ V–V exchanges, which have been identified as a bottleneck for energy-efficient $NO_x$ production in GA plasma. The reduced temperature during the interpulse period also decreases the $O_2$–$O_2$ V–T relaxation rate, curbing the vibrational energy sink and reducing energy loss to heat. This interplay ensures more efficient utilization of plasma energy for $NO_x$ production.

Pulsed plasmas not only promote the vibrational excitation of $N_2$ molecules but also mitigate the reverse reactions of the Zeldovich mechanism (**L4** and **L5** in Table 8.1), thereby preventing $NO_x$ destruction. In continuous plasmas, which maintain constant temperatures, both forward and reverse reactions are enhanced at elevated temperatures. However, their temperature dependencies differ. Specifically, while the rate coefficients of both the forward and reverse reactions of R1 increase with temperature, the reverse reaction (**L1**) is much more sensitive to temperature changes. The temperature drop during the interpulse period in pulsed plasmas significantly limits the reverse reaction (**L1**). Simultaneously, the forward reaction of **R1** benefits from the rise in the $N_2(v)$ population during this period, causing O atoms to participate in NO production rather than NO loss preferentially. As a result, the rate of **R1** remains positive even during the interpulse period, as shown in Figure 8.9.

For **R2**, while temperature changes do not favor one direction over the other, the reaction is strongly influenced by the availability of reactive species such as O atoms and $N_2$ (ground and vibrational states). Although a high temperature would theoretically benefit **R2** based solely on rate coefficients, the reaction is also strongly promoted by a high $N_2(v)$ population, which is more pronounced at lower temperatures. Thus, achieving a balance between these factors is crucial for optimizing pulsed plasma systems to maximize energy-efficient $NO_x$ production.

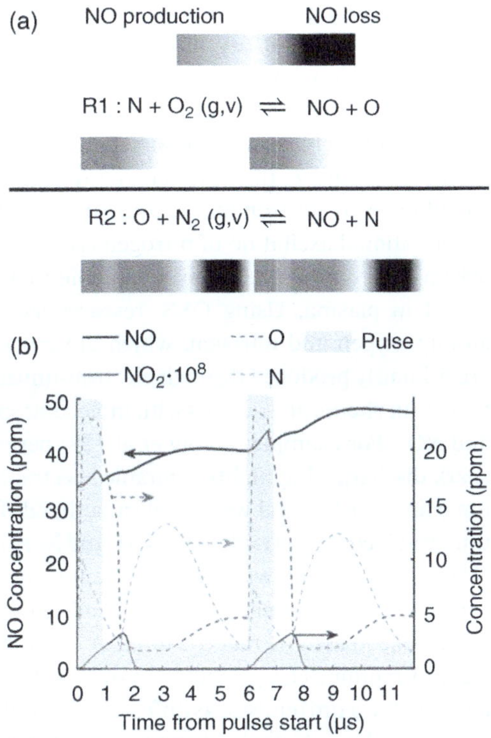


**Figure 8.9** (a) The significance of **R1** and **R2** over time during two pulses and the interpulse periods, as represented on a gradient scale, gray for forward reactions (net NO production), black for backward reactions (net NO loss), and white for no net change. Darker shades indicate higher net rates in either direction. (b) The corresponding concentrations of NO, $NO_2$, N, and O during the same two pulses. *Source:* Ref. [55] / The Royal Society of Chemistry.

In general, as illustrated in Figure 8.9, the two steps of the nonthermal Zeldovich mechanism occur at high rates during the pulses. During the interpulse period, **R1** becomes negligible, and **R2** may proceed in the reverse direction. Nevertheless, the overall effect is a steady increase in NO concentration over time, highlighting the advantages of pulsed plasma operation.

### 8.2.4 Mechanistic Insights from Experimental Studies

Experimental studies have provided valuable insights into the mechanisms underlying plasma-driven $NO_x$ synthesis, elucidating the complex interaction of reactions and energy transfer processes. These studies highlight the critical roles of plasma parameters such as electron temperature, gas composition, and pressure in influencing reaction pathways and product yields.

Experimental evidence demonstrates that electron temperature varies across different plasma types. Kim et al. [57] used OES to measure electric field distributions and average electron energies in pulsed corona and DBD in atmospheric air. Their results indicate that electron energies in a pulsed corona discharge range from

10 to 12 eV along the streamers, while time-resolved measurements show slightly lower energies of 9–10 eV in DBD streamers. Zhu et al. [58] examined the translational, rotational, vibrational, and electron temperatures of a GA reactor, revealing the order $T_e > T_v > T_r > T_t$, indicative of a highly nonequilibrium state. By simultaneously measuring plasma column length, discharge voltage, and translational temperature, they estimated a reduced electric field of around 6 Td, corresponding to an electron temperature of ~1.5 eV. This electron temperature is ideal for vibrational excitation of nitrogen ($N_2$) molecules, a key mechanism in plasma-driven $NO_x$ synthesis. OES is a valuable tool for identifying reactive species generated in plasma. Using OES, researchers have detected key radicals such as atomic oxygen and nitrogen, which are crucial for $NO_x$ formation. These species are primarily produced through electron-impact dissociation and subsequently participate in chemical reactions with molecular species, leading to the formation of NO and $NO_2$. For example, Zhang et al. [28] performed time-integrated OES of an air spark discharge, highlighting atomic spectral lines for N, O, Fe, and H, along with ionic lines for $N^+$, and weak molecular bands for $N_2$ and $N_2{}^+$. They also conducted time-resolved OES of excited O, N, H, and $N^+$. The results showed that the intensity of $N^+$ (500.1 nm) aligns with the absolute current amplitude, reflecting a time delay due to electron-impact excitation and the radiative lifetime of the upper excited state. It is reasonable to assume that the number of positive ions matches the number of electrons to maintain charge neutrality in NTP. However, the evolution of O I (777.2 nm) and N I (821.5 nm) emissions differs significantly from $N^+$, especially during the post-discharge phase. Notably, distinct intensity peaks for O I (777.2 nm) and N I (821.5 nm) emerge around 1,500 ns, whereas $N^+$ shows almost no such peaks. This behavior suggests that O and N atoms are generated through chemical reactions rather than electron-impact processes during this stage. These findings strongly support the occurrence of chain reactions such as $O\cdot + N_2(v) \rightarrow NO + N\cdot$ and $N\cdot + O_2(v) \rightarrow NO + O\cdot$, highlighting the critical role of vibrational molecules in $NO_x$ production.

A recent study [59] demonstrated enhanced $NO_x$ production using RGA by integrating rapid quenching techniques in the plasma reactor, effectively improving overall process efficiency. This study indicates that reducing the gas heating, especially in quasi-thermal plasmas, limits V–T relaxation, increasing the vibrational population of $N_2$ and promoting the efficiency of $NO_x$ synthesis.

Modeling studies have significantly advanced the understanding of plasma-driven $NO_x$ synthesis mechanisms, providing valuable insights into optimizing reactor designs and operational parameters. However, the full realization of these advancements critically depends on robust experimental investigations to validate and refine the models. Experimental studies are pivotal in bridging the gap between theoretical predictions and practical applications. They provide direct evidence of reaction pathways, plasma characteristics, and system performance under various operating conditions, offering crucial data to enhance the accuracy of kinetic models. Without experimental validation, the reliability of modeling assumptions and calculated parameters remains uncertain.

One of the key challenges in modeling plasma-assisted $NO_x$ production is the scarcity of accurate cross-section data, especially for vibrational and electronically excited species. Currently, these data are often approximated through techniques such as cross section shifting with threshold energy adjustments [60, 61] and detailed balancing methods [62, 63]. Moreover, experimental studies enable the direct determination of reaction rate coefficients, which can then be used to calculate the rate coefficients for the vibrational-state-involved reactions through the Fridman-Macheret $\alpha$-model [24]. Special attention has also been given to the role of metastable nitrogen species in optimizing reaction pathways. Experimental investigations into these species are critical for understanding their behavior and interactions, which are often difficult to predict purely through modeling. By combining experimental findings with modeling efforts, researchers can achieve a more comprehensive understanding of $NO_x$ synthesis mechanisms and identify strategies to enhance efficiency. In summary, experimental studies are not merely complementary but integral to advancing plasma-assisted $NO_x$ production. They provide the empirical foundation necessary to refine models, validate assumptions, and ultimately translate theoretical insights into practical solutions.

## 8.3 Plasma-Catalytic $NO_x$ Synthesis

Catalysts impact plasma reactions through both physical and chemical mechanisms. Physically, they modify key discharge properties, including the regime, electron density, temperature, and electric field distribution. Chemically, the active sites on the catalyst surface generate highly reactive species that can alter the reaction pathways, enhancing the overall process efficiency. Catalysts can be integrated into plasma systems in various configurations. They may be directly placed within the discharge gaps or coated onto electrode surfaces in the plasma region, where they directly interact with the plasma to enhance reaction rates. Alternatively, catalysts can be positioned outside the plasma discharge region. This configuration minimizes undesired changes in plasma properties and avoids potential instability caused by direct plasma–catalyst interactions. This flexibility in catalyst placement allows for tailored approaches to optimize plasma-assisted processes, balancing the benefits of direct interaction with the need for stable and efficient operation.

Placing the catalyst bed within the DBD plasma field enables direct interaction between the plasma and the catalysts, a setup referred to as in-plasma catalysis. This configuration is commonly employed in plasma-catalytic chemical reactions [64–66], including plasma-catalytic $NO_x$ production [4, 67]. The possible synergistic effects in this arrangement are both significant and complex, as the physicochemical properties of the plasma and the catalysts dynamically influence and modify each other. Using the in-plasma catalysis approach, Patil et al. [4] highlighted the benefits of micro-discharges for $NO_x$ production in a packed-bed DBD reactor. Their study revealed that packing materials can enhance micro-discharging formation, leading to an increased generation of high-energy species essential for NO formation. This effect was especially notable with quartz wool, whose fine, fibrous structure and rigid, sharp edges facilitated stronger micro-discharges.

While packing materials primarily influence the electric field distribution and electron temperatures, the main function of catalysts is believed to be adsorption on active sites. However, in plasma-catalysis, the discharge alteration might have a bigger impact on $NO_x$ production than this adsorption-enabled catalytic effect. For instance, in DBD-assisted plasma-catalytic $NO_x$ production, packing $\gamma$-$Al_2O_3$ doubled the $NO_x$ yield, while loading different metal oxides onto $\gamma$-$Al_2O_3$ showed only a minor (<10%) impact on $NO_x$ yield [4]. Among the catalysts tested, $WO_3$/$\gamma$-$Al_2O_3$ exhibited the highest activity, whereas well-known oxygen activation catalysts such as $PbO$/$\gamma$-$Al_2O_3$ and $Co_3O_4$/$\gamma$-$Al_2O_3$ provided only modest improvements of less than 5%. These findings suggest that the primary role of these catalysts in DBD-assisted $NO_x$ production lies in facilitating micro-discharge formation, rather than chemically interacting with plasma-activated species.

Bayer et al. [68] recently explored the effects of surface reactions, gas-phase reactions, and mass transport on plasma-catalytic NO formation using a DBD-based plasma jet. The system configuration, illustrated in Figure 8.10a, involves feeding $O_2$ into the afterglow region and positioning an Ag wire catalyst 5 mm downstream from the union point of the plasma jet. This arrangement facilitates the quantification of species densities through molecular beam mass spectrometry (BMS), enabling a detailed examination of the pathways and timescales involved in plasma-assisted $N_2$–$O_2$ reactions. By analyzing the consumption of plasma-derived N and the formation of NO, the study provides insights into both gas-phase and surface reaction dynamics.

As shown in Figure 8.10b, gas-phase reactions deplete N but do not selectively produce NO. The introduction of the non-porous Ag wire catalyst increases the rate of N· conversion to NO, primarily driven by mass transfer rather than surface reactions. However, when NO density surpasses a critical threshold, N atoms cannot diffuse to the catalyst surface quickly enough to keep up with its consumption in the gas phase through reactions with NO. As a result, the effectiveness of

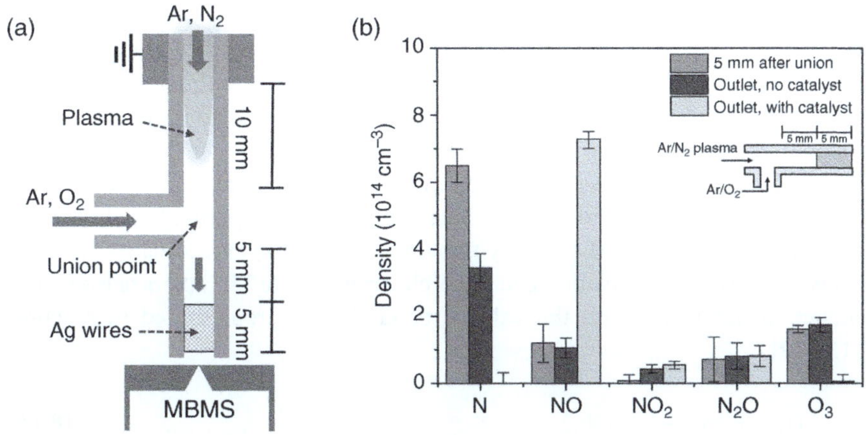


**Figure 8.10** (a) Schematic of the plasma-catalytic system using DBD-based plasma jet and Ag wires as the catalyst. (b) Densities of N, NO, $NO_2$, $N_2O$, and $O_3$ at 5 mm from the union (5 mm after union), at the outlet without a catalyst (Outlet, no catalyst), and at the outlet with Ag catalyst in the final 5 mm of the reactor (Outlet, with catalyst). *Source:* Ref. [68] / with permission from Elsevier.

heterogeneous catalysts in enhancing plasma-assisted $NO_x$ production is constrained to conditions where the diffusive transport of N atoms from the gas phase to the catalyst surface outpaces its consumption in gas-phase reactions involving NO.

Catalyst supports, typically regarded as non-catalytic in thermal reactions, can exhibit catalytic behavior under plasma conditions. For example, $\gamma$-$Al_2O_3$ has demonstrated activity in several studies [4, 69, 70]. Patil et al. reported that packing $\gamma$-$Al_2O_3$ in a DBD reactor enhanced $NO_x$ yield and altered $NO_2$ selectivity in the products [4]. This suggests that NO species may react with surface-adsorbed oxygen on $\gamma$-$Al_2O_3$, resulting in increased $NO_2$ production. Pei et al. [69] proposed a mechanism to explain how catalyst supports enhance $NO_x$ production in atmospheric-pressure air plasma. Long-lived species within the discharge plasma channel, such as excited nitrogen molecules ($N_2^*$) and oxygen atoms, diffuse to the active alumina surface and adsorb onto it. These surface-bound intermediates significantly increase the likelihood of the $N_2^* + O\cdot \rightarrow NO + N\cdot$ reaction, thereby promoting higher NO yields.

Using a low-pressure MW reactor, Gicquiel et al. [71] demonstrated that depositing $MoO_3$ on the reactor wall more than doubled the NO production rate compared to plasma-only conditions. In their experiments, with pure nitrogen plasma, they observed NO formation accompanied by the reduction of $MoO_3$ to $MoO_2$. This finding highlights the importance of lattice oxygen mobility during NO synthesis.

The reduction of the $MoO_3$ catalyst is driven by interactions with vibrationally excited molecules, which leads to the formation of NO. Based on these observations, the authors proposed that the reaction follows an Eley-Rideal mechanism, where vibrationally excited $N_2$ reacts with the mobile oxygen atoms of $MoO_3$. The proposed reaction pathway is depicted in Figure 8.11.

Recently, $TiO_2$ has been investigated in GA systems for plasma-catalytic $NO_x$ production [33, 39]. The photocatalytic mechanism of $TiO_2$, particularly in its anatase phase, is based on activation by photonic energy (hv) exceeding its bandgap energy of 3.2 eV. This activation generates electron–hole ($e^-$-$h^+$) pairs between the conduction band (CB) and the valence band (VB), as represented in reaction (8.15) [72]:

$$TiO_2 + hv \rightarrow e^- + h^+ \tag{8.15}$$

Nakamura et al. [5] demonstrated that plasma can activate $TiO_2$ photocatalysts by generating electron–hole pairs through electron impact on the $TiO_2$ surface, rather than relying on UV radiation. This is because plasma UV emissions are less intense than those produced by conventional UV lamps, a phenomenon reported in several prior studies [73–76]. Additionally, Whitehead et al. suggested that electron energies in the range of 3–4 eV are sufficient to activate $TiO_2$ photocatalysts, as shown in [77, 78]:

$$TiO_2 + e^-(>3.2\,\text{eV}) \rightarrow e^- + h^+ \tag{8.16}$$

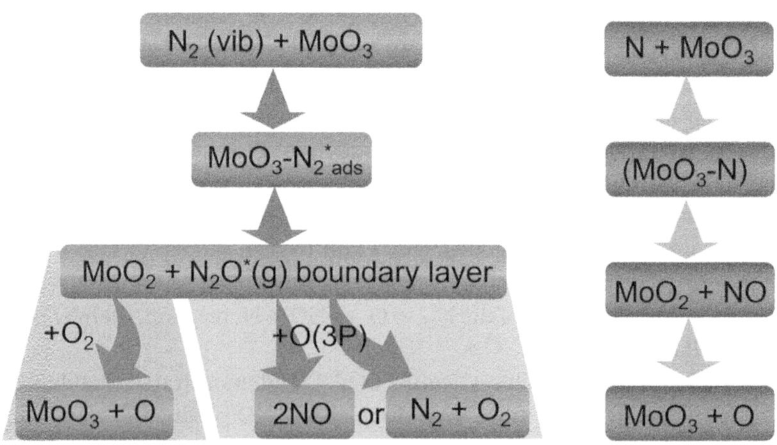


**Figure 8.11** Proposed mechanisms for plasma-catalytic $NO_x$ production in a MW plasma with $MoO_3$ as the catalyst. *Source:* Ref. [71] / IOP Publishing.

A potential mechanism for the plasma-photocatalytic process is illustrated below. Initially, high-energy electrons and ions from the plasma generate electron–hole ($e^-$–$h^+$) pairs and create oxygen vacancies ($V_o$) on the $TiO_2$ surface. These oxygen defects activate the surface, facilitating the conduction band mechanism and enhancing $NO_x$ production. This process begins with the adsorption of $O_2$ molecules at the oxygen vacancies on the $TiO_2$ surface. These adsorbed $O_2$ molecules capture photo-induced electrons and free electrons from the oxygen vacancy states, leading to the formation of superoxide radical species, such as $O_2^-$ (as shown in Eq. (8.17)) [33]. The generated $O_2^-$ species subsequently interact with $N_2$ molecules, ultimately producing NO (as described in Eq. (8.18)).

$$O_2 + e^- \rightarrow O_2^- \tag{8.17}$$

$$O_2^- + N_2 \rightarrow e^- + 2NO \tag{8.18}$$

The plasma-catalytic system involves intricate interactions between plasma and catalysts, as discussed in detailed reviews [44, 79]. These interactions include the influence of catalysts on plasma properties and the effects of plasma discharge on the catalysts. When catalysts are present within the plasma, their local surface curvature enhances the electric field, producing a packed-bed effect in the reactor. This physical phenomenon directly affects the electron energy distribution function, as well as the rates of electron-impact dissociation and ionization, thereby altering the chemical composition of the plasma [80]. Additionally, the intensified electric field promotes the formation of micro-discharges within the catalyst pores or near sharp edges [81], creating discharge characteristics that differ from those in the bulk gas [82]. This results in varying production and depletion rates of different plasma species, further influencing the overall reaction dynamics. Apart from altering plasma discharge characteristics, catalysts also play a key role in modifying reaction pathways through surface-mediated processes. The catalyst surface provides active sites for adsorption and reaction, which can lead to the selective activation of plasma-generated species. For example, oxygen vacancies on catalyst surfaces can enhance the adsorption and dissociation of gas-phase molecules, such as $O_2$ and $N_2$, thereby promoting $NO_x$ formation or other desired reactions. Moreover, the synergistic interaction between plasma and catalysts can create unique chemical environments. Plasma-generated reactive species, such as radicals, ions, and excited molecules, interact with adsorbed intermediates on the catalyst surface, enabling reaction pathways that are not accessible under conventional thermal or catalytic conditions. This synergy can also improve selectivity and energy efficiency by steering reactions toward specific products while minimizing side reactions.

Overall, the performance of plasma-catalytic systems for $NO_x$ production is driven by the interplay between plasma physics and catalyst design. This combination may enhance their efficiency and selectivity, making them a promising technology for sustainable $NO_x$ generation. Despite the proposed mechanisms for the synergistic effects between plasma and catalysts, many aspects remain unclear. For example, it is still uncertain which plasma-activated $N_2$ species are dominant in $NO_x$ synthesis across different plasma-catalytic systems, which active sites on the catalyst contribute to plasma-enhanced catalytic $NO_x$ production, and the relative significance of Langmuir-Hinshelwood vs. Eley-Rideal mechanisms. To address these gaps, advanced characterization techniques of plasma–catalyst systems are essential. In situ diagnostics, such as Fourier transform infrared (FTIR) and X-ray photoelectron spectroscopy (XPS), could offer valuable insights into the molecular interactions between plasma and the catalyst. Moreover, the effects of various metals and their combinations on plasma-catalytic $NO_x$ synthesis have not been thoroughly explored. Additionally, the combination of GA with catalysts remains under-researched.

## 8.4 Conclusion and Outlook

The plasma-based synthesis of $NO_x$ has emerged as a promising avenue for sustainable nitrogen fixation due to its ability to operate under ambient conditions and its potential for high-energy efficiency. This chapter has systematically explored mechanistic insights into $NO_x$ formation and loss processes, the role of various plasma types, and the influence of operational parameters such as feed gas composition, pressure, and pulsed discharges. The findings elucidate the critical reaction pathways and highlight the Zeldovich mechanism as the most energy-efficient pathway for $NO_x$ production in NTP systems. Furthermore, the chapter underscores the importance of minimizing $NO_x$ loss processes, optimizing vibrational kinetics, and understanding the interplay between plasma parameters to enhance the overall energy efficiency of the process.

The detailed analysis of plasma properties reveals that NTP, such as GA, offers superior energy efficiency compared to thermal plasmas due to their nonequilibrium characteristics. These plasmas enable selective excitation of molecular vibrational states, facilitating the dissociation of $N_2$ and $O_2$ molecules with less energy loss to gas heating. However, challenges remain in optimizing the electron energy distribution and mitigating energy losses associated with vibrational–translational relaxation and reverse Zeldovich reactions.

Gas composition significantly impacts the reaction pathways and product distribution in plasma-enabled $NO_x$ synthesis. The optimal $N_2/O_2$ ratio, typically around 70%, maximizes the efficiency of the Zeldovich mechanism by balancing the availability of N and O atoms. The addition of $H_2O$ as a feed gas modifier has shown potential in enhancing $NO_x$ yields through the formation of reactive OH and NH radicals, although excessive $H_2O$ levels can reduce energy efficiency. Pressure and pulsed discharge conditions also play crucial roles in determining $NO_x$ production rates and energy consumption. Elevated pressures and pulsed plasmas have demonstrated the ability to enhance $NO_x$ yields by increasing gas density and promoting nonequilibrium conditions.

The integration of catalysts into plasma systems represents a transformative approach to further improving $NO_x$ synthesis. Plasma-catalytic process can sometimes enhance reaction selectivity, modify discharge characteristics, and introduce new reaction pathways via active sites on the catalyst surface. The ability of catalysts to influence plasma parameters and their interaction with plasma-generated species offers exciting opportunities for tailoring the $NO_x$ production process. However, understanding the complex interaction between plasma and catalysts remains an ongoing challenge, necessitating further experimental and theoretical investigations.

Looking ahead, several research directions can drive advancements in plasma-based $NO_x$ synthesis:

**Mechanistic studies:** Comprehensive studies using advanced diagnostic tools and modeling techniques are needed to elucidate the dynamics of plasma-activated species and their interactions with catalysts. Understanding the role of intermediate species and reaction kinetics will enable the design of more efficient systems.

**Catalyst development:** The development of novel catalyst materials with enhanced stability under plasma conditions can significantly improve $NO_x$ yields and energy efficiency. Exploration of materials that resist deactivation and maintain high activity in harsh plasma environments will be critical.

**System integration:** Scaling up plasma-based systems for industrial applications requires innovative reactor designs that maximize energy efficiency and product yield. Key design innovations include modular plasma reactors for enhanced scalability, advanced electrode configurations to optimize gas flow dynamics, and novel cooling systems to manage thermal loads in high-power operations. Integration with renewable energy sources and coupling with downstream processes can enhance the sustainability of the nitrogen fixation process.

**Operational optimization:** Fine-tuning operating parameters, including optimizing the feed gas composition, operating at elevated pressures, and employing pulsed power supply, can further improve process performance. Techniques such as fast quenching and in situ NO removal offer pathways to minimize energy losses and maximize $NO_x$ yields.

In conclusion, plasma-based $NO_x$ synthesis holds immense potential as a sustainable alternative to conventional nitrogen fixation processes. By addressing the current challenges through multidisciplinary research and innovation, this technology can pave the way for a greener and more energy-efficient approach to meeting global nitrogen demand.

## References


1 Zhou, D., Zhou, R., Zhou, R. et al. (2021). Sustainable ammonia production by non-thermal plasmas: status, mechanisms, and opportunities. *Chem. Eng. J.* 421: 129544. https://doi.org/10.1016/j.cej.2021.129544.

2 Starikovskiy, A. and Aleksandrov, N. (2013). Plasma-assisted ignition and combustion. *Prog. Energy Combust. Sci.* 39: 61–110. https://doi.org/10.1016/j.pecs.2012.05.003.

3 Wang, W., Patil, B., Heijkers, S. et al. (2017). Nitrogen fixation by gliding arc plasma: better insight by chemical kinetics modelling. *ChemSusChem* 10: 2145–2157. https://doi.org/10.1002/cssc.201700095.

4 Patil, B., Cherkasov, N., Lang, J. et al. (2016). Low temperature plasma-catalytic $NO_x$ synthesis in a packed DBD reactor: effect of support materials and supported active metal oxides. *Appl Catal B* 194: 123–133. https://doi.org/10.1016/j.apcatb.2016.04.055.

5 Nakamura, I., Negishi, N., Kutsuna, S. et al. (2000). Role of oxygen vacancy in the plasma-treated TiO2 photocatalyst with visible light activity for NO removal. *J. Mol. Catal. A Chem.* 161: 205–212. https://doi.org/10.1016/S1381-1169(00)00362-9.

6 Helmke, A., Gerling, T., and Weltmann, K.-D. (2018). Plasma Sources for Biomedical Applications. In: *Comprehensive Clinical Plasma Medicine: Cold Physical Plasma for Medical Application* (ed. H.R. Metelmann, T. Von Woedtke, and K.D. Weltmann), 23–41. Springer https://doi.org/10.1007/978-3-319-67627-2_2.

7 Weltmann, K.-D. and Von Woedtke, T. (2011). Basic requirements for plasma sources in medicine. *EPJ Appl. Phys.* 55: 13807. https://doi.org/10.1051/epjap/2011100452.

8 Birkeland, K. (1906). On the oxidation of atmospheric nitrogen in electric arcs. *Trans. Faraday Soc.* 2: 98–116.

9 Eyde, S. (1912). Oxidation of atmospheric nitrogen and development of resulting industries in Norway. *Ind. Eng. Chem.* 4: 771–774.

10 Kaufman, F. and Kelso, J.R. (1955). Thermal decomposition of nitric oxide. *J. Chem. Phys.* 23: 1702–1707. https://doi.org/10.1063/1.1742415.

11 Vervloessem, E., Aghaei, M., Jardali, F. et al. (2020). Plasma-based $N_2$ fixation into $NO_x$: insights from modeling toward optimum yields and energy costs in a gliding arc plasmatron. *ACS Sustain. Chem. Eng.* 8: 9711–9720. https://doi.org/10.1021/acssuschemeng.0c01815.

**12** Ishikawa, K. (2021). Perspectives on functional nitrogen science and plasma-based in situ functionalization. *Jpn. J. Appl. Phys.* 61: SA0802. https://doi.org/10.35848/1347-4065/ac3558.

**13** Nicholson, D.R. and Nicholson, D.R. (1983). *Introduction to Plasma Theory*, vol. 28. New York: Wiley.

**14** Li, S., van Raak, T., Kriek, R. et al. (2023). Gliding arc reactor under AC pulsed mode operation: spatial performance profile for $NO_x$ synthesis. *ACS Sustain. Chem. Eng.* 11: 12821–12832. https://doi.org/10.1021/acssuschemeng.3c03832.

**15** Zhang, H., Li, L., Li, X. et al. (2018). Warm plasma activation of $CO_2$ in a rotating gliding arc discharge reactor. *J $CO_2$ Util* 27: 472–479. https://doi.org/10.1016/j.jcou.2018.08.020.

**16** Kim, H.H. (2004). Nonthermal plasma processing for air-pollution control: a historical review, current issues, and future prospects. *Plasma Process. Polym.* 1: 91–110. https://doi.org/10.1002/ppap.200400028.

**17** Gangoli, S., Gutsol, A., and Fridman, A. (2010). A non-equilibrium plasma source: magnetically stabilized gliding arc discharge: I. Design and diagnostics. *Plasma Sources Sci. Technol.* 19: 065003. https://doi.org/10.1088/0963-0252/19/6/065003.

**18** Liu, J.L., Gao, Y., Sun, B. et al. (2022). Mechanism study on gliding arc (GA) plasma reforming: a combination approach of experiment and modeling. *Plasma Process. Polym.* 19: 2200077. https://doi.org/10.1002/ppap.202200077.

**19** Tang, Y., Sun, J., Shi, B. et al. (2021). Extension of flammability and stability limits of swirling premixed flames by AC powered gliding arc discharges. *Combust. Flame* 231: 111483. https://doi.org/10.1016/j.combustflame.2021.111483.

**20** Chen, C.-K., Wei, T.-C., Collins, L.R., and Phillips, J. (1999). Modelling the discharge region of a microwave generated hydrogen plasma. *J. Phys. D. Appl. Phys.* 32: 688. https://doi.org/10.1088/0022-3727/32/6/015.

**21** Kosarev, I., Khorunzhenko, V., Mintoussov, E. et al. (2012). A nanosecond surface dielectric barrier discharge at elevated pressures: time-resolved electric field and efficiency of initiation of combustion. *Plasma Sources Sci. Technol.* 21: 045012. https://doi.org/10.1088/0963-0252/21/4/045012.

**22** Bibinov, N., Rajasekaran, P., Mertmann, P. et al. (2011). Basics and Biomedical Applications of Dielectric Barrier Discharge (DBD). In: *Biomedical Engineering, Trends in Materials Science* (ed. A. Laskovski). IntechOpen.

**23** Malik, M.A., Jiang, C., Heller, R. et al. (2016). Ozone-free nitric oxide production using an atmospheric pressure surface discharge–a way to minimize nitrogen dioxide co-production. *Chem. Eng. J.* 283: 631–638. https://doi.org/10.1016/j.cej.2015.07.092.

**24** Fridman, A. (2008). *Plasma Chemistry*. Cambridge University Press.

**25** Van Alphen, S., Vermeiren, V., Butterworth, T. et al. (2019). Power pulsing to maximize vibrational excitation efficiency in N2 microwave plasma: a combined experimental and computational study. *J. Phys. Chem. C* 124: 1765–1779. https://doi.org/10.1021/acs.jpcc.9b06053.

**26** Guerra, V., Tejero-del-Caz, A., Pintassilgo, C.D., and Alves, L.L. (2019). Modelling $N_2$–$O_2$ plasmas: volume and surface kinetics. *Plasma Sources Sci. Technol.* 28: 073001. https://doi.org/10.1088/1361-6595/ab252c.

**27** Vervloessem, E., Aghaei, M., Jardali, F. et al. (2020). Plasma-based $N_2$ fixation into $NO_x$: insights from modeling toward optimum yields and energy costs in a gliding arc plasmatron. *ACS Sustain. Chem. Eng.* 8: 9711–9720. https://doi.org/10.1021/acssuschemeng.0c01815.

**28** Zhang, S., Zong, L., Zeng, X. et al. (2022). Sustainable nitrogen fixation with nanosecond pulsed spark discharges: insights into free-radical-chain reactions. *Green Chem.* 24: 1534–1544. https://doi.org/10.1039/D1GC03859A.

**29** Pei, X., Gidon, D., Yang, Y.-J. et al. (2019). Reducing energy cost of $NO_x$ production in air plasmas. *Chem. Eng. J.* 362: 217–228. https://doi.org/10.1016/j.cej.2019.01.011.

**30** Liu, K., Geng, W., Zhou, X. et al. (2023). Transition mechanisms between selective $O_3$ and $NO_x$ generation modes in atmospheric-pressure plasmas: decoupling specific discharge energy and gas temperature effects. *Plasma Sources Sci. Technol.* 32: 025005. https://doi.org/10.1088/1361-6595/acb814.

**31** K. Li, H. Javed, G. Zhang, Calculation of ozone and $NO_x$ production under AC corona discharge in dry air used for faults diagnostic, 2015 2nd International Conference on Machinery, Materials Engineering, Chemical Engineering and Biotechnology, Atlantis Press, 2015, pp. 461–466.

**32** Kelly, S. and Bogaerts, A. (2021). Nitrogen fixation in an electrode-free microwave plasma. *Joule* 5: 3006–3030. https://doi.org/10.1016/j.joule.2021.09.009.

**33** Lei, X., Cheng, H., Nie, L. et al. (2021). Plasma-catalytic $NO_x$ production in a three-level coupled rotating electrodes air plasma combined with nano-sized TiO2. *J. Phys. D. Appl. Phys.* 55: 115201. https://doi.org/10.1088/1361-6463/ac3f0e.

**34** Lei, X., Cheng, H., Nie, L., and Lu, X. (2022). Nitrogen fixation as $NO_x$ enabled by a three-level coupled rotating electrodes air plasma at atmospheric pressure. *Plasma Chem. Plasma Process.* 42: 211–227. https://doi.org/10.1007/s11090-021-10222-2.

**35** Zhang, Y., Liu, B., Luo, J. et al. (2023). Research on key influencing factors and mechanisms of improved nitrogen fixation efficiency in magnetic-driven gliding arc. *J. Phys. D. Appl. Phys.* 57: 125204. https://doi.org/10.1088/1361-6463/ad13c8.

**36** Ono, R. and Oda, T. (2001). OH radical measurement in a pulsed arc discharge plasma observed by a LIF method. *IEEE Trans. Ind. Appl.* 37: 709–714. https://doi.org/10.1109/28.924749.

**37** Srivastava, N. and Wang, C. (2011). Effects of water addition on OH radical generation and plasma properties in an atmospheric argon microwave plasma jet. *J. Appl. Phys.* 110: 053304. https://doi.org/10.1063/1.3632970.

**38** Wang, C. and Wu, W. (2013). Simultaneous measurements of OH (A) and OH (X) radicals in microwave plasma jet-assisted combustion of methane/air mixtures around the lean-burn limit using optical emission spectroscopy and cavity ringdown spectroscopy. *J. Phys. D. Appl. Phys.* 46: 464008. https://doi.org/10.1088/0022-3727/46/46/464008.

39 Lei, X., Cheng, H., Nie, L., and Lu, X. (2022). Nitrogen fixation as $NO_x$ enabled by a three-level coupled rotating electrodes air plasma at atmospheric pressure. *Plasma Chem. Plasma Process.* 42: 211–217. https://doi.org/10.1007/s11090-021-10222-2.

40 Sarani, A., Nikiforov, A.Y., and Leys, C. (2010). Atmospheric pressure plasma jet in Ar and Ar/H2O mixtures: optical emission spectroscopy and temperature measurements. *Phys. Plasma* 17: 063504. https://doi.org/10.1063/1.3439685.

41 Zhang, H., Zhu, F., Li, X. et al. (2016). Rotating gliding arc assisted water splitting in atmospheric nitrogen. *Plasma Chem. Plasma Process.* 36: 813–834. https://doi.org/10.1007/s11090-016-9700-y.

42 Hibert, C., Gaurand, I., Motret, O., and Pouvesle, J.M. (1999). [OH (X)] measurements by resonant absorption spectroscopy in a pulsed dielectric barrier discharge. *J. Appl. Phys.* 85: 7070–7075. https://doi.org/10.1063/1.370514.

43 Smith, C., Hill, A.K., and Torrente-Murciano, L. (2020). Current and future role of Haber–Bosch ammonia in a carbon-free energy landscape. *Energy Environ. Sci.* 13: 331–344. https://doi.org/10.1039/C9EE02873K.

44 Bogaerts, A., Tu, X., Whitehead, J.C. et al. (2020). The 2020 plasma catalysis roadmap. *J. Phys. D. Appl. Phys.* 53: 443001. https://doi.org/10.1088/1361-6463/ab9048.

45 Benilov, M. (2008). Understanding and modelling plasma–electrode interaction in high-pressure arc discharges: a review. *J. Phys. D. Appl. Phys.* 41: 144001. https://doi.org/10.1088/0022-3727/41/14/144001.

46 Fulcheri, L., Rohani, V., Fabry, F., and Traisnel, N. (2010). Experimental electrical characterization of a low-current tip–tip arc discharge in helium atmosphere at very high pressure. *Plasma Sources Sci. Technol.* 19: 045010. https://doi.org/10.1088/0963-0252/19/4/045010.

47 Gueye, P., Cressault, Y., Rohani, V., and Fulcheri, L. (2019). MHD modeling of rotating arc under restrike mode in 'Kvaerner-type' torch: II. Dynamics and stability at 20 bar pressure. *J. Phys. D. Appl. Phys.* 52: 145202. https://doi.org/10.1088/1361-6463/aaff45.

48 Rad, R.H., Brüser, V., Schiorlin, M. et al. (2023). Enhancement of $CO_2$ splitting in a coaxial dielectric barrier discharge by pressure increase, packed bed and catalyst addition. *Chem. Eng. J.* 456: 141072. https://doi.org/10.1016/j.cej.2022.141072.

49 Rohani, V., Iwarere, S., Fabry, F. et al. (2011). Experimental study of hydrocarbons synthesis from syngas by a tip–tip electrical discharge at very high pressure. *Plasma Chem. Plasma Process.* 31: 663–679. https://doi.org/10.1007/s11090-011-9316-1.

50 Iwarere, S., Rohani, V., Ramjugernath, D. et al. (2014). Hydrocarbons synthesis from syngas by very high pressure plasma. *Chem. Eng. J.* 241: 1–8. https://doi.org/10.1016/j.cej.2013.12.003.

51 Zhu, B., Li, X.-S., Shi, C. et al. (2012). Pressurization effect on dry reforming of biogas in kilohertz spark-discharge plasma. *Int. J. Hydrog. Energy* 37: 4945–4954. https://doi.org/10.1016/j.ijhydene.2011.12.062.

52 Zhang, J.-q., Yang, Y.-j., Zhang, J.-s. et al. (2002). Non-oxidative coupling of methane to $C_2$ hydrocarbons under above-atmospheric pressure using pulsed microwave plasma. *Energy Fuel* 16: 687–693. https://doi.org/10.1021/ef010217u.

53 Eliasson, B., Kogelschatz, U., Xue, B., and Zhou, L.-M. (1998). Hydrogenation of carbon dioxide to methanol with a discharge-activated catalyst. *Ind. Eng. Chem. Res.* 37: 3350–3357. https://doi.org/10.1021/ie9709401.

54 Tsonev, I., O'Modhrain, C., Bogaerts, A., and Gorbanev, Y. (2023). Nitrogen fixation by an arc plasma at elevated pressure to increase the energy efficiency and production rate of $NO_x$. *ACS Sustain. Chem. Eng.* 11: 1888–1897. https://doi.org/10.1021/acssuschemeng.2c06357.

55 Vervloessem, E., Gorbanev, Y., Nikiforov, A. et al. (2022). Sustainable $NO_x$ production from air in pulsed plasma: elucidating the chemistry behind the low energy consumption. *Green Chem.* 24: 916–929. https://doi.org/10.1039/D1GC02762J.

56 Li, J., Yao, S., and Wu, Z. (2020). $NO_x$ production in plasma reactors by pulsed spark discharges. *J. Phys. D. Appl. Phys.* 53: 385201. https://doi.org/10.1088/1361-6463/ab946a.

57 Kim, Y., Hong, S.H., Cha, M.S. et al. (2003). Measurements of electron energy by emission spectroscopy in pulsed corona and dielectric barrier discharges. *J. Adv. Oxid. Technol.* 6: 17–22. https://doi.org/10.1515/jaots-2003-0103.

58 Zhu, J., Ehn, A., Gao, J. et al. (2017). Translational, rotational, vibrational and electron temperatures of a gliding arc discharge. *Opt. Express* 25: 20243–20257. https://doi.org/10.1364/OE.25.020243.

59 Majeed, M., Iqbal, M., Altin, M. et al. (2024). Effect of thermal gas quenching on $NO_x$ production by atmospheric pressure rotating arc plasma: a pathway towards eco-friendly fertilizer. *Chem. Eng. J.* 485: 149727. https://doi.org/10.1016/j.cej.2024.149727.

60 Thomas, S., Mozetic, M., Cvelbar, U. et al. (2018). *Non-thermal Plasma Technology for Polymeric Materials: Applications in Composites, Nanostructured Materials, and Biomedical Fields*. Elsevier.

61 Thorsteinsson, E. and Gudmundsson, J. (2010). The low pressure $Cl_2/O_2$ discharge and the role of ClO. *Plasma Sources Sci. Technol.* 19: 055008. https://doi.org/10.1088/0963-0252/19/5/055008.

62 Lieberman, M.A. and Lichtenberg, A.J. (2005). *Principles of Plasma Discharges and Materials Processing*. John Wiley & Sons.

63 Thorsteinsson, E. and Gudmundsson, J.T. (2009). A global (volume averaged) model of a nitrogen discharge: I. Steady state. *Plasma Sources Sci. Technol.* 18: 045001. https://doi.org/10.1088/0963-0252/18/4/045001.

64 Wang, W., Ma, Y., Chen, G. et al. (2022). Enhanced hydrogen production using a tandem biomass pyrolysis and plasma reforming process. *Fuel Process. Technol.* 234: 107333. https://doi.org/10.1016/j.fuproc.2022.107333.

**65** Wang, Y., Yang, W., Xu, S. et al. (2022). Shielding protection by mesoporous catalysts for improving plasma-catalytic ambient ammonia synthesis. *J. Am. Chem. Soc.* 144: 12020–12031. https://doi.org/10.1021/jacs.2c01950.

**66** Wang, Y., Craven, M., Yu, X. et al. (2019). Plasma-enhanced catalytic synthesis of ammonia over a $Ni/Al_2O_3$ catalyst at near-room temperature: insights into the importance of the catalyst surface on the reaction mechanism. *ACS Catal.* 9: 10780–10793. https://doi.org/10.1021/acscatal.9b02538.

**67** Ma, Y., Wang, Y., Harding, J., and Tu, X. (2021). Plasma-enhanced $N_2$ fixation in a dielectric barrier discharge reactor: effect of packing materials. *Plasma Sources Sci. Technol.* 30: 105002. https://doi.org/10.1088/1361-6595/ac2412.

**68** Bayer, B.N., Bruggeman, P.J., and Bhan, A. (2024). NO formation by $N_2/O_2$ plasma catalysis: the impact of surface reactions, gas-phase reactions, and mass transport. *Chem. Eng. J.* 482: 149041. https://doi.org/10.1016/j.cej.2024.149041.

**69** Pei, X., Li, Y., Luo, Y. et al. (2024). Nitrogen fixation as $NO_x$ using air plasma coupled with heterogeneous catalysis at atmospheric pressure. *Plasma Process. Polym.* 21: 2300135. https://doi.org/10.1002/ppap.202300135.

**70** Zhang, T.-Q., Li, X.-S., Liu, J.-L. et al. (2023). Plasma nitrogen fixation: $NO_x$ synthesis in $MnO_x/Al_2O_3$ packed-bed dielectric barrier discharge. *Plasma Chem. Plasma Process.* 43: 1907–1919. https://doi.org/10.1007/s11090-023-10345-8.

**71** Gicquel, A., Cavadias, S., and Amouroux, J. (1986). Heterogeneous catalysis in low-pressure plasmas. *J. Phys. D. Appl. Phys.* 19: 2013. https://doi.org/10.1088/0022-3727/19/11/003.

**72** Schneider, J., Matsuoka, M., Takeuchi, M. et al. (2014). Understanding $TiO_2$ photocatalysis: mechanisms and materials. *Chem. Rev.* 114: 9919–9986. https://doi.org/10.1021/cr5001892.

**73** Guaitella, O., Thevenet, F., Puzenat, E. et al. (2008). $C_2H_2$ oxidation by plasma/$TiO_2$ combination: influence of the porosity, and photocatalytic mechanisms under plasma exposure. *Appl Catal B* 80: 296–305. https://doi.org/10.1016/j.apcatb.2007.11.032.

**74** Chen, G., Britun, N., Godfroid, T. et al. (2017). An overview of $CO_2$ conversion in a microwave discharge: the role of plasma-catalysis. *J. Phys. D. Appl. Phys.* 50: 084001. https://doi.org/10.1088/1361-6463/aa5616.

**75** Wu, Z., Zhu, Z., Hao, X. et al. (2018). Enhanced oxidation of naphthalene using plasma activation of $TiO_2$/diatomite catalyst. *J. Hazard. Mater.* 347: 48–57. https://doi.org/10.1016/j.jhazmat.2017.12.052.

**76** Sano, T., Negishi, N., Sakai, E., and Matsuzawa, S. (2006). Contributions of photocatalytic/catalytic activities of $TiO_2$ and $\gamma$-$Al_2O_3$ in nonthermal plasma on oxidation of acetaldehyde and CO. *J. Mol. Catal. A Chem.* 245: 235–241. https://doi.org/10.1016/j.molcata.2005.10.002.

**77** Whitehead, J.C. (2010). Plasma catalysis: a solution for environmental problems. *Pure Appl. Chem.* 82: 1329–1336. https://doi.org/10.1351/PAC-CON-10-02-39.

**78** Wallis, A.E., Whitehead, J.C., and Zhang, K. (2007). Plasma-assisted catalysis for the destruction of CFC-12 in atmospheric pressure gas streams using $TiO_2$. *Catal. Lett.* 113: 29–33. https://doi.org/10.1007/s10562-006-9000-x.

**79** Whitehead, J.C. (2019). Plasma-catalysis: is it just a question of scale? *Front. Chem. Sci. Eng.* 13: 264–273. https://doi.org/10.1007/s11705-019-1794-3.

**80** Neyts, E. and Bogaerts, A. (2014). Understanding plasma catalysis through modelling and simulation—a review. *J. Phys. D. Appl. Phys.* 47: 224010. https://doi.org/10.1088/0022-3727/47/22/224010.

**81** Chen, H.L., Lee, H.M., Chen, S.H., and Chang, M.B. (2008). Review of packed-bed plasma reactor for ozone generation and air pollution control. *Ind. Eng. Chem. Res.* 47: 2122–2130. https://doi.org/10.1021/ie071411s.

**82** Hensel, K., Martišovitš, V., Machala, Z. et al. (2007). Electrical and optical properties of AC microdischarges in porous ceramics. *Plasma Process. Polym.* 4: 682–693. https://doi.org/10.1002/ppap.200700022.

# Part IV

# Environmental and Economic Viability

# 9

# Environmental Impact and Sustainability Aspects of Plasma-Based Nitrogen Fixation

*Nam Nghiep Tran*[1], *Nguyen Van Duc Long*[1], *Muhammad Yousaf Arshad*[1], *Jose Luis Osorio Tejada*[2], *and Volker Hessel*[1,2]

[1] *School of Chemical Engineering, Adelaide University, Adelaide, SA, Australia*
[2] *School of Engineering, University of Warwick, Coventry, UK*

## 9.1 Introduction

Integrating sustainability principles [1] into chemical engineering design has emerged as a substantial challenge for modern society [2–5]. Achieving sustainability in the context of chemical production involves improving energy efficiency [6, 7], reducing greenhouse gas (GHG) emissions and environmental impact [8], integrating renewable energy [9], promoting localized production [10], improving resource efficiency [11], and optimizing waste resource efficiency through the Circular Economy [12, 13]. Despite efforts to lower GHG emissions within the chemical industry, some chemical products still rely heavily on non-renewable natural resources [14]. One representative example of an industrial chemical process that faces substantial sustainability challenges is the nitrogen fixation process [15, 16].

Nitrogen fixation is vital for all forms of life, as nitrogen is essential for biosynthesizing the fundamental building blocks of organisms, plants, and other life forms. Currently, nitrogen fixation can be categorized into two main processes, including ammonia synthesis reaction by nitrogen reduction and the direct production of $NO_x$ or $NO_3^-/NO_2^-$ by oxidation method [17–20]. The Haber-Bosch (HB) process, the primary method of artificial nitrogen fixation, has been essential to the agriculture and chemical industries since the 1910s. However, its significant drawbacks, such as high-energy consumption and substantial emissions, present critical challenges [17]. Particularly, operating at 450–600 °C and 150–350 bar, the HB process is centralized and continuous, producing at least 2,000 tons of $NH_3$ per

*Plasma-Assisted Nitrogen Fixation for Sustainable Process Industries*, First Edition. Edited by Volker Hessel, Annemie Bogaerts, Gabriele Centi, Evgeny Rebrov, and Nguyen Van Duc Long.

Published 2026 by John Wiley & Sons, Inc.

day [21]. Additionally, ammonia synthesis relies heavily on natural gas as a feedstock, resulting in considerable $CO_2$ emissions (up to 1.5 kg of $CO_2$ per kilogram of $NH_3$ produced) [22–24]. Furthermore, the efficiency of the HB process decreases upon downsizing and upon intermittent operation, which is one of the bottlenecks in matching with renewable energy sources [25, 26]. Due to its reliance on fossil feedstocks, global $NH_3$ production is concentrated in a few countries with access to affordable natural gas or coal [27, 28]. This centralization increases the carbon footprint as ammonia must be shipped internationally to meet global demand. Large-scale HB plants are designed for continuous operation, often taking hours or even days to reach a steady state, making it challenging to connect with intermittent renewable sources like solar or wind power [29, 30]. Recent trade disruptions, including the Suez Canal blockage, COVID-related lockdowns, and container shortages, have further raised international shipping costs and times, raising concerns about the sustainability of global trade [27].

Therefore, developing a sustainable nitrogen fixation process has been a critical area of research. Enhancing the energy efficiency of nitrogen fixation would bring significant economic and environmental benefits, especially in the context of the rapidly growing population and dwindling natural resources. As a result, current research into sustainable nitrogen fixation techniques focuses on several key objectives, including improved energy efficiency, reduced emissions, and minimized environmental impact [31].

Among the various technologies, plasma-assisted nitrogen fixation (PANF) technology is especially promising due to its small scale, mild reaction conditions, and flexible parameters [32]. This approach has the potential to work effectively with intermittent energy sources, such as wind and solar, which are characterized by rapid load fluctuations [33]. For nitrogen reduction, ammonia synthesis via low-temperature plasma using nitrogen and hydrogen or water, combined with renewable energy, offers a pathway to zero-emission nitrogen fixation [17]. In plasma oxidation, the direct oxidation of air or a nitrogen–oxygen mixture can reduce the need for high-purity hydrogen and nitrogen. Generally, low-temperature plasmas are categorized into two types: thermal (equilibrium) plasma (TP) and nonthermal (nonequilibrium) plasma (NTP) [9]. The main drawbacks of TPs, including low excitation selectivity, very high gas temperature, significant quenching requirements, and electrode issues, result in limited energy efficiency and reduce the applicability of TP sources, making them less suitable for $NH_3$ synthesis [34].

In contrast, NTPs such as low-pressure glow and radio frequency, microwave discharges, dielectric barrier discharges (DBDs), and laser-produced plasmas have been utilized due to their high selectivity in plasma chemical reactions [9]. These plasmas operate effectively at low temperatures without the need for quenching. NTPs are characterized by an extremely high electron temperature, while the bulk of the gas remains at a mild temperature. These nonequilibrium properties enable thermodynamically unfavorable chemical reactions under mild conditions such as atmospheric pressure and low temperature. Additionally, NTP offers an opportunity to use catalysts to benefit surface reactions and increase selectivity toward the desired product. Combination of NTP with a catalyst (plasma catalysis) often yields synergetic effects, with the efficiency of the combined system exceeding that of the constituent parts. Figure 9.1 shows the developments of plasma reactors, highlighting $NH_3$ yield and energy yield [35–44].

Besides, significant efforts have also been made toward the synthesis of $NO_x$ molecule, which can be classified into three different categories: TP, cold plasma or NTP, and warm plasma [45]. Most of the early research studies reported the use of TPs, such as the electric arc [46, 47] or the spark discharge [48], though these methods have high theoretical energy consumption limitations [49]. NTP is operated at close to room temperature and has a lower energy density than TP. However, to date, high-energy efficiencies have not been reported under NTP conditions, leading some researchers to question its suitability for $NO_x$ production [49, 50]. Warm plasmas, combining TP and NTP conditions, have been used for $NO_x$ generation, and progress regarding energy efficiency and product selectivity has been recently reported in the literature [51].

## 9.2 Environmental Benefits of PANF

PANF significantly reduces carbon emissions and enhances industrial nitrogen output compared to the HB process, particularly when combined with renewable energy sources [52]. The carbon footprint of the HB process is about 1.5 kg $CO_2$/kg$NH_3$ generated, mostly attributable to the natural gas used in the hydrogen generation phase [53]. The process has several environmental advantages, including decreased energy demand, reduced resource use, and significant mitigation of GHG emissions, especially when combined with renewable energy sources [14, 54]. Figure 9.2 illustrates the global nitrogen fixation process. Biological nitrogen fixation (BNF) contributes $58 \pm 50\%$ Tg N year$^{-1}$ on land and $140 \pm 50\%$ Tg N year$^{-1}$ in oceans, while human activities, like the HB process for fertilizer production, add $120\% \pm 10$ Tg N year$^{-1}$, combustion adds $30\% \pm 10$ Tg N year$^{-1}$, and agricultural BNF adds $60 \pm 30$ Tg N year$^{-1}$. Anthropogenic nitrogen input, totaling 210% Tg N year$^{-1}$, not only increases food production but also leads to environmental challenges like eutrophication, pollution, and disruptions in the nitrogen cycle, highlighting the need for sustainable nitrogen management. Annual nitrogen stands at 413 Tg N year$^{-1}$ [55, 56].

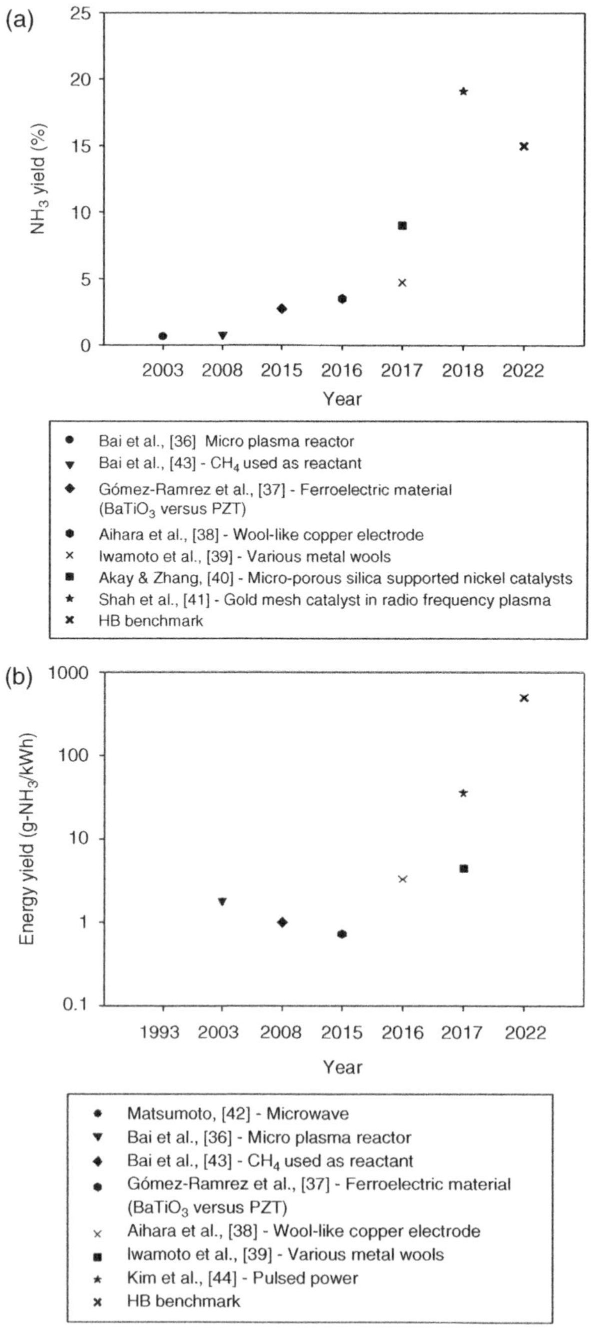


**Figure 9.1** Reported (a) $NH_3$ yield and (b) energy yield for different plasmas [35–44].

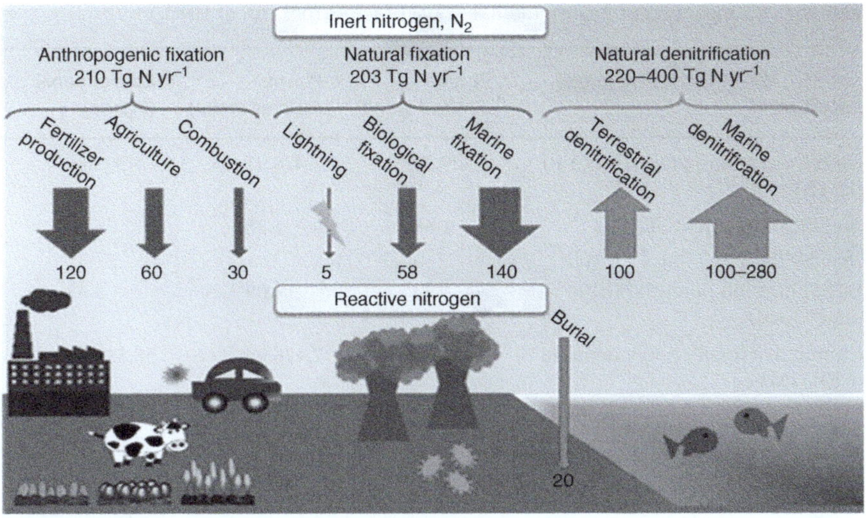


**Figure 9.2** Nitrogen availability in atmospheric layers. *Source:* [55] / American Chemical Society.

### 9.2.1 Carbon Footprint Analysis

PANF is notable for its capacity to reduce the carbon footprint of nitrogen manufacturing methods [57]. Life cycle assessment (LCA) in the study literature highlights that plasma-assisted manufacturing significantly decreases global warming potential (GWP) relative to traditional approaches [58]. The GWP of plasma-assisted processes powered by renewable energy ranges from 478.97 to 529.34 kg $CO_2$-equivalent per ton of $HNO_3$, depending on whether wind or solar energy is used. In stark contrast, conventional nitric acid production using fossil fuels presents a GWP of 796.80 kg $CO_2$-equivalent per ton of $HNO_3$, with emissions escalating dramatically to 5,399.89 kg $CO_2$-equivalent per ton when lignite is employed [14, 59].

The integration of renewable energy sources with PANF is crucial for attaining these reductions [52]. The process's dependence on electricity instead of fossil fuels for plasma generation facilitates a shift to wind or solar energy, therefore eliminating the carbon emissions usually linked to nitrogen fixation [57, 60, 61]. This capacity to use renewable energy immediately facilitates the decarbonization of industrial nitrogen production [62]. A plasma-assisted processes powered by renewable energy have a carbon footprint that can be 19% lower when using renewable electricity, such as solar or wind. Furthermore, plasma-assisted systems are highly adaptable to decentralized production models, which can be established near agricultural fields or industrial applications [14, 53, 57]. By producing

**Table 9.1** Comparison of environmental impact factors (per ton of $HNO_3$).

| Impact factor | Plasma-assisted (wind) | Plasma-assisted (solar) | Conventional (lignite) |
|---|---|---|---|
| Global warming potential (GWP) (kg $CO_2$-eq) | 529.34 | 478.97 | 5,399.89 |
| Acidification potential (AP) (kg $SO_2$-eq) | 1.68 | 1.38 | 6.37 |
| Eutrophication potential (EP) (kg $NO_x$-eq) | 1.23 | 1.09 | 5.43 |
| Cumulative fossil energy demand (CED) (MJ-eq) | 7,974.27 | 7,426.98 | 58,586.86 |

*Source:* Adapted from [58, 63, 64].

nitrogenous compounds such as ammonia or nitric acid closer to their point of use, the system reduces the carbon emissions that would otherwise result from transporting these chemicals over long distances [54].

Table 9.1 demonstrates that PANF, when fueled by renewable energy sources, significantly surpasses traditional methods for GWP, acidification potential, eutrophication potential, and cumulative fossil energy consumption. The decreases in GWP and CED are especially significant when lignite serves as the energy source in traditional processes [52, 57, 64].

### 9.2.2 Comparison with the HB Process

The HB process has long been the gold standard for ammonia production, but its environmental costs are substantial [57]. This process consumes approximately 1–2% of global energy annually and contributes over 300 million metric tons of $CO_2$ to the atmosphere each year [14, 53, 65]. Plasma-assisted methods may be scaled down to satisfy localized demand – something that the HB process finds challenging to accomplish – they are not only suitable for huge, centralized operations [14, 53, 57, 66]. Because of this decentralization, ammonia may be produced in smaller, renewable energy-powered facilities, avoiding the carbon footprint associated with the logistics and transportation of large-scale HB plants.

Even though PANF is still being refined, existing systems have energy efficiencies between 1.9 and 17.2 g $NH_3$/kWh, which is lower than the HB process's by 8–12 g $NH_3$/kWh efficiency under ideal circumstances [54]. But in decentralized systems, plasma systems offer the benefit of using intermittent renewable energy sources, which increases their adaptability and environmental friendliness [17]. The substantial benefits of PANF, especially when combined with renewable

**Table 9.2** Reduction in greenhouse gas emissions, energy consumption and reduction potential (per ton of $NH_3$).

| Process | GHG emissions (kg $CO_2$-eq) | Reduction potential (%) | Energy consumption ($MJ\ mol^{-1}$) |
|---|---|---|---|
| PANF | 0 (with renewable energy) | Up to 86% | 4.2–8 |
| Haber-Bosch process | 1,800–2,100 | – | 0.48 |

energy, are shown in Table 9.2. The total removal of $CO_2$ emissions in renewable-powered plasma systems makes it a preferable option to the HB process, despite the fact that its energy consumption per mole is presently very high [53].

### 9.2.3 Energy Efficiency and Consumption

In its current configuration, the PANF process is less energy efficient than the highly optimized HB process; yet, it has a significant opportunity for advancement [57, 65]. Plasma reactors consume approximately 4.2–8 MJ per mole of $NO_x$, depending on the specific configuration and operating conditions. This is higher than the energy consumption of the HB process, which uses 1.7 MJ per mole of nitrogen [54]. In contrast to the HB process, which is mostly dependent on fossil fuels, plasma-assisted systems use renewable electricity, thereby mitigating their elevated energy requirements. It provides an ideal technology for both surplus renewable energy utilization and major consumer of nitrogen feedstocks and end-products [52, 67].

The energy consumption of various nitrogen fixation methods varies drastically. Biological fixation is the most efficient, consuming just above 1 $MJ\ mol^{-1}$. The HB process uses 8–12 $MJ\ mol^{-1}$, with a theoretical limit of 1.7 $MJ\ mol^{-1}$. The cyanamide process consumes 6–10 $MJ\ mol^{-1}$, while TP fixation ranges from 20 to 100 $MJ\ mol^{-1}$, with a theoretical limit near 2 $MJ\ mol^{-1}$. NTP fixation consumes 5–50 $MJ\ mol^{-1}$, with a similar limit, and metallo-complex fixation is around 1–2 $MJ\ mol^{-1}$, close to its theoretical efficiency. These values highlight the energy-saving potential of emerging methods compared to traditional ones [17, 65].

As advancements in plasma technology continue, improvements in reactor design and energy recovery systems, such as the recycling of tail gases and heat, are expected to enhance the overall energy efficiency of plasma-assisted processes [53, 62]. For example, tail gas recycling can capture the excess heat and gases produced during nitrogen fixation, allowing these byproducts to be reused, thus further reducing the energy consumption of the system [65]. Intensification and sustainability assessments are expected to drive a paradigm shift for major ammonia producers and fuel cell technology providers.

### 9.2.4 Reduction in GHG Emissions

PANF offers a significant benefit by enabling the reduction of GHG emissions. The conventional HB process generates substantial $CO_2$ emissions due to its dependence on steam methane reforming (SMR) for hydrogen production from natural gas, a method characterized by high carbon intensity. On average, each ton of ammonia produced using the HB process emits 1.8–2.1 tons of $CO_2$ [53, 68]. PANF using renewable energy may attain almost zero $CO_2$ emissions [57]. This procedure utilizes air as a nitrogen supply and water as a hydrogen source via electrolysis, so completely removing the need for fossil fuels [52]. The decentralized character of plasma-assisted systems enables their placement at the point of use, hence reducing emissions linked to the transportation of ammonia and other nitrogenous compounds from centralized manufacturing facilities to distant sites as a decentralized production model [17, 67].

As indicated in Table 9.2, PANF can reduce the GHG emissions by up to 86% compared to the HB process when powered by renewable energy sources. This reduction is a crucial contribution to achieving international climate goals, such as those set by the Paris Agreement [17].

### 9.2.5 Integration with Renewable Energy Sources

The capacity to seamlessly integrate with renewable energy sources is a crucial distinguishing factor for PANF [57]. In contrast to the HB process, which needs a constant and reliable energy supply from fossil fuels, plasma systems are adept at functioning intermittently, making them suitable for integration with fluctuating renewable energy sources, such as wind, solar, or hydropower [17, 52]. PANF devices can rapidly initiate and terminate operations, making them very adaptable to renewable energy networks that encounter supply volatility [69].

Research indicates that plasma-assisted processes, when energized by wind energy, may decrease cumulative energy demand (CED) by 84%, and solar-powered systems can cut CED by 77% in comparison to traditional techniques [68]. These reductions highlight the potential for PANF to become a more sustainable and energy-efficient process as renewable energy technologies continue to improve. When integrated with renewable energy, the plasma-assisted process can achieve up to 30% improvements in energy efficiency [53]. Figure 9.3 highlights such roles of plasma-based nitrogen fixation technologies in the evolving landscape of remote and on-site production, particularly in the context of cost, energy use, and farm-level production, alongside emerging commercial and

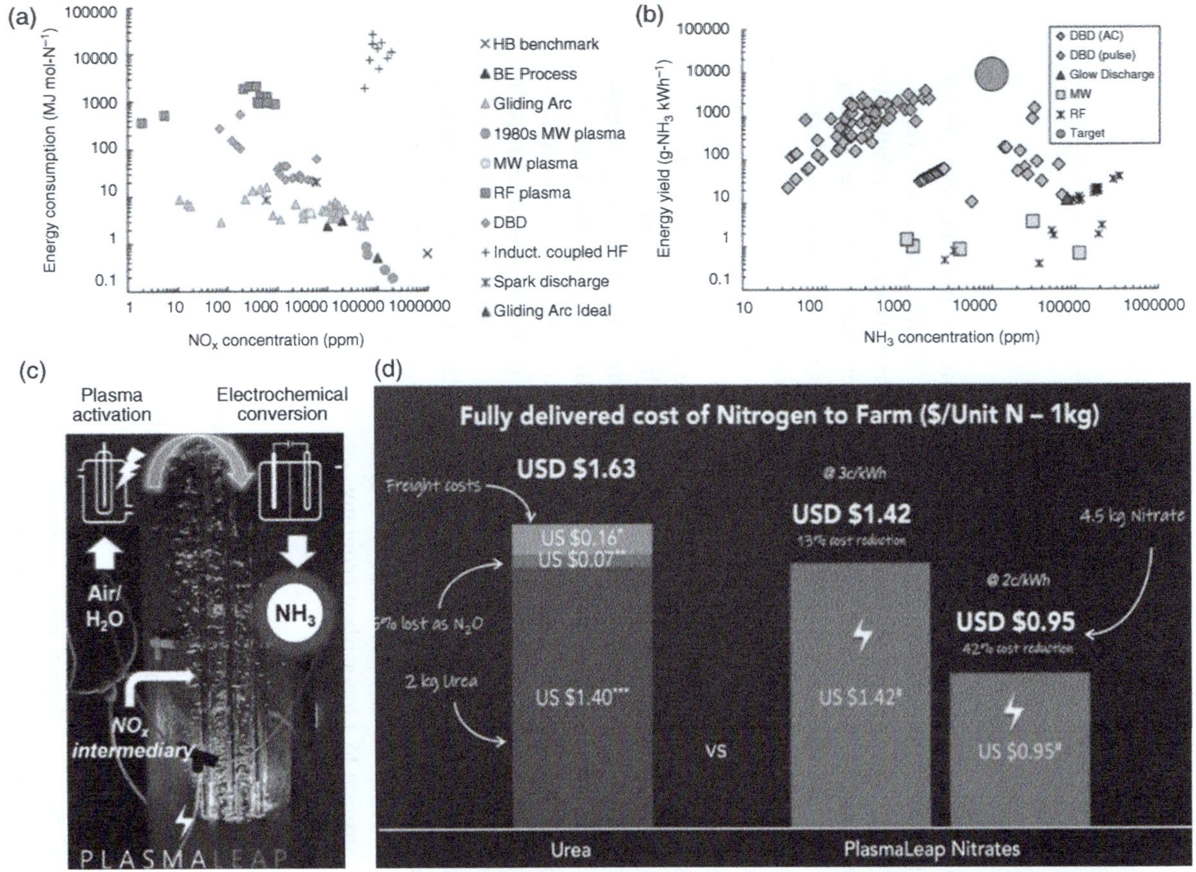


**Figure 9.3** (a) The power consumption of plasma-based reactors utilized for NO production: (b) A comparison with regard to energy yield in plasma-assisted and ammonia production studies (c) PlasmaLeap Pty. Ltd.'s zero-emission ammonia production technology, and (d) PlasmaLeap's nitrogen fixation technology's commercial prospects compared to conventional urea fertilizer. *Source:* [70–72] / Elsevier / CC BY 4.0.

**Table 9.3** Integration of renewable energy into PANF (per ton of $NH_3$).

| Energy Source | Global warming potential (GWP) (kg $CO_2$-eq) | Cumulative energy demand (CED) (MJ-eq) |
|---|---|---|
| Wind energy | 529.34 | 7,974.27 |
| Solar energy | 478.97 | 7,426.98 |
| Conventional (lignite) | 5,399.89 | 58,586.86 |

*Source:* Adapted from [54, 73].

financial opportunities. Four different panels are merged into one scenario for the requirements and performance of a farm-level plasma system. Figure 9.3a examines the power consumption of various plasma-based reactors relative to nitrogen concentration, reflecting ongoing advancements in energy efficiency. Figure 9.3b evaluates energy yield across plasma-assisted ammonia production methods, illustrating differences in process performance. Figure 9.3c outlines Plasma-Leap's zero-emission ammonia synthesis, integrating plasma activation with electrochemical conversion to present an alternative to conventional large-scale HB systems. Figure 9.3d considers the cost of nitrogen delivery, indicating how plasma-derived ammonia could align with or improve upon the economics of conventional urea fertilizers under favorable energy conditions. These insights contribute to the broader discussion on the feasibility of decentralized ammonia production, particularly in relation to evolving market dynamics, energy costs, and the integration of renewable resources in nitrogen fixation.

Table 9.3 underscores the significant environmental advantages of integrating renewable energy into PANF [17, 52]. Both wind and solar energy dramatically reduce the GWP and CED of the process, positioning PANF as a key technology in the transition to a low-carbon economy.

## 9.3 Circular Economy Considerations

Several benefits that improve sustainability and add to a circular economy are offered by PANF, which is of particular interest to enterprises that are trying to conform to ESG principles [53, 74]. In this part, we will look at the ways in which processes that use plasma may integrate into closed-loop systems, optimize resource utilization, and decrease waste [68, 75].

### 9.3.1 PANF Within the Circular Economy Model

Industrial systems may be made more efficient and less wasteful by following the circular economy concept, which aims to reuse resources and energy [54, 76]. In keeping with these ideas, PANF allows for the localized and decentralized generation of nitrogenous compounds, while simultaneously reducing the requirement for large-scale, centralized plants [57]. Reduced emissions and energy consumption from transportation may be achieved by having production of ammonia or nitric acid carried out closer to the point of consumption via decentralized production facilities [17, 52].

Furthermore, nitrogen compounds may be used as inputs for various industrial applications using plasma-assisted processes that are part of circular production cycles. One way to improve resource efficiency is by applying nitrogenous chemicals on-site as fertilizers or chemical feedstocks [54, 77]. This reduces dependency on external sources. Plasma reactors may be run more sustainably with the use of renewable energy sources [53, 78]. Solar photovoltaic (PV)-electrolysis yields 5.95–7.38 $mgH_2\ (Wh)^{-1}$, with solar electrochemical ammonia synthesis reaching 732 $mgNH_3\ (Wh)^{-1}$. Arshad and Azam [16] highlight that decentralized production for 100 ha has a GWP of solar urea routes ranging from 0.091 to 1.083 kg $CO_2$-eq/kg urea. One way to make manufacturing more sustainable is to use renewable energy and energy recovery technologies to minimize waste [17]. PANF helps with this by reducing the need for fossil fuel input [57, 78]. The use of PANF lends credence to the circular economy model's claims of resource efficiency and waste minimization. Reduced waste, improved resource efficiency, and decreased operating environmental impact are all benefits of the system's ability to recycle unreacted hydrogen and nitrogen gases.

### 9.3.2 Resource Utilization and Waste Minimization

One of the key advantages of PANF is its reduced resource demand compared to conventional methods. Natural gas and other fossil fuels are rendered unnecessary by plasma systems, which primarily use air and water as their feedstocks [79]. Less emissions and a lessening of the environmental impact are direct results of this cutback in resource use [54]. Furthermore, there is a decrease in resource demand due to the fact that plasma-assisted processes may function at room temperature and pressure, requiring less energy-intensive equipment [52, 80, 81]. In terms of waste minimization, PANF produces fewer emissions and byproducts than the HB process. The integration of energy recovery systems – such as tail gas recycling and heat recovery – captures excess energy and gases produced during nitrogen fixation, which can then be reused to enhance overall efficiency and further reduce waste [67, 82]. An on-site decentralized clean fuel and ammonia can be produced

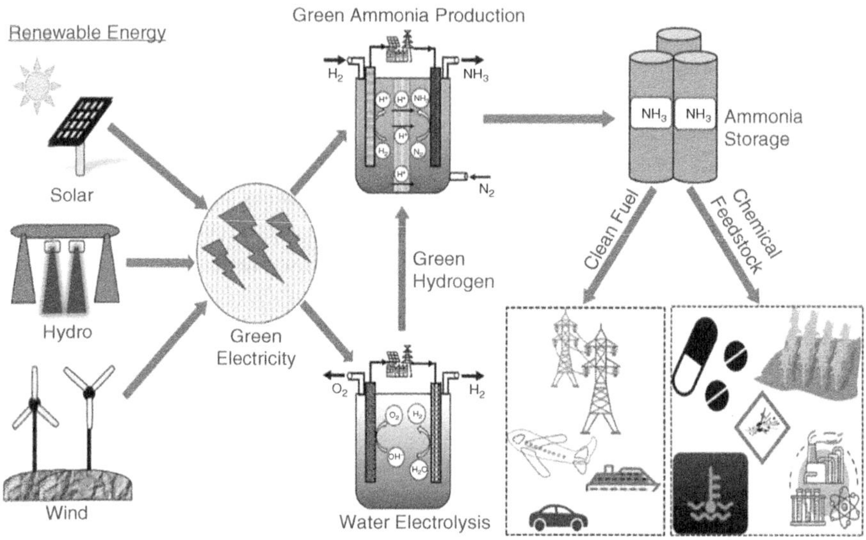


**Figure 9.4** An environmentally friendly method of producing ammonia from water and air that uses electrocatalysis, thermal catalysis, photoelectric catalysis, and nonthermal plasma fueled by renewable energy sources is the Haber-Bosch process. *Source:* [83] / with permission of Elsevier.

using green electricity. Figure 9.4 highlights that renewable energy (solar, wind, hydro) generates green electricity, which powers water electrolysis, resulting in green hydrogen. Hydrogen is converted to green $NH_3$, which is stored and used as a clean fuel or chemical feedstock for industries and agriculture and stands as an example for the PANF system [84–86]. However, another production schematic of an extended version of such conventional atmospheric pressure processes and highlighted in Figure 9.5, and both entail green ammonia on-site production.

### 9.3.3 Closed-Loop Systems and Recycling Opportunities

There are prospects for incorporating PANF into closed-loop systems, especially in sectors that rely on nitrogen compounds as feedstocks [57]. Industries may decrease their reliance on external nitrogen supplies and maximize waste reduction by permitting on-site manufacture of nitrogen compounds like ammonia or nitric acid. This creates a self-sustaining production cycle that reuses these compounds inside the system [53, 68, 87]. Figure 9.5 shows a small-scale, atmospheric-pressure processes like plasma and electrochemical technologies offer a sustainable alternative to the energy-intensive HB process. One way to make plasma-assisted processes more suitable for the circular economy is to power them with

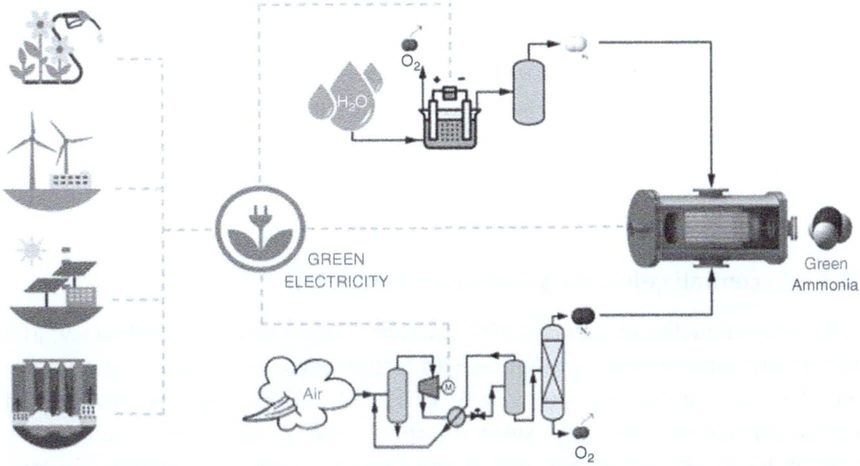


**Figure 9.5** Renewable energy and plasma-supported nitrogen fixation process for pure ammonia production. *Source:* [64] / with permission of Elsevier.

renewable energy [53]. Energy recovery systems capture and reuse waste energy, which reduces overall energy consumption and improves resource efficiency, contributing to a more sustainable and resilient production model [52, 79]. Another example, explained schematically, Figure 9.6, is where nitrogenous fertilizers can be generated on demand at the field level using NTP and associated

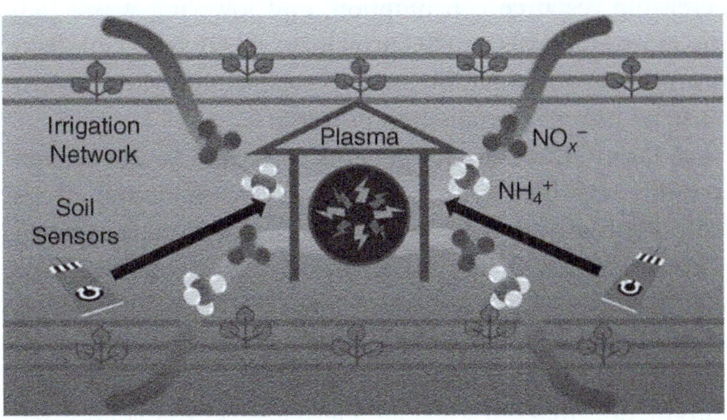


**Figure 9.6** A diagram of the recommended on-demand production of nitrogenous fertilizers at the field level, utilizing nonthermal plasma in conjunction with combined electrochemical soil sensors. Real-time data from sensors in the soil facilitate the adaptable plasma-activated synthesis of aqueous rN species, which can be distributed through present irrigation systems in fields of agriculture. *Source:* [55] / American Chemical Society.

electrochemical soil sensors. Real-time soil sensor data facilitate the dynamic, plasma-activated synthesis of aqueous reactive nitrogen (rN) species, which can be integrated into existing agricultural irrigation systems [55]. This implementation of a closed-loop and sensor-supported system further lessens nitrogen fixation and environmental burden through efficient process and better resource management.

### 9.3.4 Decentralization via Small-Scale Production

PANF offers benefits to reduce GHG emissions, improve energy efficiency, and lower resource consumption, aligning with corporate goals of minimizing environmental impact and conserving resources [3, 74, 79]. The decentralization of plasma-assisted systems also supports social and governance objectives by enabling localized production, which can improve access to essential nitrogen compounds in remote or underserved regions [88, 89]. The decentralized nature of PANF makes it adaptable for smaller scale production facilities, which can be integrated into local supply chains [17]. This improves the resilience of nitrogen supply chains and reduces the environmental impact associated with transporting nitrogen compounds over long distances [52, 53]. Furthermore, the ability to produce nitrogen compounds on-site reduces the need for large infrastructure investments, making the technology accessible to smaller industries [89].

PANF is an innovative method that diminishes the environmental impact of nitrogen synthesis [57]. Plasma-assisted processes may significantly contribute to the shift toward a more sustainable and resilient nitrogen industry by using renewable energy, decreasing resource consumption, and lowering waste [53]. With companies placing more emphasis on ESG principles, PANF is a viable solution for attaining environmental sustainability, social responsibility, and governance excellence in industrial nitrogen production [79, 88].

Table 9.4 supports that PANF offers several benefits compared to the conventional HB process, especially regarding sustainability and environmental effects. Functioning in ambient circumstances with renewable energy sources substantially decreases energy consumption, GHG emissions, and reliance on fossil fuels. The energy efficiency of plasma-assisted procedures may fluctuate, although its carbon impact is significantly reduced, particularly when using renewable energy sources. Moreover, its substantial compatibility with renewable energy and capacity for decentralized production make it a viable alternative to the centralized, energy-intensive HB process, which is largely dependent on natural gas and generates considerable $CO_2$ emissions. PANF provides a cleaner, more adaptable method for ammonia synthesis, with enhanced potential for incorporation into future sustainable energy systems.

**Table 9.4** PANF comparison with Haber-Bosch process for sustainability and process parameters.

| Characteristic | PANF | Haber-Bosch process |
|---|---|---|
| Operating conditions | Ambient temperature and atmospheric pressure | 400–600 °C, 150–300 atm |
| Energy source | Renewable (wind, solar, hydropower) | Fossil fuels (natural gas) |
| Energy efficiency (g $NH_3$/kWh) | 1.9–17.2 g $NH_3$/kWh | 8–12 g $NH_3$/kWh |
| Energy consumption (MWh/ton $NH_3$) | 28 MWh/ton $NH_3$ | 9–12 MWh/ton $NH_3$ |
| GHG emissions | Lower with renewables, 40% reduction compared to HB | 2.1 tons of $CO_2$ per ton of $NH_3$ produced |
| Carbon footprint | 19% lower when powered by renewables | High due to reliance on fossil fuels |
| Recycling opportunities | Unreacted gases ($N_2$, $H_2$) can be recaptured and reused | Limited to steam reuse |
| Resource utilization | Air and renewable energy, reduced natural gas dependency | Heavy reliance on natural gas |
| Integration with renewables | High compatibility with wind, solar, and hydropower | Difficult to integrate with intermittent renewables |
| Waste production | Minimal, potential for closed-loop recycling | Significant $CO_2$ and waste byproducts |
| Decentralized production | Allows for localized production, reducing transport needs | Requires centralized, large-scale facilities |

*Source:* Adapted from [53, 83, 88].

## 9.4 Life Cycle Assessment

### 9.4.1 LCA of PANF: Overview

When analyzing the potential environmental impacts of emerging technologies, it is essential to consider not only carbon emissions throughout the value chain affecting climate change but also other environmental impact categories, including emissions of gases released during feedstock production and product distribution to end-users. For example, when electrifying an industrial process – where a conventional fossil fuel such as natural gas is replaced by grid electricity – the environmental impacts can extend to acidification, ozone formation, and water eutrophication, particularly in regions such as Australia, China, India, Poland,

and other countries where most electricity is generated from fossil sources (Cozzi et al. [90]; ETH [91]; Ritchie and Rosado [92]). In these regions, the predominance of coal and lignite plants contributes airborne pollutants, including nitrogen oxides ($NO_x$), which can lead to acid rain and deposit onto land and water bodies, increasing nutrient loads and potentially causing harmful algal blooms, oxygen depletion, and negative effects on aquatic ecosystems and human health (HH) [26, 93].

Other sectors, such as transport, face a similar dilemma: new technologies can reduce dependence on fossil fuels, decreasing GHG emissions while simultaneously increasing impacts related to water use and human toxicity, particularly from battery disposal [94]. Therefore, strategies to reduce carbon emissions need to be evaluated from a more comprehensive perspective, such as through the LCA methodology [95].

For these reasons, emerging nitrogen fixation technologies that utilize electricity are being promoted with renewable energy sources, leveraging the recent decline in electricity costs from solar and wind energy [96]. Plasma-based plants could also operate at small scales, allowing them to be installed in rural areas off the national grid. In this context, it is possible to evaluate the environmental performance of these plants under assumptions of dedicated renewable electricity generation. Producing nitrogen-based products, such as fertilizers, at these distributed plants could substantially reduce carbon emissions. However, from an LCA perspective, even renewable energy sources like solar PV are not without impact, as significant fuel and electricity are used during silicon mining, other minerals used in inverters, and installations also contribute to terrestrial and human toxicity [27].

This comprehensive analysis of emerging technologies underscores the importance of methodologies like LCA. For instance, Osorio-Tejada et al. [10] performed an LCA comparing the conventional SMR-HB process with a low-temperature plasma reactor for ammonia production, employing a cradle-to-gate scope. This assessment considers the impacts of feedstock extraction, production, and transportation to the ammonia factory, as well as the impacts of factory construction, ammonia synthesis, and purification, without including product storage and distribution. Therefore, a cradle-to-site, or cradle-to-distribution, approach would provide a more complete picture by incorporating the impacts of storage and transportation to the point of use.

### 9.4.2 Benchmarking Against the HB Process

A core aspect of using LCA is the necessity of comparing at least two processes producing a similar product or service to contextualize the results. Evaluating potential environmental impacts individually (e.g. climate change in kg of $CO_2$-eq, terrestrial eutrophication in kg of N-eq) often makes it challenging to determine whether a process is environmentally friendly [97]. Thus, a primary approach is to benchmark new processes against a baseline process to determine whether environmental impacts have increased or decreased across specific impact categories. However, practitioners may intentionally select highly polluting processes for comparison to emphasize substantial emission reductions in the new process; this approach can overstate environmental benefits. Therefore, it is essential to benchmark the new process against a widely used, conventional process that incorporates the best available techniques.

For ammonia production, the conventional HB process is the industry standard, accounting for around 90% of global ammonia production and relying primarily on fossil resources [98, 99]. Since its development in 1908, the HB process has significantly improved its energy efficiency [100], which has also reduced environmental impacts such as climate change by decreasing fossil fuel use per ammonia unit. Despite these efficiencies, the HB process's overall energy consumption remains high, at approximately 28 MJ $kg^{-1}$ $NH_3$. A fair comparison for a plasma-based ammonia synthesis plant would, therefore, involve a high-efficiency HB plant, ideally one equipped with $CO_2$ capture systems. Additionally, selecting a conventional HB process that uses a less environmentally intensive feedstock, such as natural gas rather than coal, would further enhance fairness, as coal-based HB plants can produce twice the emissions of those using natural gas [101].

In this context, emerging technologies for nitrogen-based chemicals like ammonia should be benchmarked against natural gas-supplied SMR-HB plants, offering a realistic comparison for assessing environmental benefits and challenges.

### 9.4.3 Environmental Impact Analysis (Including $CO_2$ Emissions and Pollutants)

An environmental LCA study, following ISO 14040-44 standards [95], involves four essential steps: (a) defining the goal and scope, (b) conducting an inventory analysis, (c) performing an impact assessment, and (d) interpreting the results.

In this case study, the goal of the LCA was to evaluate the environmental impacts of producing 1 ton of anhydrous ammonia using a DBD plasma reactor, with hydrogen supplied by proton exchange membrane (PEM) water electrolyzers and nitrogen produced in pressure swing absorption (PSA) units. The results were compared against a conventional method of producing the equivalent product in a large-scale natural gas-supplied SMR-HB plant. The scope of the LCA was cradle-to-site in Australia and included refrigerated national storage, pressurized regional and local storage, and the transportation of ammonia over significant distances. Specifically, ammonia was transported over a national distance of 4,000 km by diesel-electric train and over a regional distance of up to 800 km by articulated and rigid tanker trucks. This scenario represents the optimal supply chain for the national plant, utilizing train transport as much as possible. For plasma-based plants located closer to end-users, transportation was assumed to be by rigid tanker trucks. Additionally, this study examined ammonia production using a novel high-temperature plasma (HTP) technology to generate hydrogen, which could potentially replace the SMR process in supplying HB systems at the regional scale. A total of three pathways were analyzed: (i) a national-scale SMR-HB plant supplying 1.6 million tons of ammonia annually across distances up to 4,800 km; (ii) a regional-scale HTP-HB plant supplying 320,000 tons of ammonia annually up to 800 km; and (iii) a local-scale electrolyzer-NTP plant producing 10,650 tons of ammonia annually, located 30 km from end-use sites. The system expansion approach was applied to account for co-products (e.g. carbon black from HTP, oxygen from electrolysis) and byproducts (e.g. heat or steam from HB) generated in each pathway.

For the life cycle inventory analysis, the different feedstocks, energy, and emissions balances for $NH_3$ production and storage were developed from literature data, lab experiments, and process simulation software. Australian-specific datasets for materials and energy flows were sourced from Eco-invent 3.8 [102]. PV electricity was only assumed to supply the local plants, as the high electricity demand of the regional and national plants would require large-scale solar PV farms that would be difficult to implement. The electricity consumption for the DBD reactor was assumed to be 28.0 MWh $t^{-1}$ $NH_3$ [44], plus an additional 1.65 MWh $t^{-1}$ required for the post-separation system with a recycling loop for unreacted $N_2/H_2$, given an ammonia concentration of 1.0 mol% [72] and product compression requirements. Additionally, given that the current maximum energy efficiency of DBD plasma reactors for $NH_3$ synthesis is still low for commercial feasibility, a sensitivity analysis was performed to account for potential improvements in energy efficiency, which could increase by up to 25 times [72]. A hypothetical plant with the highest NTP energy efficiency was termed pathway 3x.

Transport emissions were modeled specifically for Australian conditions, considering various truck types and emission standards, as well as emissions from truck manufacturing, fuel production, infrastructure construction, and end-of-life activities. This comprehensive, life cycle approach to transport services [10, 27, 89, 103] also included specific fuel compositions, such as sulfur, carbon, hydrogen, and oxygen content, for low-sulfur diesel B5 (10 ppm sulfur with a 5% v/v content of biodiesel). For the life cycle impact assessment step, specific methods have been developed to address relevant environmental impact categories across different regions. The ReCiPe method [104] includes 18 impact categories applicable in a global context. By contrast, the Environmental Footprint (EF) method, designed by the European Commission for consistent environmental footprinting [105], includes 16 midpoint impact categories relevant to policy and regulatory needs in Europe.

Beyond the importance of location in selecting an impact assessment method, the sector being analyzed is also crucial, as some methods may exclude impact categories relevant to the chemical industry. Maranghi and Brondi [103] analyzed numerous LCA studies for chemical production, from which ten critical impact categories were identified based on their frequency of use and the significant environmental impacts observed in various results. These categories include climate change, freshwater eutrophication, freshwater ecotoxicity, terrestrial acidification, terrestrial ecotoxicity, ozone formation – HH, ozone formation – terrestrial ecosystems (TEc), human carcinogenic toxicity, human non-carcinogenic toxicity, and fossil resource scarcity. Given the need for global representativeness and the inclusion of the ten most relevant categories for chemical production, the ReCiPe 2016 method [106] was chosen as the impact assessment method. The analysis was conducted using the SimaPro 9.2 software [107] under the ReCiPe hierarchical (100-year) perspective, excluding long-term emissions. The impact assessment results are presented in Figure 9.7.

The LCA results of Figure 9.7 show that, although electrolyzer-plasma-based plants have lower impacts than natural gas-based plants in the climate change category, this advantage does not hold across 5 of the 10 analyzed impact categories. For instance, in terrestrial ecotoxicity and human non-carcinogenic toxicity, the net impact of the current Elect-NTP (pathway 3) is 30- and 3.8-fold higher than that of the SMR-HB plant. These high impacts are primarily associated with the use of solar PV electricity, as particles of silver, copper, and other metals are released into the environment during the manufacturing of PV modules, inverters, and electrical installations. However, using electricity from the national grid can result in even greater impacts in categories, such as climate change, ozone formation, terrestrial acidification, and freshwater eutrophication. This is evident in pathway 2, HTP-HB, which – despite relying more on electricity instead of fossil fuels for feedstock production and the heat and cooling needed in the synthesis process than the SMR-HB plant – performs poorly due to the high consumption of coal and lignite in grid power generation.

In the future scenario for pathway 3x, a neutral impact could potentially be achieved across six impact categories, provided the oxygen co-product is sold. The reversal of the emissions balance is primarily driven by oxygen sales, as conventional oxygen production requires significant grid electricity, thereby reducing the market demand for conventionally produced oxygen with a higher environmental impact. It is also notable that in this pathway, most of the impacts would be associated with electricity consumption for feedstock production, primarily for hydrogen from the electrolyzer, as shown in the light blue bars.

The reduced transport distances for decentralized plants also play a crucial role in lowering supply chain impacts. Together with storage requirements, transportation accounts for approximately 17% of the net carbon footprint of centrally

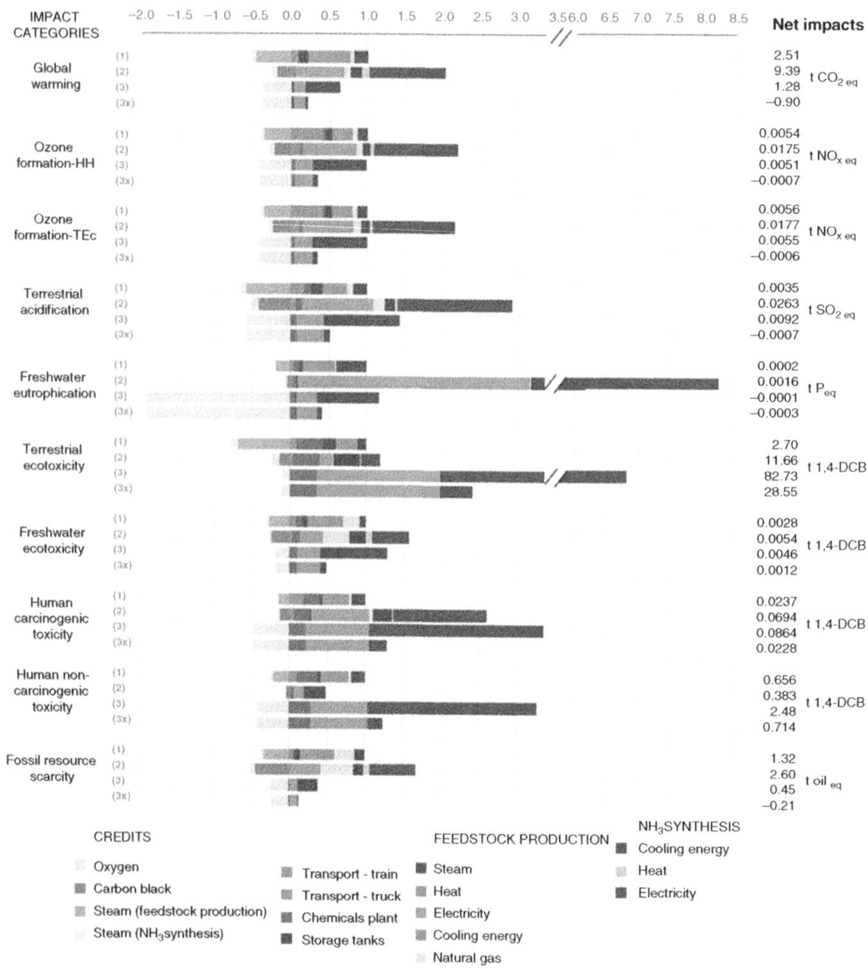


**Figure 9.7** Midpoint results for conventional and alternative $NH_3$ production pathways: (1) SMR-HB, (2) HTP-HB, and (3) electrolysis-NTP and (3x) electrolysis-NTPx. *Notes:* Inputs and outputs with contributions lower than 1% across all impact categories were excluded from the chart. The results for the SMR-HB pathway were normalized to 1.0 on the horizontal axis, with the results for other pathways representing their variation relative to the SMR-HB pathway.

produced $NH_3$. The environmental impact of $NH_3$ distribution is exacerbated by the significant distances between conventional $NH_3$ plants and farms, as well as factors unique to Australia, such as load characteristics, trucking fleets, road infrastructure, and energy sources. Additionally, the production of low-sulfur diesel B5, containing 5% v/v vegetable oil methyl ester, generates about 17% more carbon emissions than 100% fossil diesel, due to the considerable carbon footprint of palm oil cultivation in Southeast Asia [108–111]. Diesel-electric trains in Australia, which also use diesel B5, experience emissions increase from 0.052 to 0.060 kg $CO_2$-eq/tkm [27]. Transport also plays an important role in ozone formation due to $NO_x$ emissions from older trucks, and in freshwater eutrophication due to electricity consumption for vehicle and road maintenance. This electricity production releases phosphates into water because of the extensive use of hard coal and lignite in power plants. Oxygen credits also have a notable impact on freshwater eutrophication in electrolyzer-NTP plants due to the reduced market demand for oxygen produced via fossil-based electricity.

## 9.5 Perspectives for Sustainable Plasma-Based Nitrogen Fixation

The shift toward plasma-based nitrogen fixation offers significant promise for reducing GHG emissions, particularly when integrated with renewable energy sources. However, to ensure these systems achieve true sustainability, several strategic improvements and innovations should be prioritized.

- **Optimizing energy sources and efficiency:** The LCA results indicate that, in the future scenario (pathway 3x), most environmental impacts will primarily arise from the electricity consumed during feedstock production, with a notably minimal impact from the electricity used in ammonia synthesis itself. This imbalance is largely due to the substantial energy demands of water electrolysis, which remains a critical process in hydrogen production for plasma-based ammonia synthesis. Consequently, further improvements in the environmental performance of plasma-based ammonia production will depend on advancements in the energy efficiency of water electrolysis.
- **Expanding local production and supply chains:** Decentralized plasma-based ammonia production plants located near end-users (e.g. farms) could drastically reduce transportation emissions. This decentralized model would also lower reliance on fossil-derived fuels in transport and mitigate infrastructure-related impacts. To realize this potential, future developments should promote the expansion of infrastructure for distributed energy sources, such as small-scale solar or wind installations, to support these local plants.
- **Utilizing other mature renewable energy sources and regional focus in LCA:** The impact of the electricity source is also a critical consideration, as LCA studies have shown that wind energy can have lower impacts across multiple categories, including climate change, compared to solar PV energy [10]. Conducting LCA studies in regions with a greater potential for wind energy

integration could provide additional insights into regional variations in environmental impacts [112]. To better capture these differences, future studies should compare environmental impacts across renewable sources to optimize energy portfolios, potentially combining solar and wind resources to offset intermittency and reduce impacts.

- **Market development for co-products:** The findings here highlight the role of oxygen and carbon black co-production in offsetting environmental impacts. By expanding markets for these co-products, and especially high purity oxygen, the process can achieve better net impact reduction. Therefore, policies that encourage co-product utilization, particularly in regions with high demand for industrial oxygen, could provide economic and environmental incentives.
- **Advancing LCA methodology and benchmarks:** As plasma technology continues to evolve, LCA approaches should also advance to capture the full spectrum of impacts accurately. The ongoing refinement of impact assessment methods, especially in emerging markets, is essential. Future studies could focus on establishing consistent benchmarks with updated datasets that reflect current technology, resource extraction practices, and region-specific environmental factors. Moreover, adapting LCA to internalize the environmental impacts into social and economic analyses could benefit these decentralized systems, providing a holistic assessment of their sustainability.

In conclusion, plasma-based nitrogen fixation is a promising avenue toward sustainable nitrogen fixation, but achieving widespread adoption and sustainability will require technological, infrastructural, and policy advancements.

## 9.6 Conclusion and Outlook

PANF will not replace the HB process in the near future, and certainly not by aiming alone for better energy efficiency. PANF-LCA does not provide arguments strong enough to compensate current economic and scale-up disadvantages of PANF. As a way out of this dilemma, the window of opportunity for PANF is to capitalize use of renewable energy and to establish a business in distributed small-scale production, as well as to establish recycling, circularity, and cost-internalization. Therefore, the key leverage of PANF is to spark a change in the business model of N-fixation and to leverage a societal-governance switch in the fertilizer, agricultural, and food production.

This review provides critical key information to those levers. The renewable energy transition has started with strong societal-governance impetus. This, in principle, could propel plasma technology but alone has turned out not to be sufficient. Competing electrolyzer technology is progressing better and has found broad acceptance of investment among stakeholders and shareholders. The scale-up of plasma technology beyond lower pilot scale has not been achieved or even tackled, which limits trust and public awareness of this emerging technology branch. The economics of hypothetical scale-ups are simply not convincingly promising. The scale-up milestone has been achieved by the competing electrolyzers and that kind of industrial-societal trust is lacking for PANF. Therefore, the window of opportunity for PANF is the distributed production, which can be achieved with current demonstrated PANF scale-ups, to satisfy local needs with a quality and time of response not achievable by centralized world-scale plants. Issues of modularity in small-scale production would then be the key to a distributed production. Those need to be tackled much more in future PANF-process designs. In addition, the circularity of PANF needs more insight via respective holistic assessments, including the capability of separation of products compared to the HB process, and new opportunities for recycling.

The fate of PANF is to properly address societal-governance needs and decision-making. Yet, while PANF needs to be proactive, it is also receptive to global megatrends. The overall economic development in terms of cost internalizations decides about the PANF fate. Sustainable cash-flow investments are a reality in carbon dioxide removal; they need to be a reality in the prevention of carbon dioxide formation, which is the business case for PANF. Then, a game change in nitrogen fixation would not be out of reach.

## References


1 Patil, B.S., Wang, Q., Hessel, V., and Lang, J. (2015). Plasma N2-fixation: 1900–2014. *Catal. Today* 256: 49–66. https://doi.org/10.1016/j.cattod.2015.05.005.

2 Chang, F., Zhang, X., Zhan, G. et al. (2021). Review of methods for sustainability assessment of chemical engineering processes. *Ind. Eng. Chem. Res.* 60 (1): 52–66. https://doi.org/10.1021/acs.iecr.0c04720.

3 Yousaf, A.M. and Aqsa, R. (2023). Integrating circular economy, SBTI, digital LCA, and ESG benchmarks for sustainable textile dyeing: a critical review of industrial textile practices. *Glob. NEST J* 25: 39–51.

4 Yu, L., Keilani, A., Tran, N.N. et al. (2025). ESG assessment methodology for emerging technologies: plasma-versus conventional technology for ammonia production. *RSC Sustain.* 3 (3): 1102–1113.

5 Escriba-Gelonch, M., Osorio-Tejada, J., Vertongen, R. et al. (2025). Sustainability assessment of plasma-based and electrolytic $CO_2$ conversion to CO. *J. Clean. Prod.* 488: 144578.

6 Van Duc Long, N. and Lee, M. (2012). Dividing wall column structure design using response surface methodology. *Comput. Chem. Eng.* 37: 119–124. https://doi.org/10.1016/j.compchemeng.2011.07.006.

7 Arshad, M.Y., Jaffer, H., Tahir, M.W. et al. (2024). Maximizing hydrogen-rich syngas production from rubber wood biomass in an updraft fluidized bed gasifier: an advanced 3D simulation study. *J. Taiwan Inst. Chem. Eng.* 156: 105365.

8 Long, N.V.D., Lee, D.Y., Kwag, C. et al. (2021). Improvement of marine carbon capture onboard diesel fueled ships. *Chem. Eng. Process. Process Intensif.* 168: 108535. https://doi.org/10.1016/j.cep.2021.108535.

9 Van Duc Long, N., Al-Bared, M., Lin, L. et al. (2022). Understanding plasma-assisted ammonia synthesis via crossing discipline borders of literature: a critical review. *Chem. Eng. Sci.* 263: 118097. https://doi.org/10.1016/j.ces.2022.118097.

10 Osorio-Tejada, J., van't Veer, K., Long, N.V.D. et al. (2022). Sustainability analysis of methane-to-hydrogen-to-ammonia conversion by integration of high-temperature plasma and non-thermal plasma processes. *Energy Convers. Manag.* 269: 116095. https://doi.org/10.1016/j.enconman.2022.116095.

11 Chen, W., Liu, X., and Wei, Y. (2023). Promoting sustainability by resource efficiency and green energy: policy recommendation for green growth. *Res. Policy* 87: 104246. https://doi.org/10.1016/j.resourpol.2023.104246.

12 Elroi, H., Zbigniew, G., Agnieszka, W.-C., and Piotr, S. (2023). Enhancing waste resource efficiency: circular economy for sustainability and energy conversion. *Front. Environ. Sci.* 11: 1303792. https://doi.org/10.3389/fenvs.2023.1303792.

13 Arshad, Y., Rashid, A., Gul, H. et al. (2022). Optimization of acid-assisted extraction of pectin from banana (Musa Acuminata) peels by central composite design. *Glob. NEST J* 24: 752–756.

14 Anastasopoulou, A., Keijzer, R., Butala, S. et al. (2020). Eco-efficiency analysis of plasma-assisted nitrogen fixation. *J. Phys. D. Appl. Phys.* 53 (23): 234001.

15 Appl, M. (2011. https://doi.org/10.1002/14356007.o02_o11). Ammonia, 2. Production Processes. In: *Ullmann's Encyclopedia of Industrial Chemistry* (ed. C. Ley). Weinheim, Germany: Wiley-VCH.

16 Arshad, M.Y. and Azam, M. (2021). Environmental Friendly Specialty Chemical Plants for Developing World: A Roadmap for Economic Development and Sustainability with Waste Reduction. In: *Book of Abstracts* (ed. Farooq), 294. Serbia: **AIDASCO Publishing**.

17 Chen, H., Yuan, D., Wu, A. et al. (2021). Review of low-temperature plasma nitrogen fixation technology. *Waste Disposal Sustain Energy* 3 (3): 201–217. https://doi.org/10.1007/s42768-021-00074-z.

18 Arshad, M.Y., Rashid, A., Mahmood, F. et al. (2023). Metal (II) triazole complexes: synthesis, biological evaluation, and analytical characterization using machine learning-based validation. *Eur. J. Chem.* 14: 155–164.

**19** Rafique, M., Kiran, S., Ashraf, A. et al. (2022). Effective removal of direct orange 26 dye using copper nanoparticles synthesized from Tilapia fish scales. *Glob. NEST J* 24: 311–317.

**20** Rafique, M.A., Kiran, S., Jamal, A. et al. (2022). Green synthesis of copper nanoparticles using Allium cepa (onion) peels for removal of disperse yellow 3 dye. *Desalin. Water Treat.* 272: 259–265.

**21** Rouwenhorst, K.H.R., Kim, H.-H., and Lefferts, L. (2019). Vibrationally excited activation of N2 in plasma-enhanced catalytic ammonia synthesis: a kinetic analysis. *ACS Sustain. Chem. Eng.* 7 (20): 17515–17522. https://doi.org/10.1021/acssuschemeng.9b04997.

**22** Pfromm, P.H. (2017). Towards sustainable agriculture: Fossil-free ammonia. *J. Renew. Sustain. Energy* 9 (3): 034702. https://doi.org/10.1063/1.4985090.

**23** Mehmood, A., Tahir, M.W., Saeed, M.A. et al. (2023). Optimization of gasifying agents in 3D downdraft gasification for enhanced gas composition, combustion, and $CO_2$ utilization. *Fire* 6 (9): 361.

**24** Saeed, M.A., Niedzwiecki, L., Arshad, M.Y. et al. (2022). Combustion and explosion characteristics of pulverised wood, valorized with mild pyrolysis in pilot scale installation, using the modified ISO 1 m3 dust explosion vessel. *Appl. Sci.* 12 (24): 12928.

**25** Rouwenhorst, K.H.R., Van der Ham, A.G.J., Mul, G., and Kersten, S.R.A. (2019). Islanded ammonia power systems: technology review & conceptual process design. *Renew. Sust. Energ. Rev.* 114: 109339. https://doi.org/10.1016/j.rser.2019.109339.

**26** Arshad, M.Y., Saeed, S., Raza, A. et al. (2023). Integrating life cycle assessment and machine learning to enhance black soldier fly larvae-based composting of kitchen waste. *Sustainability* 15 (16): 12475.

**27** Osorio-Tejada, J., Tran, N.N., and Hessel, V. (2022). Techno-environmental assessment of small-scale Haber-Bosch and plasma-assisted ammonia supply chains. *Sci. Total Environ.* 826: 154162. https://doi.org/10.1016/j.scitotenv.2022.154162.

**28** Arshad, M.Y., Ahmad, A.S., Mularski, J. et al. (2024). Pioneering the future: a trailblazing review of the fusion of computational fluid dynamics and machine learning revolutionizing plasma catalysis and non-thermal plasma reactor design. *Catalysts* 14 (1): 40.

**29** Long, N.N.T.N.V.D., Osorio-Tejada, J., and Hessel, V. (2022). Local ammonia production – fertiliser made for local, social and environmental benefits. *Chemi. Industr Digest* 28: 1590–1603.

**30** Yar, A., Arshad, M.Y., Asghar, F. et al. (2022). Machine learning-based relative performance analysis of monocrystalline and polycrystalline grid-tied PV systems. *Int. J. Photoenergy* 2022 (1): 3186378.

31 Klimek, A. and Piercey, D.G. (2024). Nitrogen fixation via plasma-assisted processes: mechanisms, applications, and comparative analysis—a comprehensive review. *Processes* 12 (4): 786.

32 Van Duc Long, N., Pourali, N., Lamichhane, P. et al. (2024). Catalytic ammonia formation in a microreaction chamber with electrically intensified arc plasma. *ChemCatChem* 16 (13): e202400005. https://doi.org/10.1002/cctc.202400005.

33 Brandenburg, R., Bogaerts, A., Bongers, W. et al. (2019). White paper on the future of plasma science in environment, for gas conversion and agriculture. *Plasma Process. Polym.* 16 (1): 1700238. https://doi.org/10.1002/ppap.201700238.

34 D'Angola, A., Colonna, G., and Kustova, E. (2022). Editorial. Thermal and non-thermal plasmas at atmospheric pressure. *Front. Phys.* 10: 852905. https://doi.org/10.3389/fphy.2022.852905.

35 Mindong, B., Zhitao, Z., Xiyao, B. et al. (2003). Plasma synthesis of ammonia with a microgap dielectric barrier discharge at ambient pressure. *IEEE Tran. Plasma Sci.* 31 (6): 1285–1291. https://doi.org/10.1109/TPS.2003.818761.

36 Bai, M., Zhang, Z., Bai, M. et al. (2008). Synthesis of ammonia using $CH_4/N_2$ plasmas based on micro-gap discharge under environmentally friendly condition. *Plasma Chem. Plasma Process.* 28 (4): 405–414. https://doi.org/10.1007/s11090-008-9132-4.

37 Gómez-Ramírez, A., Cotrino, J., Lambert, R.M., and González-Elipe, A.R. (2015). Efficient synthesis of ammonia from $N_2$ and $H_2$ alone in a ferroelectric packed-bed DBD reactor. *Plasma Sources Sci. Technol.* 24 (6): 065011. https://doi.org/10.1088/0963-0252/24/6/065011.

38 Aihara, K., Akiyama, M., Deguchi, T. et al. (2016). Remarkable catalysis of a wool-like copper electrode for $NH_3$ synthesis from $N_2$ and $H_2$ in non-thermal atmospheric plasma. *Chem. Commun.* 52 (93): 13560–13563. https://doi.org/10.1039/C6CC06752B.

39 Iwamoto, M., Akiyama, M., Aihara, K., and Deguchi, T. (2017). Ammonia synthesis on wool-like Au, Pt, Pd, Ag, or Cu electrode catalysts in nonthermal atmospheric-pressure plasma of $N_2$ and $H_2$. *ACS Catal.* 7 (10): 6924–6929. https://doi.org/10.1021/acscatal.7b01624.

40 Akay, G. and Zhang, K. (2017). Process intensification in ammonia synthesis using novel coassembled supported microporous catalysts promoted by nonthermal plasma. *Ind. Eng. Chem. Res.* 56 (2): 457–468. https://doi.org/10.1021/acs.iecr.6b02053.

41 Shah, J., Wang, W., Bogaerts, A., and Carreon, M.L. (2018). Ammonia synthesis by radio frequency plasma catalysis: revealing the underlying mechanisms. *ACS Appl Energy Mater* 1 (9): 4824–4839. https://doi.org/10.1021/acsaem.8b00898.

42 Matsumoto, O. (1988). Plasma catalytic reaction in ammonia synthesis in the microwave discharge. *J. Phys.* 8: Pr7-411. https://doi.org/10.1051/jp4:1998736.

**43** Bai, M., Zhang, Z., Bai, M. et al. (2008). Conversion of methane to liquid products, hydrogen, and ammonia with environmentally friendly condition-based microgap discharge. *J. Air Waste Manag. Assoc.* 58 (12): 1616–1621. https://doi.org/10.3155/1047-3289.58.12.1616.

**44** Kim, H.-H., Teramoto, Y., Ogata, A. et al. (2017). Atmospheric-pressure nonthermal plasma synthesis of ammonia over ruthenium catalysts. *Plasma Process. Polym.* 14 (6): 1600157. https://doi.org/10.1002/ppap.201600157.

**45** Patil, B.S., Cherkasov, N., Lang, J. et al. (2016). Low temperature plasma-catalytic NOx synthesis in a packed DBD reactor: effect of support materials and supported active metal oxides. *Appl. Catal. B Environ.* 194: 123–133. https://doi.org/10.1016/j.apcatb.2016.04.055.

**46** Hu, H., Liang, H., Li, J. et al. (2007). Study on production of inhaled nitric oxide for medical applications by pulsed discharge. *IEEE Tran. Plasma Sci.* 35 (3): 619–625. https://doi.org/10.1109/TPS.2007.896782.

**47** Namihira, T., Katsuki, S., Hackam, R. et al. (2002). Production of nitric oxide using a pulsed arc discharge. *IEEE Tran. Plasma Sci.* 30 (5): 1993–1998. https://doi.org/10.1109/TPS.2002.807502.

**48** Dobrynin, D., Fridman, A., and Starikovskiy, A.Y. (2012). Reactive oxygen and nitrogen species production and delivery into liquid media by microsecond thermal spark-discharge plasma jet. *IEEE Tran. Plasma Sci.* 40 (9): 2163–2171. https://doi.org/10.1109/TPS.2012.2204780.

**49** Li, S., Medrano, J.A., Hessel, V., and Gallucci, F. (2018). Recent progress of plasma-assisted nitrogen fixation research: a review. *Processes* 6 (12): 248.

**50** Malik, M.A. (2016). Nitric oxide production by high voltage electrical discharges for medical uses: a review. *Plasma Chem. Plasma Process.* 36 (3): 737–766. https://doi.org/10.1007/s11090-016-9698-1.

**51** Janda, M., Martišovitš, V., Hensel, K., and Machala, Z. (2016). Study of transient spark discharge focused at $NO_x$ generation for biomedical applications. *J. Phys. Conf. Ser.* 768 (1): 012009. https://doi.org/10.1088/1742-6596/768/1/012009.

**52** Peng, P., Chen, P., Addy, M. et al. (2018). *Current Situation of Non-Thermal Plasma (NTP) Ammonia Synthesis and Its Potential to Build a Sustainable Nitrogen Fixation Industry.* United Kingdom: Elsevier.

**53** Ahmed, H.S., Yahya, Z., Khan, W.A., and Faraz, A. (2024). Sustainable pathways to ammonia: a comprehensive review of green production approaches. *Clean Energy* 8 (2): 60–72. https://doi.org/10.1093/ce/zkae002.

**54** Sharma, R.K., Patel, H., Mushtaq, U. et al. (2020). Plasma activated electrochemical ammonia synthesis from nitrogen and water. *ACS Energy Lett.* 6 (2): 313–319.

**55** Butler, C., Fan, Y., Grewal, S., and Winter, L.R. (2023). At-field and on-demand nitrogenous fertilizer synthesis. *ACS Sustain. Chem. Eng.* 11 (15): 5803–5818. https://doi.org/10.1021/acssuschemeng.2c06551.

**56** Bradu, C., Kutasi, K., Magureanu, M. et al. (2020). Reactive nitrogen species in plasma-activated water: generation, chemistry and application in agriculture. *J. Phys. D. Appl. Phys.* 53 (22): 223001. https://doi.org/10.1088/1361-6463/ab795a.

**57** Sun, J., Alam, D., Daiyan, R. et al. (2021). A hybrid plasma electrocatalytic process for sustainable ammonia production. *Energy Environ. Sci.* 14: 865–872. https://doi.org/10.1039/D0EE03769A.

**58** Anastasopoulou, A., Butala, S., Lang, J. et al. (2016). Life cycle assessment of the nitrogen fixation process assisted by plasma technology and incorporating renewable energy. *Ind. Eng. Chem. Res.* 55 (29): 8141–8153. https://doi.org/10.1021/acs.iecr.6b00145.

**59** Li, D., Zan, L., Chen, S. et al. (2022). Direct conversion of $N_2$ and $O_2$: status, challenge and perspective. *Natl. Sci. Rev.* 9 (12): nwac042. https://doi.org/10.1093/nsr/nwac042.

**60** Arshad, M.Y., Saeed, M.A., Tahir, M.W. et al. (2023). Role of experimental, modeling, and simulation studies of plasma in sustainable green energy. *Sustainability* 15 (19): 14193.

**61** Tahir, M., Tahir, M.W., Arshad, M.Y. et al. (2025). Modelling and simulation of an integrated coupled reactor for hydrogen production and carbon dioxide utilisation in an integrated fuel cell power system. *J. Taiwan Inst. Chem. Eng.* 167: 105857. https://doi.org/10.1016/j.jtice.2024.105857.

**62** Lamichhane, P., Pourali, N., Rebrov, E., and Hessel, V. (2023). Energy intensified nitrogen fixation through fast modulated gas discharge from pyramid-shaped micro-electrode. *Plasma Chem. Plasma Process.* 44: 1–24. https://doi.org/10.1007/s11090-023-10376-1.

**63** Anastasopoulou, A., Butala, S., Patil, B. et al. (2016). Techno-economic feasibility study of renewable power systems for a small-scale plasma-assisted nitric acid plant in Africa. *Processes* 4 (4): 54.

**64** Samipour, S. and Rahimpour, M.R. (2024). Chapter 2 – Nonthermal Plasma-Assisted Ammonia Synthesis Technologies. In: *Progresses in Ammonia: Science, Technology and Membranes* (ed. A. Basile and M.R. Rahimpour), 33–62. Elsevier https://doi.org/10.1016/B978-0-323-88516-4.00015-9.

**65** Vervloessem, E., Gorbanev, Y., Nikiforov, A. et al. (2022). Sustainable $NO_x$ production from air in pulsed plasma: elucidating the chemistry behind the low energy consumption. *Green Chem.* 24 (2): 916–929. https://doi.org/10.1039/D1GC02762J.

**66** Saeed, S., Arshad, M.Y., Raza, A. et al. (2023). Influence of thermal and chemical treatment on biosorbent from rice husk and its application in removal of resorcinol from industrial wastewater. *Processes* 11 (12): 3344.

**67** Chaudhary, R., van Rooij, G., Li, S. et al. (2020). Low-temperature, atmospheric pressure reverse water-gas shift reaction in dielectric barrier plasma discharge, with outlook to use in relevant industrial processes. *Chem. Eng. Sci.* 225: 115803. https://doi.org/10.1016/j.ces.2020.115803.

**68** Anastasopoulou, A., Keijzer, R., Patil, B. et al. (2020). Environmental impact assessment of plasma-assisted and conventional ammonia synthesis routes. *J. Ind. Ecol.* 24 (5): 1171–1185. https://doi.org/10.1111/jiec.12996.

**69** Cherkasov, N., Ibhadon, A.O., and Fitzpatrick, P. (2015). A review of the existing and alternative methods for greener nitrogen fixation. *Chem. Eng. Process. Process Intensif.* 90: 24–33. https://doi.org/10.1016/j.cep.2015.02.004.

**70** Panchal, D., Lu, Q., Sakaushi, K., and Zhang, X. (2024). Advanced cold plasma-assisted technology for green and sustainable ammonia synthesis. *Chem. Eng. J.* 498: 154920. https://doi.org/10.1016/j.cej.2024.154920.

**71** Rouwenhorst, K.H., Jardali, F., Bogaerts, A., and Lefferts, L. (2021). From the Birkeland–Eyde process towards energy-efficient plasma-based NO*x* synthesis: a techno-economic analysis. *Energy Environ. Sci.* 14 (5): 2520–2534.

**72** Rouwenhorst, K.H., Engelmann, Y., van't Veer, K. et al. (2020). Plasma-driven catalysis: green ammonia synthesis with intermittent electricity. *Green Chem.* 22 (19): 6258–6287.

**73** Hessel, V., Anastasopoulou, A., Wang, Q. et al. (2013). Energy, catalyst and reactor considerations for (near)-industrial plasma processing and learning for nitrogen-fixation reactions. *Catal. Today* 211: 9–28. https://doi.org/10.1016/j.cattod.2013.04.005.

**74** Saeed, S., Arshad, M.Y., and Ahmed, A.S. (2023). Advancing circular economy in industrial chemistry and environmental engineering: principles, alignment with United Nations sustainable development goals, and pathways to implementation. *Eur. J. Chem.* 14 (3): 414–428.

**75** Arshad, M.Y., Lewis, D., and Tran, N.N. (2025). Empowering Sustainability: Integrating ESG Principles for a Net-Zero Future Through Decentralised Process Plant Design. In: *International Conference on Environmental, Social, and Governance (ICESG 2024)*, 60–64. Atlantis Press.

**76** Arshad, M.Y., Saeed, M.A., Tahir, M.W. et al. (2023). Advancing sustainable decomposition of biomass tar model compound: machine learning, kinetic modeling, and experimental investigation in a non-thermal plasma dielectric barrier discharge reactor. *Energies* 16 (15): 5835.

**77** Kelly, S. and Bogaerts, A. (2021). Nitrogen fixation in an electrode-free microwave plasma. *Joule* 5 (11): 3006–3030. https://doi.org/10.1016/j.joule.2021.09.009.

**78** Srivastava, N., Saquib, M., Rajput, P. et al. (2023). Prospects of solar-powered nitrogenous fertilizers. *Renew. Sust. Energ. Rev.* 187: 113691. https://doi.org/10.1016/j.rser.2023.113691.

**79** Winter, L.R. and Chen, J.G. (2021). N2 fixation by plasma-activated processes. *Joule* 5 (2): 300–315. https://doi.org/10.1016/j.joule.2020.11.009.

**80** Hussain, H., Tahir, M.W., Mehmood, A., and Arshad, M.Y. (2024). Kinetic and thermodynamic parameter optimization of an anode-supported solid oxide fuel cell: a computational fluid dynamics study, journal of electrochemical. *Sci. Technol.* 16 (2): 197–215.

**81** Pham, T.K., Nguyen, T.T., Long, N.V.D. et al. (2024). Developing microfluidic purification techniques for biodiesel production from recycled grease trap waste. *Sustain. Energy Fuel.* 8 (24): 5806–5817.

**82** Hessel, V. (2015). Special issue: design and engineering of microreactor and smart-scaled flow processes. *Processes* 3 (1): 19–22.

**83** Santhosh, C.R. and Sankannavar, R. (2023). A comprehensive review on electrochemical green ammonia synthesis: from conventional to distinctive strategies for efficient nitrogen fixation. *Appl. Energy* 352: 121960. https://doi.org/10.1016/j.apenergy.2023.121960.

**84** Gul, H., Arshad, M.Y., and Tahir, M.W. (2023). Production of $H_2$ via sorption enhanced auto-thermal reforming for small scale applications—A process modeling and machine learning study. *Int. J. Hydrog. Energy* 48 (34): 12622–12635.

**85** Tahir, F., Arshad, M.Y., Saeed, M.A., and Ali, U. (2023). Integrated process for simulation of gasification and chemical looping hydrogen production using Artificial Neural Network and machine learning validation. *Energy Convers. Manag.* 296: 117702.

**86** Saleem, F., Abbas, A., Rehman, A. et al. (2022). Decomposition of benzene as a biomass gasification tar in $CH_4$ carrier gas using non-thermal plasma: parametric and kinetic study. *J. Energy Inst.* 102: 190–195.

**87** Modzelewska, A., Jackowski, M., Boutikos, P. et al. (2024). Sustainable production of biohydrogen: feedstock, pretreatment methods, production processes, and environmental impact. *Fuel Process. Technol.* 266: 108158.

**88** Li, J., Lan, C., Nie, L. et al. (2023). Distributed plasma-water-based nitrogen fixation system based on cascade discharge: generation, regulation, and application. *Chem. Eng. J.* 478: 147483. https://doi.org/10.1016/j.cej.2023.147483.

**89** Osorio-Tejada, J.L., Rebrov, E., and Hessel, V. (2023). Internalisation of environmental costs of decentralised nitrogen fertilisers production. *Int. J. Life Cycle Assess.* 28 (11): 1590–1603. https://doi.org/10.1007/s11367-023-02187-5.

**90** Cozzi, L., Gould, T., Bouckart, S. et al. (2019). World energy outlook 2020. *Energy* 2020: 30.

**91** Hug, G., Demiray, T., Guidati, G., McKenna, R., Oswald, K., Patt, A., Saar, M.O., Sansavini, G., Schaffner, C., Schwarz, M. and Steffen, B., 2022. Steps to Fossil-Fuel Independence for Switzerland: Expert Group "Security of Supply"–Policy Brief. ETH Zurich.

**92** Ritchie, H., 2020. You want to reduce the carbon footprint of your food? Focus on what you eat, not whether your food is local. Our world in data.

**93** Smith, C. and Torrente-Murciano, L. (2021). The potential of green ammonia for agricultural and economic development in Sierra Leone. *One Earth* 4 (1): 104–113. https://doi.org/10.1016/j.oneear.2020.12.015.

**94** Engerer, H. and Horn, M. (2010). Natural gas vehicles: an option for Europe. *Energy Policy* 38 (2): 1017–1029.

**95** Tarí, J.J., Molina-Azorín, J.F., and Heras, I. (2012). Benefits of the ISO 9001 and ISO 14001 standards: a literature review. *J. Indust. Eng. Manage. JIEM* 5 (2): 297–322.

**96** Fasihi, M. and Breyer, C. (2024). Global production potential of green methanol based on variable renewable electricity. *Energy Environ. Sci.* 17 (10): 3503–3522.

**97** Arshad, M.Y. and Halog, A. (2025). Life Cycle Assessment of Various Process Routes Including Biological Processes for Renewable Fuel Production. In: *Sustainable and Green Catalytic Processes for Renewable Fuel Production with Net-Zero Emissions*, 377–428. Elsevier.

**98** Guerra, C.F., Reyes-Bozo, L., Vyhmeister, E. et al. (2020). Technical-economic analysis for a green ammonia production plant in Chile and its subsequent transport to Japan. *Renew. Energy* 157: 404–414.

**99** McCollum, D.L., Zhou, W., Bertram, C. et al. (2018). Energy investment needs for fulfilling the Paris agreement and achieving the sustainable development goals. *Nat. Energy* 3 (7): 589–599.

**100** Erisman, J.W., Sutton, M.A., Galloway, J. et al. (2008). How a century of ammonia synthesis changed the world. *Nat. Geosci.* 1 (10): 636–639.

**101** Brightling, J. (2018). Ammonia and the fertiliser industry: the development of ammonia at Billingham. *Johnson Matthey Technol. Rev.* 62 (1): 32–47.

**102** Tiong, Y.W., Sharma, P., Tian, H. et al. (2023). Startup performance and microbial communities of a decentralized anaerobic digestion of food waste. *Chemosphere* 318: 137937. https://doi.org/10.1016/j.chemosphere.2023.137937.

**103** Osorio-Tejada, J.L., Ferlin, F., Vaccaro, L., and Hessel, V. (2023). The sustainability impact of Nobel Prize Chemistry: life cycle assessment of C–C cross-coupling reactions. *Green Chem.* 25 (23): 9760–9778. https://doi.org/10.1039/D3GC01896B.

**104** Valente, A., Tulus, V., Galán-Martín, Á. et al. (2021). The role of hydrogen in heavy transport to operate within planetary boundaries. *Sustain. Energy Fuel* 5 (18): 4637–4649.

**105** Bürgin, A. (2023). The European Commission: A Climate Policy Entrepreneur. In: *Handbook on European Union Climate Change Policy and Politics*, 23–37. Edward Elgar Publishing.

**106** Maranghi, S. and Brondi, C. (2020). *Life Cycle Assessment in the Chemical Product Chain, Life Cycle Assessment in the Chemical Product Chain*. Siena: Springer.

**107** Consultants, Pré (2022). *SimaPro*. Amersfoort: *Pré Consultants*.

**108** Efe, E.P. Climate change and productivity in Sub Saharan Africa. *Africa J. Environ.* 2 (1): 1–10.

**109** Sutar, D.D. and Jadhav, S.V. (2022). Life cycle assessment of methanol production by natural gas route. *Mater Today Proc* 57: 1559–1566. https://doi.org/10.1016/j.matpr.2021.12.164.

**110** Mustafa, A., Faisal, S., Singh, J. et al. (2024). Converting lignocellulosic biomass into valuable end products for decentralized energy solutions: a comprehensive overview. *Sustain Energy Technol Assess* 72: 104065. https://doi.org/10.1016/j.seta.2024.104065.

**111** Mishra, K., Siwal, S.S., Nayaka, S.C. et al. (2023). Waste-to-chemicals: green solutions for bioeconomy markets. *Sci. Total Environ.* 887: 164006.

**112** Portillo, F., Alcayde, A., Garcia, R.M. et al. (2024). Life cycle assessment in renewable energy: solar and wind perspectives. *Environments* 11 (7): 147.

# 10

# Industrial Applications and Economic Viability of Plasma-Based Nitrogen Fixation

*Magnus Nyvold[1,2] and Rune Ingels[1]*

[1] *Faculty of Environmental Sciences and Natural Resource Management, Norwegian University of Life Sciences, Ås, Norway*
[2] *N2 Applied AS, Asker, Norway*

## 10.1 Introduction

Plasma-based nitrogen fixation (PNF) is applicable in a range of industries and has the potential to reduce their reliance on finite fossil resources. A widespread adoption of PNF has to date been restrained by its high EC compared to conventional nitrogen fixation (CNF): An optimized CNF, based on the Haber-Bosch process, produces reactive nitrogen at an EC of 0.5 MJ $molN^{-1}$, whereas the best PNF technology currently operates in the range of 2.0–2.4 MJ $molN^{-1}$ [1]. Despite the high current EC, PNF has gained attention for its low theoretical EC at 0.2 MJ $molN^{-1}$, which could, in principle, halve the energy directed to industrial nitrogen fixation. At present, the extent to which the theoretical limit can be approached is unknown, but the innovation potential, and consequently the potential for the industry, is significant.

Although the EC is important for all industrial applications of PNF, its economic viability does not solely depend on the amount of energy used. In an industrial setting, it can be equally important to leverage the unique properties of plasma technology. These properties are as follows:

- Direct conversion of air to reactive nitrogen with electricity and air as the only input.
- Flexible operation – ramp-up and ramp-down in milliseconds.
- Low capital cost at small-to-medium scale (0.3–30 MW).
- High-temperature heat as a by-product.

*Plasma-Assisted Nitrogen Fixation for Sustainable Process Industries*, First Edition. Edited by Volker Hessel, Annemie Bogaerts, Gabriele Centi, Evgeny Rebrov, and Nguyen Van Duc Long.

Published 2026 by John Wiley & Sons, Inc.

This set of advantages will hold different weight for different industrial applications, but a common denominator is the simplicity of the process. Electricity generation used to be a detour in nitrogen fixation, as it was generated mainly by hydrocarbon combustion. Now that an increasing share of electricity generation stems from non-hydrocarbon sources [2], the direct use of electricity has become a shortcut (Figure 10.1). In addition, an increasing share of electricity generation is spatially distributed and intermittent, which is suitable for PNF processes, and by extension, the distributed end-users (composed mainly of the widespread food producers).

The electric conversion of air to reactive nitrogen holds other particular advantages. As can be seen in Figure 10.2, PNF is a direct path from the nitrogen ($N_2$) and oxygen ($O_2$) in air to nitrate ($NO_3^-$). Since it does not involve carbon, it does not emit $CO_2$. Nor is the Ostwald process, which oxidizes ammonia and emits the potent greenhouse gas (GHG) nitrous oxide ($N_2O$), necessary. Consequently, PNF can reduce the GHG emissions from nitrogen fixation, provided the electricity originates from low-emitting sources.

The electricity-driven process also enables ramp-up and ramp-down duration in the order of milliseconds. Exploiting either variable electricity production or variable electricity prices is, therefore, possible. Electricity price variability is expected to rise in the coming years, with regions where there is a large influx of intermittent electricity production predicting negative prices for up to 30% of the year by 2030 [2, 3]. Industrial processes that are capable of uncompromised intermittent operation will benefit in these increasingly common situations.

Successful intermittent operation depends on the capital expenditure (capex) of the industrial plant. A PNF process placed in a hypothetical region where

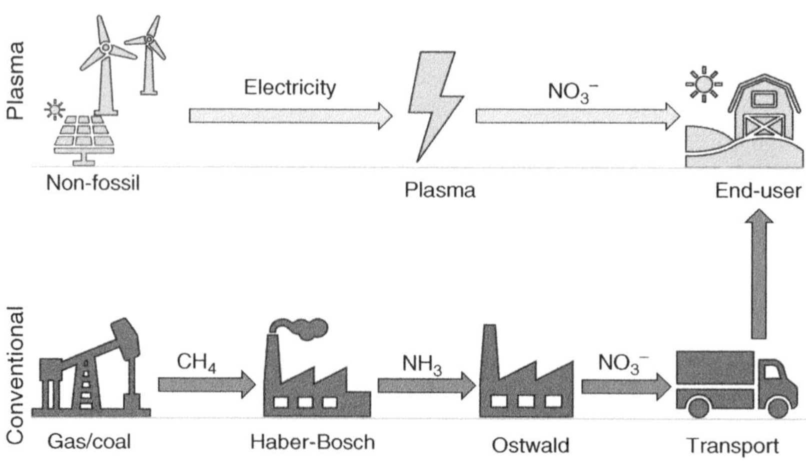


**Figure 10.1** The value chain of plasma-based nitrogen fixation compared to conventional nitrogen fixation.

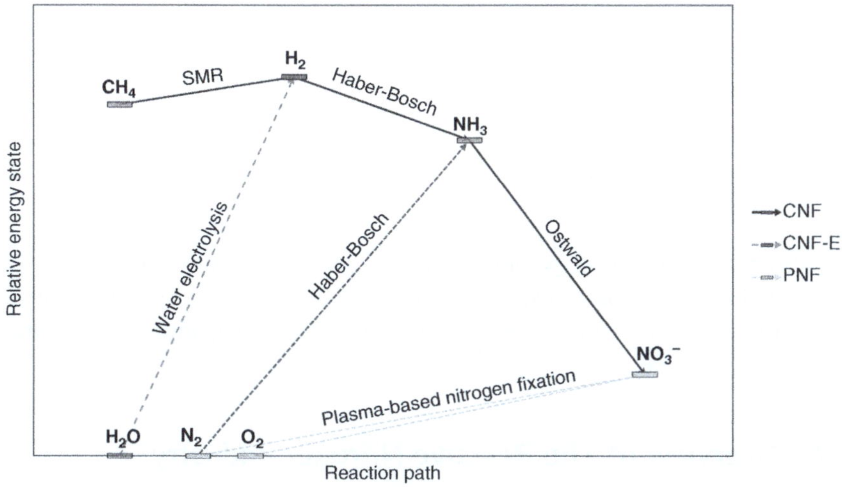


**Figure 10.2** Simplified energy diagram of three nitrogen fixation methods. In conventional nitrogen fixation (CNF), methane ($CH_4$) is used as energy and hydrogen source in the steam methane reforming (SMR) method, whereas electrolysis-based nitrogen fixation (CNF-E) uses electricity to generate hydrogen ($H_2$). Plasma-based nitrogen fixation (PNF) is a direct electric conversion of oxygen ($O_2$) and nitrogen ($N_2$) to nitric oxide (NO). The plasma process will require more energy than NO's enthalpy of formation (0.09 MJ $molNO^{-1}$).

electricity prices are near-zero 30% of the time could, in principle, produce reactive nitrogen at zero cost. However, the capex per unit product will be more than triple in such a situation, which may offset the potential benefit. Ensuring a sufficiently low capex is, therefore, paramount to a viable intermittent operation. The simplicity of PNF, in which there is no need for high-pressure, multistage, catalyzed reactions, has the potential for low capex (detailed in Section 10.4). If the potential is realized, the technology will be highly attractive in future energy systems where electricity prices fluctuate.

Any nitrogen fixation method will generate substantial amounts of heat. Even the optimized CNF process loses 80% of its energy as heat when producing $NO_3^-$. The heat generated in a plasma system has the advantage of high exergy. Thermal plasmas commonly operate at average temperatures of 1,500–5,000 K, which enables high-temperature heat to be recovered from the system. This heat can be used favorably either in the process itself or by providing the heat (as hot water or pressurized steam) to other users. This will improve the economic feasibility of PNF.

To summarize, PNF may already have a place in industrial nitrogen fixation despite its high EC, if applied in a manner that exploits the abovementioned advantages. In the sections that follow, we provide an overview of today's reactive nitrogen industry and describe the conventional ammonia and nitrate production process, through both the steam methane reforming (SMR) and the electrolysis pathway. We present key metrics for the two conventional methods and use these metrics as benchmarks for a set of specific PNF applications, namely:

- Stand-alone nitric acid process
- Integrated nitric acid process
- Nitrate enrichment of organic substrates

We also present two other potential applications that we outline in less detail, given their lower *technology readiness level* and lack of industrial performance data. Lastly, we estimate the economic viability of PNF, today and in the future.

## 10.2 Overview of The Reactive Nitrogen Industry

As of 2022, industrial nitrogen fixation amounted to 149 megatons ammonia–nitrogen per year (MT $NH_3$-N year$^{-1}$) [4]. Fertilizers constitute the majority of end-use at 114 MT N year$^{-1}$, of which urea holds the dominant share (Figure 10.3) [5]. Approximately 10% of global $NH_3$ is used to produce nitrate ($NO_3^-$) in the nitric acid ($HNO_3$) process [6]. The $HNO_3$ is primarily combined with $NH_3$ to produce ammonium nitrate (a fertilizer and an important precursor in other industrial processes), but it does have other direct industrial uses (e.g. in adipic acid and nitrobenzene production). This also applies to $NH_3$ and urea, both of which have a myriad of industrial applications [8]. Given nitrogen's important and wide-ranging uses, the reactive nitrogen industry is and will remain an integral part of modern society.

The rate of global nitrogen fixation is expected to rise in the coming decades. The International Energy Agency (IEA) [4] foresees an increase to more than 200 MT N year$^{-1}$ for fertilizer and other industrial uses in 2050 in their *Stated Policies* scenario (Figure 10.4). In addition, $NH_3$ has been touted as a potential carbon-free fuel for the maritime shipping sector and as fuel for electric power generation.

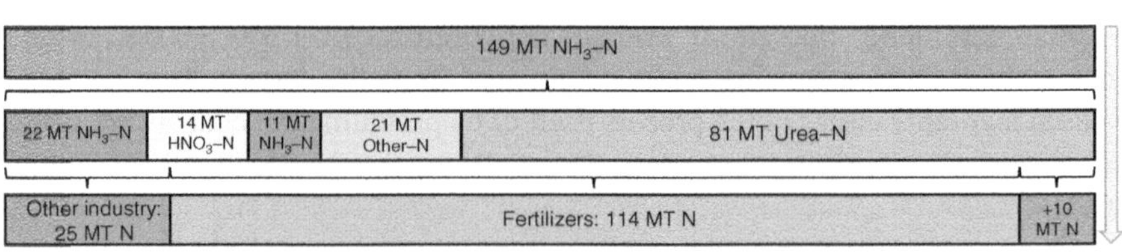


**Figure 10.3** Schematic representation of global reactive nitrogen production and application. All reactive nitrogen starts as ammonia $NH_3$, which serves as a basis for other nitrogen products. Fertilizers are the main end-use at 114 MT year$^{-1}$, while the remaining 35 MT year$^{-1}$ is applied in other industries. Based on estimates from [4–7].

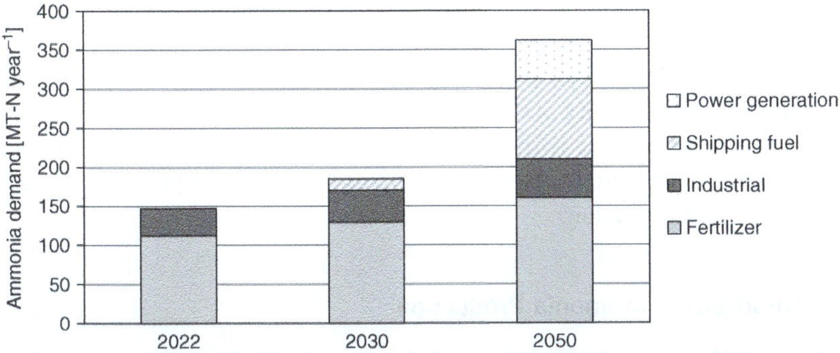


**Figure 10.4** Current, short-term, and long-term estimates of global ammonia ($NH_3$-N) production. There is significant uncertainty tied to the shipping fuel and power generation estimate. Adapted from [4, 9–12].

There are various estimates on how this will impact global $NH_3$ production. We base our 152 MT $NH_3$-N year$^{-1}$ in 2050 on a set of these estimates [9–12].

Today, more than 99% of the $NH_3$ production depends on finite fossil fuels: 5–6% of global natural gas extraction is devoted to $NH_3$ synthesis [13]. In addition, the reactive nitrogen sector emits more than 1.5% of global anthropogenic GHGs. The industry is, therefore, motivated to improve its long-term sustainability. Carbon capture and storage is a favored approach, given the high concentration of the $CO_2$ emitted from the $NH_3$ process, but it does not alleviate the fossil-fuel dependency, and it adds cost. Both electrolysis and PNF could shift nitrogen fixation away from fossil-fuel dependency, but these technologies currently do not play a major role in the *Stated Policies* scenario (<2% of global N fixation in 2050, as per the IEA) [14]. In the *Sustainable Development* scenario, however, the share of electricity-based nitrogen fixation is close to 30% in 2050.

## 10.3 Conventional Nitrogen Fixation

CNF begins with the synthesis of ammonia ($NH_3$). Since PNF is ideally used to produce nitrate ($NO_3^-$), we describe the entire conventional $NO_3^-$ production process to derive an applicable benchmark for PNF. Because the hydrogen ($H_2$) required for the $NH_3$ synthesis can be preceded by either the conversion of fossil fuels to $H_2$ or the electric conversion of water ($H_2O$) to $H_2$, we list the performance metrics for both routes. The conversion of $NH_3$ to $NO_3^-$ is the same for both CNF routes: $NH_3$ is oxidized to nitric oxide (NO), which is absorbed in water as nitric acid ($HNO_3$) – an $NO_3^-$ source.

To ensure a representative comparison between technologies that differ internally in their performance (i.e., there are significant variations in the energy efficiency of $NH_3$ plants), we employ two scenarios for mature fossil-based $NH_3$ technology: (i) the *best available technology* (BAT), defined as the performance necessary to be part of the 10% best plants and (ii) the global average, defined as the average operating performance of the world's $NH_3$ production. For the less mature electrolysis route, we operate with a *current* scenario that represents the performance obtainable today, and the *potential* scenario that states the potential of electrolysis-based $NH_3$ production.

### 10.3.1 Fossil-Based Ammonia Production

Conventional $NH_3$ production is commonly based on SMR (72% of global production), followed by coal (22%) and heavy oil (4%) [15]. We limit the upcoming process description to the SMR process, which is considered the BAT, but we present key metrics for both the BAT and the current global average. Reaction (10.1) summarizes the overall SMR $NH_3$ synthesis.

$$8N_2 + 2O_2 + 10H_2O + 7CH_4 \rightarrow 7\,CO_2 + 16NH_3 \tag{10.1}$$

As shown in Figure 10.5, Reaction (10.1) is not a single-step process. The $NH_3$ synthesis begins with desulphurization of the natural gas ($CH_4$) feedstock at

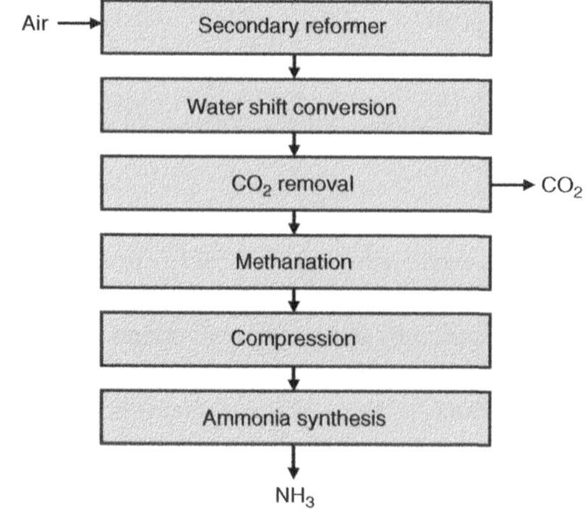


**Figure 10.5** Process diagram of the conventional fossil-based ammonia process.

350–400 °C, in which sulfur is hydrogenated to $H_2S$ in a reaction catalyzed by molybdenum oxide and subsequently adsorbed on pelletized zinc oxide, preventing downstream sulfur poisoning of the catalysts. In the primary reformer, steam ($H_2O$) and the desulfurized $CH_4$ are converted to a mixture consisting of $H_2$, carbon monoxide (CO), and $CO_2$ over a nickel-containing reforming catalyst [16]. The reaction is endothermic, and additional $CH_4$ is used as fuel to maintain a temperature of 500–600 °C in the reformer. The conversion of $CH_4$ in the primary reformer will be incomplete and is followed by a secondary reformer, where air is introduced to combust the remaining $CH_4$, yielding CO and $H_2$. This step also introduces the nitrogen ($N_2$) needed in the $NH_3$ synthesis. The excess heat from the combustion in the secondary reformer is used to generate high-pressure steam. The secondary reformer is followed by a water gas shift conversion of CO and $H_2O$ to $CO_2$ and $H_2$ over a catalyst (catalyst composition varies). The gas is cooled, and an alkaline aqueous solution consisting of either amines or potassium hydroxide is used to remove the $CO_2$. Small amounts of remaining and problematic CO and $CO_2$ can be converted to the inert $CH_4$ in the methanation step before the process gas, now consisting primarily of $N_2$ and $H_2$, enters the $NH_3$ synthesis operating at 100–250 bar and 350–550 °C. The process generates sufficient heat to supply high-pressure steam for turbines that power the system's pumps and compressors. Some steam may also be exported to external users.

Key metrics for the process are presented in Table 10.1. The EC for the BAT is calculated based on Reaction (10.1) with an additional 29% $CH_4$ needed as fuel to heat the primary reformer [15]. Its EC is consequently 0.50 MJ $molN^{-1}$. The global average EC is based on the share of $CH_4$, coal, and heavy oil used globally and their respective EC values listed by Rouwenhorst et al. [17]. Since Rouwenhorst et al. operate with BAT numbers that do not account for process inefficiencies, we adjust the calculated number to match the IEA's estimate of 0.64 MJ $molN^{-1}$ as a global average [4].

The GHG intensity is estimated in the same manner. Reaction (10.1) and the additional fuel requirements will provide a lower emission estimate of 1.9 $tonCO_2$ $tonNH_3\text{-}N^{-1}$, which is similar to the stated BAT GHG intensity [4]. Estimating the

**Table 10.1** Key metrics for the conventional fossil-based ammonia production.

| Parameter | BAT | Global average | References |
|---|---|---|---|
| Energy consumption [MJ $molN^{-1}$] | 0.50 | 0.64 | [4] [15–17] |
| Capex [EUR $tonNH_3\text{-}N^{-1}$ capacity] | 1800 | 1500 | [4] [18–20] |
| Fixed opex [EUR $tonNH_3\text{-}N^{-1}$] | 60 | 100 | [4] |
| GHG intensity [$tonCO_2$ $tonNH_3\text{-}N^{-1}$] | 1.9 | 3.6 | [4, 17] |

global average based on the fuel and feedstock distribution will again yield an estimate that does not account for the inefficiencies prevalent in the global $NH_3$ sector. We, therefore, adopt Yara's [5] estimate of 3.6 ton$CO_2$ ton$NH_3$-$N^{-1}$, as that estimate also results in the 0.45 GT $CO_2$ global emissions from $NH_3$ production (2006) listed by IEA [21] when multiplied by the global $NH_3$ production volumes in 2006 (125 MT $NH_3$-N year$^{-1}$).

The capital expenditure (capex) estimate is based on information from various source [4] [18–20]. Costs related to a new industrial plant and site (*greenfield*) were used, and all identified estimates were inflation adjusted to reflect 2024 prices. Table 10.1 summarizes the key metrics.

Other operational expenditures such as maintenance and labor costs (fixed opex) were estimated to be 60 EUR ton$NH_3$-$N^{-1}$ produced [4]. For coal-based production, fixed opex was above 216 EUR ton$NH_3$-$N^{-1}$, which led to a global average of 100 EUR ton$NH_3$-$N^{-1}$ produced.

### 10.3.2 Electricity-Based Ammonia Production

Ammonia ($NH_3$) production can overcome its dependency on fossil fuels by using water electrolysis to generate hydrogen ($H_2$). The generated $H_2$ is used as feedstock in the $NH_3$ synthesis (Reaction (10.3)), rendering the process fully electric. As less high-temperature heat is generated in the process than in a fossil-fuel-based process, electricity, not steam, is used to drive the system utilities. The lack of oxygen consumption in the now-obsolete reformers necessitates air separation to ensure that only $N_2$ and $H_2$ enter the $NH_3$ reactor. Figure 10.6 provides a process overview.

$$3H_2O \rightarrow 3H_2 + 1.5O_2 \qquad (10.2)$$

$$3H_2 + N_2 \rightarrow 2NH_3 \qquad (10.3)$$

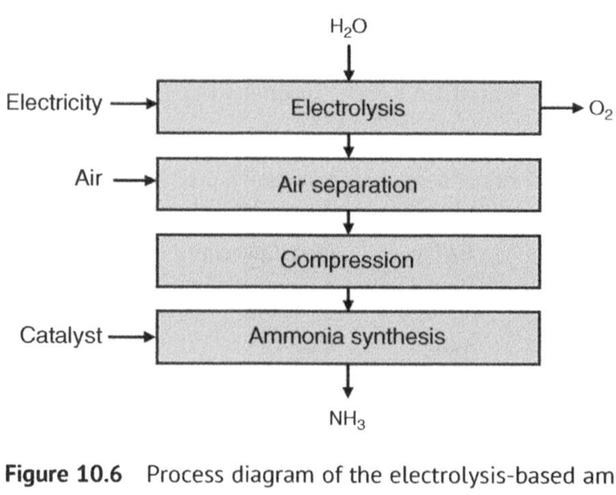


**Figure 10.6** Process diagram of the electrolysis-based ammonia process.

**Table 10.2** Key metrics for the conventional electrolysis-based ammonia production.

| Parameter | Current | Potential | References |
|---|---|---|---|
| Energy consumption [MJ $molN^{-1}$] | 0.63 | 0.56 | [15] |
| Capex [EUR $tonNH_3$-$N^{-1}$ capacity] | 2340 | 2250 | [4, 17, 18, 22] |
| Fixed opex [EUR $tonNH_3$-$N^{-1}$] | 100 | 100 | [4] |
| GHG intensity [$tonCO_2$ $tonNH_3$-$N^{-1}$] | 0 | 0 | |

Table 10.2 lists key metrics for conventional electrolysis-based ammonia production. The EC is based on a higher heating value (HHV) electrolyzer efficiency of 70%, resulting in a hydrogen EC of 0.406 MJ $molH_2^{-1}$. The $H_2$ requirement is given by (10.3), and the resultant EC is 0.60 MJ $molN^{-1}$. An additional 5% of electric energy is used to power compressors and air separation units, resulting in a current EC of 0.63 MJ $molN^{-1}$ [15]. For the technology potential, we set the electrolyzer efficiency to 80%.

The electrolyzers increase the capital expenditure (capex) of the ammonia plant. Based on averaging various estimates, the capex is set to 1.3 times the cost of a conventional ammonia plant [4, 17, 18]. The cost of electrolyzers has been predicted to fall, and the potential is herein set to 1.2 times the cost of an SMR $NH_3$ plant [22].

The fixed opex increases compared to the fossil-based method because of higher maintenance requirements for electrolyzers. For grid-connected electrolysis, the fixed opex was estimated to be 100 EUR $tonNH_3$-$N^{-1}$ produced [4]. Increased electrolyzer efficiency requires more complex technologies, and consequently, we do not foresee a potential reduction in fixed opex. If the electrolyzers were to be adapted to variable renewable energy (VRE), fixed opex would increase to 200 EUR $tonNH_3$-$N^{-1}$ [4].

Any GHGs emitted from electricity generation are considered to be outside the process boundaries. The GHG intensity is, therefore, set to zero. However, GHG emission from electricity generation is an important factor for the overall impact of the technology, and access to low-emission electricity generation is key to the success of the technology.

### 10.3.3 Nitric Acid Production

Both the fossil-fuel-based and the electric ammonia ($NH_3$) synthesis require the same final conversion step to produce nitrate ($NO_3^-$). The production of $NO_3^-$ is composed of the catalytic conversion of $NH_3$ to nitric oxide (NO) in the Ostwald process and the absorption of nitrogen oxides ($NO_x$) in water to produce nitric acid ($HNO_3$). Reactions (10.4) and (10.5) show the simplified overall two-stage reaction, which involves several intermediate steps [23, 24]. The oxidation of $NH_3$ to NO is highly exothermic and the released energy is used to power all process utilities, eliminating the need for additional energy input. An overview of the process is shown in Figure 10.7.

$$4NH_3 + 5O_2 \rightarrow 4\,NO + 6H_2O \tag{10.4}$$

$$4NO + 3O_2 + 2H_2O \rightarrow 4HNO_3 \tag{10.5}$$

The conversion of $NH_3$ to NO is commonly 96% efficient [23]. The $NH_3$ that does not react to NO is converted to either $N_2$ or $N_2O$. Besides being an efficiency loss, the $N_2O$ is particularly problematic, given its potency as a GHG (273 times more potent than $CO_2$ [25]). Up to 2.5% of the $NH_3$-N is converted to $N_2O$, resulting in a GHG emission intensity of 10.7 ton$CO_2$eq ton$HNO_3$-$N^{-1}$, which corresponds well with the estimates of Wood and Cowie [26]. To counter these emissions, new nitric acid plants employ various catalytic systems to decompose $N_2O$, which reduces $N_2O$ emissions by 80–95% [23, 27]. For the average plant with an $N_2O$ decomposition system, we assume a 90% emission reduction. To calculate the world average emissions, we estimate that 45% of the world's nitric acid plants operate without $N_2O$ removal. The global average emissions are consequently 5.4 ton$CO_2$ ton$HNO_3$-$N^{-1}$.

Since all new nitric acid plants will include $N_2O$ removal, we only list the capital cost of a modern plant. By evaluating the cost of nitric acid plants commissioned in recent years and adjusting for inflation, the capex of a nitric acid plant is set to 2,100 EUR ton N capacity$^{-1}$ [28, 29].

The fixed opex of a nitric acid plant is considered to be lower than that of the ammonia plant, given the lower number of catalyzed, high-temperature reaction stages. Regular maintenance and replacement of the platinum–rhodium gauze is necessary to ensure sufficiently high conversion. The fixed opex is set to 15 EUR ton$NH_3$-$N^{-1}$ (approximately 15% of the capex contribution).

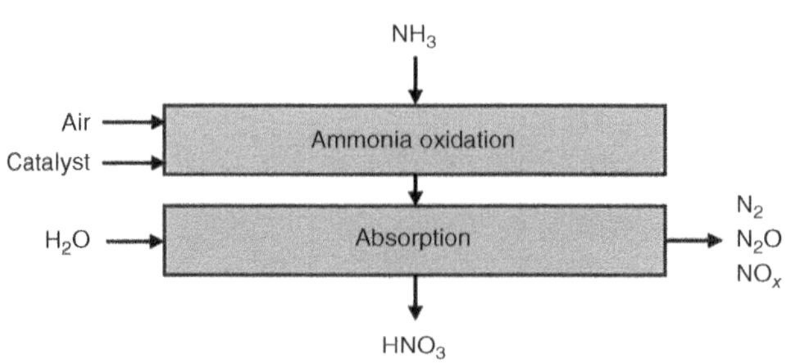


**Figure 10.7** Process diagram of the conventional nitric acid process.

**Table 10.3** Key metrics for the Ostwald process and nitric acid production.

| Parameter | $N_2O$ removal | World average | References |
|---|---|---|---|
| Conversion efficiency ($HNO_3$-N/$NH_3$-N) | 96% | 96% | [23] |
| Capex [EUR ton$HNO_3$-$N^{-1}$ capacity] | 2100 | NA | [28] |
| Fixed opex [EUR ton$NH_3$-$N^{-1}$] | 15 | 15 | |
| GHG intensity [ton$CO_2$ ton$NH_3$-$N^{-1}$] | 1.1 | 5.4 | [23, 26, 30] |

**Table 10.4** Key metrics for combined ammonia and nitric acid production. BAT $HNO_3$: best available technology steam methane reforming. The current electrolysis performance is shown in this table.

| Parameter | BAT $HNO_3$ | Electrolysis $HNO_3$ | Global average $HNO_3$ |
|---|---|---|---|
| Energy consumption [MJ mol$N^{-1}$] | 0.52 | 0.66 | 0.67 |
| Capex [EUR ton$NO_3$-$N^{-1}$ capacity] | 3900 | 4350 | 3600 |
| Fixed opex [EUR ton$NO_3$-$N^{-1}$] | 75 | 115 | 115 |
| GHG intensity [ton$CO_2$ ton$NO_3$-$N^{-1}$] | 3.0 | 0 | 9.0 |

### 10.3.4 Overall Performance of Nitrate Production

To estimate the overall performance of nitrate production, the ammonia and nitric acid production must be combined. Table 10.4 lists key metrics for nitrate production based on the BAT for the steam methane reformation (BAT $HNO_3$) and electrolysis (electrolysis $HNO_3$), as well as the current global average (global average $HNO_3$). The EC is higher for $HNO_3$ production than for $NH_3$ production because of the incomplete conversion (Tables 10.3 and 10.4).

## 10.4 Plasma-Based Nitrate Production

PNF begins with an electric conversion of air ($N_2$ and $O_2$) to nitric oxide (NO). A set of reactive species is generated in the plasma, and these species vary in their influence on nitrogen fixation. A simplified reaction sequence is shown as follows:

$$N_2 + O_2 \xrightarrow{\text{energy}} N_2^* + O_2^* \tag{10.6}$$

$$N_2^* + O_2^* \rightleftharpoons 2NO \tag{10.7}$$

A plasma system that enables (10.6) and (10.7) consists of a plasma generator (PG), power supply unit (PSU), and air supply unit (ASU) (Figure 10.8). The PG's main constituents are two electrodes and a heat exchanger to cool the plasma gas. The PSU contains a rectifier to convert alternating current (AC) to direct current (DC), a transformer to supply the correct current/voltage ratio, and an output filter to ensure a steady current. The cost of power electronics is inelastic to scale and the cost potential for the required components is set to 30 EUR per kW installed power, while the current cost is set to 100 EUR $kW^{-1}$ [31]. The ASU is assumed to be a compressor, although fans could be used if the pressure drop through the generator is limited. Both the PSU and ASU require additional energy, the share of which is listed in Table 10.5.

For all *current* systems, we employ a rotating gliding arc (RGA) due to its simplicity and relatively low specific EC (2.4 MJ $molN^{-1}$) at high NO concentration (5.4%) [32] and an additional 14% in auxiliary energy (Table 10.5). We are agnostic to the plasma technology that will be most efficient in the future, but we set its *medium-term* (defined as 5–10 years from now) EC to 1.25 MJ $molN^{-1}$ (including auxiliary energy), which is a significant reduction relative to current EC, but still five times higher than its theoretical potential. If the medium-term EC and the cost potential are realized, the resultant capex will be 214 EUR ton-N $capacity^{-1}$.

In the sections that follow, the plasma system is applied in various industrial processes. Key metrics are estimated for each process, and these metrics are used in the economic evaluation (Section 10.5).

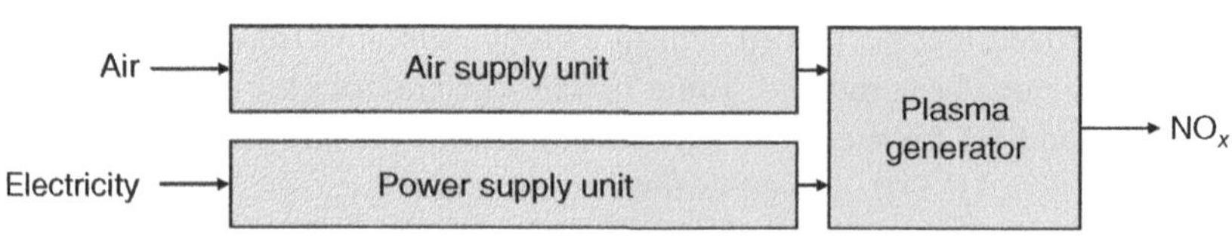


**Figure 10.8** Process diagram of plasma system generating nitrogen oxides ($NO_x$).

**Table 10.5** Distribution of energy consumption and cost potential for plasma-based nitrogen fixation.

| Parameter | Energy consumption (%) | Cost potential [EUR $kW^{-1}_{plasma}$] |
|---|---|---|
| Plasma generator | 100 | 20 |
| Power supply unit | +4 | 30 |
| Air supply unit | +10 | 50 |

### 10.4.1 Stand-alone Nitric Acid Process

In a stand-alone nitric acid process, the plasma system is used to generate a reactive nitrogen gas ($NO_x$) that is absorbed in water as nitric acid ($HNO_3$). There is no longer a need for the ammonia ($NH_3$) process, and because the oxidation of $NH_3$ to $NO_x$ is unnecessary, a large proportion of the equipment and catalysts required in the nitric acid process (Section 10.3.3) is obsolete. This simplifies the overall process, as shown in Figure 10.9.

Given the simplification of the nitric acid process, we estimate the cost of $NO_x$ *absorption* to constitute one third of the entire conventional nitric acid process (Table 10.3). This cost is added to the cost of the plasma system (Table 10.5). For the plasma system, an additional capex of 30% is included for system integration. As with the electrolysis-based nitrogen fixation, GHG emissions from electricity generation are outside the scope of the analysis. The fixed opex is estimated to be 100 EUR tonN$^{-1}$ based on electrode replacement and compressor and heat exchanger maintenance, and these are reduced as efficiency and electrode lifetime are improved. Table 10.6 summarizes the key metrics for the stand-alone plasma-based nitric acid process.

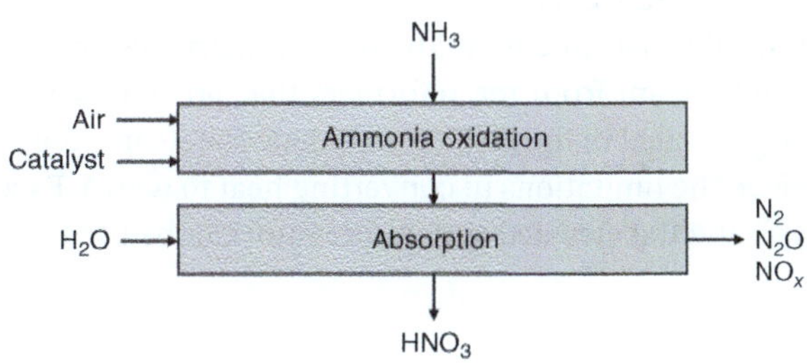


**Figure 10.9** Process diagram for the plasma-based nitric acid process.

**Table 10.6** Key metrics for the plasma-based nitric acid process.

| Parameter | Current | Medium term |
|---|---|---|
| SEC [MJ molN$^{-1}$] | 2.70 | 1.25 |
| Capex [EUR ton$NO_3$-N$^{-1}$ capacity] | 2,160 | 1,140 |
| Fixed opex [EUR ton$NO_3$-N$^{-1}$] | 100 | 50 |
| GHG intensity [ton$CO_2$ ton$NO_3$-N$^{-1}$] | 0 | 0 |

### 10.4.2 Integrated Nitric Acid Process

There are several ammonia ($NH_3$) plants in the world that are not connected to a nitric acid ($HNO_3$) plant. In these cases, PNF can favorably integrate into the conventional $NH_3$ process to make ammonium nitrate ($NH_4NO_3$). PNF then serves as both an instrument to produce more reactive nitrogen and a means to upgrade $NH_3$ to the more valuable $NH_4NO_3$. Moreover, the heat from PNF can be utilized to the benefit of the overall process, reducing energy costs. Other co-benefits also exist. In this section, we describe three processes in which PNF can be favorably integrated.

Integration into an existing $NH_3$ plant is the simplest of the three. In this case, the existing plant exclusively produces $NH_3$, but there is a local demand for $NH_4NO_3$ fertilizer (Figure 10.10). This demand may be significantly lower than the full production and too low for an investment in a nitric acid plant. PNF can be used to cover the nitrate demand, and a portion of the ammonia is used in the $NH_4NO_3$ process. Such an integration is of particular interest where there is a variable supply of low-cost electricity, since PNF can operate intermittently without complications. The process can be a tool to reduce the emissions from parts of global fertilizer production and ensure local access to $NH_4NO_3$.

Another option for integration is an existing ammonia plant that no longer uses methane ($CH_4$) as a feedstock and fuel, but has switched to electricity and electrolysis (Figure 10.11). In this scenario, PNF will enable nitrate production without an Ostwald investment. With the reformers obsolete, less heat is generated in the plant. Steam from the reformers that previously powered the ASUs can instead be generated in the PNF (we estimate the steam value to hold 50% electricity value, given the limitations in converting heat to work). Excess oxygen from the electrolysis and the elevated system pressure can be used to reduce the EC of the PNF to

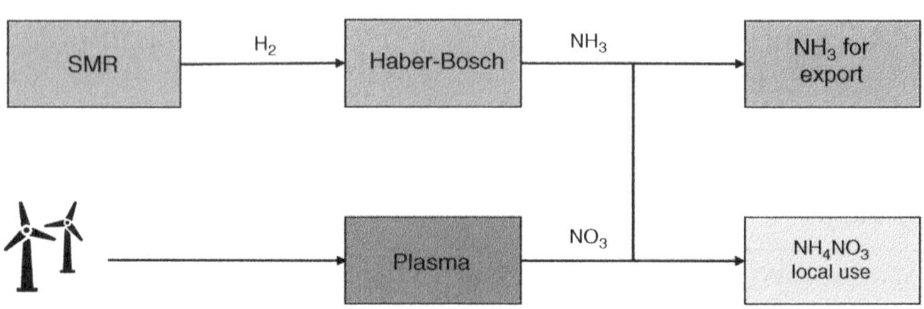


**Figure 10.10** Partial electrification of nitrogen fixation in an existing ammonia plant. Plasma is used to generate nitric acid for local ammonium nitrate consumption with local intermittent renewable energy.

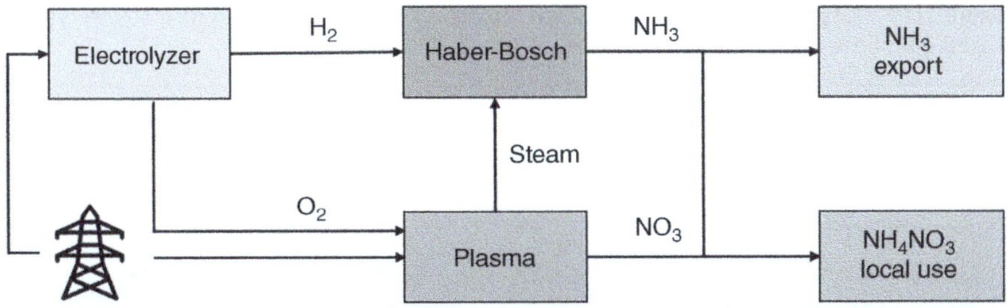


**Figure 10.11** Integration of plasma-based nitrogen fixation in an ammonia plant upgraded with electrolyzers. Approximately 10% of the ammonia is upgraded to ammonium nitrate with nitric acid from the plasma system. The plasma system generates steam to power the utilities of the entire plant. Oxygen-enriched air and elevated pressure improve the efficiency of the plasma process.

1.8 MJ $molN^{-1}$ [33]. If 90% of the energy is recovered as steam, the net PNF EC reaches a net value of 1.0 MJ $molN^{-1}$.

PNF can also integrate into a fully electric electrolysis-based ammonia process. Again, PNF replaces the Ostwald step and enables low-emission production of $HNO_3$. The system compressors are now powered by electricity, and the heat from the plasma is instead used in the $NH_4NO_3$ production (Figure 10.12). The integration still benefits from the available pressure and oxygen.

To evaluate the performance potential of the integrated approach, scenario 2 (Figure 10.11) is used. The EC is set to 1.8 $molN^{-1}$ (all auxiliary energy other than PSU is covered by process heat) for the current scenario, and a 50% reduction as the potential (0.9 $molN^{-1}$). As for the process heat, 90% is recovered as steam at 50% of the electricity value, resulting in a steam credit of 45%. We estimate a brownfield (existing industrial plant) advantage for the capex, eliminating the ASU cost and

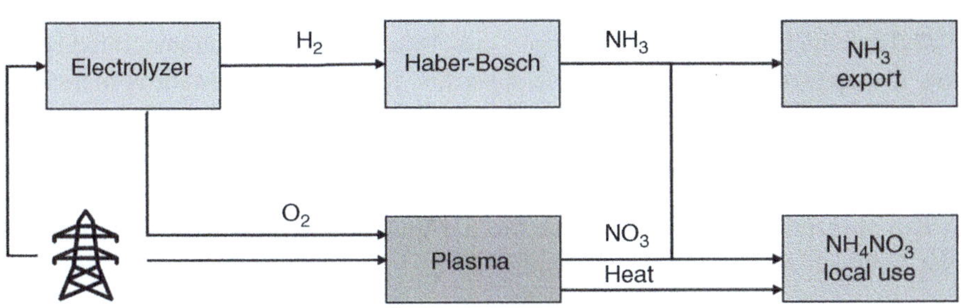


**Figure 10.12** A fully electric ammonia process uses a plasma system instead of an Ostwald process to produce ammonium nitrate. Oxygen-enriched air and elevated pressure improve the efficiency of the plasma process.

**Table 10.7** Key metrics for plasma-based nitrogen fixation integrated into electrolysis-based ammonia production.

| Parameter | Current | Medium term |
|---|---|---|
| Energy consumption [MJ $molN^{-1}$] | 1.8 | 0.9 |
| Steam credit (%)* | 45 | 45 |
| Capex [EUR $tonNO_3$-$N^{-1}$ capacity] | 1560 | 960 |
| Fixed opex [EUR $tonNO_3$-$N^{-1}$] | 100 | 50 |
| GHG intensity [$tonCO_2$ $tonNO_3$-$N^{-1}$] | 0 | 0 |

*Affects net energy consumption, but not the production capacity and therefore not capex per ton N.

reducing the system integration cost (Section 10.4.1) to 15%. The capex per ton N capacity is also reduced due to the reduced EC, whereas maintenance requirements stay the same. Table 10.7 summarizes the key metrics.

### 10.4.3 Nitrate Enrichment of Organic Substrates

PNF can also be used to enrich organic fertilizers with nitrate ($NO_3^-$) and thereby improve their fertilizer performance. Organic fertilizers such as livestock manure contain significant amounts (103 MT-N $year^{-1}$) of reactive nitrogen due to the inefficient conversion of livestock-feed into meat and milk [34]. Organic substrates are consequently considered a valuable resource in food production. This value is diminished by the high pH of these organic substrates, which causes the ammoniacal nitrogen to volatilize and be lost to air at a rate ranging from 30 to 70% depending on storage and application conditions [35]. The lost nitrogen must in turn be replaced by purchased synthetic fertilizer to prevent agronomic yield losses.

PNF can be used to prevent nitrogen loss. By absorbing plasma-generated reactive nitrogen directly in the organic substrate, the nitrogen content is increased and the nitrogen losses reduced, which is shown as

$$2NO + O_2 + H_2O \rightarrow 2H^+ + NO_3^- + NO_2^- \tag{10.8}$$

The absorption in (10.8) reduces the pH and shifts (10.9) from $NH_3$ toward $NH_4^+$, which remains in the substrate.

$$NH_3 + H^+ \leftrightarrow NH_4^+ \tag{10.9}$$

The direct absorption occurs at a moderate pH level (pH 4–6), which improves the absorption efficiency substantially compared to absorption in concentrated

$HNO_3$. Since the absorption occurs at moderate pH, a share of the nitrogen will be absorbed as nitrite ($NO_2^-$), which can be further oxidized to nitrate ($NO_3^-$).

Because of the impact on pH, every unit of nitrogen produced in the plasma system will prevent the loss of ammoniacal nitrogen. The value of the plasma-generated reactive nitrogen is, therefore, increased in accordance with Eq. (10.10):

$$\text{Added nitrogen value} = \frac{\text{loss } NH_3}{NO_3 : NH_4} \tag{10.10}$$

where loss $NH_3$ is the share of ammoniacal nitrogen lost (set to 50% [35]), and $NO_3 : NH_4$ is the molar ratio of nitrate and ammonium required to reach the target pH (the ratio is set to 1.5 for the *current* and 0.8 for the *medium-term* scenario). The increased nitrogen value is consequently 33% for the *current* and 63% for the *medium-term*, resulting in a *de facto* EC of 2.0 and 0.74 MJ $molN^{-1}$, respectively. We assume an additional 14% in auxiliary power needed for the *current* scenario (Table 10.5).

In addition to the increased nitrogen content and reduced nitrogen losses, the absorption of reactive nitrogen confers antimicrobial properties [36, 37]. The process, therefore, inhibits the formation of methane ($CH_4$) from stored organic substrates [38]. Stored organic substrates represent a major source of GHG, and an estimated 108 kg $CO_2$eq as $CH_4$ is emitted per ton of stored organic substrate in a temperate region [35]. Under the assumption that 2 $kgNO_3$-N is added per ton substrate in the plasma process, the GHG reduction is 54 kg $CO_2$eq $kgNO_3$-$N_{pasma}{}^{-1}$.

An early commercial version of the process is currently manufactured by the GEA Group based on a technology license from $N_2$ Applied AS. The current first-generation unit cost was estimated to be EUR 200,000 for a 50 kW unit, resulting in a capex of 23,000 EUR $tonNO_3$-$N^{-1}$ capacity [39]. We consider the technology to have the same capex potential as the non-organic concepts (Table 10.8).

**Table 10.8** Key metrics for nitrate enrichment of organic substrates.

| Parameter | Current | Medium term |
|---|---|---|
| Energy consumption* [MJ $mol^{-1}$] | 2.3 | 0.74 |
| Capex [EUR $tonNO_3$-$N^{-1}$ capacity] | 23 000 | 1140 |
| Fixed opex [EUR $tonNO_3$-$N^{-1}$] | 100 | 50 |
| GHG intensity [$tonCO_2$ $tonNO_3$-$N^{-1}$] | −54 | −54 |

*See Eq. (10.10).

### 10.4.4 Other Avenues

The industrial application of PNF is not limited to those described in the previous sections (10.4.1–10.4.3). Several other industrial applications of PNF exist, albeit at a lower technology readiness level than the industrial plasma-based nitrate production invented in the early twentieth century. These applications include, but are not limited to, plasma-based ammonia ($NH_3$) production and plasma-activated water (PAW).

Plasma-based $NH_3$ synthesis is separated into three main pathways. In the first step, nitrogen ($N_2$) is combined with hydrogen ($H_2$) in a plasma catalytic process to form $NH_3$. The reactants are the same as in the conventional Haber-Bosch process, but the $NH_3$ synthesis can occur under reduced pressure and temperature. This pathway requires $H_2$ formation prior to the plasma process, which increases the energy cost. The conversion is often low (0.1%) at acceptable EC, which requires recycling of the reactants. Consequently, the overall EC of these systems is generally above 4 MJ $molN^{-1}$ when considering the cost of $H_2$ production, recycling, and air separation [40]. To circumvent the $H_2$ production, ammonia can be produced from $N_2$ and $H_2O$ directly in a plasma process. However, the conversion is generally low (~200 ppm) and the EC is higher than the $H_2$-based plasma-assisted ammonia synthesis [41]. In a water-vapor non-equilibrium situation, Vervloessem et al. [42] reached an EC of 22 MJ $molN^{-1}$ at 15% $NH_3$ selectivity (the remaining N fixed as $HNO_2$). The final method involves the production of $NO_x$ in an air plasma followed by adsorption and catalytic reduction of the $NO_x$ to $NH_3$. The process is estimated to reach an EC of 5.6 MJ $molN^{-1}$ with current technology, and its EC is largely tied to the EC of the $NO_x$ and $H_2$ production [43]. In summary, the main motivation for plasma-based ammonia synthesis will be the avoidance of the capital investment required for electrolysis-based conventional $NH_3$ production.

Compared to plasma-based ammonia synthesis, PAW is not primarily considered an application of PNF, but rather a more general application of plasma technology. However, its properties do depend on reactive nitrogen and oxygen species that will eventually form stable reactive nitrogen compounds such as nitrate. The industrial application of PAW is mainly related to its antimicrobial properties [44]. Since water and electricity are the only input factors, other chemicals can be replaced. Most of the reactive species are also short-lived, potentially lessening the waste-handling cost. PAW has several applications ranging from decontaminating hospital equipment to reducing pesticide use. The economic viability of PAW, therefore, primarily depends on the substances it can replace and the required plasma treatment time (the determinant of the EC). PAW remains a promising plasma application in need of more research and industrial development [45].

## 10.5 Economic Comparison

A widespread industrial adoption of PNF demands a certain economic performance. The economic performance of PNF depends in large part not only on the performance of the technology but also on the conditions under which it operates. To assess the technology performance, we use the *current* technology status (defined as what is already shown to be obtainable) and the *medium-term* potential (defined as what is assumed obtainable within 5–10 years) listed in Sections 10.3 and 10.4 to compare PNF with CNF. Two scenarios that capture important external factors, such as electricity and natural gas price, are used in the assessment (Table 10.9).

Our economic assessment of nitrate ($NO_3^-$) production has certain limitations. The scope is limited to include the capital investment (capex), other operational expenses (fixed opex), energy and feedstock costs (variable opex), and the cost of GHG emissions. The cost of transport is omitted, primarily because the transportation costs depend on, among several other factors, the placement of the factory. In addition, the factors that we do list will vary considerably depending on the facility's location and the technology used. Nevertheless, we consider the comparisons to be valid for the given scenarios, as the main cost drivers are included.

The cost of energy (natural gas or electricity) and feedstock (natural gas for the SMR) is commonly the main contributor to the production cost. As of October 2024, the EU natural gas price was 40 EUR $MWh^{-1}$. The price of electricity is set to the same value, 40 $MWh^{-1}$. Although this is higher than current average electricity prices in most regions of the world, it is a price obtainable in specific regions and in certain long-term industrial electricity contracts [2]. Further reductions in electricity prices require intermittent operation to exploit VRE. We, therefore, include a VRE scenario where the capacity factor is reduced from the standard 90% to 50%, which increases the capex per fixed nitrogen, but reduces electricity cost to 15 EUR $MWh^{-1}$, which is still above the lower estimate of 10 EUR $MWh^{-1}$ used by IEA [2–4]. For the plasma system, 90% of the plasma

**Table 10.9** Parameters defining the standard and the variable renewable energy (VRE) scenario.

| Parameter | Standard | VRE |
|---|---|---|
| Natural gas price [EUR $MWh^{-1}$] | 40 | 40 |
| Electricity price [EUR $MWh^{-1}$] | 40 | 15 |
| Factory lifespan [years] | 27 | 27 |
| Capacity factor [%] | 90 | 50 |
| Carbon price [EUR ton $CO_2$eq] | 70 | 70 |

energy can be captured as heat. This heat is considered to hold a value at one third of the electricity price and therefore reduces the energy cost of PNF by 30% (no heat capture is included from the auxiliary energy used for the compressor and power supply). For the integrated PNF, a value of 45% is used (as described in Section 10.4.2). In the VRE scenario, no heat value is included.

The capex values listed in Sections 10.3 and 10.4 are converted to a cost per ton nitric acid nitrogen ($HNO_3$-N – the $NO_3^-$ precursor) produced in a factory with a lifespan of 27 years and a capacity factor of 90% for all systems (except for the VRE scenario where the capacity factor is set to 50%). The lifespan does not account for the shorter lifespan of the electrolyzer stack. The lack of long-term industrial PNF with modern technology makes its lifespan estimate uncertain, but the replaceable electrodes are assumed to be the main component exposed to significant stress.

For the cost of GHG emissions, we neglect the intricacies of carbon trading systems and set a fixed GHG price for emissions occurring within the system boundaries. The GHG intensities for each technology are listed in Sections 10.3 and 10.4. The European Emission Trading System (ETS) trades carbon permits at 70 EUR $tonCO_2^{-1}$ in October 2024, which is the value used to determine the cost of GHG emissions.

The current technology status under the standard scenario results in the *World average $HNO_3$* production having the lowest production cost when combining capex and fuel costs, but losing its advantage when the high GHG emissions are included (Figure 10.13). Given the reduction in GHG emissions resulting from nitrogen enrichment of organic fertilizers, this option could, in principle, pay for itself. However, agricultural emissions are currently not a part of the ETS. The *Plasma $HNO_3$ integrated* results in lower $HNO_3$ production cost than the world average but higher than *Electrolysis $HNO_3$*, while the stand-alone *Plasma $HNO_3$* has the highest net production cost.

In the medium-term scenario, all the technologies have an improved performance, as detailed in Sections 10.3 and 10.4. The BAT $HNO_3$ is used for the conventional processes. Under these conditions, the integrated plasma approach and nitrogen-enriched organic fertilizer provide the best economic performance, followed by electrolysis-based $HNO_3$ (see Figure 10.14).

The above scenarios operate under steady conditions (capacity factor = 90%). By operating intermittently, lower electricity prices can be obtained. In Figure 10.15, both Electrolysis $HNO_3$ and Plasma $HNO_3$ are operated intermittently (VRE scenario in Table 10.9) at the medium-term technology performance. The BAT $HNO_3$ is kept as a reference case. The low capex of Plasma $HNO_3$ results in a lower overall production cost despite the higher specific EC.

The production cost of $HNO_3$ is highly sensitive to the price of energy, either as natural gas or as electricity. In the above scenarios, the price of the two energy carriers was equal. Commonly, the two fluctuate and can differ substantially. Figure 10.16 shows the sensitivity of *Plasma $HNO_3$* to the electricity price under

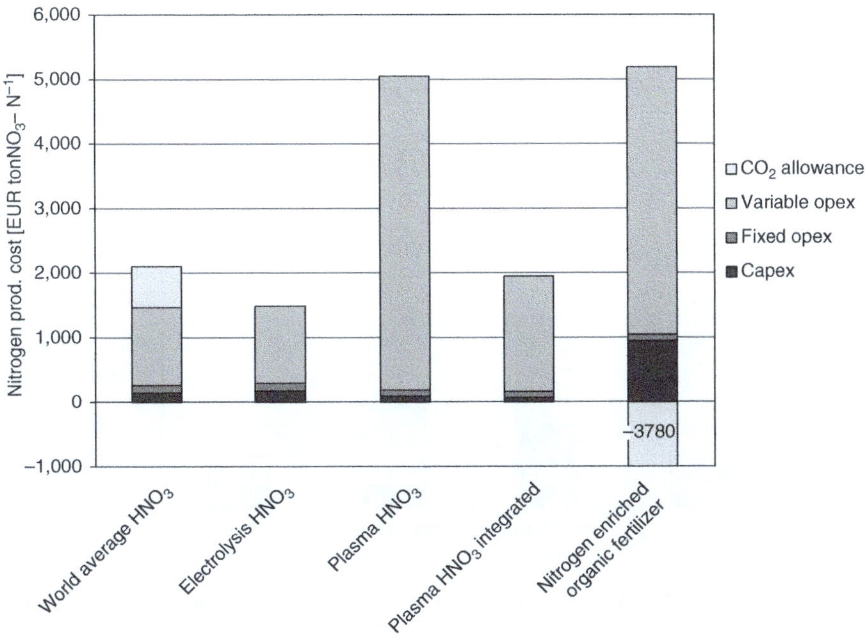


**Figure 10.13** Nitrogen production cost with current technology under standard conditions (Table 10.9).

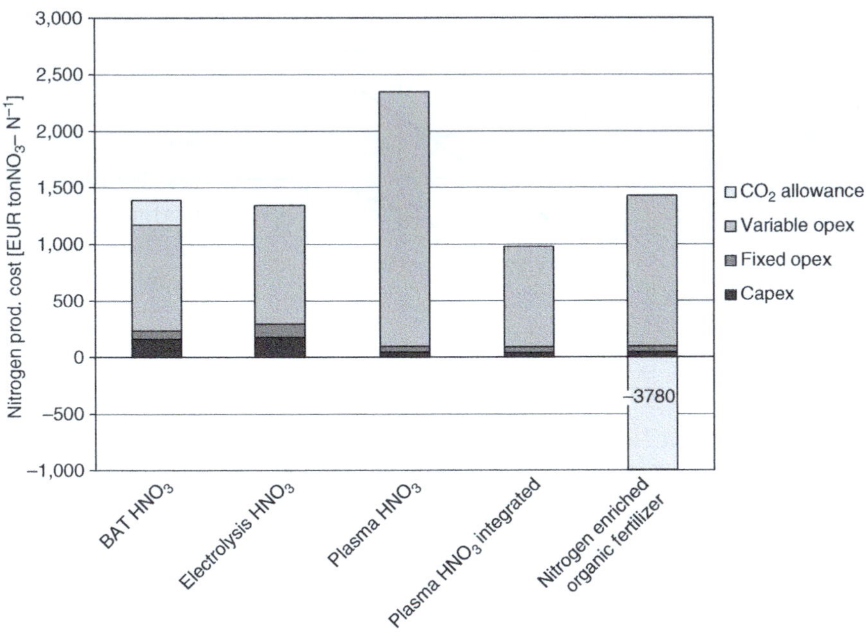


**Figure 10.14** Nitrogen production cost with medium-term technology under standard conditions (Table 10.9).

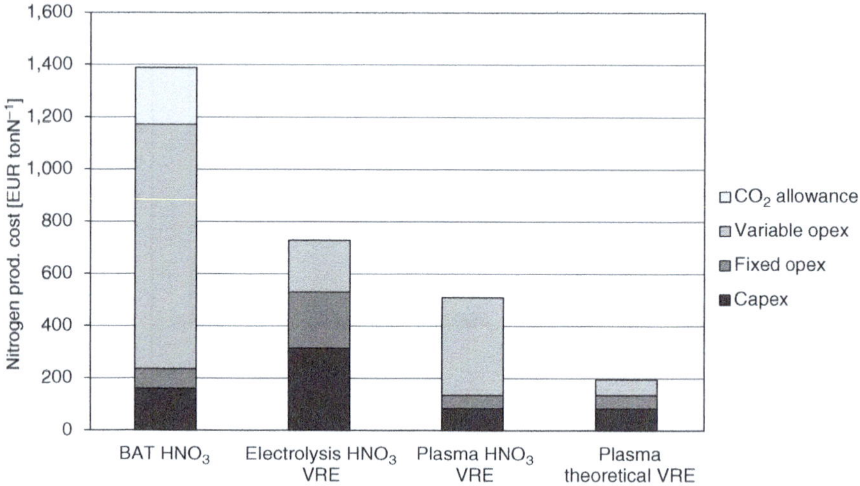


**Figure 10.15** Nitrogen production cost with medium-term technology under the variable renewable energy (VRE) scenario (Table 10.9). No heat credit is included in Plasma $HNO_3$ VRE. "Plasma theoretical VRE" shows the nitrogen cost at the theoretical energy consumption of 0.2 MJ molN$^{-1}$ in the VRE scenario.

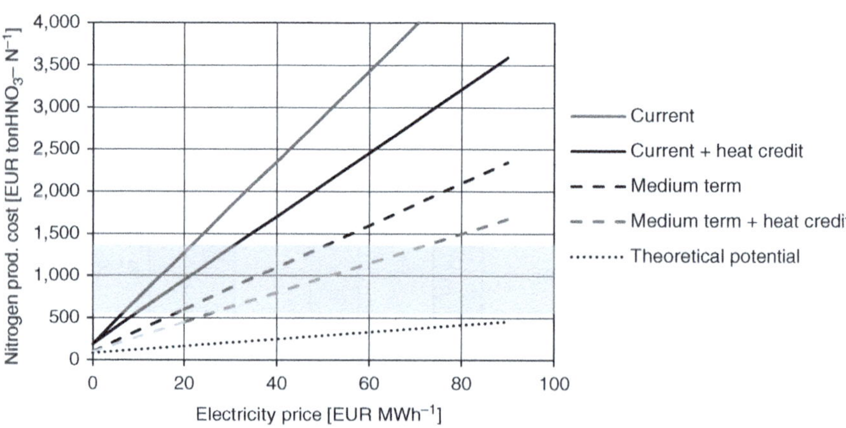


**Figure 10.16** The nitrogen production cost sensitivity to electricity price for plasma-based nitrogen fixation under three different technology scenarios, with and without 30% heat credit (no heat credit included for the theoretical potential). The gray region shows the production cost of nitrate with the conventional method for natural gas prices ranging from 5 to 90 EUR kWh$^{-1}$.

three different technology scenarios: the current, medium-term, and theoretical potential (0.2 MJ $molN^{-1}$). The sensitivity is presented with and without heat credit (30% heat credit). We also include the cost region of conventional fossil-based $HNO_3$ production for natural gas prices ranging from 5 to 90 EUR $MWh^{-1}$ (capex and $CO_2$ allowance are included).

As Figure 10.16 indicates, the economic viability of PNF depends on the cost of energy as electricity compared to its cost as natural gas. This ratio will be a major determinant of how widespread its adoption will be. In addition, technological advances and improved integration will play an important role. Lastly, the desire for smaller, regional production facilities closer to the consumer can drive technology adoption, as it eliminates many costly intermediaries.

## 10.6 Competitive Landscape

Several companies have entered the PNF market in recent years (Figure 10.17). The accelerating number of companies indicates that the technology's relevance is in motion. The increased attention is likely driven by the desired departure from fossil-fuel-based industries to a more widespread use of electricity in chemical processes, as well as a renewed focus on food security and reduced reliance on sensitive global supply chains. The companies shown in Figure 10.17 are approaching these central themes from different angles.

$N_2$ Applied AS specializes in technology for nitrogen fixation and technology necessary to valorize a reactive nitrogen gas. The company holds patents on efficient nitrogen fixation and various application technologies, some of which cover nitrogen-enrichment of organic substrates [46–48]. The latter process has been licensed to the GEA Group and is now marketed as the *ProManure E2950 Manure Enricher* [49]. $N_2$ Applied has also piloted nitrogen fixation technology for the production of mineral fertilizers to greenhouse growers [50].

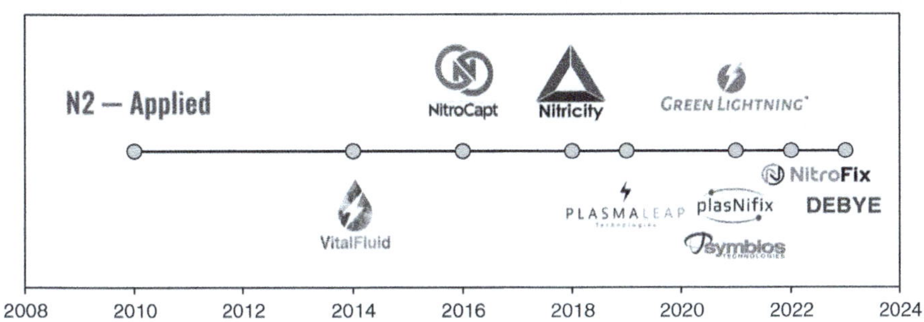


**Figure 10.17** Overview of companies focusing on plasma-based nitrogen fixation and application.

VitalFluid BV focuses primarily on the production of PAW for food production, aiming to reduce pathogenic activity, improve germination, and supply reactive nitrogen to the crops. The company holds a patent on a multi-electrode AC PG and a combined thermal/non-thermal system for PAW production [51, 52]. Three PAW units are listed in VitalFluid's product catalog, ranging from 0.1 to 15 kW nominal power [53].

NitroCapt AB aims to build production facilities to produce nitrogen fertilizer. The company's patent portfolio comprises a high-pressure (10–100 bar) microwave- or radio-frequency plasma system and a gas-permeable reaction chamber [54, 55].

Nitricity Inc. develops plasma-based systems that produce liquid mineral fertilizers. The company targets local fertilizer production with renewable energy and holds patents on a power supply for intermittent operation, a microwave plasma system, and a version of the nitrophosphate process [56–58]. Nitricity is currently producing calcium nitrate in partnership with Olam Food Ingredients for the fertilization of almonds in California [59].

The more recently founded companies listed in Figure 10.17 are all pursuing the development of PNF, but limited public information exists on their state of progress. As the potential of PNF becomes more evident, the number of companies and initiatives are expected to increase.

## 10.7 Conclusion

CNF is currently significantly more energy efficient than PNF. By improving the efficiency of the plasma-based approach, the rate of industrial adoption would increase, and a realization of its theoretical potential would make plasma the favored nitrate production technology. However, improved efficiency is not the only pathway to economic viability. The way in which the advantages of plasma technology are leveraged and how it is adapted to future energy and food systems are equally important. We foresee four scenarios that would lead to a general economic viability of PNF:

1) The capex targets for PNF are met, and variable electricity production results in prices below 20 EUR $MWh^{-1}$ for more than 30% of the year.
2) There is a premium placed on reduced $CO_2$ emissions in the agricultural sector of 50 EUR $tonCO_2^{-1}$.
3) There is an increased demand for smaller-scale regional nitrogen fixation to improve food security and reduce dependency on global supply chains.
4) The EC of PNF is reduced by 50%.

Either of the above scenarios would warrant major industrial investment in PNF.

## References


1 Abdelaziz, A.A., Komuro, A., Teramoto, Y. et al. (2024). Atmospheric-pressure plasmas for NOx production; short review on current status. *Curr. Opin. Green Sustain. Chem.* 50: 100977. https://doi.org/10.1016/j.cogsc.2024.100977.

2 International Energy Agency (2024). *Electricity 2024 – Analysis and Forecast to 2026*. OECD.

3 Storaker, K., Østensen, E., Korneliussen, R., and Bøhnsdalen, E. (2024). Kortsiktig Markedsanalyse 2024–2029, Statnett, Sep. 2024.

4 International Energy Agency (2021). *Ammonia Technology Roadmap: Towards More Sustainable Nitrogen Fertiliser Production*. OECD https://doi.org/10.1787/f6daa4a0-en.

5 Yara (2022). *Fertilizer Industry Handbook*. Yara.

6 Levi, P.G. and Cullen, J.M. (2018). Mapping global flows of chemicals: from fossil fuel feedstocks to chemical products. *Environ. Sci. Technol.* 52 (4): 1725–1734. https://doi.org/10.1021/acs.est.7b04573.

7 Hatfield, O. (2021). Country traded ammonia logistics and storage, present and future, 2021.

8 Domene, L.A.F. and Ayres, R.U. (2001). Nitrogen's role in industrial systems. *J. Ind. Ecol.* 5 (1): 77–103. https://doi.org/10.1162/108819801753358517.

9 International Energy Agency (2023). *Net Zero Roadmap: A Global Pathway to Keep the 1.5 °C Goal in Reach - 2023 Update*. OECD.

10 Verschuur, J., Salmon, N., Hall, J., and Bañares-Alcántara, R. (2024). Optimal fuel supply of green ammonia to decarbonise global shipping. *Environ. Res.: Infrastruct. Sustain.* 4 (1): 015001. https://doi.org/10.1088/2634-4505/ad097a.

11 International Energy Agency (2021). *Global Hydrogen Review*. OECD.

12 Potential of ammonia as a shipping fuel | LR. Available: https://www.lr.org/en/knowledge/horizons/april-2021/decarbonising-shipping--could-ammonia-be-the-fuel-of-the-future/ (accessed October 24, 2024)

13 International Energy Agency (2023). *Medium-Term Gas Report 2023: Including the Gas Market Report, Q4-2023*. OECD https://doi.org/10.1787/5da099e5-en.

14 Executive Summary – Ammonia Technology Roadmap – Analysis, IEA. https://www.iea.org/reports/ammonia-technology-roadmap/executive-summary (accessed October 24, 2024)

15 Liu, X., Elgowainy, A., and Wang, M. (2020). Life cycle energy use and greenhouse gas emissions of ammonia production from renewable resources and industrial by-products. *Green Chem.* 22 (17): 5751–5761. https://doi.org/10.1039/D0GC02301A.

16 European Fertilizer Manufacturers Association (2000). *Best Available Techniques for Pollution Prevention and Control in the European Fertilizer Industry*. European Fertilizer Manufacturers Association.

17 Rouwenhorst, K.H.R., Krzywda, P.M., Benes, N.E. et al. (2021). Ammonia Production Technologies. In: *Techno-Economic Challenges of Green Ammonia as an*

*Energy Vector* (ed. A. Valera-Medina and R. Banares-Alcantara), 41–83. Elsevier https://doi.org/10.1016/B978-0-12-820560-0.00004-7.

**18** Nosherwani, S.A. and Neto, R.C. (2021). Techno-economic assessment of commercial ammonia synthesis methods in coastal areas of Germany. *J Energy Storage* 34: 102201. https://doi.org/10.1016/j.est.2020.102201.

**19** Yara and Enbridge to Develop and Construct a Low-Carbon Blue Ammonia Project at Enbridge Ingleside Energy Center | Yara International, Yara None. https://www.yara.com/corporate-releases/yara-and-enbridge-to-develop-and-construct-a-low-carbon-blue-ammonia-project-at-enbridge-ingleside-energy-center/ (accessed October 19, 2024)

**20** The Capital Intensity of Small-Scale Ammonia Plants, Ammonia Energy Association. https://ammoniaenergy.org/articles/the-capital-intensity-of-small-scale-ammonia-plants/ (accessed October 19, 2024)

**21** International Energy Agency (2007). *Tracking Industrial Energy Efficiency and $CO_2$ Emissions. OECD https://doi.org/10.1787/9789264030404-en.*

**22** L. Bertuccioli and A. Chan, Development of Water Electrolysis in the European Union, Fuel Cells and Hydrogen Joint Undertaking, Feb. 2014. https://web.archive.org/web/20160310213740/http:/www.fch.europa.eu/sites/default/files/study%20electrolyser_0-Logos_0_0.pdf (accessed October 20, 2024)

**23** Pérez-Ramírez, J., Kapteijn, F., Schöffel, K., and Moulijn, J.A. (2003). Formation and control of N2O in nitric acid production. *Appl. Catal. B Environ.* 44 (2): 117–151. https://doi.org/10.1016/S0926-3373(03)00026-2.

**24** Loutet, K.G., Mahecha-Botero, A., Boyd, T. et al. (2011). Experimental measurements and mass transfer/reaction modeling for an industrial $NO_x$ absorption process. *Ind. Eng. Chem. Res.* 50 (4): 2192–2203. https://doi.org/10.1021/ie100436p.

**25** Intergovernmental Panel on Climate Change (2014). *Climate Change 2014 Mitigation of Climate Change: Working Group III Contribution to the Fifth Assessment Report of the Intergovernmental Panel on Climate Change*. Cambridge: Cambridge University Press https://doi.org/10.1017/CBO9781107415416.

**26** Wood, S. and Cowie, A. A Review of Greenhouse Gas Emission Factors for Fertiliser Production.

**27** Wiesenberger, H. (2001). *State-of-the-Art for the Production of Nitric Acid with Regard to the IPPC Directive*, vol. no. Bd. 150. Wien: Umweltbundesamt.

**28** next_develop, The Art of Building a Nitric Acid Plant in Uzbekistan • Casale, Casale. https://casale.ch/the-art-of-building-a-nitric-acid-plant/ (accessed October 20, 2024)

**29** Omnia's New Nitric Acid Complex on Time and Below Budget – Omnia Holdings Limited. https://www.omnia.co.za/investors-and-media/press-releases/2012/105-omnias-new-nitric-acid-complex-on-time-and-below-budget-09-may-2012 (accessed October 26, 2024)

**30** European Commission, Update of Benchmark Values for the Years 2021–2025 of Phase 4 of the EU ETS, European Commission, 2021.

31 van Rooij, G.J., Akse, H.N., Bongers, W.A., and van de Sanden, M.C.M. (2017). Plasma for electrification of chemical industry: a case study on $CO_2$ reduction. *Plasma Phys. Control. Fusion* 60 (1): 014019. https://doi.org/10.1088/1361-6587/aa8f7d.

32 Jardali, F., Van Alphen, S., Creel, J. et al. (2021). $NO_x$ production in a rotating gliding arc plasma: potential avenue for sustainable nitrogen fixation. *Green Chem.* 23 (4): 1748–1757. https://doi.org/10.1039/D0GC03521A.

33 Tsonev, I., O'Modhrain, C., Bogaerts, A., and Gorbanev, Y. (2023). Nitrogen fixation by an arc plasma at elevated pressure to increase the energy efficiency and production rate of $NO_x$. *ACS Sustain. Chem. Eng.* 11 (5): 1888–1897. https://doi.org/10.1021/acssuschemeng.2c06357.

34 Warren, D.R., Nicola, F.T., Turner, D., and Serrano, J.M. (2018). *Nitrogen Inputs to Agricultural Soils from Livestock Manure*. Food and Agriculture Organization of the United Nations.

35 Eggleston, H. S., Buendia, L., Miwa, K., Ngara, T., and Tanabe, K. (2006). IPCC Guidelines for National Greenhouse Gas Inventories, Jul. 2006, https://www.osti.gov/etdeweb/biblio/20880391 (accessed January 25, 2022)

36 Nicol, M.J., Brubaker, T.R., Honish, B.J. et al. (2020). Antibacterial effects of low-temperature plasma generated by atmospheric-pressure plasma jet are mediated by reactive oxygen species. *Sci. Rep.* 10 (1): 3066. https://doi.org/10.1038/s41598-020-59652-6.

37 Nannestad, Å. D., Espelund, M., Kjønnø, O., Nyvold, M. R., and Vik, E. A. (2024). Plasma-based nitrogen enrichment and the hygienisation effect on organic slurries, 2024.

38 Nyvold, M. and Dörsch, P. (2024). Complete elimination of methane formation in stored livestock manure using plasma technology. *Front. Sustain. Food Syst.* 8: 1370542. https://doi.org/10.3389/fsufs.2024.1370542.

39 New Slurry Treatment Fixes N from Air to Create Enriched Fertiliser, Farmers Weekly. https://www.fwi.co.uk/arable/crop-management/nutrition-and-fertiliser/new-slurry-treatment-fixes-n-from-air-to-create-enriched-fertiliser (accessed October 20, 2022)

40 Rouwenhorst, K.H.R. and Lefferts, L. (2020). Feasibility study of plasma-catalytic ammonia synthesis for energy storage applications. *Catalysts* 10 (9): 9. https://doi.org/10.3390/catal10090999.

41 Zhang, T., Zhou, R., Zhang, S. et al. (2023). Sustainable ammonia synthesis from nitrogen and water by one-step plasma catalysis. *Energy Environ. Mater.* 6 (2): e12344. https://doi.org/10.1002/eem2.12344.

42 Vervloessem, E., Gromov, M., De Geyter, N. et al. (2023). $NH_3$ and $HNO_x$ formation and loss in nitrogen fixation from air with water vapor by nonequilibrium plasma. *ACS Sustain. Chem. Eng.* 11 (10): 4289–4298. https://doi.org/10.1021/acssuschemeng.3c00208.

43 Hollevoet, L., Jardali, F., Gorbanev, Y. et al. (2020). Towards green ammonia synthesis through plasma-driven nitrogen oxidation and catalytic reduction.

*Angew. Chem. Int. Ed.* 59 (52): 23825–23829. https://doi.org/10.1002/anie.202011676.

44 Xiang, Q., Fan, L., Li, Y. et al. (2022). A review on recent advances in plasma-activated water for food safety: current applications and future trends. *Crit. Rev. Food Sci. Nutr.* 62 (8): 2250–2268. https://doi.org/10.1080/10408398.2020.1852173.

45 Kolbe, B.H., Coad, B., and Richter, K. (2024). Plasma-activated water's potential contribution to 'One Health'. *Microbiol. Aust.* 45 (2): 83–87. https://doi.org/10.1071/MA24024.

46 Ingels, R. (2017). Energy efficient process for producing nitrogen oxide, EP2704989B1, Jun. 28, 2017.

47 Ingels, R. (2020). Low pressure plasma reactor loop process and system, NO345196B1, Nov. 02, 2020.

48 Ingels, R. (2020). "Nitrogen enrichment of organic fertilizer with nitrate and air plasma," NO345195B1, Nov. 02, 2020.

49 Produce your own sustainable fertilizer | GEA ProManure E2950.

50 Nesje, S. D. (2024). Snarum Gartneri, N2 Applied. https://n2applied.com/2024/12/12/snarum-gartneri/ (accessed December 23, 2024)

51 Pemen, A. J. M., Hoeben, W. F. L. M., van Ooij, P. P., and Leenders, P. H. M. (2020). Plasma activated water, US10669169B2, Jun. 02, 2020.

52 van Ooij, P. P. and Beckers, F. J. C. M. (2022). Plasma activated fluid processing system, US20220159821A1, May 19, 2022.

53 Products – VitalFluid. https://vitalfluid.com/products/ (accessed December 23, 2024)

54 van Rooij, R. G. J., Forsberg, G., and Baeling, P. (2022). Method for the synthesis of nitrogen oxides in a thermal reactor, SE544603C2, Sep. 20, 2022

55 Forsberg, G., Baeling, P., and Andersson, R. (2024). Thermal reactor comprising a gas permeable cage arranged to influence a flow path of gas, US20240010493A1, Jan. 11, 2024.

56 King, W. C., Schwalbe, J. A., McEnaney, J. M., and Pinkowski, N. H. (2023). Plasma power supply using an intermittent power source, US20230191356A1, Jun. 22, 2023.

57 McEnaney, J. M., Schwalbe, J. A., Chockalingam, S., and Pinkowski, N. H. (2023). Microwave plasma system for efficiently producing nitric acid and nitrogen fertilizers, US20230219052A1, Jul. 13, 2023.

58 Pinkowski, N. H., McEnaney, J. M., Schwalbe. J. A., and Athariboroujeny, M. (2024). Sustainable plasma nitrophosphate process for production of nitrogen and phosphorous products and for carbon sequestration, US20240228287A9, Jul. 11, 2024.

59 Nitricity launches field trial of low-emissions nitrogen fertilizer on almond trees with Olam Food Ingredients, Nitricity.

# Part V

# Advanced Processes of Nitrogen Fixation

# 11

# Microplasma for Nitrogen Fixation

*Liangliang Lin*

*Key Laboratory of Synthetic and Biological Colloids, Ministry of Education, School of Chemical and Material Engineering, Jiangnan University, Wuxi, China*

## 11.1 Introduction

Plasma, the fourth state of matter, consists of a wide array of components such as electrons, ions, excited atoms and molecules, free radicals in metastable states and various types of radiations such as ultraviolet (UV) photons [1]. It is regarded as a promising tool in chemical reaction engineering, as the collisions between plasma species and other chemical compounds can generate a controlled reactive physicochemical environment that is not accessible by traditional methods [2]. Microplasma is a special category of plasma that is confined in the submillimeter range in at least one dimension [3]. It employs beneficial properties of the atmospheric pressure gas discharge in microscale geometry for various plasma-enabled processes, resulting in a new and facile branch of applied plasma science. In general, microplasma is commonly generated by electrical breakdown in microscale gaps in gases or liquids and can be ignited either by applying an overvoltage to electrodes or by focusing high-power laser radiation into the discharge medium [4]. Once the applied voltage or electric field across the microscale gap is high enough to pull electrons from gas molecules to ionize them, electrical breakdown occurs, and the gas between the gap becomes electrically conductive.

As to the gas breakdown voltage of microplasma, according to Paschen's law, it depends on the product of gas pressure and gap length between electrodes [5]. For a certain gas such as air, nitrogen, oxygen, and argon, high pressure and small size are mutually consistent. Thus, maximizing the operational pressure of microplasma requires minimizing the gap distance between electrodes. For air and

*Plasma-Assisted Nitrogen Fixation for Sustainable Process Industries*, First Edition. Edited by Volker Hessel, Annemie Bogaerts, Gabriele Centi, Evgeny Rebrov, and Nguyen Van Duc Long.

Published 2026 by John Wiley & Sons, Inc.

$N_2$ gas, the smallest gaps for the breakdown voltage at atmospheric pressure were measured to be approximately half a micrometer [6]. In such a small dimension, plasma can be operated at atmospheric or even exceeding pressures without vacuum systems and is favored for industrialization. Besides, the electrical breakdown voltage is strongly affected by many factors, such as gas composition, gas temperature, electrode geometry, electrode surface roughness, and voltage waveforms. In the past century, the effects of these parameters on the breakdown characteristics of microplasma discharges have been extensively investigated [7].

Compared to conventional plasmas, microplasma has many unique properties. Thanks to its high surface-to-volume ratio, heat and mass transfer rates are drastically increased, without excessive temperature rise. This not only makes microplasma appealing for the synthesis of temperature-sensitive materials but also prevents particle growth and agglomeration. Thus, microplasma can prepare nanomaterials with a small size and narrow size distribution. Non-equilibrium is another important feature of microplasma, where the electron temperature ($10^4$ to more than $10^5$ K) is much higher than the gas temperature (ranging from room temperature to about 2,500 K). As reported by Okubo, since electrons transfer only a small amount of energy to heavy particles in a single collision, the energy they gain from the electric field is primarily due to their own acceleration. This typically leads to the formation of non-equilibrium plasma, where only the electron temperature remains high (Figure 11.1) [8]. This allows for chemical reactions to be driven by the electron temperature, while the bulk temperature of the reaction media typically remains at or close to room temperature. In addition, microscale operation helps to reduce safety risks, especially when handling hazardous

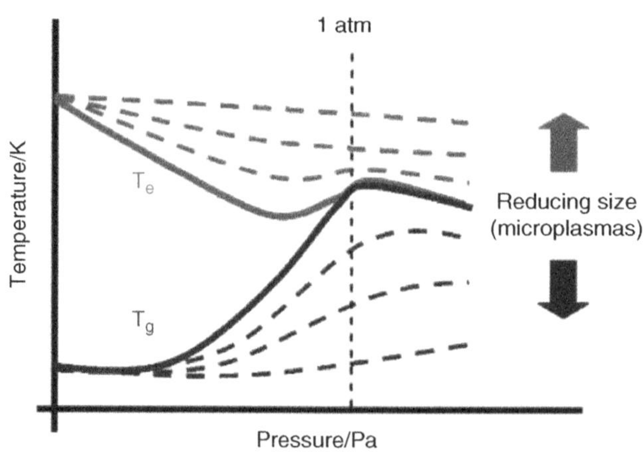


**Figure 11.1** Electron and gas temperature divergence at different gas pressures and the same current. *Source:* Reproduced from [8] / AIP Publishing.

materials or explosive reactions. Thus, from the cost-efficiency standpoint, microplasma allows saving of the remarkable costs associated with maintaining vacuum and using complex transfer chambers. From the process-efficiency perspective, microplasma is characterized by higher densities of radicals and electron energy, resulting in enhanced plasma–chemical reaction rates. Therefore, microplasma provides a simple, cheap, intrinsic-safe, and more convenient substitute for conventional plasmas.

High flexibility of microplasma configuration is another important advantage, where setups of diverse geometries have been developed. They can integrate with techniques like 3D printing to realize plasma additive manufacturing at ambient condition. For instance, Lin et al. [9] combined microplasma with a commercial 3D printer to construct a mobile microplasma jet and fed with rare earth acetylacetonates to prepare $Y_2O_3$-based nanophosphor patterns. This microplasma setup was also used to prepare Au/$SiO_2$-based surface-enhanced Raman scattering (SERS) platforms for chemical analyte detection [10] and the synthesis of gold nanoparticles for catalytic growth of CdS nanowires [11]. In addition to the flexible configuration, the process versatility of microplasma for nanomaterials synthesis and processing cannot be overstated. It is able to be stably operated using a wide range of gases (e.g. Ar, He, $N_2$, $CO_2$, and air) and diverse feedstock phases (i.e. liquid, solid, and gas). For instance, microplasma–liquid interactions are rich sources of reactive species, such as solvated electrons ($e_{aq}^-$), ·OH, O·, $HO_2$·, $H_2O_2$, and UV radiation. They can replace chemical reducing agents that are used in conventional wet-chemistry methods to avoid persistent ions (e.g. $NO_3^-$, $Cl^-$, and $SO_4^{2-}$) and complex purification steps [12]. Besides, discharges can be generated with direct current (DC), alternating current (AC), radiofrequency (RF), pulsed, or other excitations, leading to the development of different microplasma setup scales ranging from portable and benchtop devices to large-scale arrays [13]. These microplasma systems are being used in a variety of application areas and have opened new research opportunities, blending plasma science, chemistry, electrochemistry, and physics.

Fascinated by these advantages, microplasma has been applied in a wide range of fields, such as the destruction of volatile organic compounds (VOCs), synthesis of nanomaterials, regeneration of spent catalysts, surface modification of polymers, development of optical devices, ozone generation, and environmental remediation [14]. Currently, a growing number of studies have focused on microplasma-based process and setup, resulting in significant progress in both plasma configurations and their applications. Several representative applications of microplasma in recent years have been summarized and presented in Figure 11.2.

Nitrogen ($N_2$) is an extremely stable gas plentiful in air that has a very strong N≡N triple bond (bond dissociation energy of 9.79 eV) [23]. It is very difficult to be activated at normal temperature and pressure conditions. Nitrogen fixation

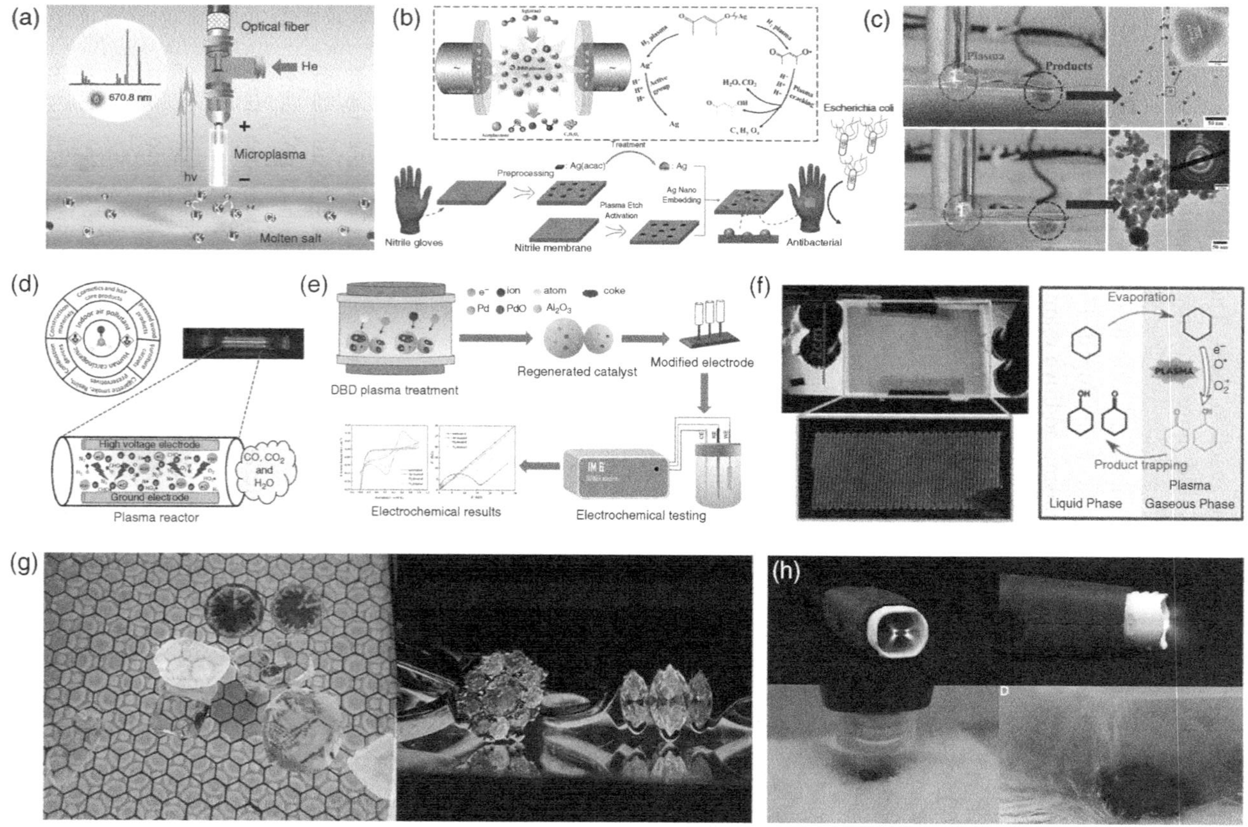


**Figure 11.2** Representative applications of microplasma: (a) *in situ* quantitative analysis of lithium element. *Source:* Reproduced from [15] / with permission of Elsevier. (b) Surface modification of nitrile membranes. *Source:* Reproduced from [16] / with permission of Elsevier. (c) Synthesis of gold nanoparticles for $Pb^{2+}$ detection. *Source:* Reproduced from [17] / with permission of Elsevier. (d) Degradation of formaldehyde. *Source:* Reproduced from [18] / with permission of Elsevier. (e) Regeneration of spent $Pd/Al_2O_3$ catalysts. *Source:* Reproduced from [19] / with permission of American Chemical Society. (f) Controlled oxidative processes for synthesis of KA oil. *Source:* Reproduced from [20] / with permission of Royal Society of Chemistry. (g) Photographs of diamonds illuminated by microplasma lamp. *Source:* Reproduced from [21] / John Wiley & Sons/CC BY 4.0. (h) Skin therapy by microplasma jet. *Source:* Reproduced from [22] / Frontiers Media S.A. / CC BY 4.0.

refers to any natural or industrial process that transforms free nitrogen into more reactive nitrogen compounds such as ammonia, nitrates, nitrites, nanomaterials, or other nitrogen-containing substances [24]. In recent years, plasma-assisted nitrogen fixation, with the ability to break N≡N bond without pressure and heating, has significantly affected nitrogen fixation research. It offers us the opportunity to convert electrical energy from renewable sources into chemical energy and has become one of the pioneering ways for industrial nitrogen fixation [25].

Miniaturization of plasma reactors proves to be an effective way to achieve process intensification for nitrogen fixation. The existence of a significant fraction of high-energy electrons (>10 eV) in microplasma renders it possible to dissociate the N≡N bond and trigger reactions at atmospheric pressure and low temperatures. As a result, a wide variety of nitrogen derivatives are produced via the collisions and reactions of the resulting species with oxygen and hydrogen [26]. Meanwhile, since microplasma is operated at ambient pressure, the densities of radicals are higher than in low-pressure plasma systems. This, in turn, leads to much enhanced collisions between nitrogen species and other radicals. Besides, the confined small dimensions ensure narrow residence time distribution (RTD) as well as reduced temperature and concentration variations, addressing the inherent discontinuity and poor reproducibility of batch processing [27].

Microplasma catalysis for nitrogen fixation is an innovative approach that merges the unique advantages of microplasma with catalytic processes to revolutionize ammonia synthesis and other nitrogen fixation processes. By leveraging microplasma's ability to generate highly reactive species (e.g. electrons, ions, and radicals) at ambient conditions, this hybrid method possesses plasma's high reactivity and catalysis's selectivity. As a result, it allows to enhance nitrogen activation significantly, which is a major bottleneck in conventional methods such as the Haber-Bosch process. Compared with other plasma-based nitrogen fixation approaches, microplasma catalysis for nitrogen fixation has several key advantages: (i) high-energy efficiency: it operates at low temperatures/pressures and consumes less energy than thermal-driven processes. (ii) Rapid reaction kinetics: the plasma–catalyst interaction boosts nitrogen dissociation and hydrogenation rates. (iii) High process flexibility: compatible with diverse catalysts (e.g. transition metals, metal oxides, and zeolites), enabling tunable selectivity for $NH_3$ or other nitrogen products.

Despite a series of studies have been performed on plasma-assisted nitrogen fixation, only some reviews are reported [28–31]. To the best of our knowledge, there is still no review focusing on microplasma nitrogen fixation. This chapter aims to introduce the state of the art of nitrogen fixation by microplasma. Representative microplasma configurations are first presented. Then, specific examples on a broad category of nitrogen derivatives obtained by microplasma are presented, such as $NO_x$, $NH_3$, nitride nanomaterials, and N-doped nanomaterials. Finally, existing challenges and perspectives of microplasma nitrogen fixation are discussed.

## 11.2 Microplasma Configurations for Nitrogen Fixation

To date, a broad range of microplasma configurations have been designed and applied for nitrogen fixation, with the process parameters being explored and optimized. These studies not only suggest the versatility of microplasma systems, but also reveal the high flexibility in process control during nitrogen fixation. Owing to the diversity of microplasma configurations, it is difficult to make a uniform classification for different microplasma types. According to the electrode geometry and plasma power source, microplasma configuration can be divided into six main categories: microwave microplasma, gliding arc microplasma, dielectric barrier discharge (DBD) microplasma, plasma jet exiting microreactor, plasma jet confined in microreactor, and needle electrode-based microplasma. It is worth mentioning that the classification of microplasma configurations is not rigid, since mixed cases can be found. Figure 11.3 presents typical microplasma configurations that have been developed for nitrogen fixation.

Microplasma generated by microwave has desirable characteristics, including good uniformity, high ionization fraction, and relatively low electron energy (typically 1–3 eV). In such a configuration, there are no internal electrodes; thus, it can produce products with high purity. Vibrational excitation is the dominant pathway for energy transfer, and the electric field will selectively heat the electrons due to their lighter mass [32]. Meanwhile, microwave plasma is capable of distributing energy rapidly among higher energy levels of $N_2$ molecules through intermolecular collisions, thus promoting the dissociation of $N_2$ considerably. Compared with other microplasma reactor designs, it yields high $N_2$ conversion, high-energy efficiency, and high product throughput. However, it is worth mentioning that this type of microplasma configuration consists of a sophisticated microwave generator and a waveguide for microwave transmission, which is commonly more expensive than other plasma sources.

Gliding arc microplasma is another commonly used configuration for nitrogen fixation. It uses two or more diverging electrodes to sustain dynamic plasma arcs with DC or AC power sources. A fast gas flow containing $N_2$ is injected near the shortest path between the electrodes and pushes the generated plasma discharges upward, forming a gliding arc. During this process, nitrogen compounds are formed by the interaction of nitrogen radicals, oxidants, and other plasma species. It is worth mentioning that the gliding arc microplasma has a dynamic behavior. As the arc ignites and rises, it will extend and propagate and finally extinguish once the gap between the electrodes becomes too large [38]. Therefore, the plasma properties vary with the development of arc, which brings uniformity in the conversion, selectivity, and energy efficiency of the nitrogen fixation process [39].

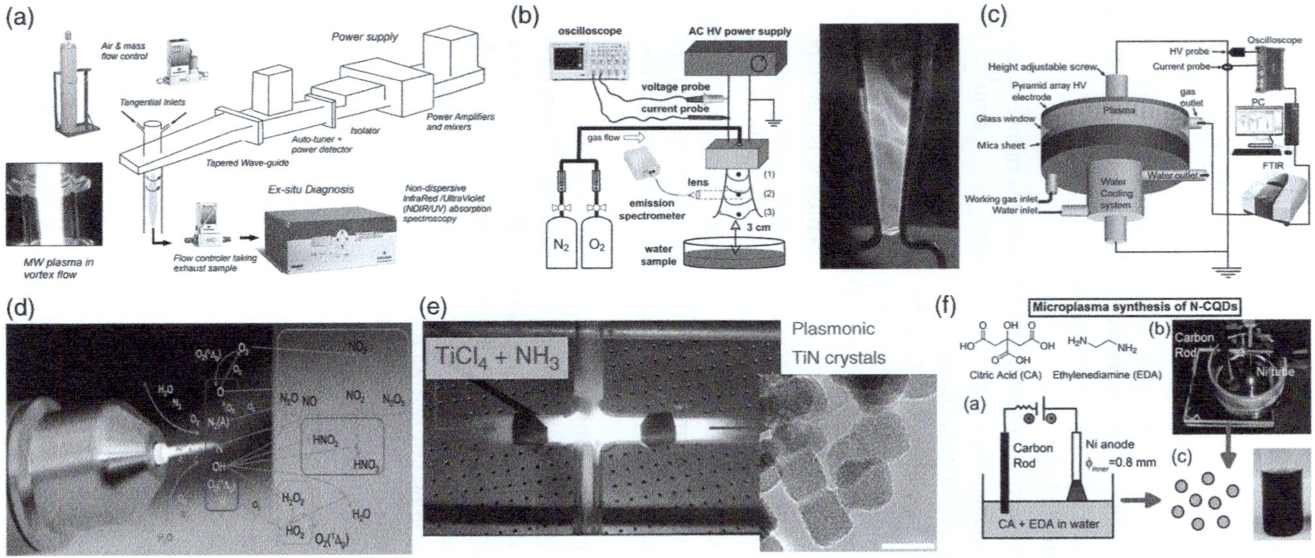


**Figure 11.3** Typical microplasma configurations for nitrogen fixation. (a) Nitrogen fixation from air into $NO_x$ by microwave plasma. *Source:* [32] / with permission of Elsevier. (b) Mini-gliding arc discharge for the formation of oxidative species in gas and liquid. *Source:* [33] / Springer Nature / CC BY 4.0. (c) $NO_x$ production by DBD microplasma with a pyramid-shaped electrode. *Source:* Reproduced from [34] / Springer Nature / CC BY 4.0. (d) Plasma jet exiting microreactor initiates strong non-equilibrium chemical processes. *Source:* [35] / IOP Publishing / CC BY 3.0.
(e) Microplasma jet confined in quartz tube for the synthesis of TiN nanoparticles. *Source:* Reproduced with permission from [36] / American Chemical Society. (f) Needle electrode-based microplasma for the synthesis of nitrogen-doped carbon quantum dots. *Source:* [37] / Royal Society of Chemistry / CC BY 3.0.

DBD microplasma is a widely used plasma that has dielectric materials covering the electrodes with micrometer-size electrode gaps. Such a configuration can generate stable and uniform plasma discharges at low temperature, atmospheric pressure, and low discharge voltages (~1 kV) and is very simple to scale up to large sizes for industrial application [40]. When $N_2$ and $O_2$ or $H_2$ flow into the DBD microplasma, nitrates, nitrites, and ammonia are formed. Thanks to the simple structure, DBD microplasma has lower operation and maintenance costs compared with other microplasma structures. Besides, catalysts can be integrated with the plasma easily to generate a synergy, with the ability to enhance the energy efficiency and product selectivity [41].

Plasma jet exiting microreactor (microplasma jet for convenience) is mostly created inside a microreactor, such as a capillary or a tube, and is emitted from the microreactor by passing a gas mixture through it. They are commonly sustained by RF powers inductively or capacitively coupled by external electrodes outside the microreactor. Diffusion of ambient species into the microplasma jet leads to the formation of diverse chemical species, such as OH, NO, $HO_2$, and $NO_2$. A well-known design is the pen-sized kINPen® plasma jet, which was developed at the Leibniz-Institute for Plasma Science and Technology, INP Greifswald, Germany, and has been used for various applications such as wound healing, sterilization, and tissue ablation [35, 42]. It is operated with inert gases or pressurized air at ambient pressure, with around 1-cm long visible plasma jet exiting the nozzle. Such a configuration is portable and easy to use, and the plasma jet can be directed onto the target area.

Plasma jet confined in microreactor is also developed for nitrogen fixation, particularly favored for the synthesis of nitride nanomaterials. This type of microplasma is often confined inside a quartz tube, with a metal plate or coil surrounding it as the external electrode. High-voltage pulse power supply is commonly used to sustain the microplasma. In some cases, the pressure inside the microreactor can be adjusted by integrating with a vacuum system. With the supply of precursors and $N_2$ gas into the plasma zone, nitrogen derivatives are produced and will be collected downstream from the plasma. This microplasma configuration has a high degree of flexibility in process control, as a wider range of operation parameters can be set, such as precursor types and ratios, residence time, and power coupling mode [43] Thus, different nitrogen-containing compounds can be produced, such as $NH_3$, $NO_x$, or nitride, carbonitride, or oxynitride nanomaterials.

In addition to microwave, AC, or pulsed potentials, microplasma can be powered using DC. A salient configuration of microplasma for nitrogen fixation is needle electrode-based microplasma. This type of microplasma often employs a hollow needle as both electrode and gas channel, while the counter electrode is a parallel plate that is separated by narrow gaps [44]. Upon the supply of high voltage to the electrodes, microplasma is formed between the electrodes. Interestingly, by immersing the plane electrode in a highly conductive liquid like water, microplasma can also be sustained between the tip of the needle and the water. Thus, stable secondary species such as $H_2O_2$, $NO_2^-$, $NO_3^-$, or other nitrogen-containing organics will be generated with the addition of $N_2$. This is a great benefit of needle electrode-based microplasma, where numerous chemical compounds can be used as the raw materials, as long as they have good solubility in liquids [45]. Furthermore, the needle electrode can be easily replaced at low cost after heavy sputtering on its tip.

## 11.3 Microplasma-Based Process for Nitrogen Fixation

In a microplasma environment, the N≡N bond of $N_2$ was first dissociated by electron collisions, followed by the excitation, dissociation, ionization, and recombination processes to generate a wide variety of nitrogen-based species. In general, the microplasma nitrogen fixation process involves the reaction of nitrogen with either hydrogen or oxygen, leading to the generation of nitrogen derivatives such as $NH_3$, nitrogen oxide ($NO_x$), or $HNO_3$. Figure 11.4 shows nitrogen derivatives (general formula: $H_xN_yO_z$) obtained from chemical reactions with nitrogen atoms, as summarized by Ishikawa [26]. In this section, we divided the nitrogen derivatives into four categories: $NO_x$, $NH_3$, nitrides or derivatives, and N-doped materials. Salient examples of the nitrogen fixation process by microplasma are also provided.

### 11.3.1 $NO_x$

As an easy-accessible, abundant, and affordable raw material, air has been widely used in microplasma nitrogen fixation to produce nitrogen oxides ($NO_x$). These compounds are soluble in water and, eventually, form nitrite ($NO_2^-$) and nitrate ($NO_3^-$). The dominant chemistry during $NO_x$ generation follows the Zeldovich mechanism, which begins with the dissociation of $N_2$ and $O_2$ by microplasma electrons to form N and O radicals. This is followed by the generation of NO from the radicals. Since NO is easily oxidized by oxygen radicals or ozone, $NO_2$ will be readily formed. When any of these $NO_x$ compounds dissolve in water, nitrous acid ($HNO_2$) or nitric acid ($HNO_3$) is finally generated. The main reactions are provided as follows [28]:

$$N_2 + e^* \rightleftharpoons 2N^* + e^- \tag{11.1}$$

$$O_2 + e^* \rightleftharpoons 2O^* + e^- \tag{11.2}$$

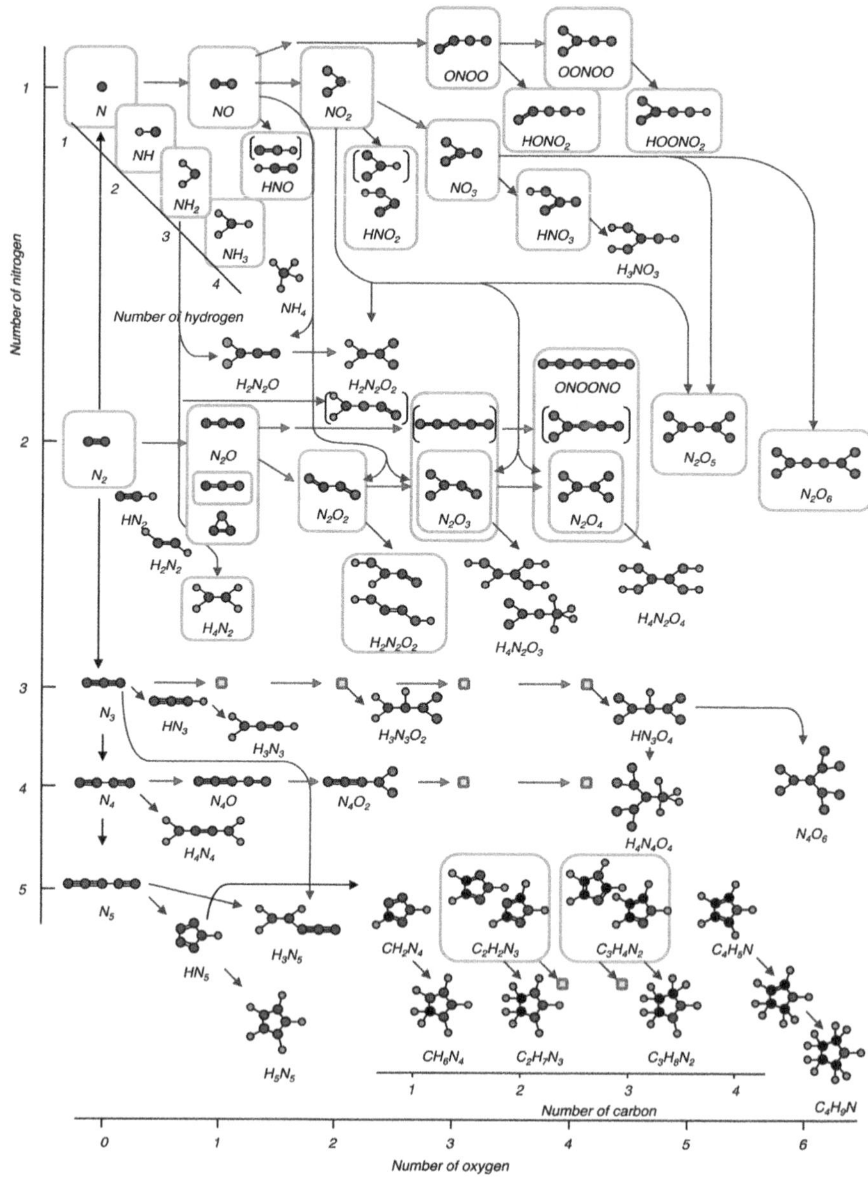


**Figure 11.4** Summarization of a wide range of nitrogen derivatives having the general formula of $H_xN_yO_z$. *Source:* Reproduced from [26] / The Japan Society of Applied Physics / CC BY 4.0.

$$O^* + N_2 \rightarrow NO + N^* \quad (11.3)$$

$$N^* + O_2 \rightarrow NO + O^* \quad (11.4)$$

$$O^* + NO \rightarrow NO_2 \quad (11.5)$$

$$O^* + O_2 \rightarrow O_3 \quad (11.6)$$

$$O_3 + NO \rightarrow NO_2 + O_2 \quad (11.7)$$

Owing to the advantages of microplasma, increasing studies have been carried out in recent years to produce $NO_x$. In the research by Yang et al. [46], a DBD plasma reactor of needle-array electrode was used for the synthesis of $NO_x$ from $N_2/O_2$ mixture, with $\alpha$-$Al_2O_3$ and $\gamma$-$Al_2O_3$ as the catalysts (Figure 11.5a). Operation parameters such as the discharge power, discharge area, pulse repetition rate, and gas composition were regulated to optimize the $NO_x$ synthesis, including $NO_x$ concentration, selectivity, and energy cost. The results demonstrated the efficient production of $NO_x$ ($NO_2$, NO, and $N_2O$) by the needle-array packed bed DBD plasma reactor, where the selectivity of $NO_2$ exceeds 90% in all cases. At the optimized condition, the $NO_x$ concentration reaches 1.1%, with a low-energy cost of 17 MJ $mol^{-1}$. To intensify the nitrogen fixation process through electric field enforcement, Hessel et al. [34] developed a DBD plasma with pyramid-shaped microelectrode to replace conventional flat electrode (Figure 11.5b). Due to an increased secondary electron and field emission rate of nitrogen molecules, higher $NO_x$ production efficiency can be obtained. Compared with flat electrodes, 58% more $NO_x$ yield efficiency at 32% less energy cost is achieved by the pyramid electrodes. Afterward, they incorporated $TiO_2$ to study the synergistic effect of plasma and catalysts. The addition of $TiO_2$ catalysts contributed to a sharp increase in $NO_x$ production. At the optimum condition, a 20% increase in $NO_x$ production along with a 23% enhancement in energy yield was realized [47].

Gliding arc microplasma is regarded as "warm" discharge compared to other microplasma types and has been used for the synthesis of $NO_x$. Bogaerts et al. [31] used a milliscale gliding arc microplasma to produce $NO_x$ at different feed ratios of $N_2/O_2$, of which the narrowest discharge gap was merely 1.3 mm. Such a structure benefits from an intensified contact of the $N_2/O_2$ gas with plasma species, thus promoting energy-delivering efficiency as well as $NO_x$ generation. It is revealed that 10 times less energy is required by the gliding arc than the thermal process of $NO_x$ formation. In a study by Talukder et al. [48], an air bubble microplasma jet was utilized to produce $NO_x$ from air. The plasma jet consists of three capillary tubes, of which the inner diameter is 1 mm. Plasma discharges were formed inside the capillary tubes and protruded 10 mm into downstream water. Meanwhile, an external magnetic field was used to enhance the production of reactive plasma species. $NO_x$ proved to be successfully produced, resulting in the formation of $NO_2^-$ and $NO_3^-$ in water. A maximum concentration of ~82% for $NO_2^-$ and ~74% for $NO_3^-$ was achieved at the magnetic field strength of

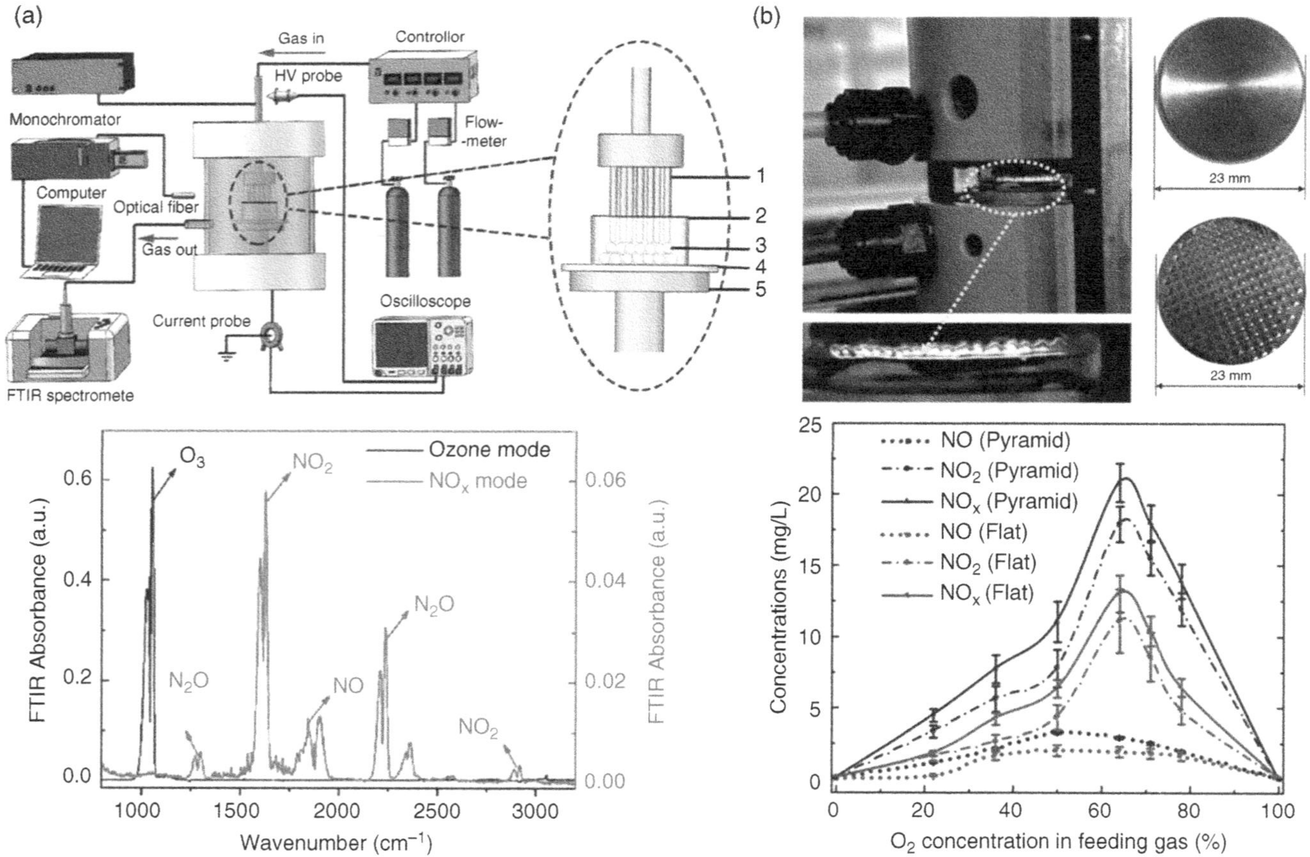


**Figure 11.5** (a) DBD plasma reactor of needle-array electrode for $NO_x$ production. *Source:* Reproduced from [46] / with permission of Elsevier. (b) DBD plasma with pyramid-shaped microelectrode for $NO_x$ production. *Source:* Reproduced from [47] / John Wiley & Sons / CC BY 4.0.

290 mT. Due to the reduced radial diffusion loss of plasma species in the presence of magnetic field, the energy cost for $NO_x$ synthesis was also decreased by ~78%.

In addition to the synthesis process of $NO_x$, the underlying mechanisms were also investigated. Pattyn et al. [49] adopted a needle-water DC microplasma system to study the fundamental mechanisms of $NO_2^-/NO_3^-$ synthesis. In this work, a stainless-steel capillary tube of 0.2-mm inner diameter was used as the high-voltage electrode, while a Pt wire, as the counter electrode, was immersed in an aqueous solution of NaCl. Microplasma was generated between the needle and the water, with the gap distance maintained at $0.7 \pm 0.1$ mm. $N_2$ and $O_2$ gas mixtures of different compositions were injected into the needle-water DC microplasma. Microplasma discharges were analyzed using optical emission spectrometry. The results suggested the $N_2$ has a vibrational temperature in the range of 3,500–5,000 K, which increases with the rise of the $O_2$ content in the gas mixture. The process energy efficiency and selectivity were affected by gas composition as well as plasma-induced water heating. The dominant process for the initial NO formation is the extended Zeldovich mechanism between N atoms and OH radicals, rather than being restricted to reactions between $O/O_2$ and $N/N_2$. Besides, different $NO_x$ species (e.g. NO, $NO_2$, and $N_2O$) were formed in plasma and can be directly dissolved in water to generate stable species such as $NO_2^-$ and $NO_3^-$. To provide a better overview of microplasma-assisted $NO_x$ synthesis, Table 11.1 shows a summary of reported microplasma configurations, operation parameters, and the obtained products.

### 11.3.2 $NH_3$

Currently, the Haber-Bosch process is still the main industrial procedure for ammonia ($NH_3$) synthesis. Conventionally, this process is carried out in the presence of iron catalysts at temperatures around 400–500 °C and pressures ranging from 150 to 250 atm. [56]. With the growing need for scalable $NH_3$ production at mild conditions and the use of renewable energy and feedstock to replace the Haber-Bosch process, numerous efforts have been devoted to microplasma-based processes, as it is capable of being stably operated at ambient pressure and low temperature. Besides, the presence of a high density of reactive species not only intensifies the process but also can use renewable energy as an energy input. Therefore, it has better energy efficiency, lower capital cost, and extended catalyst lifetime [57]. Microplasma-assisted $NH_3$ synthesis mainly includes dissociation, ionization, and excitation of $N_2$ and $H_2$ to produce radicals, followed by the formation of NH radicals, and, finally, $NH_3$ production. The main reactions are shown as follows [58]:

$$N_2 + e^* \rightleftharpoons 2N^* + e^- \tag{11.8}$$

$$H_2 + e^* \rightleftharpoons 2H^* + e^- \tag{11.9}$$

**Table 11.1** Summary of existing microplasma configurations, operation parameters, and the corresponding $NO_x$ products.

| Configuration | Reactant | Product | Pressure (atm) | Temperature (K) | Energy consumption | Power | Throughput ($NO_x$) | $NO_X$ concentration (ppm) | Ref. |
|---|---|---|---|---|---|---|---|---|---|
| Microwave microplasma | Air | NO, $NO_2$ | 1 | 1,700–4,300 | 2 MJ $mol^{-1}$ | 1.07 kW | 0.77 L $min^{-1}$ | 40,000 | [32] |
| Gliding arc microplasma | Air, $O_2$ | NO, $NO_2$ | 1 | 3,000 | 4 MJ $mol^{-1}$ | 60–90 W | | 8,000 | [31] |
| Gliding arc microplasma | $N_2$, $O_2$ | NO, $NO_2$ | 1 | 313.15 | 9.1 kWh $kg^{-1}$ | | | 6,870 | [50] |
| DBD microplasma | $N_2$, $O_2$ | $H_2O_2$, $HNO_2$, $HNO_3$ | 1 | 350–450 | 167 kWh $kg^{-1}$ | 47 W | 260 mg $L^{-1}$ | | [51] |
| DBD microplasma | $N_2$,$O_2$ | NO, $NO_2$ | 1 | 6,058 6,408 | 74 kWh $kg^{-1}$ | | 44.7 mg $L^{-1}$ | | [47] |
| DBD microplasma | $N_2$, $O_2$ | NO, $NO_2$ | 1 | 303.35–318.35 | | 43.8–144.7 W | 1,113.7 mg $m^{-3}$ | | [52] |
| Plasma jet confined in microreactor | Air | NO, $NO_2$ | 1 | | 2.9–7.7 MJ $mol^{-1}$ | 50–60 W | 167–586 μmol $min^{-1}$ | 1,600–16,600 | [53] |
| Plasma jet confined in microreactor | Air | NO, $NO_2$ | 1 | Room temperature | 1.76 MJ $mol^{-1}$ | 1.50–2.90 mW | 4.82 μmol $h^{-1}$ | | [54] |
| Plasma jet exits microreactor | Air | NO | 1 | 308.15–368.15 | | 20 W | $3.5 \times 10^{19}$ molecules $s^{-1}$ | 1.5–2.0 | [55] |
| Needle electrode based microplasma | $N_2$, $O_2$ | $HNO_2$, $HNO_3$, $H_2O_2$ | 1 | 1,400–1,800 | 167.4 MJ $mol^{-1}$ | 7.3 W | | | [49] |

$$N^* + H^* \rightarrow NH^* \tag{11.10}$$

$$H^* + NH^* \rightarrow NH_2{}^* \tag{11.11}$$

$$NH^* + H_2 \rightarrow NH_2{}^* + H^* \tag{11.12}$$

$$NH_2{}^* + H^* \rightarrow NH_3 \tag{11.13}$$

$$NH_2{}^* + H_2 \rightarrow NH_3 + H^* \tag{11.14}$$

Renner et al. [59] designed a needle electrode-based microplasma system for the synthesis of $NH_3$ from $N_2$ and water at ambient temperature and pressure, as shown in Figure 11.6a. In this work, a Pt foil was immersed in the aqueous solution of $H_2SO_4$ as the anode, while the cathode was a stainless-steel capillary 2 mm

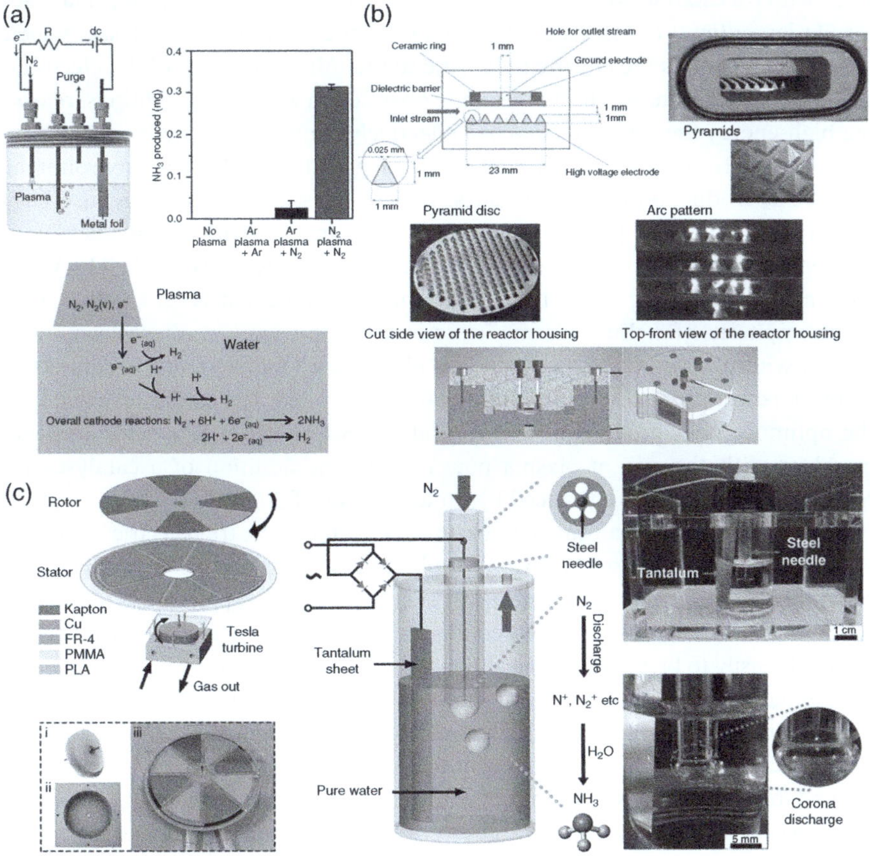


**Figure 11.6** (a) Needle electrode-based microplasma for the synthesis of $NH_3$ from $N_2$ and water. *Source:* Reproduced from [59] / American Association for the Advancement of Science. (b) Electrically intensified DBD microplasma with pyramid-shaped electrode for $NH_3$ production. *Source:* Reproduced from [60] / John Wiley & Sons / CC BY 4.0. (c) A simple and efficient self-powered ammonia synthesis system based on a Tesla turbine device was developed for $NH_3$ synthesis. *Source:* Reproduced from [61] / Springer Nature / CC BY 4.0.

above the solution. Microplasma was generated between the cathode and the solution surface. In this case, the solution not only supplied protons ($H^+$) to reduce $N_2$ but also trapped the as-synthesized $NH_3$. The results showed that solvated electrons played a crucial role in the reaction chemistry, especially the reduction of $H^+$ to H·, since $NH_3$ is mainly generated via a mechanism involving H·. However, compared to the Haber-Bosch process, the power consumption of the microplasma system is considerably larger and is estimated to be 2,270 kWh $kg^{-1}$ $NH_3$ at the studied condition. Such a high-energy consumption is due to the requirement of large electric fields to generate and sustain electrons.

In order to improve reaction performance, Hessel et al. [60] developed a DBD microplasma reactor with multi-pyramid arrays (Figure 11.6b) and then incorporated with Ru catalysts to study the synthesis of $NH_3$. This novel microplasma reactor design allows to generate discharges at the surface of catalysts, which effectively increases the electric field. Due to the miniaturization of the electrode gap and reaction volume, discharges were operated in warm arc conditions, releasing high-energy free electrons that can effectively convert low-reactive molecules. The optimal $N_2/H_2$ ratio was tested to 3:1, suggesting an $N_2$-rich environment favors the production of $NH_3$. Besides, optical emission spectroscopy also suggested a symbiosis of plasma-excited species generated with the Ru catalysts. Microplasma-catalytic synthesis of $NH_3$ from $N_2$ and $H_2$ was also studied by Vervloedt and Keudell [62], where iron, platinum, and copper catalysts, together with an atmospheric RF helium plasma, were used. In this work, a microplasma was ignited within a 1-mm gap between two electrodes. One electrode is connected to an RF power generator, and the other one is grounded. The results revealed that the optimum $NH_3$ production occurred at the 5% $N_2/(N_2 + H_2)$ ratio, and this increases with the rise of plasma power. With the addition of a catalyst, the $NH_3$ production will be enhanced by up to a factor of 2.

To achieve high mechanical energy conversion efficiency, triboelectric nanogenerators (TENGs), based on Maxwell's displacement current, have been utilized to convert mechanical energy into electrical energy for diverse applications [63]. With an appropriate design of the TENG configuration, it not only can build strong electric fields easily to break $N_2$ chemical bonds for $N_2$ fixation but also allows to make the discharge process to become self-powered. In the study by Wang et al. [61], a simple and efficient self-powered ammonia synthesis system based on a Tesla turbine device was developed for $NH_3$ synthesis (Figure 11.6c). They used a bladeless turbine to transform gas kinetic energy into high-speed rotation energy to power the TENG. $N_2$ microplasma discharges were simply generated using the high voltage of the TENG. By flowing $N_2$ and water into the microplasma, $NH_3$ was formed as a product reaction in the developed system. An $NH_3$ yield of 2.14 μg $h^{-1}$ (0.126 μmol $h^{-1}$) was achieved at a pressure of 0.12–0.13 MPa supplied by an air compressor using the self-powered system under ambient conditions. Representative examples of microplasma-assisted NH3 synthesis were summarized in Table 11.2.

**Table 11.2** Representative examples of microplasma-assisted NH3 synthesis.

| Structure | Reactant | Temperature (K) | Pressure (atm) | Power | Energy consumption | Throughput ($NH_3$) | $NH_3$ concentration (ppm) | Ref |
|---|---|---|---|---|---|---|---|---|
| Microwave microplasma | $N_2$, $H_2$ | 5,500–6,500 | 1 | 0.5–1.1 kW | | 20–50 μmol min$^{-1}$ | | [64] |
| Gliding arc Microplasma | $N_2$, $H_2O$ | | 1 | | 28 kWh kg$^{-1}$ | 1.43 mg/(cm$^2$·h$^{-1}$) | | [65] |
| DBD microplasma | $N_2$, $H_2$ | 381.15 | | 65 W | 99 MJ mol$^{-1}$ | 42 mg h$^{-1}$ | 10,000 | [66] |
| DBD microplasma | $N_2$, $H_2$ | 393.15–593.15 | 1 | 30.81 W | 0.0012 kWh kg$^{-1}$ | 1,500 μmol/(g h) | | [67] |
| DBD microplasma | $N_2$, $H_2$ | 373.15–383.15 | 1 | 5–18 W | 15.87 kWh kg$^{-1}$ | 1.4 μmol min$^{-1}$ | 151–1,788 | [68] |
| DBD microplasma | $N_2$, $H_2$ | Room temperature | 1 | 10.10 W | 130 MJ mol$^{-1}$ | 3,424 μmol/(h g$_{_cat}$) | | [69] |
| DBD microplasma | $N_2$, $H_2$ | Room temperature | 1 | 5–25 W | 421.94 kWh kg$^{-1}$ | | 2,912 | [70] |
| Plasma jet confined in microreactor | $N_2$, $H_2O$ | Room temperature | 1 | | | 4.5–11.7 mg l$^{-1}$ | | [71] |
| Plasma jet exits microreactor | $N_2$, $H_2$ | 309.85–314.35 | 1 | 0.25 W | | 1.921–2.745 mg l$^{-1}$ | | [72] |
| Needle electrode based microplasma | $N_2$, $H_2O$ | 1,473.15–1,773.15 | | 0.1 W | | 0.064 mg h$^{-1}$ | | [73] |

### 11.3.3 Nitride, Carbonitride, and Oxynitride Nanomaterials

Nitride materials refer to a class of chemical compounds in which nitride ions are combined with elements of similar or lower electronegativity, such as B, Si, or metals. Due to their novel electrochemical activities and high chemical stabilities, nitride nanomaterials have attracted widespread attention for different applications. For example, BN is an important ceramic material that has outstanding thermal conductivity, electrical insulation, and thermal properties. Group-13 nitrides such as AlN, GaN, InN, and their alloys have been widely used in producing electronic or optoelectronic devices [74]. Transition-metal nitrides such as $FeN_4$, $Co_4N$, $Mo_2N$, $Co_3Mo_3N$, and $Fe_3Mo_3N$ are useful in heterogeneous catalysis, as the incorporation of N in the lattice of transition metals can change their electron densities, making them easily chemisorb hydrogen and oxygen [75]. In general, nitride nanomaterials are prepared at high temperatures (800–1,500 °C) or high pressures. The harsh reaction conditions always result in products of large sizes and wide size distributions. To obtain uniform nanostructured nitrides, active precursors, halide raw materials, additives, or active intermediates are always introduced [76].

In addition to nitride, carbonitride, oxynitride, and oxycarbonitride nanomaterials are also of great potential for advanced applications. They are solid solutions of carbides, nitrides, and oxides containing —C—N, —N—O, and —C—N—O bonds with diverse chemical compositions. Such compounds possess improved physicochemical properties, including tunable thermal conductivity and hardness. However, the existing synthesis approaches are commonly time-intensive and energy-consuming, and the products always have a wide size distribution as well as impurities of free carbon, oxygen, and nitrogen [30].

As mentioned above, microplasma has emerged as a promising technique for nanomaterials synthesis. Indeed, there are many benefits to using microplasma for the synthesis of nitride, carbonitride, and oxynitride nanomaterials. For instance, with the help of high-density electrons with energy as high as 10 eV, it can dissociate $N_2$ molecules at low temperature, ambient pressure, and small plasma powers, thus overcoming the energy-intensive nature of the conventional Haber-Bosch process [77]. By varying operational parameters such as the applied power, reaction time, or reagent concentration, microplasma can effectively regulate the chemical composition and size of the products [78].

In the study by Hessel et al. [23], a hollow cathode microplasma was used for the synthesis of TiN nanoparticles from $TiCl_4$ and $N_2$, as seen in Figure 11.7a. Crystalline TiN nanoparticles prove to be synthesized in one step, and the addition of $H_2$ is an effective way to prevent TiN oxidation. From the optical emission spectrum, the gas temperature of the microplasma is estimated to be around 1,750 K. Regarding the reaction mechanism, microplasma-assisted TiN synthesis is a

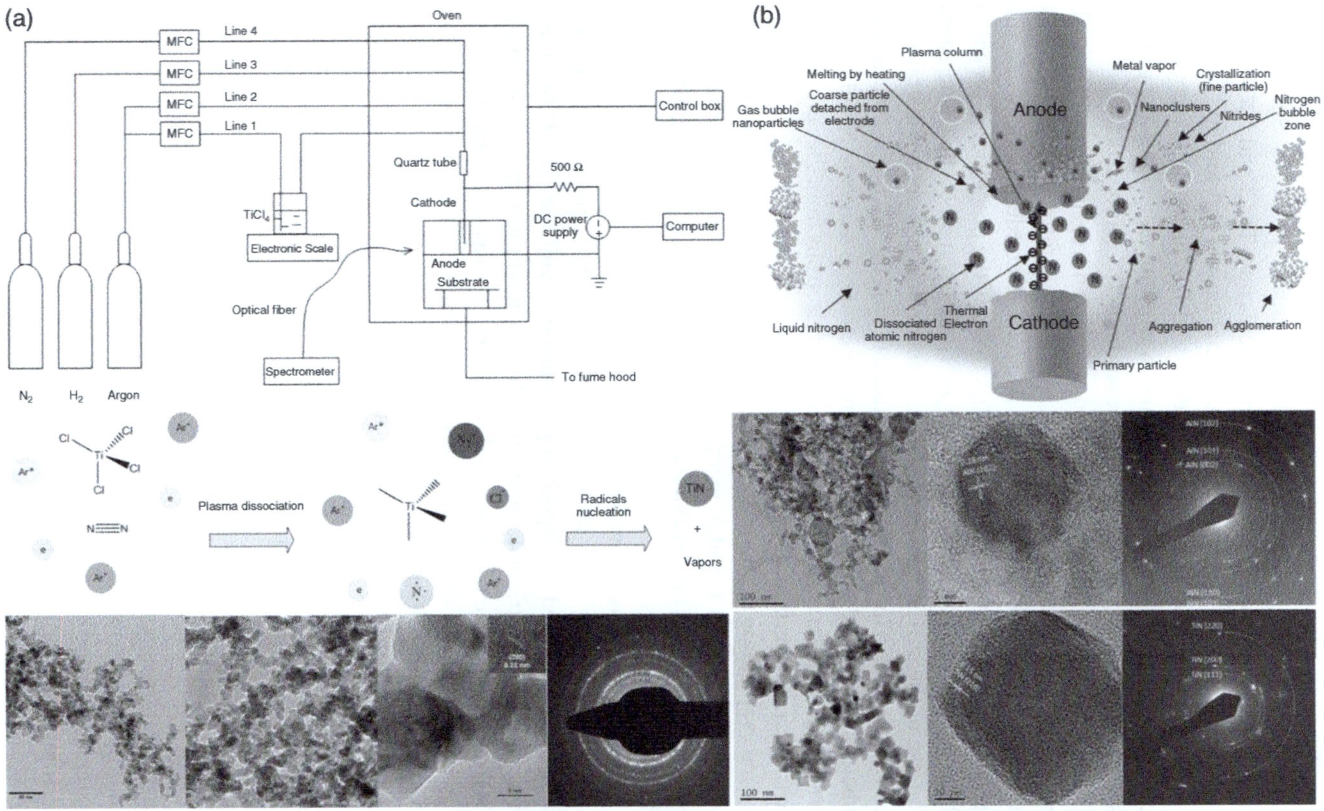


**Figure 11.7** (a) Synthesis of TiN nanoparticles using a hollow cathode microplasma. *Source:* Reproduced from [23] / with permission of Elsevier; (b) Synthesis of AlN and TiN nanoparticles using an arc microplasma in liquid nitrogen. *Source:* Reproduced from [79] / MDPI / CC BY 4.0.

complex process involving reactive radicals, ions, molecules, and electrons, initiated and driven by consecutive or parallel reactions such as the electron-impact process, Penning ionization, and charge transfer processes. Specifically, once $TiCl_4$ and $N_2$ were introduced into the microplasma, Ti–Cl and N≡N would be dissociated by high-energy species such as electrons, $Ar^*$, $N_2^*$, and $Ar^+$, leading to the formation of atomic Ti, atomic N, and ionized Ti moieties. The recombination reactions of these fragments, in conjunction with the nucleation and growth process, would eventually result in the generation of TiN nanoparticles.

Apart from $N_2$ gas, liquid nitrogen can also serve as a nitrogen source. Park et al. [79] proposed using submerged arc microplasma in liquid nitrogen to prepare AlN and TiN nanoparticles directly from Ti and Al metal electrodes. In this work, the microplasma current was tuned from 5 to 30 A to study the properties of prepared particles, suggesting the crystallinity of both AlN and TiN nanoparticles increased with the rise of the current. A hypothesized reaction mechanism is illustrated in Figure 11.7b. In brief, a lot of nitrogen gas existed during the discharge process and would be continuously dissociated by plasma electrons to generate N radicals as long as the microplasma persisted. Meanwhile, the electrons from the cathode collided with the anode surface, in which the temperature is high enough to cause the melting and evaporation of anode materials. Then, the generated N radicals would react with the evaporated Ti and Al metals to form TiN and AlN nanoparticles. The developed method opens new avenues for simple, green, and direct preparation of fine nitride nanoparticles.

The versatility of microplasma for nitrogen fixation in nanomaterials was also demonstrated by Huang et al. [80], where an atmospheric pressure air microplasma jet was employed to deposit organosilicon oxynitride ($SiO_xC_yN_z$) films on the surface of polycarbonate substrates. In this work, tetramethyldisiloxane (TMDSO) was used as the precursor, and the plasma jet was sustained at a power of 300 W and an audio frequency of 20 kHz. By injecting TMDSO into the microplasma jet, a high deposition rate up to 19.9 nm $s^{-1}$ can be achieved. The surface hardness of the polycarbonate substrates was greatly enhanced after the deposition of $SiO_xC_yN_z$ films, while the bulk substrates still retain high transparency. Table 11.3 provides typical examples on microplasma-assisted synthesis of nitride, carbonitride, and oxynitrid nanomaterials.

### 11.3.4 N-Doped Nanomaterials

Introducing heteroatoms into nanomaterials has become a hot research topic, as it represents an effective way to regulate their physical and chemical properties. Currently, atoms including boron (B), nitrogen (N), oxygen (O), fluorine (F), phosphorus (P), sulfur (S), chlorine (Cl), and iodine (I) have been used to alter the surface, interfacial, and electronic properties of nanomaterials, aiming to endow them with

**Table 11.3** Typical examples on microplasma-assisted synthesis of nitride, carbonitride, and oxynitrid nanomaterials.

| Structure | Reactant | Temperature (K) | Pressure | Power | Product | Throughput | Reference |
|---|---|---|---|---|---|---|---|
| Microwave microplasma | Ga, $N_2$ | 883.15–973.15 | 200–400 Pa | 420 W | GaN | 0.5 μm $h^{-1}$ | [81] |
| Microwave microplasma | NiCo-LDH, $N_2$, CC, | 573.15 | 14 Torr | 450 W | NiCo-N-O | | [82] |
| Gliding arc microplasma | $N_2$, MgO | 1,000–2,500 | 1 atm | | $Mg_3N_2$ | 0.15 g $kWh^{-1}$ | [83] |
| DBD microplasma | Si, $NH_3$, $N_2$ | 573.15 | 80–110 Pa | 35–200 W | $SiN_x$ | 11 nm $min^{-1}$ | [84] |
| DBD microplasma | $N_2$, $CH_4$, $C_2H_2$, $C_2H_4$ | | 25–1,000 Pa | | α—C:N:H | 0.3–106 nm $min^{-1}$ | [85] |
| Plasma jet confined in microreactor | Si, $N_2$, $NH_3$ | 10,000 – | 1 atm | 13–14 kW | $Si_3N_4$ | | [86] |
| Plasma jet confined in microreactor | $C_3H_6N_6$, $Ga(NO_3)_3{\cdot}H_2O$, $NH_3$ | | 1 atm | 12.6 kW | GaN | | [87] |
| Plasma jet exits microreactor | Ti, $N_2$ | | 1 atm | | TiN | | [88] |
| Needle electrode based Microplasma | Ethylenediamine, PDMS | 308.15 | 1 atm | 8 W | PPEDA | | [89] |

new functionalities for diverse applications ([90, 91]). Among them, nitrogen doping has received wide attention for several reasons. First, due to the existence of two lone pair electrons, doping N atoms can improve the electron density of materials, providing them with higher electrochemical activity [92]. Second, a lot of defects will be created after N-doping. These defects can alter the uniform charge distribution of materials, resulting in change in electronic structures and magnetic properties. In some cases, these defects also function as emission centers for phosphors, or active sites for catalysts [93]. Third, the formation of nitrogen functional groups, such as pyrrolic NH, pyridinic-N, graphitic-N, and quaternary-N, is believed to contribute to the electrical conductivity enhancement. As a result, N-doping allows nanomaterials to be used in many electronic applications.

Nitrogen-doping can be generated by either direct synthesis or post-treatment. The former one generates N-doping during the nucleation and growth of nanomaterials, while the latter one forms N-doping by post-treating the target materials with N-precursors [94]. Despite remarkable progress, existing methods still rely on time- and energy-consuming operations, requiring strong acids or alkalis, which tend to be energy inefficient, have multiple processes, and have adverse effects on the environment. The existence of a high density of reactive species, such as ions, electrons, free radicals in metastable states, and photons, makes microplasma attractive for N-doping, since $N_2$ can be dissociated rapidly at relatively low temperatures and power consumption. Therefore, reactions can be completed within a few minutes. Besides, no extra nitrogen sources are needed, except for the $N_2$, helping to avoid strong acids or alkalis as well as the associated purification process.

To date, many studies have been performed to prepare N-doped materials via the microplasma-assisted method. Ma et al. [95] used a hollow cathode microplasma jet to prepare N-doped carbon dots (N-CDs) from the aqueous solution of folic acid (Figure 11.8a). In this work, a stainless-steel capillary was used as the cathode and placed 1 mm above the surface of the solution, while a platinum disk, as the counter electrode, was immersed in the liquid. Microplasma discharges were formed between the capillary and the solution surface. The influence of process parameters on the surface groups and photoluminescence property of the products was studied in detail, including reaction time, operating voltage, and electrode size. The results revealed the successful fabrication of N-CDs by the microplasma method, as indicated by the presence of N–H, C–N functional groups. Prolonged plasma processing time and enhanced plasma voltage contributed to a higher carbonization degree and increased particle aggregation. A maximum N-CD yield of 0.42% was achieved at 5 kV voltage after two hours of plasma processing. A similar method was also used by Chiang et al. [96] to produce colloidal nitrogen-doped graphene quantum dots (N-GQDs) from chitosan and acetic acid (Figure 11.8b). The hydroxyl radicals and solvated electrons ($e_{aq}^-$) proved to play an important

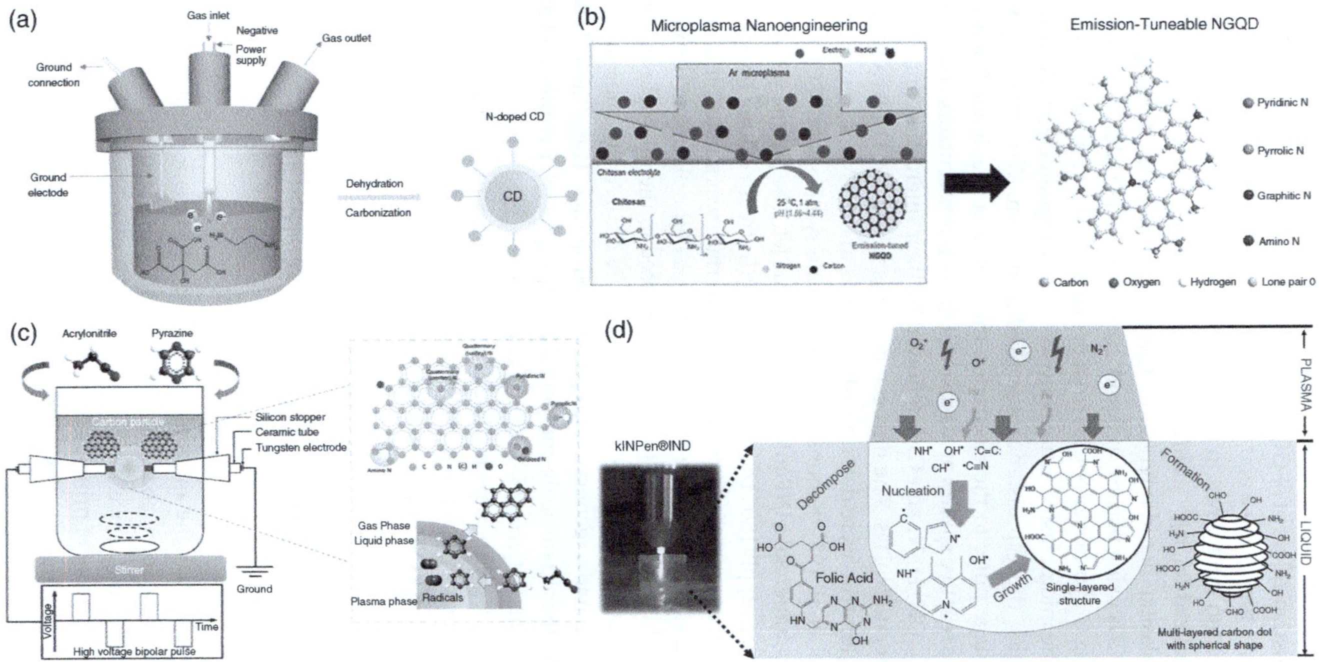


**Figure 11.8** (a) Hollow cathode microplasma for the synthesis of N-CDs from citric acid and ethylenediamine. *Source:* Reproduced from [95] / with permission of Elsevier; (b) Microplasma nanoengineering of N-GQDs using chitosan dissolved with acetic acid solution. *Source:* Reproduced from [96] / with permission of Elsevier; (c) Synthesis of N-CDs by a liquid microplasma process from pyrazine and acrylonitrile. *Source:* Reproduced from [97] / with permission of Royal Society of Chemistry; (d) kINPen®IND microplasma jet for gram-scale synthesis of N-CDs from folic acid. *Source:* Reproduced from [98] / with permission of Elsevier.

part in preparing N-GQDs during the microplasma synthesis. Furthermore, N-doping is able to change the configuration and functional groups of GQDs, resulting in tunable emission colors of N-GQDs. This characteristic allows N-GQDs to be used as nanosensors for selective and sensitive detection of $Fe^{3+}$, $Cu^{2+}$, and $Hg^{2+}$. A rather low-detection limit down to 47.9 nM with a broad linear detection range (0.5–300 μM) is realized for $Hg^{2+}$ using the microplasma-fabricated N-GQDs.

In addition to plasma–liquid interaction, there are also studies focusing on the preparation of N-CDs using plasma in liquid. Ishizaki et al. [97] developed a solution plasma process to synthesize N-CDs from pyrazine and acrylonitrile, as shown in Figure 11.8c. The microplasma setup consists of two tungsten electrodes within electrolyte, in which the gap distance was fixed at 0.5 mm. With the supply of 2.0 kV of high-voltage pulse to the electrodes, microplasma was ignited between the electrodes. The reactive species of plasma will dissociate precursors and $N_2$ into fragments (i.e., $C_2$, CH, H, N, and CN), followed by their interaction and recommendation to generate N-CDs. The nitrogen contents in the N-CDs derived from pyrazine and acrylonitrile were determined to be 8.58 and 2.97 at.%, respectively.

Recently, Hessel et al. [98] demonstrated gram-scale synthesis of N-CDs from folic acid using a commercial kINPenIND microplasma jet. The plasma was driven by a 110 V AC power supply (50 Hz), with a 1 cm length stream exiting the nozzle. The effect of microplasma jet distance on the water surface was studied, indicating that the closer the distance, the more the spreading area of the plasma in the whole gaseous reaction volume. Once the folic acid is exposed to the microplasma, it is decomposed into fragments such as CH·, ·CN, OH·, and NH·. At the same time, $N_2$ in air was also dissociated to generate nitrogen-excited species. These radicals will interact with each other to nucleate compounds such as phenyl, pyrrolic, and graphitic nitrogen radicals. Afterward, they will collide and aggregate to form graphene sheet-like structures with nitrogen defects and $-NH_2$ functional groups, before further assembling layer by layer to form N-CDs. The proposed mechanism of the N-CDs synthesis by the kINPenIND microplasma jet is shown in Figure 11.8d.

## 11.4 Challenges and Perspectives for Microplasma Nitrogen Fixation

Nitrogen fixation based on the microplasma method has demonstrated fruitful achievements. However, some challenges still remain. An apparent limitation is the rather low throughput, since most fundamental research uses only a single microplasma unit as a proof of principle to produce $NO_x$, $NH_3$, nitride nanomaterials, or N-doped nanomaterials. It should be noted that the throughput can be improved by different strategies. One solution is to maximize the operational potential of the process (e.g. microplasma–catalyst coupling, catalyst properties, or reaction parameters such as power input, frequency, and gas flow rates) to improve the productivity [99]. For instance, the inner diameter of the microplasma reactor can be increased, or the precursor and gas flow rates can be enhanced, accompanied by the appropriate rise of the plasma power. When optimizing the potential of the process, it is important to ensure that the reaction conditions are maintained at the same level. Another solution is to develop new reactor configurations with the capacity to handle larger quantities of reagents. This can be implemented by numbering up the microplasma unit. For example, increasing the number of microplasma jet (e.g. microplasma array) can significantly boost the productivity. Meanwhile, thanks to the small dimension of microplasma, the array configuration is still very compact, making it suitable for industrial production [23]. These strategies require a bit of trial and error in both the lab-scale research and pilot studies as well as collaborations across multidisciplinary teams.

Regarding the energy, currently, most nitrogen fixation processes are driven by the use of fossil fuels. Taking the Haber-Bosch process as an example, it consumes about 2% of global energy and releases around 1.5% of global $CO_2$ [100]. Microplasma-assisted nitrogen fixation process can be driven by electrical energy derived from renewable energy sources, such as solar, hydro, wind, biomass, and geothermal power. This trend is expected to continue in the coming years. Thus, building microplasma nitrogen fixation systems in regions with renewable energy sources while using cheap feedstocks such as air, water, and $N_2$ to value-added chemicals may be a promising approach to develop zero-emissions energy conversion systems. Another key advantage of the microplasma nitrogen fixation process driven by electrical energy is its fast response, in which the process can be realized instantaneously "on-and-off," just by switching the microplasma on and off. Such a characteristic renders it possible to achieve remote and digital process control.

Microplasma nitrogen fixation is a complex process involving numerous intermediates and multiple reactions. To date, the related plasma physics, plasma chemistry, and reaction kinetics are still not fully understood. Thus, data scarcity is one of the major challenges to fully understand the reaction mechanisms. Fortunately, modeling and advanced characterization techniques have allowed us to gain better insight into the plasma physics and plasma chemistry during the nitrogen fixation process. As demonstrated by Bogaerts et al. [101], modeling has successfully predicted the optimal reaction conditions driven by a gliding arc plasma for enhancing nitrogen fixation efficiency. Besides, *in situ* characterization techniques, such as Langmuir probe, optical emission spectroscopy, and interferometer, should offer a more realistic picture of the rapid transient nature of microplasma phenomena. With their help, we can identify the role of different plasma species and their excited states, as well as the underlying chemistry. On the other hand, due to the extreme transient nature of plasma dynamics, the essential evolution processes of microplasma discharges are still difficult to assess. Plasma modeling and numerical simulation is a powerful tool to investigate the physical and chemical processes occurring in the plasma in detail [102]. It is based on the combination of plasma theories and fluid models to address complex problems in plasma, which overcomes the experimental difficulties and offers insights into the space and time dependence of plasma parameters, fast decay, and complex chemistry of the plasma [103, 104]. With the development of computer technology, future investigation on the microplasma-assisted nitrogen fixation process will undoubtedly benefit from plasma modeling and numerical simulation.

## 11.5 Conclusions

As a special category of plasma that employs beneficial properties of non-thermal plasma in microscale geometry, microplasma possesses several key advantages, including non-equilibrium characteristics, microscale operation, flexible configuration, and versatile process. Compared to conventional plasma, microplasma has an enhanced density of reactive species and can be operated at ambient pressure and low temperatures. These properties provide microplasma with the ability to dissociate extremely stable $N_2$ under mild conditions, offering the opportunity to convert electrical energy from renewable sources into chemical energy. This chapter, for the first time, focuses on the state of the art of nitrogen fixation by microplasma. First, the properties, advantages, and main applications of microplasma were introduced, followed by a discussion on the benefits of using microplasma for nitrogen fixation. Then, commonly used microplasma configurations for nitrogen fixation were demonstrated, including microwave microplasma, gliding arc microplasma, DBD microplasma, plasma jet exiting microreactor, plasma jet confined in microreactor, and needle electrode-based microplasma. Afterward, four types of nitrogen derivatives produced by microplasma were reviewed, including $NO_x$, $NH_3$, nitrides or derivatives, and N-doped materials. Finally, existing challenges and perspectives for microplasma nitrogen fixation were discussed in terms of scaling-up strategy, sustainable energy, and reaction mechanisms.

## Acknowledgments

The author would like to acknowledge the financial support from the National Natural Science Foundation of China (22078125 and 52004102), China Postdoctoral Science Foundation (2023 M741472), and Undergraduate Education and Teaching Reform Research Project of Jiangnan University (1045214012240190).

## References


1 Samukawa, S., Hori, M., Rauf, S. et al. (2012). The 2012 plasma roadmap. *J. Phys. D. Appl. Phys.* 45: 253001. https://doi.org/10.1088/0022-3727/45/25/253001.

2 Nikiforov, A., Ma, C., Choukourov, A., and Palumbo, F. (2022). Plasma technology in antimicrobial surface engineering. *J. Appl. Phys.* 131 (1): 011102. https://doi.org/10.1063/5.0066724.

3 Mariotti, D. and Sankaran, R.M. (2010). Microplasmas for nanomaterials synthesis. *J. Phys. D. Appl. Phys.* 43 (32): 323001. https://doi.org/10.1088/0022-3727/43/32/323001.

4 Schoenbach, K.H. and Becker, K. (2016). 20 years of microplasma research: a status report, The. *Eur. Phys. J. D.* 70: 1–22. https://doi.org/10.1140/epjd/e2015-60618-1.

5 Mathew, P., George, J., Mathews, S., and Kurian, T.P.J. (2019). Experimental verification of modified Paschen's law in DC glow discharge argon plasma. *AIP Adv.* 9 (2): 025215. https://doi.org/10.1063/1.5086246.

6 Hourdakis, E., Simonds, B.J., and Zimmerman, N.M. (2006). Submicron gap capacitor for measurement of breakdown voltage in air. *Rev. Sci. Instrum.* 77 (3): 034702. https://doi.org/10.1063/1.2185149.

7 Fu, Y., Zhang, P., Verboncoeur, J.P., and Wang, X. (2020). Electrical breakdown from macro to micro/nano scales: a tutorial and a review of the state of the art. *Plasma Res. Express* 2: 013001. https://doi.org/10.1088/2516-1067/ab6c84.

8 Okubo, M. (2023). Fundamentals and applications of nonthermal plasma fluid flows: a review. *Plasma* 6: 277–307. https://doi.org/10.3390/plasma6020020.

9 Zhang, Z., Rui, L., and Lin, L. (2024). Direct writing of fluorescent patterns of rare-Earth-doped nanophosphors by microplasma jet. *Ind. Eng. Chem. Res.* 63 (11): 5000–5004. https://doi.org/10.1021/acs.iecr.4c00294.

10 Zhang, Z., Wang, F., Hessel, V. et al. (2024). Microplasma-printed Au-based SERS sensing platform for ultra-sensitive chemical analyte detection. *Reaction Chem. Eng.* 9 (6): 1541–1549. https://doi.org/10.1039/d4re00011k.

11 Hu, H., Zhou, J., Miao, Z. et al. (2024). Catalytic growth of CdS Nanowires by microplasma-prepared gold nanoparticles. *Ind. Eng. Chem. Res.* 64 (1): 502–510. https://doi.org/10.1021/acs.iecr.4c03906.

12 Lin, L., Li, X., Zhou, J. et al. (2021). Plasma-aided green and controllable synthesis of silver nanoparticles and their compounding with gemini surfactant. *J. Taiwan Inst. Chem. Eng.* 122: 311–319. https://doi.org/10.1016/j.jtice.2021.04.061.

13 Chiang, W.H., Mariotti, D., Sankaran, R.M. et al. (2020). Microplasmas for advanced materials and devices. *Adv. Mater.* 32 (18): 1905508. https://doi.org/10.1002/adma.201905508.

14 Becker, K.H., Zhu, W., and Lopez, J.L. (2017). Microplasmas: Environmental and Biological Applications. In: *Encyclopedia of Plasma Technology*

(ed. A.R. Hoskinson), 791–805. Boca Raton: Taylor & Francis https://doi.org/10.1081/E-EPLT-120051237.

**15** Cai, Z., Chen, H., Gao, M., and Wang, Z. (2024). Quantification of lithium in molten chlorides by optical emission spectrometry using a novel molten-salt-electrode microplasma source. *Talanta* 266: 125111. https://doi.org/10.1016/j.talanta.2023.125111.

**16** Rui, L., Wang, Y., Zhou, J. et al. (2022). Surface modification of nitrile membranes by DBD plasma and their antibacterial properties. *Chem. Eng. Res. Des.* 186: 125–135. https://doi.org/10.1016/j.cherd.2022.07.038.

**17** Li, X., Zhao, C.-X., and Lin, L. (2022). Plasma-based instant synthesis of functionalized gold nanoparticles for colorimetric detection of lead ions. *Chem. Eng. Sci.* 260: 117849. https://doi.org/10.1016/j.ces.2022.117849.

**18** Dhakal, O.B., Dahal, R., Lamichhane, P. et al. (2024). Gas phase formaldehyde degradation: continuous versus duty cycle driven plasma reactor. *Process Saf. Environ. Prot.* 190: 744–754. https://doi.org/10.1016/j.psep.2024.07.089.

**19** Lin, L., Tao, Y., Starostin, S.A. et al. (2023). Plasma regeneration of spent $Pd/Al_2O_3$ catalysts and their electrochemical performance. *Ind. Eng. Chem. Res.* 62 (17): 6634–6645. https://doi.org/10.1021/acs.iecr.3c00126.

**20** Wengler, J., Ognier, S., Zhang, M. et al. (2018). Microfluidic chips for plasma flow chemistry: application to controlled oxidative processes. *Reaction Chem. Eng.* 3 (6): 930–941. https://doi.org/10.1039/c8re00122g.

**21** Kim, J., Mironov, A.E., Cho, J.H. et al. (2022). Commercialization of microcavity plasma devices and arrays: systems for VUV photolithography and nanopatterning, disinfection of drinking water and air, and biofilm deactivation for medical therapeutics. *Plasma Process. Polym.* 19 (10): e2200075. https://doi.org/10.1002/ppap.202200075.

**22** Tan, F., Wang, Y., Zhang, S. et al. (2022). Plasma dermatology: skin therapy using cold atmospheric plasma. *Front. Oncol.* 12: 918484. https://doi.org/10.3389/fonc.2022.918484.

**23** Lin, L., Starostin, S.A., Wang, Q., and Hessel, V. (2017). An atmospheric pressure microplasma process for continuous synthesis of titanium nitride nanoparticles. *Chem. Eng. J.* 321: 447–457. https://doi.org/10.1016/j.cej.2017.03.128.

**24** Li, S., Medrano Jimenez, J., Hessel, V., and Gallucci, F. (2018). Recent progress of plasma-assisted nitrogen fixation research: a review. *Processes* 6: 248. https://doi.org/10.3390/pr6120248.

**25** Li, J., Lan, C., Nie, L. et al. (2023). Distributed plasma-water-based nitrogen fixation system based on cascade discharge: generation, regulation, and application. *Chem. Eng. J.* 478: 147483. https://doi.org/10.1016/j.cej.2023.147483.

**26** Ishikawa, K. (2022). Perspectives on functional nitrogen science and plasma-based in situ functionalization. *Jpn. J. Appl. Phys.* 61 (SA): Sa0802. https://doi.org/10.35848/1347-4065/ac3558.

**27** Li, Z., Lin, L., Shen, J. et al. (2024). Microflow synthesis of fluorescent carbon dots for selective $Co^{2+}$ detection. *Ind. Eng. Chem. Res.* 63 (10): 4420–4429. https://doi.org/10.1021/acs.iecr.3c04051.

**28** Klimek, A. and Piercey, D.G. (2024). Nitrogen fixation via plasma-assisted processes: mechanisms, applications, and comparative analysis-a comprehensive review. *Processes* 12 (4): 786. https://doi.org/10.3390/pr12040786.

**29** Chen, H., Yuan, D., Wu, A. et al. (2021). Review of low-temperature plasma nitrogen fixation technology. *Waste Disposal Sustain. Energy* 3 (3): 201–217. https://doi.org/10.1007/s42768-021-00074-z.

**30** Lin, L., Xu, H., Gao, H. et al. (2020). Plasma-assisted nitrogen fixation in nanomaterials: Fabrication, characterization, and application. *J. Phys. D. Appl. Phys.* 53 (13): 133001. https://doi.org/10.1088/1361-6463/ab5f1f.

**31** Wang, W., Patil, B., Heijkers, S. et al. (2017). Nitrogen fixation by gliding arc plasma: better insight by chemical kinetics modelling. *ChemSusChem* 10 (10): 2145–2157. https://doi.org/10.1002/cssc.201700095.

**32** Kelly, S. and Bogaerts, A. (2021). Nitrogen fixation in an electrode-free microwave plasma. *Joule* 5 (11): 3006–3030. https://doi.org/10.1016/j.joule.2021.09.009.

**33** Pawlat, J., Terebun, P., Kwiatkowski, M. et al. (2019). Evaluation of oxidative species in gaseous and liquid phase generated by mini-gliding arc discharge. *Plasma Chem. Plasma Process.* 39 (3): 627–642. https://doi.org/10.1007/s11090-019-09974-9.

**34** Lamichhane, P., Pourali, N., Rebrov, E.V., and Hessel, V. (2024). Energy intensified nitrogen fixation through fast modulated gas discharge from pyramid-shaped micro-electrode. *Plasma Chem. Plasma Process.* 44 (3): 1369–1392. https://doi.org/10.1007/s11090-023-10376-1.

**35** Reuter, S., von Woedtke, T., and Weltmann, K.-D. (2018). The kINPen-a review on physics and chemistry of the atmospheric pressure plasma jet and its applications. *J. Phys. D. Appl. Phys.* 51 (23): 233001. https://doi.org/10.1088/1361-6463/aab3ad.

**36** Barragan, A.A., Ilawe, N.V., Zhong, L. et al. (2017). A non-thermal plasma route to plasmonic TiN nanoparticles. *J. Phys. Chem. C* 121 (4): 2316–2322. https://doi.org/10.1021/acs.jpcc.6b08910.

**37** Carolan, D., Rocks, C., Padmanaban, D.B. et al. (2017). Environmentally friendly nitrogen-doped carbon quantum dots for next generation solar cells. *Sustain. Energy Fuel.* 1 (7): 1611–1619. https://doi.org/10.1039/c7se00158d.

**38** Ivanov, V., Paunska, T., Lazarova, S., and Bogaerts, A. (2023). Gliding arc/glow discharge for $CO_2$ conversion: Comparing the performance of different discharge configurations. *J $CO_2$ Util 67: 102300. https://doi.org/10.1016/j.jcou.2022.102300.*

**39** Li, S., van Raak, T., Kriek, R. et al. (2023). Gliding arc reactor under AC pulsed mode operation: spatial performance profile for $NO_x$ synthesis. *ACS Sustain. Chem. Eng.* 11 (34): 12821–12832. https://doi.org/10.1021/acssuschemeng.3c03832.

**40** Wang, F., Miao, Z., Li, C., and Lin, L. (2024). Degradation of stains from metal surfaces using a DBD plasma microreactor. *Micromachines* 15 (3): 297. https://doi.org/10.3390/mi15030297.

**41** Gao, X., Liang, J., Wu, L. et al. (2022). Dielectric barrier discharge plasma-assisted catalytic $CO_2$ hydrogenation: synergy of catalyst and plasma. *Catalysts* 12 (1): 66. https://doi.org/10.3390/catal12010066.

**42** Bekeschus, S., Schmidt, A., Weltmann, K.-D., and von Woedtke, T. (2016). The plasma jet kINPen–a powerful tool for wound healing. *Clin.Plasma Med.* 4 (1): 19–28. https://doi.org/10.1016/j.cpme.2016.01.001.

**43** Lin, L. and Wang, Q. (2015). Microplasma: a new generation of technology for functional nanomaterial synthesis. *Plasma Chem. Plasma Process.* 35: 925–962. https://doi.org/10.1007/s11090-015-9640-y.

**44** Lin, L., Starostin, S.A., Hessel, V., and Wang, Q. (2017). Synthesis of iron oxide nanoparticles in microplasma under atmospheric pressure. *Chem. Eng. Sci.* 168: 360–371. https://doi.org/10.1016/j.ces.2017.05.008.

**45** Zhao, X. and Tian, Y. (2023). Sustainable nitrogen fixation by plasma-liquid interactions. *Cell Report. Phys. Sci.* 4 (10): 101618. https://doi.org/10.1016/j.xcrp.2023.101618.

**46** Li, Y., Qin, L., Wang, H.-L. et al. (2023). High efficiency $NO_x$ synthesis and regulation using dielectric barrier discharge in the needle array packed bed reactor. *Chem. Eng. J.* 461: 141922. https://doi.org/10.1016/j.cej.2023.141922.

**47** Lamichhane, P., Pourali, N., Rebrov, E.V., and Hessel, V. (2024). Sustainable plasma-catalytic nitrogen fixation with pyramid shaped μ-electrode DBD and titanium dioxide. *ChemistrySelect* 9 (24): e202401076. https://doi.org/10.1002/slct.202401076.

**48** Al-Amin, M., Sah, A.K., Roy, N.C., and Talukder, M.R. (2023). Enhancement of $NO_x$ production in water by combining an air bubble plasma jet and an external magnetic field. *Phys. Plasma* 30 (9): 093507. https://doi.org/10.1063/5.0161173.

**49** Pattyn, C., Maira, N., Remy, A. et al. (2020). Potential of $N_2/O_2$ atmospheric pressure needle-water DC microplasmas for nitrogen fixation: nitrite-free synthesis of nitrates. *Phys. Chem. Chem. Phys.* 22 (42): 24801–24812. https://doi.org/10.1039/d0cp03858j.

**50** Patil, B.S., Palau, J.R., Hessel, V. et al. (2016). Plasma nitrogen oxides synthesis in a milli-scale gliding arc reactor: investigating the electrical and process parameters. *Plasma Chem. Plasma Process.* 36 (1): 241–257. https://doi.org/10.1007/s11090-015-9671-4.

**51** Fujera, J., Hoffer, P., Prukner, V. et al. (2024). Surface DBD in moist air for nitrogen fixation: a comparative study of pulsed versus amplitude-modulated AC powered discharge. *J. Phys. D. Appl. Phys.* 57 (40): 405205. https://doi.org/10.1088/1361-6463/ad5f99.

**52** Han, Y., Wen, S., Tang, H. et al. (2018). Influences of frequency on nitrogen fixation of dielectric barrier discharge in air. *Plasma Sci. Technol.* 20 (1): 014001. https://doi.org/10.1088/2058-6272/aa947a.

**53** Pei, X., Li, Y., Luo, Y. et al. (2024). Nitrogen fixation as $NO_x$ using air plasma coupled with heterogeneous catalysis at atmospheric pressure. *Plasma Process. Polym.* 21 (1): 2300135. https://doi.org/10.1002/ppap.202300135.

**54** Wang, J., Zhang, B., Liu, Y. et al. (2023). Efficient $N_2$ fixation in air enabled triboelectric plasma jet. *Chem Catalysis* 3 (7): 100647. https://doi.org/10.1016/j.checat.2023.100647.

**55** Pipa, A.V., Bindemann, T., Foest, R. et al. (2008). Absolute production rate measurements of nitric oxide by an atmospheric pressure plasma jet (APPJ). *J. Phys. D-Appl. Phys.* 41 (19): 194011. https://doi.org/10.1088/0022-3727/41/19/194011.

**56** Smith, C., Hill, A.K., and Torrente-Murciano, L. (2020). Current and future role of Haber-Bosch ammonia in a carbon-free energy landscape. *Energy Environ. Sci.* 13 (2): 331–344. https://doi.org/10.1039/c9ee02873k.

**57** Hong, J., Prawer, S., and Murphy, A.B. (2018). Plasma catalysis as an alternative route for ammonia production: status, mechanisms, and prospects for progress. *ACS Sustain. Chem. Eng.* 6 (1): 15–31. https://doi.org/10.1021/acssuschemeng.7b02381.

**58** Nguyen Van Duc, L., Al-Bared, M., Lin, L. et al. (2022). Understanding plasma-assisted ammonia synthesis via crossing discipline borders of literature: a critical review. *Chem. Eng. Sci.* 263: 118097. https://doi.org/10.1016/j.ces.2022.118097.

**59** Hawtof, R., Ghosh, S., Guarr, E. et al. (2019). Catalyst-free, highly selective synthesis of ammonia from nitrogen and water by a plasma electrolytic system. *Sci. Adv.* 5 (1): eaat5778. https://doi.org/10.1126/sciadv.aat5778.

**60** Long, N.V.D., Pourali, N., Lamichhane, P. et al. (2024). Catalytic ammonia formation in a microreaction chamber with electrically intensified arc plasma. *ChemCatChem* 16 (13): e202400005. https://doi.org/10.1002/cctc.202400005.

**61** Han, K., Luo, J., Chen, J. et al. (2021). Self-powered ammonia synthesis under ambient conditions via $N_2$ discharge driven by tesla turbine triboelectric nanogenerators. *Microsyst. Nanoeng.* 7 (1): 7. https://doi.org/10.1038/s41378-020-00235-w.

**62** Vervloedt, S.C.L. and von Keudell, A. (2024). Ammonia synthesis by plasma catalysis in an atmospheric RF helium plasma. *Plasma Sources Sci. Technol.* 33 (4): 045005. https://doi.org/10.1088/1361-6595/ad38d6.

**63** Shao, J. and Wang, Z.L. (2023). *Displacement Current Theory of Triboelectric Nanogenerators, Handbook of Triboelectric Nanogenerators*, 1–64. Springer https://doi.org/10.1007/978-3-031-05722-9_5-1.

**64** Nakajima, J. and Sekiguchi, H. (2008). Synthesis of ammonia using microwave discharge at atmospheric pressure. *Thin Solid Films* 516 (13): 4446–4451. https://doi.org/10.1016/j.tsf.2007.10.053.

**65** Wu, A., Yang, J., Xu, B. et al. (2021). Direct ammonia synthesis from the air via gliding arc plasma integrated with single atom electrocatalysis. *Appl. Catal B-Environ.* 299: 120667. https://doi.org/10.1016/j.apcatb.2021.120667.

**66** Gorbanev, Y., Engelmann, Y., van't Veer, K. et al. (2021). $Al_2O_3$-supported transition metals for plasma-catalytic $NH_3$ synthesis in a DBD plasma: metal activity and insights into mechanisms. *Catalysts* 11 (10): 1230. https://doi.org/10.3390/catal11101230.

**67** Liu, Y., Wang, C.W., Xu, X.F. et al. (2022). Synergistic effect of Co-Ni bimetal on plasma catalytic ammonia synthesis. *Plasma Chem. Plasma Process.* 42 (2): 267–282. https://doi.org/10.1007/s11090-021-10223-1.

**68** Zhu, X., Liu, J., Hu, X. et al. (2022). Plasma-catalytic synthesis of ammonia over Ru-based catalysts: insights into the support effect. *J. Energy Inst.* 102: 240–246. https://doi.org/10.1016/j.joei.2022.02.014.

**69** Li, K., Chen, S., Li, M. et al. (2024). Plasma catalytic ammonia synthesis in a fluidized-bed DBD reactor over $M/Al_2O_3$: enhancement of plasma-catalyst surface interaction. *J. Colloid Interface Sci.* 683 (Pt 1): 652–662. https://doi.org/10.1016/j.jcis.2024.12.055.

**70** Liu, J., Zhu, X., Zhou, C. et al. (2022). Plasma-catalytic ammonia synthesis over $BaTiO_3$ supported metal catalysts: process optimization using response surface methodology. *Vacuum* 203: 111205. https://doi.org/10.1016/j.vacuum.2022.111205.

**71** Lamichhane, P., Paneru, R., Nguyen, L.N. et al. (2020). Plasma-assisted nitrogen fixation in water with various metals. *Reaction Chem. Eng.* 5 (11): 2053–2057. https://doi.org/10.1039/d0re00248h.

**72** Kosumsupamala, K., Thana, P., Palee, N. et al. (2022). Air to $H_2$-$N_2$ pulse plasma jet for in-vitro plant tissue culture process: source characteristics. *Plasma Chem. Plasma Process.* 42 (3): 535–559. https://doi.org/10.1007/s11090-022-10228-4.

**73** Gorbanev, Y., Vervloessem, E., Nikiforov, A., and Bogaerts, A. (2020). Nitrogen fixation with water vapor by nonequilibrium plasma: toward sustainable ammonia production. *ACS Sustain. Chem. Eng.* 8 (7): 2996–3004. https://doi.org/10.1021/acssuschemeng.9b07849.

**74** Fang, L., Wang, L., Jiang, L. et al. (2020). Low-temperature chemical synthesis of nanostructured indium nitride from indium hydroxide. *J. Ceram. Soc. Jpn.* 128 (1): 42–45. https://doi.org/10.2109/jcersj2.19165.

**75** Dongil, A.B. (2019). Recent progress on transition metal nitrides nanoparticles as heterogeneous catalysts. *Nano* 9 (8): 1111. https://doi.org/10.3390/nano9081111.

**76** Ma, Y., Xiong, L., Lu, Y. et al. (2021). Advanced inorganic nitride nanomaterials for renewable energy: a mini review of synthesis methods. *Front. Chem.* 9: 638216. https://doi.org/10.3389/fchem.2021.638216.

**77** Lin, L., Starostin, S.A., Ma, X. et al. (2019). Facile synthesis of lanthanide doped yttria nanophosphors by a simple microplasma-assisted process. *Reaction Chem. Eng.* 4 (5): 891–898. https://doi.org/10.1039/c8re00357b.

**78** Li, X., Lin, L., Chiang, W.-H. et al. (2022). Microplasma synthesized gold nanoparticles for surface enhanced Raman spectroscopic detection of methylene blue. *Reaction Chem. Eng.* 7 (2): 346–353. https://doi.org/10.1039/d1re00446h.

**79** Park, Y.S., Kodama, S., and Sekiguchi, H. (2021). Preparation of metal nitride particles using arc discharge in liquid nitrogen. *Nano* 11 (9): 2214. https://doi.org/10.3390/nano11092214.

**80** Lin, Y.-S., Weng, M.-S., Chung, T.-W., and Huang, C. (2011). Enhanced surface hardness of flexible polycarbonate substrates using plasma-polymerized organosilicon oxynitride films by air plasma jet under atmospheric pressure. *Surf. Coat. Technol.* 205 (13–14): 3856–3864. https://doi.org/10.1016/j.surfcoat.2011.01.060.

**81** Ozawa, T., Dohi, M., Matsuura, T., and Hayakawa, Y. (2008). Synthesis of GaN bulk crystals and melt growth of GaN layers under nitrogen plasma. *J. Cryst. Growth* 310 (7–9): 1785–1789. https://doi.org/10.1016/j.jcrysgro.2007.11.138.

**82** Kong, X., Gao, Q., Bu, S. et al. (2021). Plasma-assisted synthesis of nickel-cobalt nitride-oxide hybrids for high-efficiency electrochemical hydrogen evolution. *Materials Today Energy* 21: 100784. https://doi.org/10.1016/j.mtener.2021.100784.

**83** N. Jiang, J. Yang, L. Qin, C. Li, J. Li, J. Li, Plasma-assisted synthesis of $Mg_3N_2$ in gas-powder rotating gliding arc discharge plasma. Available at SSRN 4734645. https://doi.org/10.2139/ssrn.4743440.

**84** Li, Q., Liu, J., Dai, Y. et al. (2016). Fabrication of $SiN_x$ thin film of micro dielectric barrier discharge reactor for maskless nanoscale etching. *Micromachines* 7 (12): 232. https://doi.org/10.3390/mi7120232.

**85** Niu, J., Zhang, L., Zhang, Z. et al. (2010). Deposition of hydrogenated amorphous carbon nitride films by dielectric barrier discharge plasmas. *Appl. Surf. Sci.* 256 (22): 6887–6892. https://doi.org/10.1016/j.apsusc.2010.04.106.

**86** Kim, T.-H., Lee, S., and Park, D.-W. (2022). Synthesis of silicon nitride nanoparticles by upcycling silicon wafer waste using thermal plasma jets. *Materials* 15 (24): 8796. https://doi.org/10.3390/ma15248796.

**87** Kim, T.-H., Choi, S., and Park, D.-W. (2016). Thermal plasma synthesis of crystalline gallium nitride nanopowder from gallium nitrate hydrate and melamine. *Nano* 6 (3): 38. https://doi.org/10.3390/nano6030038.

**88** S. Mohsenian, J. Fathi, B. Shokri, Reactive nitrogen plasma spray coating of titanium nitride: Plasma torch design and coating analysis, *arXiv preprint arXiv:1805.05466*, 2018. https://doi.org/10.48550/arXiv.1805.05466.

**89** Bashir, M., Bashir, S., Rees, J.M., and Zimmerman, W.B. (2014). Surface coating of bonded PDMS microchannels by atmospheric pressure microplasma. *Plasma Process. Polym.* 11 (3): 279–288. https://doi.org/10.1002/ppap.201300123.

**90** Li, Z., Miao, Z., Shen, J. et al. (2024). Green synthesis of carbon dots by microflow method and their application as sunscreen agent. *Chin. J. Chem. Eng.* 77: 135–143. https://doi.org/10.1016/j.cjche.2024.10.008.

**91** Lin, L., Xia, Y., Wen, H. et al. (2023). Green and continuous microflow synthesis of fluorescent carbon quantum dots for bio-imaging application. *AIChE J* 69 (1): e17901. https://doi.org/10.1002/aic.17901.

**92** Wei, S., Wan, C., Zhang, L. et al. (2022). N-doped and oxygen vacancy-rich $NiCo_2O_4$ nanograss for supercapacitor electrode. *Chem. Eng. J.* 429: 132242. https://doi.org/10.1016/j.cej.2021.132242.

**93** Lin, L., Yin, Y., Li, Z. et al. (2022). Continuous microflow synthesis of fluorescent phosphorus and nitrogen co-doped carbon quantum dots. *Chem. Eng. Res. Des.* 178: 395–404. https://doi.org/10.1016/j.cherd.2021.12.037.

**94** Li, C., Wang, K., and Xie, D. (2022). A review of approaches for the design of high-performance electrocatalysts for ethanol electrooxidation. *Surf Interfaces* 28: 101594. https://doi.org/10.1016/j.surfin.2021.101594.

**95** Ma, X., Li, S., Hessel, V. et al. (2020). Synthesis of N-doped carbon dots via a microplasma process. *Chem. Eng. Sci.* 220: 115648. https://doi.org/10.1016/j.ces.2020.115648.

**96** Kurniawan, D., Weng, R.-J., Setiawan, O. et al. (2021). Microplasma nanoengineering of emission-tuneable colloidal nitrogen-doped graphene quantum dots as smart environmental-responsive nanosensors and nanothermometers. *Carbon* 185: 501–513. https://doi.org/10.1016/j.carbon.2021.09.050.

**97** Li, O.L., Chiba, S., Wada, Y. et al. (2016). Selective nitrogen bonding states in nitrogen doped carbon via a solution plasma process for advanced oxygen reduction reaction. *RSC Adv.* 6 (111): 109354–109360. https://doi.org/10.1039/c6ra24546c.

**98** Pho, Q.H., Lin, L., Rebrov, E.V. et al. (2023). Process intensification for gram-scale synthesis of N-doped carbon quantum dots immersing a microplasma jet in a gas-liquid reactor. *Chem. Eng. J.* 452: 139164. https://doi.org/10.1016/j.cej.2022.139164.

**99** Lin, L., Yin, Y., Starostin, S.A. et al. (2021). Microfluidic fabrication of fluorescent nanomaterials: A review. *Chem. Eng. J.* 425: 131511. https://doi.org/10.1016/j.cej.2021.131511.

**100** Li, J., Xiong, Q., Mu, X., and Li, L. (2024). Recent advances in ammonia synthesis: from Haber-Bosch process to external field driven strategies. *ChemSusChem* 17 (15): e202301775. https://doi.org/10.1002/cssc.202301775.

**101** Vervloessem, E., Aghaei, M., Jardali, F. et al. (2020). Plasma-based $N_2$ fixation into $NO_x$: insights from modeling toward optimum yields and energy costs in a gliding arc plasmatron. *ACS Sustain. Chem. Eng.* 8 (26): 9711–9720. https://doi.org/10.1021/acssuschemeng.0c01815.

**102** Van Dijk, J., Kroesen, G., and Bogaerts, A. (2009). Plasma modelling and numerical simulation. *J. Phys. D. Appl. Phys.* 42 (19): 190301. https://doi.org/10.1088/0022-3727/42/19/190301.

**103** Irimiciuc, S.A., Chertopalov, S., Lancok, J., and Craciun, V. (2021). Langmuir probe technique for plasma characterization during pulsed laser deposition process. *Coatings* 11 (7): 762. https://doi.org/10.3390/coatings11070762.

**104** Zhang, Y.-T., Gao, S.-H., and Ai, F. (2023). Efficient numerical simulation of atmospheric pulsed discharges by introducing deep learning. *Front. Phys.* 11: 1125548. https://doi.org/10.3389/fphy.2023.1125548.

# 12

# Plasma–Liquid Interaction for Nitrogen Fixation

*Tianqi Zhang, Jungmi Hong, and Patrick Cullen*

*School of Chemical and Biomolecular Engineering, The University of Sydney, Sydney, New South Wales, Australia*

## 12.1 Introduction

Lightning strikes to the ocean is a common and natural phenomenon, which has been suggested as a pathway to amino acid formation and the origin of life on early Earth [1]. Analogous to natural lightning over oceans, plasma discharge across a liquid surface mimics these powerful natural processes, enabling the formation of solvated electrons and reactive radicals. Plasma–liquid discharges are an important area of science within atmospheric non-thermal plasma research, bridging plasma chemistry, plasma physics, and liquid-phase applications. This interaction is impactful across fields such as wastewater treatment, gas conversion, and microbial inactivation, due to the generation of reactive oxygen and nitrogen species (RONS) [2].

The generation of nitrate ($NO_2^-$), nitrite ($NO_3^-$), and relative nitrogen species like peroxynitrite ($ONOO^-$), or aqueous $NO_x$ in plasma-activated water (PAW) has long been reported, with a focus on the synthesis of a broad range of reactive species, such as OH, O, and $H_2O_2$ [3]. However, the development of applying plasma–liquid interaction for nitrogen fixation application has particularly drawn attention recently [4].

Comparing the process to air or $N_2/O_2$ plasmas without liquid interaction or water content, many studies on plasma–liquid interactions focus on the generation of ammonia ($NH_3$) from water molecules ($H_2O$), an abundant and sustainable hydrogen source. Various configurations have been introduced, such as direct plasma discharge within the bulk liquid, plasma interaction with the water surface, introducing water vapor into the gas discharge, or using bubble discharges

*Plasma-Assisted Nitrogen Fixation for Sustainable Process Industries*, First Edition. Edited by Volker Hessel, Annemie Bogaerts, Gabriele Centi, Evgeny Rebrov, and Nguyen Van Duc Long.

Published 2026 by John Wiley & Sons, Inc.

to enhance the production rate of aqueous-phase products for $N_2$ fixation – such as ammonium ($NH_{4(aq)}{}^+$) as well as nitrite and nitrate – and to improve energy efficiency [5–9].

Supplying high-density reactive species from the gas-phase discharge is crucial, as the interaction between plasma and water, depending on the configuration, can significantly influence the final performance of the aqueous-phase chemistry. Section 12.2 introduces various plasma–liquid systems used for $N_2$ fixation, with the associated plasma physics and chemistry discussed in Section 12.3. In Section 12.4, the status and key challenges in diagnostics and modeling are addressed to guide further research, concluding with a summary in Section 12.5.

## 12.2 Plasma Systems for Plasma–Liquid Discharges

To generate plasma discharges with liquids, several approaches have been employed, as shown in Figure 12.1. These can be categorized into three types of reactors: (i) plasma on the liquid surface, (ii) plasma immersed in liquid, and (iii) plasma with liquid droplets. Water generally serves as both a reactant in the reactions and an absorbent for the generated products.

Direct discharge on the liquid surface discharge (Figure 12.1a) is a commonly used method, particularly in early studies, driven by the ease of system development, providing direct plasma–liquid interaction on the surface with surrounding gases. Gromov et al. investigated and discussed energy losses from water evaporation, plasma thermalization, and the quenching of excited nitrogen species by water in their pin-water configuration, compared to a pin-to-pin configuration [10]. These factors were identified as key characteristics of $H_2O$-included gas discharge, with evidence of a lower vibrational temperature of $N_2$, as supported by ref. [11].

A common plasma reactor design, known as a plasma jet (Figure 12.1b,c), incorporates a controlled gas flow to facilitate the propulsion of reactive species into an interfaced liquid. This design offers a number of advantages: enhanced interaction of reactive species with the liquid and a guided flow that helps maintain a stable gas atmosphere. In some studies, a dielectric material is used to cover the electrode, creating a dielectric barrier discharge (DBD) regime aimed at generating specific reactive species.

To further enhance the dissolution of generated $NO_x$ species, an underwater discharge system has been developed, improving mass transfer through bubbling (Figure 12.1d). Unlike discharges on the liquid surface, this approach can allow

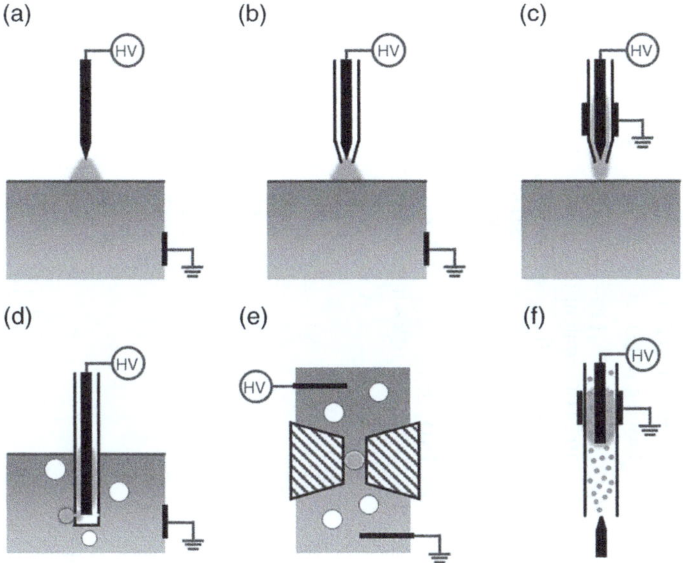


**Figure 12.1** Schematic of different discharges used in plasma/liquid interactions for nitrogen fixation. (a) Direct discharge on liquid surface. (b) Plasma jet discharge on liquid surface. (c) Plasma jet without direct contact with liquid. (d) In liquid plasma bubble discharge. (e) Discharge in bubbles. (f) Droplets/aerosol plasma discharge. Dark gray = liquid, Light gray = plasma, black = electrode.

two discharge regimes (DBD and spark) to form simultaneously. The combination of these two regimes may significantly boost $NO_x$ production [7].

In contrast to bubbles beneath the liquid surface, another approach for maximizing the plasma–liquid interface area involves aerosols or droplets in the plasma region (Figure 12.1f). This approach enables scalable, in-flow production of plasma-activated mist (PAM) containing nitrogen species like $NH_4^+$, $NO_3^-$, and $NO_2^-$ through reactions between the air plasma and water mist with micron-sized droplets. The process, which has been driven by solar-energy-assisted plasma oxidation in droplet microreactors, was demonstrated to achieve high-energy efficiency compared to conventional lab-scale plasma reactors and produce nitrogen-based nutrients for aeroponic plant growth, boosting leaf emergence and surface area by over 30%. Its compatibility with renewable energy and industrial-scale flow chemistry is regarded as offering a sustainable, low-carbon solution for nitrogen fixation [12]. Table 12.1 lists some key data reported for plasma-liquid nitrogen fixation studies.

**Table 12.1** Summary of plasma–liquid nitrogen fixation.

| Plasma source | Reactor type | Plasma power | Gas and liquid conditions | Production rate | Product selectivity | Energy efficiency | Reference |
|---|---|---|---|---|---|---|---|
| Modulated microsecond pulse AC | Plasma bubble reactor | 8.4 kV, 11.27 W | 2000 sccm air, M.Q. water | 2.92 mg $min^{-1}$ | [$NO_3^-$]: 85.3%, [$NO_2^-$]: 14.7%. (10 min, DRGSD+ Raschig rings) | 3.8 kWh $mol^{-1}$ ($NO_x$) | [7] |
| Microsecond AC pulse | Plasma jet spark discharge to water surface | 1 kV, 170 mA, 0.1 W | 0.2–1.4 L $min^{-1}$ $N_2$ (humidified), 5 mL D.I. water | 0.064 mg $h^{-1}$ | [$NH_3$]: >95% (2%–10% $H_2O$ content in $N_2$, aqueous) | 100 MJ $mol^{-1}$ ($NH_3$), 15 MJ $mol^{-1}$ ($N_2$ fixation) | [13] |
| AC plasma power supply | Pin-to-liquid discharge | 4–20 W | 1–3 SLM synthetic air ($N_2$:$O_2$ = 4:1), 0.5 L $NaH_2PO_4$ solution pumped at 13.3 mL $min^{-1}$ rate | | [$NO_3^-$]: [$NO_2^-$] = 1:18 ~ 1:8 | $16.9 \pm 2.4$ g $kWh^{-1}$ (liquid electrode system) | [10] |
| DC plasma power supply | Pin-to-liquid discharge | ~500 V, 2–8 mA | $N_2$ and Ar as purging gas, 20 mL $H_2SO_4$ solution (pH 3.5) | 0.44 mg $h^{-1}$ | | 2,270 kWh $kg^{-1}$ ($NH_3$) | [14] |
| Nanosecond pulse power supply | Plasma bubble reactor | | 1 L $min^{-1}$ air | 1.8 mmol $h^{-1}$ | | $27.0 \pm 0.1$ kWh $mol^{-1}$ ($NO_3^-$) | [15] |
| AC plasma power supply | DBD | | 1 LPM air, 23 mL DI water | 12.9 μmol $min^{-1}$ | | 20.7 MJ $mol^{-1}$ | [16] |

| | | | | | | | |
|---|---|---|---|---|---|---|---|
| AC plasma power supply | Water falling film dielectric barrier discharge | | 2,000 mL $min^{-1}$ air, 70 mL water circulating at 148 mL $min^{-1}$ | 198.3 μmol $min^{-1}$ | | 39.6 MJ $mol^{-1}$ | [17] |
| AC plasma power supply | Coaxial DBD reactor | | 5 L $min^{-1}$ air, 10 mL nebulized water aerosol droplets (~1 μm) | 24 ± 3 μmol $h^{-1}$ ($NH_3$), 8.1 ± 0.1 μmol $h^{-1}$ ($NO_2^-$), 65 ± 4 μmol $h^{-1}$ ($NO_3^-$) | [$NO_3^-$]: 66.9%, [$NO_2^-$]: 8.3%, [$NH_3$]: 24.7% (24 W, $H_2O/N_2$ plasma) | 9,500 ± 700 MJ $mol^{-1}$ | [8] |
| AC plasma power supply | In water $N_2$ plasma bubble | 8.5–11 kV, 80–130 W | 500 mL $min^{-2}$ $N_2$, | 11.2 μmol $min^{-1}$ | [$NO_3^-$]: 55%, [$NO_2^-$]: 42%, [$NH_3$]: 3% | 696.4 MJ $mol^{-1}$ | [6] |
| Nanosecond pulse power supply | Wire-plate plasma array with water mist from ultrasonic nebulizer | 3.85 W, 9–10.5 kV, | | 279.7 μmol $h^{-1}$ | [$NO_3^-$]: 78.4%, [$NO_2^-$]: 19.0%, [$NH_3$]: 3.6% (air, aqueous). [$NO_3^-$]: 77.4%, [$NO_2^-$]: 24.3%, [$NH_3$]: 5.4% ($N_2$, aqueous) | 47.79 MJ $mol^{-1}$ | [12] |

## 12.3 Mechanisms of Nitrogen Fixation in Plasma–Liquid Systems

### 12.3.1 Physical Aspects of Plasma–Liquid Interactions

Understanding the physics of plasma–liquid interactions is essential for advancing nitrogen fixation applications. Similar to other plasma-based processes, key parameters such as electron density, electron energy distribution, and its temporal evolution play significant roles in driving subsequent chemical reactions. These factors ultimately determine the efficiency of nitrogen fixation.

To optimize this process, it is critical to investigate the initiation and evolution of electrical discharges at the plasma–liquid interface. These discharges vary widely, influenced by electrode configurations, input power conditions, and liquid properties.

#### 12.3.1.1 Breakdown Mechanism of Plasma–Liquid Discharges

Plasma discharges in configurations A, B, and C of Figure 12.1 can be categorized as gas-phase discharges over a liquid, following the characterizations by Bruggeman et al. [3]. In these systems, plasma is primarily initiated and maintained in the gas phase above the liquid. However, near the plasma–liquid interface, the presence of water vapor – arising from sputtering by energetic ions or evaporation –may significantly impact the subsequent physics and chemistry. This vapor contribution should be considered to fully understand the fundamental processes occurring in the system.

The nature of plasma–liquid discharges depends on several factors, including the gap distance between the plasma electrode and the liquid, gas composition, applied voltage and power, and the liquid's conductivity, temperature, and other physicochemical properties. Discharges can range from microstreamers to sparks or even glow discharges, as noted by Bruggeman et al. [3].

Considering the different electrical aspects of the discharge, the polarity of the applied voltage is important in gaining a fundamental understanding of how plasma interacts with the water interface. As can be seen in Figure 12.2, when the high-voltage electrode is in positive polarity (Figure 12.2a), high-density positive ions ($M^+$) are accelerated toward the water interface. These ions impact the water, transferring charge to form $H_2O^+$ at a high rate, leading to a high density of $OH_{(aq)}$. In contrast, under negative polarity (Figure 12.2b, cathodic operation), electrons injected into the water become solvated and react rapidly with water, forming $H_{2(aq)}$ [18].

The discharge photographs in Figure 12.2c and d highlight the differences between anodic and cathodic glow-like discharges. Stronger emissions are

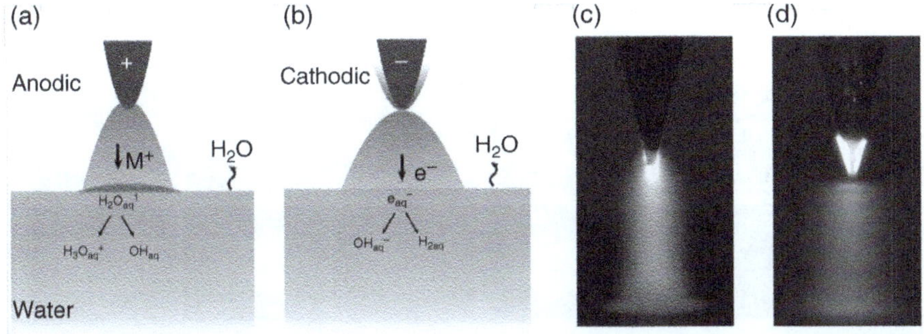


**Figure 12.2** Schematic overview of the (a) anodic and (b) cathodic configurations for the plasma formed at the surface of liquid water, with representative digital microscope images of the plasma during (c) anodic and (d) cathodic operation. *Source:* Keniley et al. [18] / IOP Publishing.

observed at the water interface under positive polarity due to the impinging high-density ions. In both cases, the current and the gap distance between the high-voltage electrode and the water surface were set to 2 mA and 1 mm, respectively.

More typical plasma–water discharges based on air or air-like gas environment requires higher voltage to initiate and maintain the discharge than inert gas discharges, such as Ar or He. This often leads to streamer or arc discharge development with higher transient current density. Hamdan et al. report a glow-like uniform discharge is maintained under low-voltage conditions in their pin-to-water air discharge configuration. However, as peak voltage and pulse width was increased, the discharge transitions into a streamer regime with higher current densities (Figure 12.3). During the positive polarity phase, high-density positive ionic species propagate as streamers at speeds approximately 10 times faster ($\sim$300 km s$^{-1}$) than those under negative polarity ($\sim$30 km s$^{-1}$). This behavior is attributed to space charge formation and electric field enhancement at the streamer head. In contrast, negative polarity streamers are primarily supported by electron drift, which counters field enhancement and results in slower propagation speeds [19].

Liu et al. reported similar findings, observing high densities of positive ions ($N^+$, $O^+$, $N_2^+$, $O_2^+$) and reactive excited species such as OH(A), $N_2$(C), $N_2^+$(B), He(3s), and O(3p) through mass spectrometry and optical emission analysis. In water, predominant aqueous species include $O_{2(aq)}^-$, $OH_{(aq)}$, and $ONOO_{(aq)}^-$, whereas negative plasma jets generate higher concentrations of nitrite ($NO_{2(aq)}^-$) and nitrate ($NO_{3(aq)}^-$). This is attributed to the generation of high-density negative ions ($OH^-$, $O_2^-$, $NO_2^-$, $NO_3^-$) in negative polarity conditions [20].

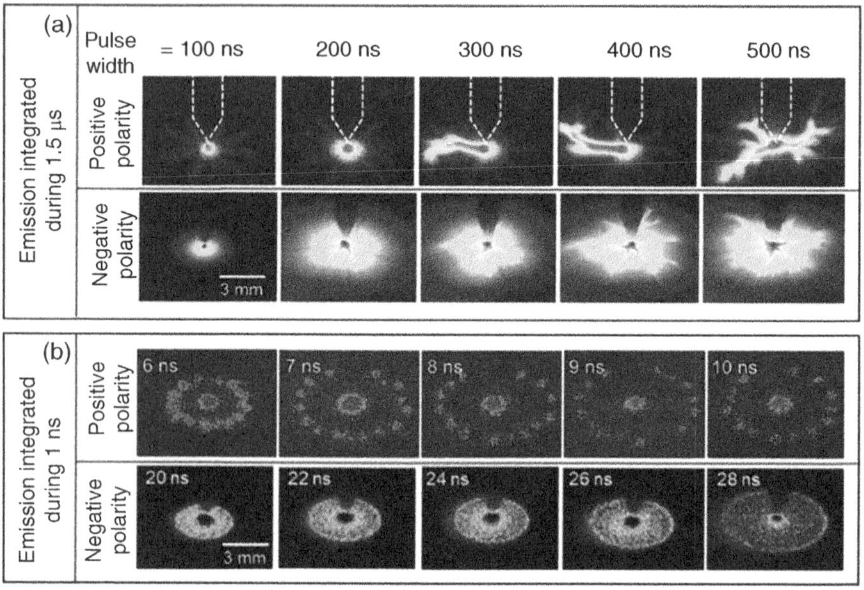


**Figure 12.3** Comparison of the pulsed discharges over the liquid surface emissions generated by positive and negative voltages: (a) 1.5 μs-integrated images at voltage amplitudes of ±20 kV and different pulse widths and (b) temporal evolution of 1 ns-integrated ICCD images recorded at the voltage amplitudes ±20 kV and at a pulse width of 500 ns. *Source:* Hamdan et al. [19] / IOP Publishing / CC BY 4.0.

The distinct behavior of positive and negative streamers at the plasma–liquid interface is considered analogous to surface ionization wave propagation over dielectric surfaces, primarily driven by space charge formation [21]. This space charge formation is also believed to play a role in the intriguing phenomenon of self-organized discharge patterns on the water surface, which are significantly influenced by the gas's electronegativity, as well as the water's conductivity and temperature [22].

While alternating current (AC) power sources are commonly used in atmospheric pressure systems, these findings suggest that tailoring the polarity of applied pulses can selectively enhance and optimize target product generation depending on the application.

Direct plasma discharge in liquids, as shown in Figure 12.1e, occurs when a high-voltage electrode is immersed in water or another liquid medium. While it is often associated with bubble formation or preexisting voids, some findings suggest that breakdown can also occur without a phase change [23]. Marinov et al. [24] discuss several related mechanisms including evaporation at the discharge channel tip due to Joule heating [25], phase transitions induced by a high electric field [26], cracking caused by dielectrophoretic forces [27], and electrostrictive forces [28]. Both dielectrophoretic and electrostrictive forces are linked to the polarization of dielectric materials. In water, the strain caused by the displacement of polarized molecules under a strong electric field is amplified by the significant electronegativity difference between hydrogen and oxygen atoms. This suggests that electromechanical forces may play a more prominent role in initiating discharge in polar media like water, where the resulting strain is substantially stronger compared to non-polar hydrocarbon-based liquid media.

Despite strong scattering effects that hinder electron multiplication in the liquid phase, the observed breakdown strength in liquids is lower than predicted by extrapolation from the pressure domain of gas discharges. This discrepancy is attributed to the absence of inelastic electron energy losses in liquids [29, 30]. Marinov et al. estimated a peak electron density of $1.3 \times 10^{26}\ m^{-3}$ with the maximal shock velocity $4.2\ km\ s^{-1}$ at 15 kV, corresponding to an initial pressure of 5.8 GPa. These values were derived from analyses of Stark and van der Waals broadening of $H_\alpha$ emission [24].

It is widely acknowledged that factors such as the temporal profile of the applied voltage pulse (e.g. rise time and duration), the shape and size of the electrodes, and various physicochemical properties of the liquid medium (e.g. conductivity, viscosity, heat capacity, temperature, and impurities) influence the initiation and evolution of the discharge [31].

The plasma–liquid discharge through bubbles (Figure 12.1d) and aerosol (Figure 12.1f) are grouped as multiphase discharges, with the former as gas-phase plasma dispersed in the gas phase (bubbles) in liquid and the latter as gas-phase plasmas dispersed in a liquid phase (aerosols).

The fundamental breakdown and time evolution of the discharge may have similar profiles to the plasma bulk liquid interaction as in the configuration of Figure 12.1a–c. Instead, the differences may lie in the dynamic transport properties of plasma species in this multiphase discharge, considering the transient nature of the plasma–liquid interface in the form of either bubbles or droplets.

#### 12.3.1.2 Solvation of Plasma Species Through Plasma–Liquid Interface

Since the final reactions of plasma–liquid systems occur in the bulk liquid, the supply rate of individual species in the aqueous phase is crucial. It is important to note that this input rate of aqueous-phase species can differ from their corresponding gas-phase species production rates due to the different solvation properties. Therefore, it is important to understand the solvation of plasma species as they traverse the plasma–liquid interface.

The solvation of electrons, ions, and reactive species may play a pivotal role in $N_2$ fixation through plasma–liquid discharge systems, where the solvated electrons are known to be the smallest primary reducing agents and highly reactive intermediates in various fields ranging from environmental chemistry to radiation biology and nuclear waste reprocessing [32–34].

Solvated electrons are characterized by fast solvation dynamics (within femtoseconds) and short penetration depths (10–20 nm for electron energies of 1–10 eV) [33, 35–38]. While some electrons exist in the vapor phase, the majority become solvated in water, as confirmed by *ab initio* molecular dynamics (MD) simulations and plasma–bubble modeling [39]. These solvated electrons drive key chemical reactions at the plasma–liquid interface, often involving multi-body interactions that differ significantly from typical gas-phase reactions.

Ions are also solvated rapidly (in 10–100 ps) and contribute to plasma–liquid chemistry through processes such as vibrational relaxation and charge transfer collisions [3]. Although low-temperature plasma systems generally provide ions with energies of only a few electron volts, simulations have demonstrated that energetic ions, such as atomic oxygen ions, can sputter water molecules and initiate interfacial reactions [40]. These interactions, along with the bidirectional transport of reactive species between liquid and gas phases, highlight the unique and complex nature of plasma–liquid interfaces. Despite the limited penetration depths and lifetimes of electrons and ions, their impact on driving chemical reactions in the bulk water and interfacial regions remains significant.

In plasma–liquid interfaces, high-density neutral yet reactive species from the gas-phase plasma play a crucial role in driving chemical reactions. These interactions resemble plasma–solid surface processes, involving scattering, adsorption, absorption, desorption, and Langmuir-Hinshelwood or Eley-Rideal mechanisms [41, 42].

MD simulations reveal how radicals and molecules penetrate the liquid interface, inducing chemical transformations. Notably, aqueous-phase reactions often differ significantly from gas-phase reactions due to the polarity of water and hydrogen bonding [43, 44] as highlighted in studies on hydroxyl radical (HO•) reactions [45]. Factors such as Henry's law constants and diffusion coefficients help explain the solubility and mass transfer of plasma-generated species into liquids [46, 47].

Here it may be worth noting that plasma–liquid interactions through dynamic bubbles and aerosols may enhance the interacting area and volume and be beneficial in increasing the mass transfer efficiency into the aqueous phase, as demonstrated with improved $N_2$ fixation rates [9, 12].

### 12.3.2 Chemical Aspect of Plasma–Liquid Interactions

While the production of $NO_x$, $NH_3$, and their related radicals in gas-phase plasma is relatively well understood [48–50], the subsequent chemical reactions involving these species in the aqueous phase or at the plasma–liquid interface remain less explored. Understanding these reactions is important for fully comprehending the chemistry of plasma–liquid interaction systems.

**Table 12.2** Species list involved in air/$N_2$ plasma–liquid system.

| | | |
|---|---|---|
| Gas phase | Positive ions | $H^+$, $H_2^+$, $H_2O^+$, $H_3O^+$, $NO^+$, $NO_2^+$, $NH^+$, $N_2H^+$, $NH_2^+$, $NH_3^+$, $NH_4^+$<br>$O^+$, $O_2^+$, $OH^+$ |
| | Negative ions | $H^-$, $NO^-$, $NO_2^-$, $O^-$, $O_2^-$, $O_3^-$, $O_4^-$, $OH^-$, $HO_2^-$ |
| | Neutrals | H, $H_2$, $H_2O$, $H_2O_2$, $HO_2$, N, $N_2(\nu)$, $N_2(A3)$, $N_2(B3)$, $N_2(a'1)$, $N_2(C3)$, $O_2(\nu)$, $O_2(a1)$ $O_2(b1)$, $H_2(\nu)$, $H_2(B3)$, $H_2(B1)$, $H_2(C3)$, $H_2(A3)$, $H_2(R)$, $H_2O(\nu)$, $N_2O$, $N_2O_5$, NO, $NO_2$, $NO_3$, $HNO_2$, $HNO_3$, O, O(1D), $O_2(a1)$, $O_3$, OH, NH, $NH_2$, $NH_3$ |
| Liquid phase | Positive ions | $H_2O^+$, $H_3O^+$, $NH_4^+$ |
| | Negative ions | $OH^-$, $H^-$, $NO^-$, $NO_2^-$, $NO_3^-$, $ONOO^-$, $O_2NOO^-$, $O^-$, $O_2^-$, $O_3^-$, $HO_2^-$, $H_2O^-$ |
| | Neutrals | H, $H_2$, $H_2O_2$, N, $N_2O$, $N_2O_5$, NO, $NO_2$, $NO_3$, $HNO_2$, $HNO_3$, ONOOH, $O_2NOOH$, O, $O_3$, OH, $HO_3$, NH, HNO |

Table 12.2 shows the possible gas- and liquid-phase species in air/$N_2$-based plasma systems, including the presence of $H_2O$ molecules, in addition to electrons in the gas phase or solvated electrons in water. A list of common positive and negative ions, along with electronically and vibrationally excited molecular and atomic species related to nitrogen, oxygen, and hydrogen, included in an air-like plasma discharge system is provided.

#### 12.3.2.1 Effect of Water Content in Gas-Phase Plasma Discharge

Introducing water into the plasma discharge not only facilitates the dissolution of $NH_3$ and $NO_x$ into the aqueous solution but also impacts the plasma properties as evaporated $H_2O$ enters the gas-phase plasma region. This affects key parameters such as electron density, electric field strength, and the densities of RONS. By various electron interactions with $H_2O$ molecule as shown in Eqs. (12.1)–(12.5), atomic H, O species as well as OH can be produced at high rates, which can boost further chemical reactions.

$$e + H_2O \rightarrow e + H + OH \tag{12.1}$$

$$e + H_2O \rightarrow O^- + H_2 \tag{12.2}$$

$$e + H_2O \rightarrow e + O + 2H \tag{12.3}$$

$$e + H_2O \rightarrow 2e + H_2O^+ \tag{12.4}$$

$$e + H_2O \rightarrow 2e + H^+ + OH \tag{12.5}$$

In an $N_2/H_2O$ plasma system aimed for $NH_3$ production, increasing the $H_2O$ content (%) in the feed gas can enhance $NH_3$ yield by providing additional hydrogen atoms as precursors [9, 13]. However, further increasing the $H_2O$ content may eventually reduce ammonia density. This decrease is likely due to the quenching of vibrationally excited $N_2(\nu)$ molecules through vibrational–translational (V–T) relaxation, which leads to a lower electron energy gain, reduced power density, and a decrease in atomic nitrogen density [49]. Additionally, other studies suggest that the presence of water can cause energy losses due to water evaporation and plasma thermalization, reducing the overall nitrogen fixation efficiency in air plasma–liquid discharge systems [10].

It is worth discussing the competitive oxidative environment created by coexisting species like O and OH, the decomposition of NH, $NH_2$, and $NH_3$ can also occur through subsequent reactions listed as follows [49]:

$$OH + NH_x \rightarrow H_2O + NH_{x-1} \tag{12.6}$$

$$OH + NH_2 \rightarrow H_2 + HNO \tag{12.7}$$

$$O + NH_x \rightarrow OH + NH_{x-1} \tag{12.8}$$

$$O + NH_2 \rightarrow H + HNO \tag{12.9}$$

$$O + NH_2 \rightarrow H_2 + NO \tag{12.10}$$

$$O + NH \rightarrow H + NO \tag{12.11}$$

$$O + NH \rightarrow OH + N \tag{12.12}$$

#### 12.3.2.2 Production and Loss of Short-lived Species in Liquid

The important aqueous hydroxyl ($OH_{(aq)}$) species are initially supplied through the solvation of gas-phase OH radicals, which are produced by the dissociation of water molecules via electrons ($e^-$), ultraviolet (UV) radiation, or metastable molecules such as $Ar^*$ or $N_2^*$ [2, 51]. These hydroxyl species, with a high oxidation potential ($E^0 = 2.85$ V), play a crucial role in reactions that form $NO_x$, both in the aqueous phase and at the plasma–liquid interface [52, 53]. However, in water, a further increase in $OH_{(aq)}$ concentration is possible through the hydronium formation reaction (12.13) shown below, where the availability of $H_2O^+$ ions increases due to rapid charge transfer (12.14) from positive ions at the interface. Additionally, aqueous-phase $O_{(aq)}$ is also recognized as an important reactant for producing $OH_{(aq)}$ in water, as shown in reaction (12.15) [47].

$$H_2O^+_{(aq)} + H_2O_{(aq)} \rightarrow H_3O^+_{(aq)} + OH_{(aq)} \tag{12.13}$$

$$M^+\left(N_2^+, O_2^+, H_2^+ \text{ etc.}\right) + H_2O_{(aq)} \rightarrow H_2O^+_{(aq)} + M \tag{12.14}$$

$$O_{(aq)} + H_2O_{(aq)} \rightarrow OH_{(aq)} + OH_{(aq)} \tag{12.15}$$

These $OH_{(aq)}$ species can be rapidly consumed to form $H_2O_{2(aq)}$ or $HO_{2(aq)}$, as shown in Eqs. (12.16) and (12.17). They also play a role in $N_2$ fixation, as demonstrated in Eqs. (12.18) and (12.19), by facilitating the conversion of dissolved $NO_{2(aq)}$ into $HNO_{3(aq)}$. This process ultimately enhances the concentration of nitrate ($NO_{3(aq)}^{-}$), which is one of the most important $N_2$ fixation products in water.

$$OH_{(aq)} + OH_{(aq)} \rightarrow H_2O_{2(aq)} \quad (12.16)$$

$$OH_{(aq)} + O_{3(aq)}^{-} \rightarrow HO_{2(aq)} + O_{2(aq)}^{-} \quad (12.17)$$

$$OH_{(aq)} + NO_{2(aq)} \rightarrow HNO_{3(aq)} \quad (12.18)$$

The hydronium ion ($H_3O_{(aq)}^{+}$) is also an important aqueous species in water as well as $H_{(aq)}$ and $OH_{(aq)}$, contributing to the production of $HO_{2(aq)}$ or $HNO_{2(aq)}$, which can then readily convert into $NO_{2(aq)}^{-}$.

$$H_3O_{(aq)}^{+} + O_{2(aq)}^{-} \rightarrow HO_{2(aq)} + H_2O_{(aq)} \quad (12.19)$$

$$H_{(aq)} + NO_{2(aq)} \rightarrow HNO_{2(aq)} \quad (12.20)$$

$$OH_{(aq)} + NO_{(aq)} \rightarrow HNO_{2(aq)} \quad (12.21)$$

#### 12.3.2.3 Important Pathway of Long-lived Species Formation in Liquid

The primary benefit of introducing water into an air or $N_2$ plasma system is that it acts as both a reactant and a solvent, storing generated fixed nitrogen species in the form of stable aqueous nitrate ($NO_{3(aq)}^{-}$), nitrite ($NO_{2(aq)}^{-}$), and ammonium ($NH_{4(aq)}^{+}$). Plasma–liquid systems for nitrogen fixation primarily aim to produce $NO_x$ or $NH_3$ in an environmentally friendly manner, using air or $N_2$ plasma with water. Nitrogen ($N_2$) is commonly selected for this purpose, while water ($H_2O_{(aq)}$) serves as a proton source and facilitates the plasma–liquid discharge. The basic reactions between $NO_{x(aq)}$ species and water under standard temperature and pressure (NTP) conditions are suggested as follows:

$$2NO_{2(aq)} + 3H_2O_{(aq)} \rightarrow NO_{3(aq)}^{-} + NO_{2(aq)}^{-} + 2H_3O_{(aq)}^{+} \quad (12.22)$$

$$HNO_{3(aq)} + H_2O_{(aq)} \rightarrow NO_{3(aq)}^{-} + H_3O_{(aq)}^{+} \quad (12.23)$$

$$NO_{2(aq)}^{-} + O_{3(aq)} \rightarrow NO_{3(aq)}^{-} + O_{2(aq)} \quad (12.24)$$

$$HNO_{2(aq)} + H_2O_{(aq)} \rightarrow NO_{2(aq)}^{-} + H_3O_{(aq)}^{+} \quad (12.25)$$

Dissolved $NO_{2(aq)}$ appears to be the key species supporting a high production rate of nitrate ($NO_{3(aq)}^{-}$) and nitrite ($NO_{2(aq)}^{-}$). $NO_{3(aq)}^{-}$ can form through interactions of $HNO_{3(aq)}$ or $NO_{2(aq)}^{-}$ with $H_2O_{(aq)}$ or $O_{3(aq)}$, respectively, as shown in Eqs. (12.23) and (12.24). For efficient $NO_{2(aq)}^{-}$ production, as shown in reactions (12.21) and (12.25), a high-density NO supply from the gas phase is crucial. This ensures a high concentration of $NO_{(aq)}$, which, in the presence of $OH_{(aq)}$, leads to

$HNO_{2(aq)}$ formation as shown in Eq. (12.21). $HNO_{2(aq)}$ then reacts further to produce $NO_{2(aq)}^-$ as described in Eq. (12.25). Therefore, an abundant NO supply is essential to enhance $NO_{2(aq)}^-$ formation in water.

An energy-efficient conversion of $NH_4^+$ has been reported through electrochemical process [7]:

$$NO_{3(aq)}^- + 10H_{(aq)}^+ + 8e^- \rightarrow NH_{4(aq)}^+ + 3H_2O_{(aq)} \tag{12.26}$$

$$NO_{2(aq)}^- + 8H_{(aq)}^+ + 6e^- \rightarrow NH_{4(aq)}^+ + 2H_2O_{(aq)} \tag{12.27}$$

It is worth commenting that these suggested mechanisms may strongly be dependent on specific plasma conditions and the electrochemical environment that follow.

### 12.3.3 Mass Transport Through the Plasma–Liquid Interface

Understanding the mass transport between the plasma and bulk liquid across the plasma–liquid interface, typically ranging from a few to dozens of nanometers, remains one of the key challenges in plasma–liquid applications due to the limited data available on the underlying mechanisms. This interface plays a critical role in governing the subsequent liquid-phase chemistry and the formation of various long-lived reactive species. As illustrated in Figure 12.4, the gas-phase discharge involves complex physico-chemical reactions, where $H_2O$ molecules participate through evaporation or sputtering. Various plasma-generated species, such as

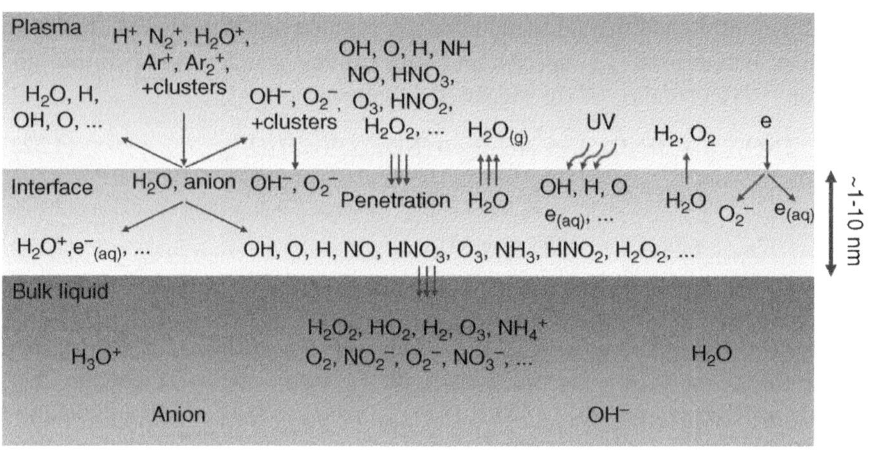


**Figure 12.4** Schematic diagram of air plasma-generated species and its mechanism in contact with water. *Source:* Bruggeman et al. [3] / IOP Publishing.

electrons, ions, and neutral radicals, interact with the liquid interface and are transported or dissolved according to their charge, diffusion properties, and solvation characteristics. Understanding how these gas-phase species transition into the aqueous phase under large local electric field environment or convective gas flow environment and contribute to further liquid-phase reactions is complex but essential for controlling plasma–liquid processes.

The plasma–liquid interface is of fundamental importance, as it serves as the primary entry point for plasma-generated species into the liquid phase, initiating secondary reactions that lead to the formation of long-lived species. The physicochemical properties of the liquid – such as pH, temperature, and conductivity – significantly influence these interactions [3]. Furthermore, chemical analysis of long-lived products in the liquid phase provides valuable insights into key reaction pathways, helping to trace the formation of target molecules. In this regard, a combination of optical and electrical plasma diagnostics, surface analysis, and modeling efforts is crucial for developing a comprehensive understanding of the intricate plasma–liquid interaction mechanisms.

## 12.4 Key Challenges

### 12.4.1 Diagnostics

Understanding reactive species in the liquid phase and at the plasma–liquid interface is crucial. However, due to the complexity of plasma reactions, the short lifetimes of many reactive species, and the challenges associated with the plasma–liquid interface, it remains difficult to directly characterize the role of these species or quantify their impact within the system. Several methods are available to study these hard-to-detect, short-lived species in liquids.

One approach involves the use of scavenger assays, which are commonly employed in plasma–liquid interaction studies to quantify the concentration of plasma-generated species and assess their specific contributions. Different types of scavengers, selected for their high reaction rates with targeted species and much lower rates with others, are used to identify the roles and impacts of specific species (Table 12.3). Controlled experiments with these scavengers allow researchers to compare results and observe the effects of particular species on reactions.

However, certain precautions are necessary when conducting scavenger experiments. The addition of scavenger compounds can alter solution properties, such as conductivity ($\sigma$), surface tension ($\gamma$), and dielectric constant ($\kappa$), potentially affecting plasma discharge characteristics and leading to variations in experimental outcomes. Another key factor is the concentration of scavengers, which influences the overall reactivity. While scavengers are highly reactive and selective for certain

**Table 12.3** Commonly used scavengers list.

| Scavengers | Target species | Reaction rates ($M^{-1}$ $s^{-1}$) | Reference |
|---|---|---|---|
| Isopropanol (IPA) | $\cdot OH$ | $1.9 \times 10^9$ | [54] |
| | $\cdot H$ | $7.4 \times 10^7$ | [54] |
| Benzoquinone (BQ) | $\cdot O_2^-$ | $1.1 \times 10^9$ | [55] |
| Sodium pyruvate (SP) | $H_2O_2$ | | [56] |
| Uranic acid (UA) | $O_3$ | $1.4 \times 10^6$ | [57] |
| | $ONOO^-$ | $1.55 \times 10^2$ | [58] |
| | $\cdot OH$ | $7.2 \times 10^6$ | [59] |
| Dimethyl sulfoxide (DMSO) | $\cdot OH$ | $6.6 \times 10^9$ | [54] |
| Mannitol | $\cdot OH$ | $2.7 \times 10^9$ | [60] |
| | $^1O_2$ | $< 10^3$ | [60] |
| Methanol (MeOH) | $\cdot OH$ | $9.7 \times 10^8$ | [54] |
| | $\cdot H$ | $2.6 \times 10^6$ | [54] |
| | $\cdot O^-$ | $7.5 \times 10^8$ | [54] |
| Ethanol (EtOH) | $\cdot OH$ | $1.9 \times 10^9$ | [54] |
| | $\cdot H$ | $1.7 \times 10^7$ | [54] |
| | $\cdot O^-$ | $1.2 \times 10^9$ | [54] |
| Carboxy-PTIO | $NO\cdot$ | $1.0 \times 10^4$ | [61] |
| Tetranitromethane (TNM) | $\cdot H$ | $5.5 \times 10^8$ | [54] |
| | $\cdot O_2^-$ | $2 \times 10^9$ | [62] |
| Sodium azide ($NaN_3$) | $^1O_2$ | $2.2 \times 10^9$ | [60] |
| | $\cdot OH$ | $1.1 \times 10^{10}$ | [60] |

radicals, the complex interconversion among radicals and reactive species means that scavenger results may not directly represent the overall importance of the scavenged species in plasma reactions.

Electron spin resonance (ESR), also known as electron paramagnetic resonance (EPR), is another technique used to detect short-lived species in liquid systems. ESR is a magnetic resonance method based on the interaction of an external magnetic field with unpaired electron spins. Due to the short lifetimes of RONS, a spin-trap reagent is often added to stabilize these highly reactive species, producing more stable by-products that can be detected by ESR over practical timescales (minutes to hours) (Table 12.4). Additionally, isotopically labeled gases or liquids can be used in ESR experiments to trace the origins of RONS generated in plasma–liquid systems [64].

**Table 12.4** Commonly used ESR spin-trap compounds.

| Spin-trapping reagents | Target species |
|---|---|
| 5,5-Dimethyl-1-pyrroline N-oxide (DMPO) | $\cdot OH$, $\cdot O_2^-$, $\cdot H$, $\cdot CH_3$ |
| *N*-tert-Butyl-α-phenylnitrone (PBN) | $\cdot OH$, $ROO\cdot$, $NO_2$ |
| 2,2,6,6-Tetramethylpiperidine (TEMP) | $^1O_2$ |
| 2-Methyl-2-nitrosopropane (MNP) | $R\cdot$ |
| 2-(4-Carboxyphenyl)-4,4,5,5-tetramethylimidazoline-1-oxyl-3-oxide (carboxy-PTIO, c-PTIO) | $NO\cdot$ |

*Source:* Davies [63] / with permission of Elsevier.

### 12.4.2 Modeling

The complex physics and chemistry at the plasma–liquid interface necessitates computational modeling to interpret the experimental results and reveal underlying mechanisms. Experimental and theoretical studies are deeply interdependent, as advanced diagnostics provide essential data to enhance model accuracy. At the molecular level, simulations have been pivotal in understanding solvation processes, including gas species and electron stabilization at the water interface. MD simulations, for example, highlight differences in hydrogen bonding between the bulk water and its interface, which affect the stability of different species [65–67]. However, the limited timescales (approximately 100 ps) and reliance on empirical force fields restrict the ability of MD to capture the full scope of chemical kinetics at the plasma–liquid interface [68].

Advanced modeling approaches, such as kinetic simulations and hydrodynamic models, have yielded detailed insights into plasma–liquid interactions. Kinetic simulations using particle-in-cell/Monte Carlo (PIC/MC) methods, combined with fluid models [35], validated experimental observations of solvated electrons [38], showing their rapid loss within nanometers of the plasma–water interface. Hydrodynamic simulations, on the other hand, have explored plasma–liquid interactions, accounting for gas- and liquid-phase kinetics influenced by factors such as applied voltage, bubble size, and liquid conductivity [69–73]. Figure 12.5 illustrates an example of the surface-hugging time evolution of plasma characteristics sustained in the bubble, showing the simulated electron density and energy profiles under different gas composition conditions [73].

Lietz and Kushner [47] conducted a computational study on plasma–liquid interactions and liquid-phase chemistry using a global model. For the given

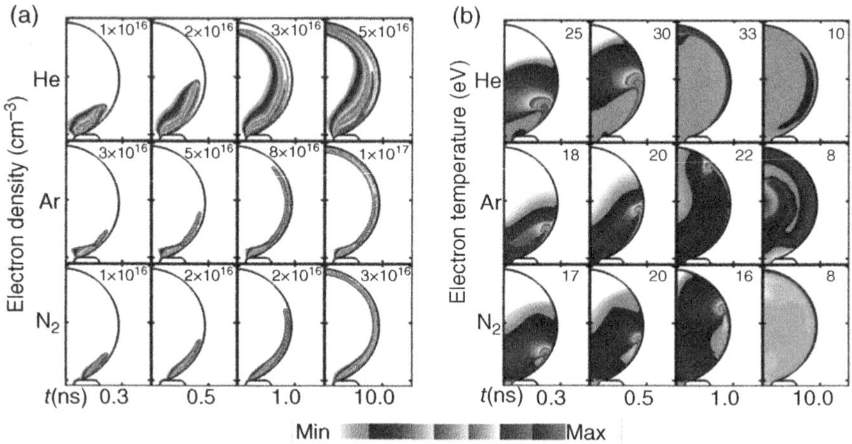


**Figure 12.5** Time evolution of plasma properties for discharges in He, Ar, and $N_2$ bubbles. (a) Electron density, (b) E/N. The discharges are surface hugging where the electric field enhancement is the largest. The contours are plotted on a log scale over three decades with maximum values for He shown in each frame. *Source:* Tian et al. [73] / IOP Publishing.

DBD plasma conditions, $O_3$ emerged as the dominant species after the plasma was turned off, while $N_2O$ persisted at a high concentration of $10^{15}\,cm^{-3}$ in the gas phase. In the liquid phase, the concentration of $NO_3^{-}{}_{(aq)}$ increased significantly, driven by reactions that generated $HNO_{x(aq)}$ from $N_xO_y$ species in the gas discharge, followed by dissociation into $NO_{x(aq)}^{-}$ and $H_3O^{+}{}_{(aq)}$. These products, detectable in liquid water samples, highlighted key reaction pathways (Figure 12.6). Their work aimed to address the wide dynamic range of timescales involved and provided a foundational framework for subsequent plasma chemistry models, particularly those focused on plasma–liquid interactions for nitrogen fixation applications.

Despite significant efforts in plasma modeling, most studies are limited to specific timescales or spatial domains. However, recent work, such as Heirman et al. [74], has extended the modeling timescale beyond plasma exposure, capturing the evolution of plasma-generated species in liquid after the interaction. Expanding hydrodynamic plasma–fluid models to bridge the gas phase and liquid volume – by incorporating diffusion and solvation reactions at the plasma–liquid interface – could further enhance visualization of both the physical and chemical processes across the entire domain. Such advancements would provide deeper insight into complex plasma–liquid interactions, driving progress in related applications [75].

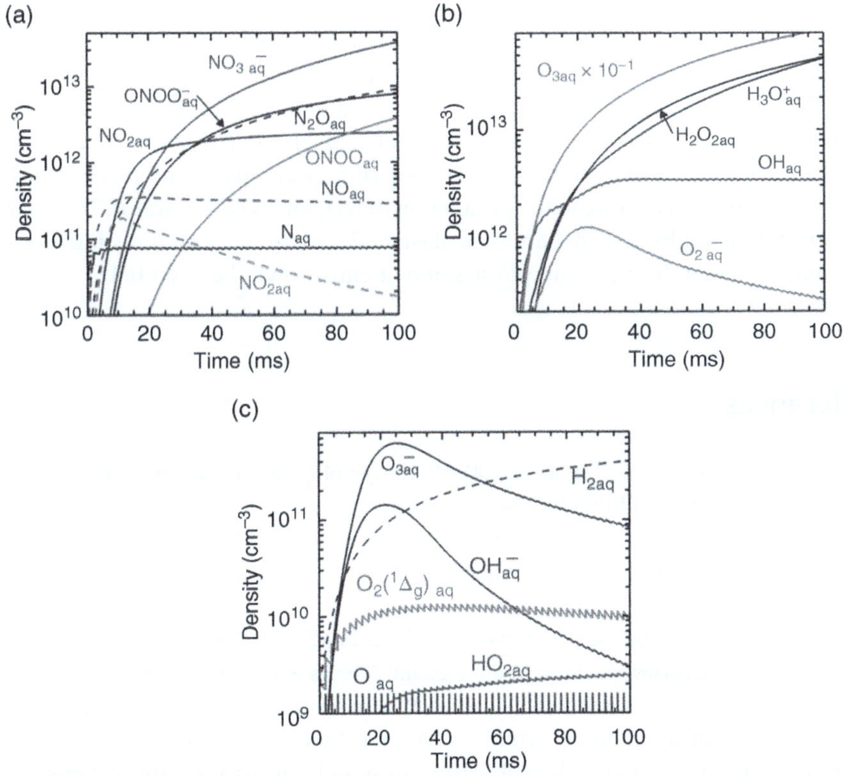


**Figure 12.6** The densities of aqueous (a) RNS and (b) and (c) ROS for the first 50 pulses of the base case. The most reactive species, including many ions and excited states, were omitted as they appear as delta functions on these timescales. Several species establish a pulsed equilibrium, while the others evolve over timescales much longer than the interpulse period. *Source:* Lietz and Kushner [47] / IOP Publishing.

## 12.5 Conclusion

Plasma–liquid interactions have demonstrated significant benefits across various fields, including water treatment, gas conversion, and healthcare, primarily due to the generation of RONS. These interactions are also beneficial for artificial nitrogen fixation applications. The introduction of water molecules in both the gas-phase discharge and liquid media within plasma–liquid systems can significantly alter plasma physics and chemistry, supplying hydrogen, oxygen, and OH radicals, which add complexity to the chemistry. The relaxation of vibrationally excited molecular species by $H_2O$, which can reduce the electron energy in the system, has been discussed as a key factor. To overcome these intrinsic challenges, various configurations have been developed and investigated. Plasma interaction through bubbles or droplets show promising potential, enhancing the transport efficiency of reactive species. Despite the existing challenges in understanding the complex physics and chemistry of this multiphase system, advanced diagnostics and multi-physics modeling can help uncover the underlying mechanisms and physical and chemical pathways, leading to the production of $NO_x$ and $NH_4^+$ in water. To better understand the multi-scale dynamics of plasma–liquid interactions, ongoing inter-disciplinary research in plasma physics and chemistry will be essential.

## References


1 Gan, D., Hong, L., Li, T. et al. (2024). Discharge plasma for prebiotic chemistry: pathways to life's building blocks. *Earth Planet. Phys.* 8 (6): 823–835.

2 Zhou, R., Zhou, R., Wang, P. et al. (2020). Plasma-activated water: generation, origin of reactive species and biological applications. *J. Phys. D. Appl. Phys.* 53 (30): 303001.

3 Bruggeman, P.J., Kushner, M.J., Locke, B.R. et al. (2016). Plasma–liquid interactions: a review and roadmap. *Plasma Sources Sci. Technol.* 25 (5): 053002.

4 Huang, Z., Xiao, A., Liu, D. et al. (2022). Plasma-water-based nitrogen fixation: status, mechanisms, and opportunities. *Plasma Process. Polym.* 19 (4): 2100198.

5 Peng, P., Chen, P., Addy, M. et al. (2018). In situ plasma-assisted atmospheric nitrogen fixation using water and spray-type jet plasma. *Chem. Commun.* 54 (23): 2886–2889. https://doi.org/10.1039/C8CC00697K.

6 Peng, P., Schiappacasse, C., Zhou, N. et al. (2019). Plasma in situ gas–liquid nitrogen fixation using concentrated high-intensity electric field. *J. Phys. D. Appl. Phys.* 52 (49): 494001.

7 Sun, J., Alam, D., Daiyan, R. et al. (2021). A hybrid plasma electrocatalytic process for sustainable ammonia production. *Energy Environ. Sci.* 14 (2): 865–872. https://doi.org/10.1039/D0EE03769A.

8 Toth, J.R., Abuyazid, N.H., Lacks, D.J. et al. (2020). A plasma-water droplet reactor for process-intensified, continuous nitrogen fixation at atmospheric pressure. *ACS Sustain. Chem. Eng.* 8 (39): 14845–14854. https://doi.org/10.1021/acssuschemeng.0c04432.

9 Zhang, T., Zhou, R., Zhang, S. et al. (2023). Sustainable ammonia synthesis from nitrogen and water by one-step plasma catalysis. *Energy Environ. Mater.* 6 (2): e12344.

10 Gromov, M., Kamarinopoulou, N., De Geyter, N. et al. (2022). Plasma-assisted nitrogen fixation: the effect of water presence. *Green Chem.* 24 (24): 9677–9689.

11 Sremački, I., Gromov, M., Leys, C. et al. (2020). An atmospheric pressure non-self-sustained glow discharge in between metal/metal and metal/liquid electrodes. *Plasma Process. Polym.* 17 (6): 1900191. https://doi.org/10.1002/ppap.201900191.

12 Gao, H., Wang, G., Huang, Z. et al. (2023). Plasma-activated mist: continuous-flow, scalable nitrogen fixation, and aeroponics. *ACS Sustain. Chem. Eng.* 11 (11): 4420–4429.

13 Gorbanev, Y., Vervloessem, E., Nikiforov, A., and Bogaerts, A. (2020). Nitrogen fixation with water vapor by nonequilibrium plasma: toward sustainable ammonia production. *ACS Sustain. Chem. Eng.* 8 (7): 2996–3004. https://doi.org/10.1021/acssuschemeng.9b07849.

14 Hawtof, R., Ghosh, S., Guarr, E. et al. (2019). Catalyst-free, highly selective synthesis of ammonia from nitrogen and water by a plasma electrolytic system. *Sci. Adv.* 5 (1): eaat5778.

15 Sun, J., Zhou, R., Hong, J. et al. (2024). Sustainable ammonia production via nanosecond-pulsed plasma oxidation and electrocatalytic reduction. *Appl. Catal. B Environ.* 342: 123426.

16 Roy, N.C., Maira, N., Pattyn, C. et al. (2023). Mechanisms of reducing energy costs for nitrogen fixation using air-based atmospheric DBD plasmas over water in contact with the electrode. *Chem. Eng. J.* 461: 141844.

17 Liu, Z., Tian, Y., Niu, G. et al. (2021). Direct oxidative nitrogen fixation from air and $H_2O$ by a water falling film dielectric barrier discharge reactor at ambient pressure and temperature. *ChemSusChem* 14 (6): 1507–1511.

18 Keniley, S., Uner, N.B., Perez, E. et al. (2022). Multiphase modeling of the DC plasma–water interface: application to hydrogen peroxide generation with experimental validation. *Plasma Sources Sci. Technol.* 31 (7): 075001.

19 Hamdan, A., Diamond, J., and Herrmann, A. (2021). Dynamics of a pulsed negative nanosecond discharge on water surface and comparison with the positive discharge. *J. Phys. Commun.* 5 (3): 035005. https://doi.org/10.1088/2399-6528/abe953.

20 Liu, Z., Xu, D., Zhou, C. et al. (2018). Effects of the pulse polarity on helium plasma jets: discharge characteristics, key reactive species, and inactivation of myeloma cell. *Plasma Chem. Plasma Process.* 38 (5): 953–968. https://doi.org/10.1007/s11090-018-9920-4.

21 Nijdam, S., Teunissen, J., and Ebert, U. (2020). The physics of streamer discharge phenomena. *Plasma Sources Sci. Technol.* 29 (10): 103001. https://doi.org/10.1088/1361-6595/abaa05.

22 Shirai, N., Uchida, S., and Tochikubo, F. (2014). Influence of oxygen gas on characteristics of self-organized luminous pattern formation observed in an atmospheric dc glow discharge using a liquid electrode. *Plasma Sources Sci. Technol.* 23 (5): 054010. https://doi.org/10.1088/0963-0252/23/5/054010.

23 Starikovskiy, A., Yang, Y., Cho, Y.I., and Fridman, A. (2011). Non-equilibrium plasma in liquid water: dynamics of generation and quenching. *Plasma Sources Sci. Technol.* 20 (2): 024003. https://doi.org/10.1088/0963-0252/20/2/024003.

24 Marinov, I., Starikovskaia, S., and Rousseau, A. (2014). Dynamics of plasma evolution in a nanosecond underwater discharge. *J. Phys. D. Appl. Phys.* 47 (22): 224017. https://doi.org/10.1088/0022-3727/47/22/224017.

25 Lisitsyn, I.V., Nomlyama, H., Katsuki, S., and Akiyama, H. (1999). Thermal processes in a streamer discharge in water. *IEEE Trans. Dielectr. Electr. Insul.* 6 (3): 351–356. https://doi.org/10.1109/94.775622.

26 Kuskova, N.I. (2001). Spark discharges in condensed media. *Tech. Phys.* 46 (2): 182–185. https://doi.org/10.1134/1.1349273.

27 Lewis, T.J. (1998). A new model for the primary process of electrical breakdown in liquids. *IEEE Trans. Dielectr. Electr. Insul.* 5: 306–315.

28 Shneider, M.N. and Pekker, M. (2013). Dielectric fluid in inhomogeneous pulsed electric field. *Phys. Rev. E* 87 (4): 043004. https://doi.org/10.1103/PhysRevE.87.043004.

29 Bonifaci, N., Denat, A., and Atrazhev, V. (1997). Ionization phenomenon in high-density gaseous and liquid argon in corona discharge experiments. *J. Phys. D. Appl. Phys.* 30 (19): 2717. https://doi.org/10.1088/0022-3727/30/19/010.

30 Atrazhev, V.M., Dmitriev, E.G., and Iakubov, I.T. (1991). The impact ionization and electrical breakdown strength for atomic and molecular liquids. *IEEE Trans. Electr. Insul.* 26 (4): 586–591. https://doi.org/10.1109/14.83675.

31 Vanraes, P. and Bogaerts, A. (2018). Plasma physics of liquids—a focused review. *Appl. Phys. Rev.* 5 (3): 031103. https://doi.org/10.1063/1.5020511.

32 Garrett, B.C., Dixon, D.A., Camaioni, D.M. et al. (2005). Role of water in electron-initiated processes and radical chemistry: issues and scientific advances. *Chem. Rev.* 105 (1): 355–390. https://doi.org/10.1021/cr030453x.

33 Sagar, D.M., Bain, C.D., and Verlet, J.R.R. (2010). Hydrated electrons at the water/air interface. *J. Am. Chem. Soc.* 132 (20): 6917–6919. https://doi.org/10.1021/ja101176r.

34 Turi, L. and Rossky, P.J. (2012). Theoretical studies of spectroscopy and dynamics of hydrated electrons. *Chem. Rev.* 112 (11): 5641–5674. https://doi.org/10.1021/cr300144z.

35 Gopalakrishnan, R., Kawamura, E., Lichtenberg, A.J. et al. (2016). Solvated electrons at the atmospheric pressure plasma–water anodic interface. *J. Phys. D. Appl. Phys.* 49 (29): 295205. https://doi.org/10.1088/0022-3727/49/29/295205.

36 Meesungnoen, J., Jay-Gerin, J.P., Filali-Mouhim, A., and Mankhetkorn, S. (2002). Low-energy electron penetration range in liquid water. *Radiat. Res.* 158 (5): 657–660. https://doi.org/10.1667/0033-7587(2002)158[0657:leepri]2.0.co;2.

37 Ottosson, N., Faubel, M., Bradforth, S.E. et al. (2010). Photoelectron spectroscopy of liquid water and aqueous solution: electron effective attenuation lengths and emission-angle anisotropy. *J. Electron Spectrosc. Relat. Phenom.* 177 (2): 60–70. https://doi.org/10.1016/j.elspec.2009.08.007.

38 Rumbach, P., Bartels, D.M., Sankaran, R.M., and Go, D.B. (2015). The solvation of electrons by an atmospheric-pressure plasma. *Nat. Commun.* 6 (1): 7248. https://doi.org/10.1038/ncomms8248.

**39** Uhlig, F., Marsalek, O., and Jungwirth, P. (2013). Electron at the surface of water: dehydrated or not? *J. Phys. Chem. Lett.* 4 (2): 338–343. https://doi.org/10.1021/jz3020953.

**40** Minagawa, Y., Shirai, N., Uchida, S., and Tochikubo, F. (2014). Analysis of effect of ion irradiation to liquid surface on water molecule kinetics by classical molecular dynamics simulation. *Jpn. J. Appl. Phys.* 53 (1): 010210. https://doi.org/10.7567/JJAP.53.010210.

**41** Chang, C.-H. and Franses, E.I. (1995). Adsorption dynamics of surfactants at the air/water interface: a critical review of mathematical models, data, and mechanisms. *Colloids Surf. A Physicochem. Eng. Asp.* 100: 1–45.

**42** Chang, C.H. and Franses, E.I. (1992). Modified Langmuir—Hinselwood kinetics for dynamic adsorption of surfactants at the air/water interface. *Colloids Surf.* 69 (2): 189–201. https://doi.org/10.1016/0166-6622(92)80230-Y.

**43** Murdachaew, G., Varner, M.E., Phillips, L.F. et al. (2013). Nitrogen dioxide at the air-water interface: trapping, absorption, and solvation in the bulk and at the surface. *Phys. Chem. Chem. Phys.* 15 (1): 204–212. https://doi.org/10.1039/c2cp42810e.

**44** Yusupov, M., Neyts, E.C., Simon, P. et al. (2014). Reactive molecular dynamics simulations of oxygen species in a liquid water layer of interest for plasma medicine. *J. Phys. D. Appl. Phys.* 47 (2): 025205. https://doi.org/10.1088/0022-3727/47/2/025205.

**45** Minakata, D., Li, K., Westerhoff, P., and Crittenden, J. (2009). Development of a group contribution method to predict aqueous phase hydroxyl radical ($HO^*$) reaction rate constants. *Environ. Sci. Technol.* 43 (16): 6220–6227. https://doi.org/10.1021/es900956c.

**46** Sander, R. (2023). Compilation of Henry's law constants (version 5.0.0) for water as solvent. *Atmos. Chem. Phys.* 23 (19): 10901–12440. https://doi.org/10.5194/acp-23-10901-2023.

**47** Lietz, A.M. and Kushner, M.J. (2016). Air plasma treatment of liquid covered tissue: long timescale chemistry. *J. Phys. D. Appl. Phys.* 49 (42): 425204. https://doi.org/10.1088/0022-3727/49/42/425204.

**48** Hong, J., Pancheshnyi, S., Tam, E. et al. (2017). Kinetic modelling of $NH_3$ production in $N_2$–$H_2$ non-equilibrium atmospheric-pressure plasma catalysis. *J. Phys. D. Appl. Phys.* 50 (15): 154005. https://doi.org/10.1088/1361-6463/aa6229.

**49** Hong, J., Zhang, T., Zhou, R. et al. (2022). Green chemical pathway of plasma synthesis of ammonia from nitrogen and water: a comparative kinetic study with a $N_2/H_2$ system. *Green Chem.* 24 (19): 7458–7468.

**50** Wang, W., Patil, B., Heijkers, S. et al. (2017). Nitrogen fixation by gliding arc plasma: better insight by chemical kinetics modelling. *ChemSusChem* 10 (10): 2145–2157.

**51** Nikiforov, A.Y., Sarani, A., and Leys, C. (2011). The influence of water vapor content on electrical and spectral properties of an atmospheric pressure plasma jet. *Plasma Sources Sci. Technol.* 20 (1): 015014.

**52** Gligorovski, S., Strekowski, R., Barbati, S., and Vione, D. (2015). Environmental implications of hydroxyl radicals (• OH). *Chem. Rev.* 115 (24): 13051–13092.

**53** Zhao, X. and Tian, Y. (2023). Sustainable nitrogen fixation by plasma-liquid interactions. *Cell Report Phys. Sci.* 4 (10).

**54** Buxton, G.V., Greenstock, C.L., Phillips Helman, W., and Ross, A.B. (1988). Critical review of rate constants for reactions of hydrated electrons, hydrogen atoms and hydroxyl radicals (·OH/·O− in Aqueous Solution. *J. Phys. Chem. Ref. Data* 17 (2): 513–886.

**55** Zhu, M., Lu, J., Hu, Y. et al. (2020). Photochemical reactions between 1,4-benzoquinone and $O_2^{\bullet -}$. *Environ. Sci. Pollut. Res.* 27: 31289–31299.

**56** Lopalco, A., Dalwadi, G., Niu, S. et al. (2016). Mechanism of decarboxylation of pyruvic acid in the presence of hydrogen peroxide. *J. Pharm. Sci.* 105 (2): 705–713.

**57** Giamalva, D., Church, D.F., and Pryor, W.A. (1985). A comparison of the rates of ozonation of biological antioxidants and oleate and linoleate esters. *Biochem. Biophys. Res. Commun.* 133 (2): 773–779.

**58** Squadrito, G.L., Cueto, R., Splenser, A.E. et al. (2000). Reaction of uric acid with peroxynitrite and implications for the mechanism of neuroprotection by uric acid. *Arch. Biochem. Biophys.* 376 (2): 333–337.

**59** Masuda, T., Shinohara, H., and Kondo, M. (1975). Reactions of hydroxyl radicals with nucleic acid bases and the related compounds in gamma-irradiated aqueous solution. *J. Radiat. Res.* 16 (3): 153–161.

**60** Basu-Modak, S. and Tyrrell, R.M. (1993). Singlet oxygen: a primary effector in the ultraviolet A/near-visible light induction of the human heme oxygenase gene. *Cancer Res.* 53 (19): 4505–4510.

**61** Akaike, T. (1976). Antagonistic action of imidazolineoxyl N-oxides against endothelium-derived relaxing factor/^ NO through a radical reaction. *J. Appl. Physiol.* 41: 764–771.

**62** Bielski, B.H., Cabelli, D.E., Arudi, R.L., and Ross, A.B. (1985). Reactivity of $HO_2/O_2^-$ radicals in aqueous solution. *J. Phys. Chem. Ref. Data* 14 (4): 1041–1100.

**63** Davies, M.J. (2016). Detection and characterisation of radicals using electron paramagnetic resonance (EPR) spin trapping and related methods. *Methods* 109: 21–30.

**64** Gorbanev, Y., Stehling, N., O'Connell, D., and Chechik, V. (2016). Reactions of nitroxide radicals in aqueous solutions exposed to non-thermal plasma: limitations of spin trapping of the plasma induced species. *Plasma Sources Sci. Technol.* 25 (5): 055017.

**65** Chalmet, S. and Ruiz-López, M.F. (2006). The structures of ozone and HOx radicals in aqueous solution from combined quantum/classical molecular dynamics simulations. *J. Chem. Phys.* 124 (19): 194502. https://doi.org/10.1063/1.2198818.

**66** Martins-Costa, M.T., Anglada, J.M., Francisco, J.S., and Ruiz-Lopez, M.F. (2012). Reactivity of atmospherically relevant small radicals at the air-water interface. *Angew. Chem. Int. Ed. Engl.* 51 (22): 5413–5417. https://doi.org/10.1002/anie.201200656.

**67** Martins-Costa, M.T.C. and Ruiz-López, M.F. (2007). Molecular dynamics of hydrogen peroxide in liquid water using a combined quantum/classical force field. *Chem. Phys.* 332 (2): 341–347. https://doi.org/10.1016/j.chemphys.2006.12.018.

**68** Bruggeman, P.J., Frontiera, R.R., Kortshagen, U.R. et al. (2021). Plasma-driven solution electrolysis. *J. Appl. Phys.* 129 (20): 200902. https://doi.org/10.1063/5.0044261.

**69** Babaeva, N.Y., Naidis, G.V., Tereshonok, D.V., and Smirnov, B.M. (2017). Streamer breakdown in elongated, compressed and tilted bubbles immersed in water. *J. Phys. D. Appl. Phys.* 50 (36): 364001. https://doi.org/10.1088/1361-6463/aa7ef1.

**70** Babaeva, N.Y., Tereshonok, D.V., and Naidis, G.V. (2015). Initiation of breakdown in bubbles immersed in liquids: pre-existed charges versus bubble size. *J. Phys. D. Appl. Phys.* 48 (35): 355201. https://doi.org/10.1088/0022-3727/48/35/355201.

**71** Kruszelnicki, J., Lietz, A.M., and Kushner, M.J. (2019). Atmospheric pressure plasma activation of water droplets. *J. Phys. D. Appl. Phys.* 52 (35): 355207. https://doi.org/10.1088/1361-6463/ab25dc.

**72** Ning, W., Lai, J., Kruszelnicki, J. et al. (2021). Propagation of positive discharges in an air bubble having an embedded water droplet. *Plasma Sources Sci. Technol.* 30 (1): 015005. https://doi.org/10.1088/1361-6595/abc830.

**73** Tian, W., Tachibana, K., and Kushner, M.J. (2014). Plasmas sustained in bubbles in water: optical emission and excitation mechanisms. *J. Phys. D. Appl. Phys.* 47 (5): 055202. https://doi.org/10.1088/0022-3727/47/5/055202.

**74** Heirman, P., Van Boxem, W., and Bogaerts, A. (2019). Reactivity and stability of plasma-generated oxygen and nitrogen species in buffered water solution: a computational study. *Phys. Chem. Chem. Phys.* 21 (24): 12881–12894. https://doi.org/10.1039/C9CP00647H.

**75** Hong, J., Zhang, T., Zhou, R. et al. (2021). Plasma bubbles: a route to sustainable chemistry. *AAPPS Bull.* 31 (1): 26. https://doi.org/10.1007/s43673-021-00027-y.

# 13

# Plasma Electrochemistry for Nitrogen Fixation

*Susanta Bera, Dimitrios Zagoraios, and Mihalis N. Tsampas*

*Dutch Institute for Fundamental Energy Research (DIFFER), Eindhoven, The Netherlands*

## 13.1 Introduction

Ammonia ($NH_3$) plays a crucial role as a platform molecule in the chemical industry and also serves as a potential fuel carrier. Currently, it is primarily produced through the Haber-Bosch process [1]. However, this process is highly energy-intensive, heavily dependent on fossil fuels, and is incompatible with the goal of achieving zero carbon emissions [2, 3]. Consequently, there is increasing global interest in developing energy-efficient and sustainable methods for synthesizing $NH_3$, using water as a hydrogen source, operating under ambient conditions, and powered by renewable energy sources [4, 5].

To achieve this long-awaited goal, various approaches have been proposed (Figure 13.1a), where electrochemistry has a prominent role, including the (i) direct electrocatalytic nitrogen ($N_2$) reduction reaction (eNRR) [4], (ii) cascade electrocatalytic nitrogen oxidation to nitrogen oxides ($NO_x$) followed by their electrocatalytic reduction (eNOR-e$NO_x$RR) [8], (iii) $Li^+$-mediated eNRR [9], (iv) plasma-enabled nitrogen reduction reaction (pNRR) [10], and (v) plasma-enabled nitrogen oxidation paired with electrocatalytic $NO_x$ reduction (pNOR-e$NO_x$RR) [11–13]. The pNOR-e$NO_x$RR involving plasma-enabled $N_2$ oxidation followed by electroreduction to $NH_3$ using renewable energy is a particularly promising pathway. It represents the reverse of the conventional nitrogen cycle based on the Haber-Bosch [1, 2] and Ostwald processes (Figure 13.1b) [14, 15]. Due to the high chemical inertness of $N_2$ [16, 17], the first two processes, i.e. eNRR and eNOR-e$NO_x$RR, face significant challenges in activating nitrogen molecules, which results in low $NH_3$ production (i.e. in the order of $10^{-1}$ and

*Plasma-Assisted Nitrogen Fixation for Sustainable Process Industries*, First Edition. Edited by Volker Hessel, Annemie Bogaerts, Gabriele Centi, Evgeny Rebrov, and Nguyen Van Duc Long.

Published 2026 by John Wiley & Sons, Inc.

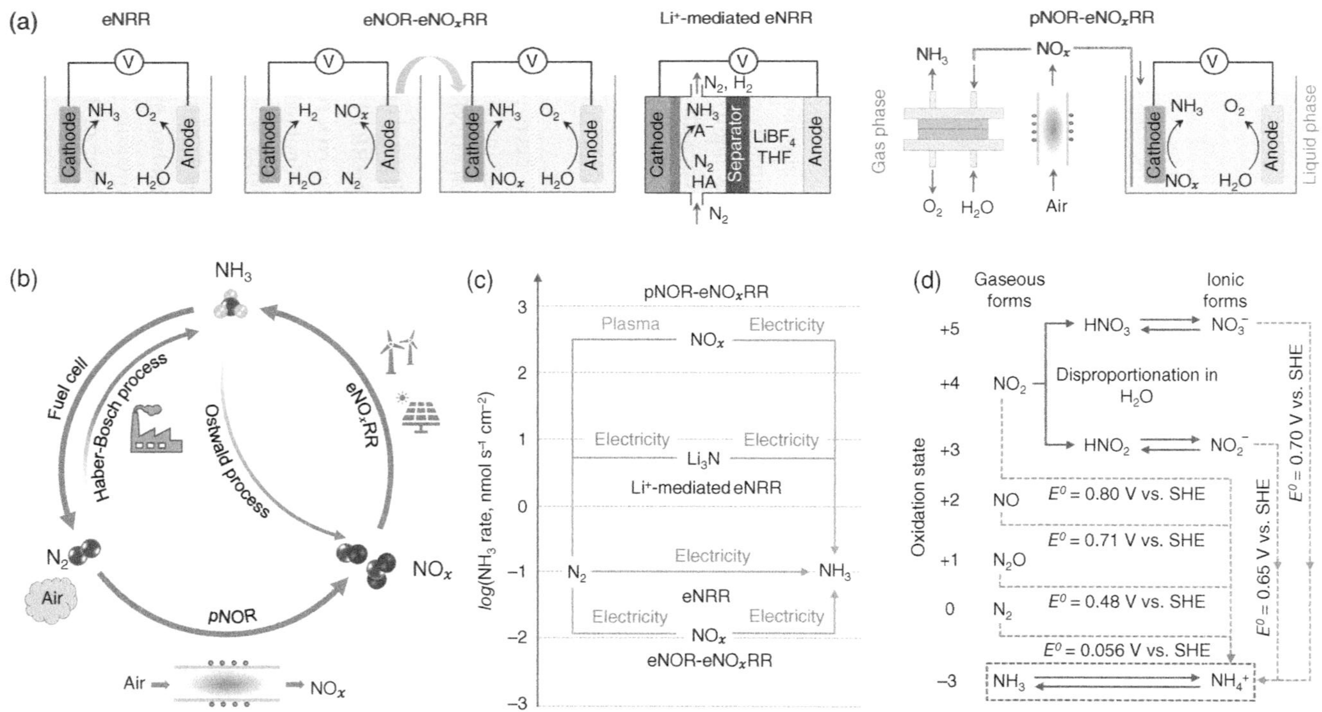


**Figure 13.1** Nitrogen fixation routes: (a) eNRR, eNOR-eNO$_x$RR, Li$^+$-mediated eNRR, and pNOR-eNO$_x$RR. (b) The conventional nitrogen cycle is based on the Haber-Bosch and Ostwald processes, and electrocatalytic reverse nitrogen cycle (anticlockwise arrows) pathways involve $N_2$ oxidation into $NO_x$ and their reverse electroreduction to $NH_3$ using renewable energy. (c) $NH_3$ production rates for the sustainable pathways of eNRR, eNOR-eNO$_x$RR, Li$^+$-mediated eNRR, and pNOR-eNO$_x$RR. (d) Pathways toward $NH_3/NH_4^+$ through the electrochemical reduction of $N_2$, and $NO_x/NO_x^-$ with standard redox potentials vs. standard hydrogen electrode (SHE). *Source:* (c) and [6] / American Chemical Society and (d) [7] Springer Nature / CC BY 4.0, respectively.

$10^{-2}$ nmol $s^{-1}$ $cm^{-2}$, respectively) (Figure 13.1c). This not only complicates reliable $NH_3$ detection but also hinders their practical applications [7, 18, 19]. The $Li^+$-mediated eNRR, on the other hand, bypasses the $N_2$ activation challenge using Li metal to spontaneously split $N_2$ molecules into $Li_3N$, achieving a higher $NH_3$ production rate (around $10^1$ nmol $s^{-1}$ $cm^{-2}$). However, the method's significant overpotential requirement (i.e. minimum of 3 V for the $Li^+$-mediated eNRR) makes the process inherently energy-intensive. Furthermore, challenges such as low system stability, the need for costly ultra-dry and oxygen-free organic solvents, pure nitrogen and hydrogen feedstocks, and lithium metal requirements are major limitations of this pathway [5, 20, 21].

Inspired by the natural process of nitrogen fixation in the Earth's atmosphere driven by lightning, plasmas have gained attention due to their ability to activate $N_2$ molecules [22, 23]. Two main pathways have been explored in this direction: (i) the reduction of plasma-activated $N_2$ by $H_2$ (generated either *in situ* or *ex situ* by a water electrolyzer) to $NH_3$ and (ii) the plasma-activated $N_2$ oxidation to $NO_x$ followed by electrochemical reduction. Studies in pNRR primarily focus on the plasma conversion of $N_2$ and $H_2$ mixtures to $NH_3$, where water splitting serves as the hydrogen production approach to maintain the green character of the process. Recent efforts also include integrating the two processes, i.e. water electrolysis and plasma-enabled nitrogen reduction, into a single system [12, 13]. The pNRR is addressed in Chapter 6 of this book, so the core of the discussions of this chapter focuses on the pNOR-eNO$_x$RR.

This chapter critically assesses the advantages and limitations of the various plasma-electrochemical pathways, offering insights into their scalability. It also includes an analysis of the pure electrochemical pathways, elucidating the role of key factors in improving Faradaic efficiency (FE) and energy efficiency. These factors are crucial for the development of more efficient, sustainable, and economically viable electrochemical ammonia synthesis methods. Finally, the synergy between fundamental research and engineering advancements in this field is emphasized, providing a roadmap for future research endeavors and industrial applications.

## 13.2 Motivation for Plasma-Enabled $N_2$ Oxidation Followed by Electrochemical Reduction

Plasma can increase the reactivity of the stable $N_2$ molecules and can convert them (in the presence of $O_2$ and/or $H_2O$, see also Chapters 6 and 12) into more reactive $NO_x$ species like nitric oxide (NO) and nitrogen dioxide ($NO_2$) [24]. These active $NO_x/NO_x^-$ species are soluble and can be electrochemically converted into $NH_3$ with higher reaction rates (i.e. in the order of $10^2$ to $10^3$ nmol $s^{-1}$ $cm^{-2}$) (Figure 13.1c) and greater selectivity compared to the eNRR process [6, 11].

From a thermodynamic point of view, the standard reduction potential (E°) of $NO_3^-$RR or $NO_2^-$RR (0.70 V and 0.65 vs. SHE, respectively) is more positive than that of eNRR (0.056 V vs. SHE), meaning that e$NO_x$RR is thermodynamically easier to carry out than the eNRR (Figure 13.1d). Moreover, the gap of standard reduction potential between e$NO_x$RR and the hydrogen evolution reaction (HER) (~0.4–0.7 V) is much larger than that between eNRR and HER (0.056 V), which indicates that the e$NO_x$RR can theoretically proceed in a wider reduction potential range than eNRR without the interference of competing HER side reaction, contributing to the high selectivity [6, 11].

The generation and utilization of $NO_x$ as an intermediate offers a practical approach to addressing the challenges of $N_2$ conversion. However, in industry, nitrites and nitrates are typically produced from ammonia through the Ostwald process [14], necessitating the development of alternative pathways for sustainable $NO_x$ generation. Plasma conversion is a promising technology for $NO_x$ production.

Plasma processes can be divided into two broad areas, thermal and nonthermal plasmas (see also Chapter 6), with significant differences in their technology and energy efficiency. One of the earliest industrial methods, i.e. the Birkeland-Eyde process [25, 26] for producing nitrogen-containing compounds from air, involved the use of thermal plasmas. This process uses air activated by a thermal plasma arc discharge, followed by quenching with water and multiple adsorption stages, yielding nitric acid with an energy consumption (EC) of approximately 0.76–0.98 kWh per mole of $HNO_3$ [24]. This process was eventually replaced by the Ostwald process, in which ammonia produced by Haber-Bosch is converted to $HNO_3$, due to its lower EC [27].

Nonthermal plasmas allow for the activation of atmospheric dinitrogen molecules while maintaining lower reaction temperatures and EC compared to thermal plasmas. The lowest EC for $NO_x$ production at atmospheric operation is 0.117 kWh mol $N^{-1}$, and it was realized via pulsed power operation [28]. The theoretical energy limit for $NO_x$ generation via nonthermal plasma is 0.056 kWh mol $N^{-1}$ (in vacuum) [27]. This value is 3–10 times lower than the EC of the conventional (i.e. hydrogen produced by steam methane reforming) Haber-Bosch process (0.13–0.30 kWh per mole of $NH_3$) and green (i.e. hydrogen produced by water electrolysis) Haber-Bosch process (0.38–0.48 kWh per mole of $NH_3$) [27, 29]. It is important to highlight that the Haber-Bosch EC increases significantly as the scale of the industrial plant decreases, e.g. by a factor of 2–2.5 times when moving from 10 MW to 10 kW units [30].

Thus, the utilization of nonthermal plasma for the $NO_x$ production offers the possibility of lower EC than Haber-Bosch, leaving room for additional energy use in the electrochemical conversion of $NO_x$ to $NH_3$. This suggests that the nonthermal plasma-based pNOR-e$NO_x$RR process could potentially compete with the Haber-Bosch, not only in scale (i.e. localized solutions) but also in energy efficiency.

## 13.3 Conventional and Plasma-Enabled $NO_x$ Feedstock

In the $eNO_xRR$, the sources of $NO_x$ can be divided into two categories: conventional and plasma-enabled feedstock. Conventional sources of $NO_x$ include wastewater, agricultural runoff, rainwater originating from industrial chemicals, and environmental $NO_x$ from air produced by lightning [31–34]. In this case, $NO_x$ is present in the liquid streams in the form of $NO_3^-$ and $NO_2^-$, which is why the majority of the electrochemical conversion studies have been focused on the liquid phase.

On the other hand, plasma-enabled processes utilize $N_2$ and $O_2$ plasmas that occur either in the gas phase or through interactions with liquid electrolytes [11, 35]. In other words, these plasma processes facilitate the operation of electrochemical systems in the gas phase, enabling additional conversion pathways.

It is worth noting that, due to the dilute nitrate concentration of the majority of the conventional $NO_x$ sources, the side reaction, i.e. HER is favored, resulting in lower ammonia yield and energy efficiency [31, 32]. For this reason, the economic viability becomes questionable when low concentration $NO_x$ sources are used. Potential solutions to address this challenge include (i) focusing on existing wastewater with high $NO_x^-$ concentration and (ii) utilizing plasmas to increase the concentration of the liquid streams. In the latter approach, the overall EC should be assessed.

## 13.4 Definition of Performance Metrics

In electrochemical reactions, FE quantifies the efficiency of the transferred charge (electrons) in a system. For the electrochemical $NO_x$ conversion to $NH_3$, FE is defined as the ratio of the charge utilized for the $NH_3$ production ($q$) to the total input charge of an electrochemical process ($Q$). It is also equivalent to the ratio of the $NH_3$ production rate, $r_{NH3}$, to the electrochemical rate $I/zF$, where $I$ is the applied current, $z$ is the number of electrons exchanged in the electrochemical reaction, and F is the Faraday's constant. This definition assumes operation in a galvanostatic mode (constant current). However, in systems operating in potentiostatic mode (constant potential), where the current varies, the definition must incorporate time-dependent factors. In such cases, the FE is determined using the integrals of the $NH_3$ production rate and applied current over time. This adjustment accounts for the dynamic current in potentiostatic conditions, ensuring accurate calculation of FE (Eq. 13.1).

$$FE_{NH_3} = \frac{q}{Q} = \frac{rNH_3}{\frac{I}{zF}} = \frac{\int rNH_3 \, dt}{\int \frac{I}{zF} \, dt} \tag{13.1}$$

The $r_{NH3}$ for electrochemical systems can be expressed as either mol $s^{-1}$ or mol $s^{-1}$ $cm_{geo}^{-2}$, with the latter normalized to the geometrical surface area of the electrode ($A_{geo}$). When using the area-normalized rate, the current density ($j = I/A_{geo}$) should be used in the FE calculation instead of the total current ($I$). Depending on the operation mode – either batch or continuous – the rate of $NH_3$ production can be calculated using the appropriate equations, incorporating conversion factors to ensure consistency in units (Eq. 13.2).

$$r_{NH_3} = \frac{\text{Concentration of } NH_3 \times \text{Reactor Volume}}{\text{Time of reaction} \times \text{Electrode geometric area}}$$
$$\text{or } \frac{\text{Concentration of } NH_3 \times \text{Volumetric flow rate}}{\text{Electrode geometric area}} \tag{13.2}$$

Another important index for any conversion process is EC. In the case of e$NO_x$RR into $NH_3$, EC is often expressed as kWh $mol_{NH3}^{-1}$ or kWh $kg_{NH3}^{-1}$ to facilitate comparisons with scaled-up systems. It is calculated as the ratio of power input for plasma and electrochemical conversion to the $r_{NH3}$ expressed in the appropriate units. This implies that there are two contributions to the EC: one from plasma ($EC_{PL}$) and another from the electrochemical system ($EC_{EL}$) (Eq. 13.3). It is worth noting that, in many literature studies, the reported $EC_{EL}$ does not account for the entire process, as it often considers model electrochemical systems. Instead, it typically reflects only the EC at the cathode (where e$NO_x$RR occurs) denoted as $EC_{CATH}$ (Eq. 13.4). This approach overlooks the energy losses (also called overpotential losses) [36] in the other components of the electrochemical cell, such as the anode and the membrane.

$$EC_{TOT} = \frac{\text{plasma power consumption} + \text{electrochemical cell power consumption}}{NH_3 \text{ production rate}}$$
$$= EC_{PL} + EC_{EL} \tag{13.3}$$

$$EC_{CATH} = \frac{\text{electrochemical cell power consumption only at the cathode}}{NH_3 \text{ production rate}} \tag{13.4}$$

## 13.5 Electrochemical Conversion of $NO_x$ to $NH_3$ – e$NO_x$RR

Before elaborating on the status of the integrated pNOR-e$NO_x$RR studies, it is important to discuss the available electrochemical conversion approaches to underscore the possibilities in the field. These conversion approaches primarily draw from water electrolyzer technologies, both in terms of reactor design and electrode materials screening strategies. A brief overview is provided here (Figure 13.2), as it is crucial to define the systems used in the $eNO_xRR$ studies. For a more detailed analysis of the water electrolyzers, the reader is encouraged to refer to the literature [36, 37].

Figure 13.2a,c–f presents the five main electrolysis technologies based on different materials and reactor designs that allow for operation across a broad range of chemical environments and temperatures. Alkaline, anion exchange membrane (AEM), and proton exchange membrane (PEM) water electrolyzers are low-temperature technologies (operation up to 80 °C), while proton ceramic electrolysis cells (PCECs) and solid oxide electrolysis cells (SOECs) operate at elevated temperature (400–600 °C and 600–900 °C, respectively) [36].

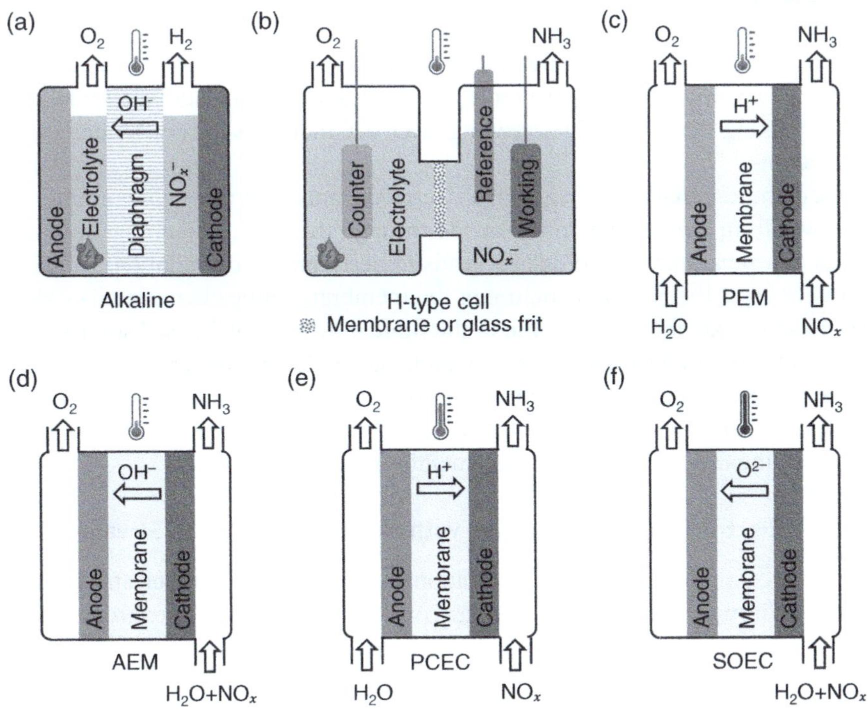


**Figure 13.2** Schematic representation of the electrochemical conversion of $NO_x$ to $NH_3$ approaches based on (a), (c)–(f) alkaline, proton exchange membrane, anion exchange membrane, protonic ceramic electrolysis cells, and solid oxide electrolysis cells. (b) H-type cell used for low-temperature material screening studies.

Alkaline electrolyzers utilize a porous diaphragm as a gas separator, whereas AEM and PEM employ ionic conducting polymeric membranes, and PCEC and SOEC ionic conducting ceramic membranes. The design of these systems is appropriate for addressing scaling considerations. It is worth noting that due to the operating temperature and reactor designs, these systems can be fed with either liquid and/or gas reactants. Specifically:

- PCEC and SOEC operate exclusively with gaseous reactants.
- AEM and PEM are versatile, accommodating both gaseous and liquid streams.
- Alkaline is restricted to liquid operation.

Simpler model electrochemical systems, such as single- or dual-chamber reactors with liquid electrolytes, are also used to accelerate electrode material discovery (Figure 13.2b). Their model nature lies in the fact that electrochemical systems focus on evaluating a single electrode (known as the working electrode) rather than the entire electrochemical system. The primary distinction between the dual-chamber reactors (also known as H-type cells) and the single-chamber systems is the presence of a diaphragm or an ionic conducting membrane that separates the two-electrode compartments. These systems operate across a broad pH range, making them ideal for screening electrode materials for low-temperature applications.

One of the key challenges in alkaline electrolysis and/or H-type cells is ammonia crossover from the cathode to the anode through the porous diaphragm, which serves as a separator. This crossover leads to performance degradation. In AEM and PEM cells, the polymeric nature of the membrane mitigates ammonia crossover to some extent. However, due to the limited number of focused studies, the magnitude of this effect remains poorly understood. Further research is needed to assess the necessity of developing alternative membranes or ammonia-tolerant anodes. In contrast, PCECs and SOECs are less prone to ammonia crossover, thanks to the ceramic nature of their membranes.

### 13.5.1 Electrochemical Conversion with $NO_x$ in Liquid-Phase Stream

The most commonly utilized configurations are the single-chamber and H-type cells containing three-electrode systems consisting of a working, reference, and counter electrodes. In these configurations, the pH of the aqueous electrolyte (acidic, neutral, and alkaline) significantly impacts electrocatalytic performance [38, 39]. However, single-chamber and H-type cells are primarily electrochemical systems used for material screening and are not intended for scaling up. Fewer studies have focused on AEM and PEM cells, where scaling considerations have been taken into account in their design, and thus, promising findings at this level can lead to practical future solutions.

The following section summarizes the studies using single-chamber, H-type, AEM, and PEM cells for $NO_x$ conversion when it is present in the aqueous electrolytes. Tables 13.1 and 13.2 summarize the most promising results, and the following paragraphs provide a brief analysis of studies with the best performance metrics across various chemical environments.

Table 13.1 highlights recent key studies conducted in H-type cells, where NO gas is introduced (by bubbling) into solution to replenish the $NO_x$ content and saturate the electrolyte solution or $NO_3^-$-based salts (e.g. $NaNO_3$, $KNO_3$) are added to the electrolytes at specific concentration.

Regarding the literature, the majority of the $eNO_xRR$ studies in H-type cells were usually conducted under neutral pH condition. In this line, Kim et al. [46] demonstrated the $eNO_xRR$ on a nanostructured Ag electrode that exhibited approximately 100% $FE_{NH3}$ with a current density of 34 mA $cm^{-2}$ at a potential of −0.165 V vs. reversible hydrogen electrode (RHE). $r_{NH3}$ of 63.6 nmol $s^{-1}$ $cm^{-2}$ was observed with an $EC_{CATH}$ of ~1.44 kWh $kg_{NH3}^{-1}$. Under neutral conditions, Chen et al. [47] designed atomically dispersed Ir confined in amorphous $MoO_3$, as a high-performance electrocatalyst for $eNO_xRR$, which delivered a maximum $r_{NH3}$ and $FE_{NH3}$ of 122 nmol $s^{-1}$ $cm^{-2}$ and 93.2% at −0.47 V vs. RHE, respectively. The $EC_{CATH}$ was ~4 kWh $kg_{NH3}^{-1}$. In this study, a mechanistic understanding of the reaction pathway was revealed using detailed atomic characterizations and theoretical investigations.

Under alkaline conditions, Wu et al. [41] developed a $CoNiO_x$@Cu catalyst to improve the NO adsorption and lower the reaction energy barrier. The synergistic effect of the CoNi oxide was achieved near 100% $FE_{NH3}$ with a maximum $r_{NH3}$ of 326 nmol $s^{-1}$ $cm^{-2}$ at −0.68 V vs. RHE. Moreover, Zhang et al. [40] developed $FeB_2$ for the $eNO_xRR$, showing a maximum $FE_{NH3}$ of 96.8% with a corresponding $r_{NH3}$ of 417 nmol $s^{-1}$ $cm^{-2}$ at −0.6 V vs. RHE, and $EC_{CATH}$ of 7.6 kWh $kg_{NH3}^{-1}$.

On the other hand, Zhang et al. [44] presented single atomic Ce sites anchored on nitrogen-doped hollow carbon spheres that are capable of electro-catalyzing NO reduction to $NH_3$ in an acidic solution, achieving $FE_{NH3}$ values up to 91% at −0.7 V vs. RHE, with a high $EC_{CATH}$ of 85.5 kWh $kg_{NH3}^{-1}$. However, the maximum $r_{NH3}$ in an acidic medium (105 nmol $s^{-1}$ $cm^{-2}$ at −0.6 V vs. RHE) was observed by Qi et al. [45] over a nanoporous VN film, with an $EC_{CATH}$ of 6.1 kWh $kg_{NH3}^{-1}$. The authors reported that the catalyst demonstrates outstanding performance, with only minimal decay in both current density and $NH_3$ production rate observed within 40 hours.

In the literature, the majority of the $eNO_xRR$ studies in flow cells were conducted in PEM configuration. Based on Table 13.2 in a PEM flow cell configuration, Li et al. [51] achieved a 90% $FE_{NH3}$ with a maximum $r_{NH3}$ of 444 nmol $s^{-1}$ $cm^{-2}$ and a high current density of 400 mA $cm^{-2}$ over an oxide-derived Cu mesh. On the other hand, Shao et al. [49] reported a

**Table 13.1** $eNO_xRR$ (liquid) with H-type cells.

| Cathode material | $NO_x$ source | Catholyte | $FE_{NH3}$ (%) | $r_{NH3}$ (nmol $s^{-1}$ $cm^{-2}$) | Ref. |
|---|---|---|---|---|---|
| $FeB_2$ | 0.1 M $NO_3^-$ | Ar-saturated 1 M KOH + 0.1 M $KNO_3$ (alkaline pH) | 96.8 | 417 | [40] |
| $CoNi_{(5:5)}O_x$@Cu | 99.9% $NO_{(g)}$ | 1 M KOH (alkaline pH) | ~100 | 326 | [41] |
| Strained Ru nanoclusters | 0.1 M $NO_3^-$ | 1 M KOH + 1 M $KNO_3$ (alkaline pH) | ~100 | 325 | [42] |
| $MoS_2$/graphite felt | 10% $NO_{(g)}$ | 0.1 M HCl + 0.5 mM iron(II) sodium benzoate (acidic pH) | 76.6 | 27.7 | [43] |
| Ce-single atom/ N-doped hollow carbon spheres | 99.9% $NO_{(g)}$ | 0.05 M HCl + 0.5 mM iron(II) sodium benzoate (acidic pH) | $91 \pm 2$ | 3.34 | [44] |
| Nanoporous VN film | 99.9% $NO_{(g)}$ | 0.1 M HCl + 0.5 mM iron(II) sodium benzoate (acidic pH) | 85 | 105 | [45] |
| Nanostructured Ag electrode | 99.9% $NO_{(g)}$ | PBS-EDTA-$Fe^{2+}$ metal complex (neutral pH) | ~100 | 63.6 | [46] |
| Ir/$MoO_3$ | 99.9% $NO_{(g)}$ | 0.5 M $Na_2SO_4$ (neutral pH) | 93.2 | 122 | [47] |
| CuO nanowire array | 0.0143 M $NO_3^-$ (from $NaNO_3$) | 0.5 M $Na_2SO_4$ (neutral pH) | 95.8 | 68 | [48] |

**Table 13.2** $eNO_xRR$ (liquid) using AEM and PEM cells with electrolyte recirculation.

| System | Cathode material | Cathodic feed | $FE_{NH3}$ (%) | $r_{NH3}$ (nmol $s^{-1}$ $cm^{-2}$) | Ref. |
|---|---|---|---|---|---|
| AEM cell | $Cu_6Sn_5$ alloy | NO (in 1 M KOH) | 96.9 | 2845 | [49] |
| PEM cell | Cu, Ru, Rh, and Pd supported on C | $KNO_3$ solution | 94 | — | [50] |
| PEM cell | Oxide-derived Cu mesh | (0.1 M $KNO_3$ + 0.1 M KOH) | 90 | 444 | [51] |
| PEM cell | Fe/C | 10% NO (in 0.5 M $H_2SO_4$) | 50.4 | 344 | [52] |
| PEM cell | Fe/C | 10% NO (in 0.5 M PBS) | 77 | 252 | [52] |

rationally designed copper–tin alloy cathode that achieves 96.9% $FE_{NH3}$ with a $r_{NH3}$ of 2,845 nmol $s^{-1}$ $cm^{-2}$ at −0.13 V vs. RHE in an AEM flow cell. The production of ammonia remained stable at >600 mA $cm^{-2}$ for 135 hours, while *ex situ* characterizations indicated that the $Cu_6Sn_5$ catalyst remains also stable. $EC_{EL}$ of the process is in the range of 15–18 kWh $kg_{NH3}^{-1}$.

### 13.5.2 Electrochemical Conversion with $NO_x$ in Gas-Phase or Catholyte-Free Stream

This section summarizes the studies conducted in AEM, PEM, and SOEC (flow) cells where gaseous $NO_x$ is fed into the cathodic compartment of these reactors. In this configuration, only NO gas is supplied to the cathode, replacing the liquid catholyte; hence, this operation is also known as catholyte free. As a result, in AEM and PEM systems, electrolyte circulation occurs only in the anodic compartment, maintained at a constant flow rate. Table 13.3 highlights the most promising results, and the following paragraphs provide a brief analysis of studies, achieving the best performance metrics for each reactor type.

Conducting a literature review, the catholyte-free (gas-phase) $eNO_xRR$ studies in flow cells are rare. Ramu et al. [55] designed a catholyte-free PEM cell for gas-phase electrochemical NO to $NH_3$ using $K_2[Ni(CN)_4]$-coated carbon felt electrode as an electrocatalyst at room temperature. This electrode exhibited excellent $eNO_xRR$ performance, with a high $r_{NH3}$ of 313 nmol $s^{-1}$ $cm^{-2}$, an excellent 89.5% $FE_{NH3}$ at −1 V vs. quasi reference Ag, and enhanced long-term $eNO_xRR$ durability for more than 10 hours. The $EC_{EL}$ of this study was 3.12 kWh $kg_{NH3}^{-1}$.

**Table 13.3** $eNO_xRR$ (gas or catholyte free) with AEM, PEM, and SOEC cells.

| System | Cathode material | Cathodic feed | $FE_{NH3}$ (%) | $r_{NH3}$ (nmol $s^{-1}$ $cm^{-2}$) | Ref. |
|---|---|---|---|---|---|
| AEM cell | Cu/C | 0.05% NO | 85.6 | 14.8 | [53] |
| PEM cell | Ru/C | 5.2% NO in He | 93 | 142 | [54] |
| PEM cell | $K_2[Ni(CN)_4]$/carbon felt | 10% NO | 89.5 | 313 | [55] |
| SOEC | $SrTi_{0.3}Fe_{0.55}Co_{0.15}O_{3-\delta}$-$Ce_{0.9}Gd_{0.1}O_{2-\delta}$ (STFC-GDC) | Ar balanced (NO + $H_2O$) | 34.8 | 523 | [56] |

Moreover, Kim et al. [53] utilized PED (pulse-electrodeposition) Cu/C as a model electrocatalyst, achieving a high $FE_{NH3}$ of 85.6% and $r_{NH3}$ up to 14.8 nmol $s^{-1}$ $cm^{-2}$) starting from 500 ppm NO in an AEM cell. They calculated an $EC_{EL}$ (including the electrical energy + energy required for acidic absorption) of 8.6 kWh $kg_{NH3}{}^{-1}$, which is lower than the conventional Haber-Bosch. Kwon et al. [56] presented a novel method for synthesizing $NH_3$ through the electrochemical reduction of NO and $H_2O$, utilizing a SOEC. More specifically, at 650 °C, a maximum electrochemical $r_{NH3}$ of 523 nmol $s^{-1}$ $cm^{-2}$ was observed, accompanied by a $FE_{NH3}$ of 34.8% at 1.5 V. Additionally, a minimal thermal $NH_3$ decomposition rate of 0.16% was observed, with $EC_{EL}$ of 93.6 kWh $kg_{NH3}{}^{-1}$.

### 13.5.3 Overview of the $eNO_xRR$ Studies

In summary, the $eNO_xRR$ in H-type cells where $NO_x$ are present in the aqueous electrolytes, $NH_3$ production rates are in the range of 3.34–417 nmol $s^{-1}$ $cm^{-2}$ depending on the $NO_x$ source and pH of the solution, whereas $EC_{CATH}$ values vary from 1.44 to 85.5 kWh $kg_{NH3}{}^{-1}$. In the case of flow cells, when $NO_x$ are present in the aqueous electrolytes, $NH_3$ production rates are in the range of 252–2845 nmol $s^{-1}$ $cm^{-2}$ and the reported $EC_{EL}$ values are in the range of 15–18 kWh $kg_{NH3}{}^{-1}$ vs. the applied potential. On the contrary, when $NO_x$ is contained in the gas phase (catholyte-free flow cells), the $NH_3$ production rates are in the range of 14.8–523 nmol $s^{-1}$ $cm^{-2}$, while $EC_{EL}$ values vary from 8.6 to 93.6 kWh $kg_{NH3}{}^{-1}$.

These conclusions are supported by a recent study from Kim et al. [53], who evaluated three different cell configurations (aqueous-phase cell, porous electrode cell, and catholyte-free cell). The study concluded that the catholyte-free cell configuration demonstrates higher $FE_{NH3}$ and lower $EC_{EL}$ compared to the other two configurations. This result is attributed to the catholyte-free cell, enabling the production of high-purity ammonium sulfate without requiring post-purification, thereby providing significant energy savings and minimizing losses associated with product back-diffusion. Additionally, the catholyte-free cell prevents issues such as electrocatalyst dissolution and corrosion during long-term e$NO_x$RR operation. The water-free environment in the cathodic compartment further suppresses undesirable phenomena such as flooding and HER, leading to enhanced $NH_3$ production. Considering the findings of Kim et al., when e$NO_x$RR is performed in the gas phase (i.e. in catholyte-free flow cells), the overall process performance is relatively higher than that of aqueous-phase cells. Therefore, future studies should prioritize and focus on optimizing the catholyte-free cells for e$NO_x$RR applications.

## 13.6 Plasma-Enabled Electrochemical Studies for $NO_x$ Conversion to $NH_3$ – pNOR-e$NO_x$RR

The potential of e$NO_x$RR, as highlighted in the Section 13.5, has driven recent studies to focus on plasma-integrated electrochemical approaches. In these studies, plasma is employed to generate $NO_x$ species and facilitate their accumulation in the electrolytes of the electrochemical cells. Figure 13.3a–d illustrates the most common approaches reported in the literature [6, 11, 35, 57–65] combining

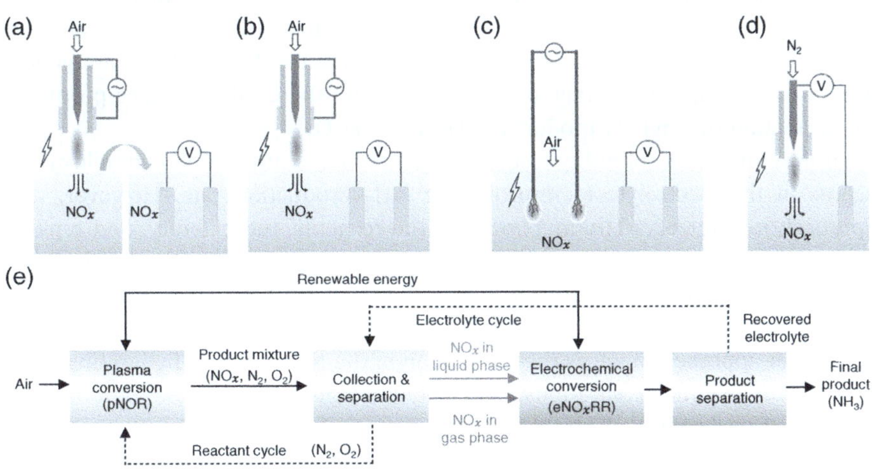


**Figure 13.3** Schematic illustration of typical reactor configurations for pNOR-e$NO_x$RR systems: (a) Gas-phase $NO_x$ production followed by dissolving in solution (as an intermediary) and used for the next e$NO_x$RR step, (b) the former process mentioned is combined in one reactor, (c) $NO_x$ generation inside liquid electrolyte followed by e$NO_x$RR step and (d) hybrid plasma-electrolytic system. (e) Illustration (adapted from [6]) for the proposed pNOR-e$NO_x$RR process integration system via liquid (right hand side) and gas phases (left hand side). Reactant and electrolyte recovery cycles are depicted with dotted lines.

plasmas with electrochemical systems, which can be categorized as follows. For simplicity, gas-phase plasma stimulation is demonstrated using the jet configuration, although this is not restrictive.

Figure 13.3a shows an approach where $NO_x$ species are generated in the gas phase, absorbed (or dissolved) in a solution as an intermediary step, and later used as the electrolyte for an electrochemical cell, whereas Figure 13.3b represents an integrated approach within a single reactor. The primary product mixture from the plasma reactor consists of $NO_x$, $N_2$, and $O_2$ (and, in some cases, $O_3$, depending on the plasma source).

It is crucial to separate the unreacted $N_2$ and $O_2$ from the product mixture, particularly $O_2$, since oxygen consumes the transported protons (from the anode compartment) to form water via oxygen reduction reaction (ORR). As the product mixture bubbles through the electrolyte, the majority of $N_2$ and $O_2$ are separated due to their low solubility in aqueous media ($N_2$: 0.63 mM, $O_2$: 1.27 mM) [66]. However, to enhance electrochemical conversion toward $NH_3$, primary importance must be placed on dissolving the $NO_x$ species ($NO_2$, NO). Nevertheless, the poor solubility of $NO_x$ species in electrolytes at ambient conditions (i.e. NO: 1.95 mM, $NO_2$: 12 mM) imposes a limitation in the e$NO_x$RR process. Although nitrous oxide ($N_2O$) has a higher solubility in water (25 mM), it is predominantly electrochemically reduced into $N_2$ rather than $NH_3$ [66]. Therefore, from energy-efficiency point of view, plasma operation should focus on minimizing $N_2O$ production. The NO concentration in the electrolyte significantly influences product selectivity. At low NO coverage, N–N coupled products ($N_2$ and $N_2O$) are generated, whereas higher coverage of NO coverage on the catalyst surface promotes the formation of single N products ($NH_3$ and $NH_2OH$) [67].

To overcome this solubility issue, researchers have utilized NO complexation agents in the electrolyte to improve the $NH_3$ production rate. However, this approach introduced additional costs for the reagents, recyclability, and separation. The performance bottleneck caused by poor $NO_x$ solubility can only be mitigated when e$NO_x$RR is performed in a flow cell configuration utilizing high surface area porous electrodes [38].

Another approach involves the generation of the $NO_x^-$ species by igniting plasmas directly in aqueous solutions, as depicted in Figure 13.3c. These solutions can either be transferred to electrochemical systems or be coupled with them. In this method, similar $NO_x$ species are generated in distinct regions of the plasma reactor (plasma-glow, plasma-spark, and at the interface between the plasma-spark and liquid) [11]. The glow discharge region primarily activates $N_2$ and $O_2$, forming reactive species such as N, $N_2$ (vib), O, $O_2$, and $O_3$. These initiate a reaction to produce small amounts of $NO_x$. The spark discharge region creates bubbles with high energy, facilitating a highly reactive environment that enhances $NO_x$ generation. In the after-glow region, located at the plasma–liquid interface, hydroxyl radical generation is favored, further assisting the production of reactive species ($NO_x$ and $NO_x^-$) in solutions. The presence of hydroxyl radicals can influence the solubility of the $NO_x$ species, but the exact magnitude remains unclear [61] and requires further investigations.

Figure 13.3d depicts an integrated system for direct $NH_3$ production from nitrogen and water using a plasma-electrolytic system, where the plasma reactor is positioned just above the liquid level [65].

The general operation schemes of the pNOR-e$NO_x$RR for the various operations are summarized in Figure 13.3e. In this scheme, pNOR involves the plasma activation of $N_2$ and $O_2$ molecules to form $NO_x$, with unreacted molecules recycled back to the plasma reactor. The generated $NO_x$ is absorbed into an aqueous electrolyte, converting the NO and $NO_2$ into $NO_2^-$ and $NO_3^-$, thereby enriching the electrolyte with $NO_x$ before its introduction into the e$NO_x$RR system [6]. Another possibility, not yet realized, involves coupling $N_2$ and $O_2$ plasmas with electrochemical cells in gaseous feeding configuration. This concept is elaborated in the following Section 13.6.1.

### 13.6.1 pNOR-e$NO_x$RR Integration Approaches

Plasma-enabled $NO_x$ production from the air generates a mixture of NO, $NO_2$, and $N_2O$, along with unreacted $N_2$ and $O_2$. However, only NO and $NO_2$ participate in electrochemical reduction to produce $NH_3$. The integration challenges and requirements for this process can be divided into two main categories.

#### 13.6.1.1 Liquid-Phase-Based pNOR-e$NO_x$RR

Studies of plasma generation in the liquid phase or at the air–liquid interface have demonstrated that purging the entire gas mixture from the plasma chamber into a liquid electrolyte can dissolve $NO_x$ and separate unreacted gases through bubbling. This separation is critical because the presence of unreacted $O_2$ in the cathode of the electrochemical system leads to the ORR, a thermodynamically more favorable reaction than e$NO_x$RR, reducing $NH_3$ selectivity.

It is worth noting that complete $O_2$ removal through bubbling is not feasible due to its partial solubility [63], which imposes a limitation on liquid-based operation. Furthermore, limited discussion exists in the literature regarding the solubility of $NO_x$ species in the various electrolytes used in pNOR-e$NO_x$RR systems, hindering accurate assessment of $NO_x$ losses from the reactor. These systems must develop efficient recirculation cycles (see Figure 13.3) to ensure that all plasma-produced $NO_x$ is fully utilized in the electrochemical systems.

Plasma conversion systems should be tuned to minimize $N_2O$ production during nitrogen oxidation, as $N_2O$ is a potent greenhouse gas (with ~300 times more impact on climate change than $CO_2$) whose production and emission must be minimized for environmental reasons. Efforts should focus on integrating plasma-enabled $N_2$ oxidation with processes for separating unreacted $O_2$, purifying NO and $NO_2$ from the $NO_x$ mixture, and facilitating their electrochemical reduction into $NH_3$.

#### 13.6.1.2 Gas-Phase-Based pNOR-e$NO_x$RR

In gas-phase nitrogen oxidation, plasma conversion must aim not only to minimize or eliminate $N_2O$ production but also to achieve high $O_2$ conversion efficiencies. Additionally, unreacted $O_2$ should be separated from the plasma reactor effluent to avoid the side reaction of ORR that reduces the $NH_3$ selectivity. Potential methods for oxygen separation include electrochemical oxygen pumps [68, 69], zeolites, and metal–organic frameworks (MOFs) [70, 71]. The resulting oxygen-free stream could then be directly fed to the cathode of AEM, PEM, and SOEC cells, facilitating efficient e$NO_x$RR.

### 13.6.2 Alternative Plasma Electrochemical Systems

There are only a few plasma-enabled electrochemical studies conducted beyond pNOR-e$NO_x$RR. These studies are noteworthy since they explore alternative reactor integration designs and hold the potential to contribute to alternatives to the aforementioned two integration approaches. The following paragraph aims to presents alternative designs beyond the conventional pNOR-e$NO_x$RR that could influence future developments.

Kumari et al. explored the integration of a plasma jet into the cathode of a PEM water electrolyzer to provide energetically activated $N_2$ species to the electrocatalyst surface [12]. At an applied bias of ~3.5 V across the electrolyzer, plasma-assisted operation was observed to produce 47% more ammonia than the combination of plasma-without-bias and bias-without-plasma conditions at room temperature. Patel et al. integrated an inductively coupled radio frequency plasma into the anode of a SOEC water electrolyzer, where water is reduced to hydrogen in the cathode, and during plasma operation, NO is produced in the anode by the reaction of oxygen ions (transported via the electrolyte) and plasma-activated nitrogen species [72]. Nitric oxide production rates and FEs are observed up to 3.15 nmol of NO $s^{-1}$ $cm^{-2}$ and 93%, respectively. The advantage of this system is the simultaneous co-generation of hydrogen in the SOEC cathode. Sharma et al. utilized the same integration approach, but at this time, plasma integration concerned the cathode of a PCEC water electrolyzer aiming for the production of ammonia [13]. In this system, hydrogen species produced by water oxidation over the anode are transported through the proton conducting membrane to the cathode where they react with the plasma-activated nitrogen toward ammonia. $r_{NH3}$ and $FE_{NH3}$ up to of 26.8 nmol $s^{-1}$ $cm^{-2}$ and 88%, respectively, were achieved.

### 13.6.3 Experimental pNOR-e$NO_x$RR Studies

A summary of the different plasma-enabled electrochemical $N_2$ fixation toward $NH_3$ is presented in Tables 13.4 and 13.5 where studies are divided into two main categories based on how $NO_x$ production is realized, i.e. in the gas or liquid phase. The majority of literature studies focus on reporting the $r_{NH3}$ and $FE_{NH3}$, while EC assessment does not always encompass the entire system.

#### 13.6.3.1 Gas-Phase $NO_x$ Generation

In the $NO_x$ production in air, $N_2$ and $O_2$ molecules initially interact with energetic electrons and become vibrationally excited and then they can further react with the atomic species to produce $NO_x$. The produced $NO_x$ should be first separated from the unreacted $N_2$ and $O_2$ (or $O_3$) and can be supplied directly to the gas-phase flow reactor or into the electrolyte of liquid cells (H-type or flow reactor) for electrochemical reduction into $NH_3$. Yang et al. recently presented a tandem air–$NO_x$–$NO_x^-$–$NH_3$ system that combines spark discharge nonthermal plasma-enabled $NO_x$ production followed by electrochemical $NO_x^-$ reduction into $NH_3$ in a flow mode using $Ni(OH)_x$/Cu electrode [35]. The plasma reactor consisted of a "T"-shaped glass tube and two stainless-steel tubes for gas inlet and plasma generation.

**Table 13.4** Plasma-enabled electrochemical conversion where the pNOR step occurs in the gas phase.

| e$NO_x$RR system | Cathode | Electrolyte | $FE_{NH3}$ (%) | $r_{NH3}$ (nmol $s^{-1}$ $cm^{-2}$) | EC (kWh $kg_{NH3}^{-1}$) | Ref. |
|---|---|---|---|---|---|---|
| H-type cell | Cu nanoparticles | 0.1 M KOH | 90 | 40 | ~26 ($EC_{CATH}$) | [6] |
| H-type cell | $Ni_3B@NiB_{2.74}$ | 0.1 M KOH | ~100 | 55 | ~6 ($EC_{CATH}$) | [59] |
| H-type cell | $Cu_2Pd$/ carbonized bacterial cellulose | 0.5 M KOH | ~94 | 32 | ~2 ($EC_{CATH}$) | [60] |
| H-type cell | $NiMoO_4$/ CuO nanowire | 1 M KOH | 98.8 | 228 | ~3.3 ($EC_{CATH}$) | [73] |
| H-type cell | N-$MoS_2$/ graphene | 0.1 M KOH | ~70 | 119 | 39.3 ($EC_{PL}$ + $EC_{CATH}$) | [58] |
| AEM cell | $Ni(OH)_x$/Cu | 1 M KOH | 92 | 347 | 300 ($E_{PL}$ + $EC_{EL}$) | [35] |

**Table 13.5** Plasma-enabled electrochemical conversion where the pNOR step occurs in the liquid phase.

| $eNO_xRR$ system | Cathode | Electrolyte | $FE_{NH3}$ (%) | $r_{NH3}$ (nmol $s^{-1}$ $cm^{-2}$) | EC (kWh $kg_{NH3}{}^{-1}$) | Ref. |
|---|---|---|---|---|---|---|
| H-type cell | Cu | Plasma-activated water (pH = 2) | 85 | 30.3 | 188 ($EC_{PL} + EC_{CATH}$) | [61] |
| H-type cell | Cu | 0.1 M KOH | 83.2 | 49 | 7203 ($EC_{PL} + EC_{CATH}$) | [62] |
| H-type cell | Co/N-doped carbon | Plasma-activated 0.1 M KOH | ~100 | ~49 | 51.8 ($EC_{PL} + EC_{CATH}$) | [63] |
| PEM cell | Cu | 0.01 M $H_2SO_4$ | 87 | 42.1 | 253 ($EC_{TOT}$) | [11] |
| Single cell ($eN_2RR$/$eNO_xRR$) | Catalyst free | Dilute $H_2SO_4$ (pH = 3.5) | 100 | 720 | 2270 ($EC_{PL} + EC_{CATH}$) | [65] |

The system delivers a steady current density of ~240 mA $cm^{-2}$ and $r_{NH3}$ of ~347 nmol $s^{-1}$ $cm^{-2}$ with an $EC_{TOT}$ of 300 kWh $kg_{NH3}{}^{-1}$.

#### 13.6.3.2 Liquid-Phase $NO_x$ Generation

$NO_x$ is directly generated inside the electrolyte by plasmas followed by electrochemical $NO_x^-$ reduction into $NH_3$ in H-type cells or circulating electrolyte with concentrated $NO_x^-$ in the flow cells. Amal et al. developed a nonthermal plasma bubble column reactor to generate $NO_x$ intermediaries and their electrochemical reduction toward $NH_3$ using a scalable electrolyzer, which operates at 1.4 V with current densities of over 50 mA $cm^{-2}$ [11]. They calculated $r_{NH3}$ of 42.1 nmol $s^{-1}$ $cm^{-2}$ and an $EC_{TOT}$ of 253 kWh $kg_{NH3}{}^{-1}$.

For the $NO_x$ production in liquid, there is also another hybrid electrolytic system consisting of a gaseous plasma electrode (as a cathode), placed just slightly above the liquid level, which facilitates plasma-activated electrochemical $N_2$ reduction toward ammonia without any material surface [65]. Renner et al. demonstrated a plasma electrolytic system that can produce $NH_3$ with high $FE_{NH3}$ (up to 100%). The system achieved $r_{NH3}$ of 720 nmol $s^{-1}$ $cm^{-2}$ with EC of 2,270 kWh $kg_{NH3}{}^{-1}$ [65].

In summary, when the pNOR step is involved in the gas phase, the $r_{NH3}$ and $EC_{CATH}$ are in the range of 32–228 nmol s$^{-1}$ cm$^{-2}$ and 2–39 kWh $kg_{NH3}{}^{-1}$, respectively, for H-type cells, whereas the $r_{NH3}$ was found to increase up to 347 nmol s$^{-1}$ cm$^{-2}$ with $EC_{TOT}$ of 300 kWh $kg_{NH3}{}^{-1}$ in the flow cell (Table 13.4). On the other hand, during the pNOR step in the liquid phase, the $r_{NH3}$ and $EC_{TOT}$ are in the range of 30–49 nmol s$^{-1}$ cm$^{-2}$ and 52–7200 kWh $kg_{NH3}{}^{-1}$, respectively, for H-type cell, and for a hybrid plasma-electrolytic single cell, these values reached 720 nmol s$^{-1}$ cm$^{-2}$ and 2,270 kWh $kg_{NH3}{}^{-1}$, respectively (Table 13.4). The reported $r_{NH3}$ (42 nmol s$^{-1}$ cm$^{-2}$) in the flow cell was slightly lower than that of the H-type cell (Table 13.5).

### 13.6.4 Overview of pNOR-eNO$_x$RR Systems

The field of pNOR-eNO$_x$RR is still in its nascent stages. Consequently, much of the existing literature focuses on proof-of-concept studies and efforts to achieve higher ammonia production rates than those demonstrated by alternative "state-of-the-art" electrochemical technologies. A comprehensive evaluation of the current status of this field requires a thorough analysis of ammonia production rates, FE, and EC. However, assessing EC remains particularly challenging, as it is often not explicitly reported in many studies.

The evaluation of EC presents intrinsic complexities for both plasma and electrochemical systems. In plasma-based systems, EC is heavily influenced by power supply losses, and the values reported typically reflect apparent EC rather than true system efficiency. For electrochemical systems, EC assessment is more straightforward when device-level reactors are employed. However, this is not the case for H-type cells, which are predominantly used in pNOR-eNO$_x$RR studies. These systems are primarily designed for material screening purposes, and the reported performance metrics typically pertain to single electrodes. While this approach facilitates material comparisons, it underestimates the total EC at the device level, which must account for the contributions of both electrodes and energy losses within the electrolyte.

Figure 13.4 provides an overview of the EC of pure and plasma-coupled eNO$_x$RR systems under the aforementioned conditions, categorized by system type and operational mode. The reported performance metrics for these systems, in terms of $r_{NH3}$, range from $10^2$ to $10^3$ nmol s$^{-1}$ cm$^{-2}$, which is significantly higher than those achieved by alternative electrochemical pathways (refer to Figure 13.1).

Insights derived from the extensive body of the literature on pure eNO$_x$RR reveal notable trends in FE and EC. For low-temperature electrochemical systems, $FE_{NH3}$ values range from 50% to 100%. H-type cells, in particular, exhibit higher $FE_{NH3}$ values (77%–100%), as their design inherently limits operation to low current densities, which promotes higher efficiency. At the device level, systems such

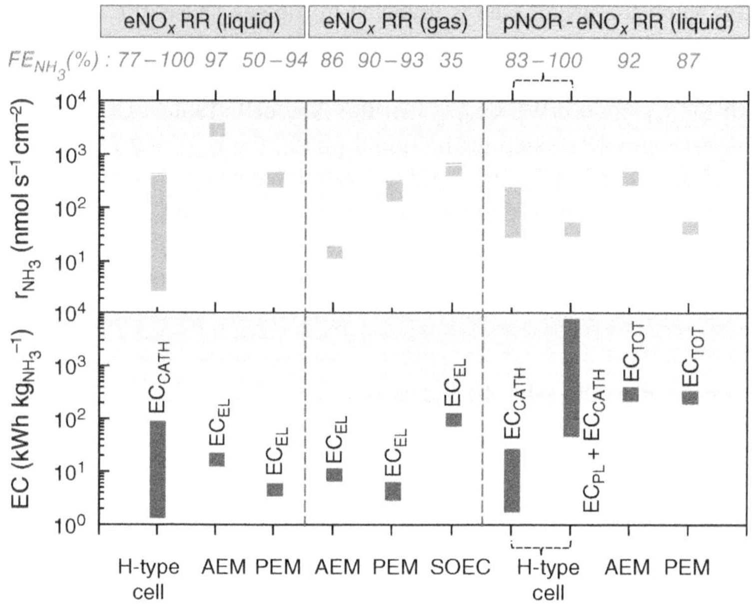


**Figure 13.4** Comparison of the $r_{NH3}$, EC, and $FE_{NH3}$ for the previously reported work covering various pathways of $eNO_xRR$ (liquid-phase), $eNO_xRR$ (gas-phase), and pNOR-$eNO_xRR$ (liquid-phase) with different cell configurations.

as AEM and PEM reactors achieve promising $FE_{NH3}$ values in the range of 50%–94% for both the liquid-phase and gas-phase operations.

The EC of AEM and PEM systems ranges from 9 to 18 and 3 to 6 kWh $kgNH_3^{-1}$, respectively, reflecting the higher technological readiness level (TRL) of PEM reactors. However, as these values are derived from laboratory-scale experiments, they underscore the significant potential for optimizing the integration of these systems into pNOR-$eNO_xRR$ processes. Current overall EC for pNOR-$eNO_xRR$ systems based on AEM and PEM reactors is in the range of 253–300 kWh $kgNH_3^{-1}$, highlighting the inefficiencies in the integration of plasma and electrochemical reactors. This also indicates that the EC associated with current plasma conversion pathways remains far from optimal.

To date, only one study has reported on the performance of $eNO_xRR$ systems employing SOECs, which operate at elevated temperatures. In this case, the reported $r_{NH3}$ is 523 nmol s$^{-1}$ cm$^{-2}$, with an $FE_{NH3}$ of 35% and EC of 94 kWh $kgNH_3^{-1}$. These results are promising, especially considering the potential for improvement through advanced material research and the possibility of integrating SOECs with plasma systems for enhanced thermal efficiency. This promising avenue warrants further investigation to unlock the full potential of SOEC-based pNOR-$eNO_xRR$ systems.

## 13.7 Implementation at Industrial Level

Despite the challenges, the potential of pNOR-eNOxRR systems has facilitated their implementation at the industrial level. Notably, Plasmaleap, an Australian company, is advancing this technology at the pilot plant stage https://www.plasmaleap.com/efuels-chemicals. The process involves activating atmospheric air and water in a plasma-bubble reactor, followed by the electrochemical conversion of these activated species into nitrate and ammonia derivatives. Moreover, $N_2$-Applied, a Norwegian company, is focused on transforming agricultural waste into sustainable fertilizers through innovative nitrogen recovery technologies. The process involves the use of plasma to create reactive nitrogen gas from air, which is then absorbed into organic materials to enhance plant-available nitrogen, stabilize ammonium, reduce nitrogen losses, inhibit methane formation, and support long-term soil health https://n2applied.com/.

Finally, several companies are focusing specifically on the pNOR step, with primary applications in fertilizers and disinfectants. For example, VitalFluid, a Dutch company, manufactures portable devices for such purposes https://vitalfluid.com/products/. These industrial-scale initiatives leverage process integration and intensification, and their role in the future of pNOR-eNOxRR systems warrants careful evaluation.

## 13.8 Key Challenges and Future Outlook

Plasma-activated nitrogen fixation for $NH_3$ production is a highly promising, sustainable, and green approach. The performance metrics analyzed in the Section 13.6 highlight its potential while also emphasizing the need for further optimization. Future research should primarily focus on decreasing EC to levels comparable to the Haber-Bosch process and in parallel address process intensification challenges. The main strategies for achieving this include:

i) *utilizing advanced nonthermal plasma systems, as detailed in Chapters 6 and 12,*
ii) *implementing advanced electrochemical systems with improvements in materials, components, and operating conditions, and*
iii) *exploring the synergistic integration of plasma and electrochemical systems, particularly regarding process and heat management.*

### 13.8.1 Electrochemical Systems

The choice of materials plays a central role in plasma-enabled ammonia production via the pNOR-e$NO_x$RR pathway. AEM, PEM, PCEC, and SOEC introduce diverse material requirements. Despite significant advancements in each field, critical challenges remain in material development, particularly concerning their selectivity, productivity, and stability.

PEM systems, which are at a high TRL, feature zero-gap designs – one of the reasons behind their superior performance. However, due to the acidic nature of their membranes, they require the use of expensive and scarce electrode materials. AEM systems, although similar in design to PEM systems, employ alkaline membranes, enabling the use of earth-abundant electrode materials, akin to alkaline electrolysis. However, their TRL is lower than that of PEM systems due to stability concerns related to their membranes and the need for efficient electrocatalysts.

SOEC systems, which are electrochemical systems with a TRL slightly below PEM, utilize abundant electrode materials. However, their main limitation lies in the long-term instability of materials due to interactions between the electrode and component materials at high operating temperatures. PCEC systems, the proton-conducting version of SOECs, operate at intermediate temperatures. This emerging and highly promising technology has a lower TRL than AEM systems and faces material stability challenges.

Achieving a balance between performance enhancement and cost-effectiveness is critical, necessitating the use of abundant, inexpensive materials. In addition to addressing challenges related to electrode materials, it is crucial to ensure the durability of membranes and other electrochemical reactor components, such as interconnects and bipolar plates.

### 13.8.2 Operational Considerations

In e$NO_x$RR, a major contribution to the energy demand is the anodic oxygen evolution reaction, although byproduct oxygen is often not utilized [74]. This motivates the search for alternative anodic oxidation reactions with lower energy demand. In this direction, biomass has been proposed as a promising reagent for the anode. The sustainability is further enhanced if both electrode reactions, oxidation at the anode and reduction at the cathode, give valuable products. Moreover, the main energy losses in plasma conversion arise from heat dissipation; hence, integrating plasma with high-temperature electrochemical systems, such as PCEC and SOEC, could enhance overall system efficiency.

### 13.8.3 Product Separation

Product separation and collection are critical steps that require attention for future-scale installations, as they will impact EC. Ammonia purification from liquid electrolyte streams remains challenging. The most widely investigated method, air-stripping followed by acid adsorption, has been recognized as a well-established approach for $NH_3$ capture over the past decade. Capturing $NH_3$ from gas-phase e$NO_x$RR offers conceptual advantages in terms of separation

[43], such as the use of zeolites, MOFs, or liquefaction. However, further investigation is needed to assess these methods' energy contributions in relation to pNOR-eNO$_x$RR devices.

### 13.8.4 Scalability and Process Integration

Transitioning ammonia production via pNOR-eNO$_x$RR from the laboratory to industrial-scale plants poses significant challenges. Bridging the gap between lab-scale and pilot-scale applications necessitates cost-effective, scalable approaches for synthesizing high-quality catalysts with optimal compositions and nanostructures, as well as large-area deposition techniques and continuous-flow reactors. Developing high-capacity reactors with favorable architecture is essential to ensure (i) efficient coupling between the plasma and the electrochemical conversion steps and (ii) compatibility with scaled-up catalysts, enabling uniform reactant distribution and efficient mass transfer under continuous-flow conditions. Additionally, it is crucial to design systems for easy maintenance, reliable operation under practical conditions, and performance validation based on empirical results. Process integration should ensure the matching of the rates of plasma and the electrochemical conversion steps while taking into account the minimization of nitrous oxide in the plasma reactor effluent and oxygen separation before the introduction of the electrochemical conversion step. Finally, the approach must emphasize low EC, economic viability, and environmental sustainability [17].

## 13.9 Conclusions

Plasma-enabled electrochemical systems represent a transformative opportunity for nitrogen fixation, offering significant environmental and energy benefits over conventional ammonia synthesis pathways. However, the field remains at an early stage, requiring substantial advancements to realize its full potential. Key conclusions include:

1) **Performance potential**: pNOR-eNO$_x$RR demonstrates significantly higher $NH_3$ production rates and FEs compared to traditional electrocatalytic approaches. $FE_{NH3}$ values of up to 100% have been achieved in laboratory-scale systems, although EC remains high.
2) **Integration challenges**: The integration of plasma systems with electrochemical reactors is hampered by inefficiencies in plasma power supply, low NO$_x$ solubility, and poor thermal and material compatibility. These factors contribute to EC far exceeding competitive benchmarks like the Haber-Bosch process.

3) **Material development**: Advanced electrocatalysts, membranes, and reactor components are critical to improving selectivity, stability, and scalability. PEM systems exhibit high TRL but rely on costly materials, while AEM, PCEC, and SOEC systems show promise with abundant materials but require durability improvements.
4) **Scalability and system design**: Transitioning from laboratory to industrial scales will necessitate innovations in reactor architecture, catalyst synthesis, and continuous operation. Emerging designs, such as catholyte-free configurations and hybrid plasma-electrolytic systems, demonstrate the potential for reduced energy losses and simplified separation processes.
5) **Future directions**: Achieving EC levels comparable to Haber-Bosch will require process intensification, including heat integration, advanced plasma technologies, and optimized electrochemical systems. Additionally, coupling $NH_3$ production with low-energy anodic oxidation reactions or biomass valorization offers opportunities for further sustainability.

In summary, while plasma-enabled electrochemical systems for $NH_3$ synthesis face significant challenges, their unique advantages position them as a viable alternative for decentralized, renewable-driven ammonia production. Continued interdisciplinary research and engineering innovation are essential to bridge the gap between laboratory results and scalable industrial applications.

## References

**1** Haber, F. and Le Rossignol, R. (1913). Über die technische Darstellung von Ammoniak aus den Elementen. *Z. Elektrochem. Angew. Phys. Chem.* 19 (2): 53–72.

**2** Kyriakou, V., Garagounis, I., Vourros, A. et al. (2020). An electrochemical Haber-Bosch process. *Joule* 4 (1): 142–158.

**3** Tan, X., Yu, C., Ren, Y. et al. (2021). Recent advances in innovative strategies for the $CO_2$ electroreduction reaction. *Energy Environ. Sci.* 14 (2): 765–780.

**4** Qing, G., Ghazfar, R., Jackowski, S.T. et al. (2020). Recent advances and challenges of electrocatalytic N2 reduction to ammonia. *Chem. Rev.* 120 (12): 5437–5516.

**5** Andersen, S.Z., Čolić, V., Yang, S. et al. (2019). A rigorous electrochemical ammonia synthesis protocol with quantitative isotope measurements. *Nature* 570 (7762): 504–508.

**6** Ren, Y., Yu, C., Wang, L. et al. (2022). Microscopic-level insights into the mechanism of enhanced $NH_3$ synthesis in plasma-enabled cascade $N_2$ oxidation-electroreduction system. *J. Am. Chem. Soc.* 144 (23): 10193–10200.

**7** Choi, J., Suryanto, B.H.R., Wang, D. et al. (2020). Identification and elimination of false positives in electrochemical nitrogen reduction studies. *Nat. Commun.* 11 (1): 1–10.

8 Wang, Y., Yu, Y., Jia, R. et al. (2019). Electrochemical synthesis of nitric acid from air and ammonia through waste utilization. *Natl. Sci. Rev.* 6 (4): 730.

9 Suryanto, B.H.R., Matuszek, K., Choi, J. et al. (2021). Nitrogen reduction to ammonia at high efficiency and rates based on a phosphonium proton shuttle. *Science* 372 (6547): 1187–1191.

10 Muzammil, I., Kim, Y.N., Kang, H. et al. (2021). Plasma catalyst-integrated system for ammonia production from $H_2O$ and $N_2$ at atmospheric pressure. *ACS Energy Lett.* 6: 3004–3010.

11 Sun, J., Alam, D., Daiyan, R. et al. (2021). A hybrid plasma electrocatalytic process for sustainable ammonia production. *Energy Environ. Sci.* 14 (2): 865–872.

12 Kumari, S., Pishgar, S., Schwarting, M.E. et al. (2018). Synergistic plasma-assisted electrochemical reduction of nitrogen to ammonia. *Chem. Commun.* 54 (95): 13347–13350.

13 Sharma, R.K., Patel, H., Mushtaq, U. et al. (2021). Plasma activated electrochemical ammonia synthesis from nitrogen and water. *ACS Energy Lett.* 6 (2): 313–319.

14 Thiemann, M., Scheibler, E., and Wiegand, K.W. (2000). Nitric acid, nitrous acid, and nitrogen oxides. *Ullmann's Encycl. Ind. Chem* 24: 1–49.

15 W. Ostwald, Improvements in the manufacture of nitric acid and nitrogen oxides. GB190200698A, issued Jan. 9, 1902.

16 Van Der Ham, C.J.M., Koper, M.T.M., and Hetterscheid, D.G.H. (2014). Challenges in reduction of dinitrogen by proton and electron transfer. *Chem. Soc. Rev.* 43 (15): 5183–5191.

17 van Langevelde, P.H., Katsounaros, I., and Koper, M.T.M. (2021). Electrocatalytic nitrate reduction for sustainable ammonia production. *Joule* 5 (2): 290–294.

18 Bera, S., van der Breggen, R., Kunturu, P.P. et al. (2024). Electrocatalytic nitrogen reduction in continuous-flow cell via water oxidation at ambient conditions: Promising for ammonia or diazene? *Electrochem. Commun.* 166: 107794.

19 Rezaie, F., Læsaa, S., Sahin, N.E. et al. (2023). Low-temperature electrochemical ammonia synthesis: measurement reliability and comparison to Haber–Bosch in terms of energy efficiency. *Energ. Technol.* 11 (10): 2300410.

20 Lazouski, N., Chung, M., Williams, K. et al. (2020). Non-aqueous gas diffusion electrodes for rapid ammonia synthesis from nitrogen and water-splitting-derived hydrogen. *Nat. Catal.* 3 (5): 463–469.

21 Maheshwari, S. and Janik, M.J. (2019). Kinetics of Li-mediated N2 electroreduction. *Joule* 3 (4): 915–916.

22 Navarro-González, R., McKay, C.P., and Mvondo, D.N. (2001). A possible nitrogen crisis for Archaean life due to reduced nitrogen fixation by lightning. *Nature* 412 (6842): 61–64.

23 Winter, L.R. and Chen, J.G. (2021). N2 fixation by plasma-activated processes. *Joule* 5 (2): 300–315.

**24** Rouwenhorst, K.H.R., Jardali, F., Bogaerts, A., and Lefferts, L. (2021). From the Birkeland–Eyde process towards energy-efficient plasma-based $NO_x$ synthesis: a techno-economic analysis. *Energy Environ. Sci.* 14 (5): 2520–2534.

**25** Birkeland, K.R. (1906). On the oxidation of atmospheric nitrogen in electric arcs. *Trans. Faraday Soc.* 2 (December): 98–116.

**26** Eyde, S. (1912). Oxidation of atmospheric nitrogen and development of resulting industries in norway. *Ind. Eng. Chem.* 4 (10): 771–774.

**27** Cherkasov, N., Ibhadon, A.O., and Fitzpatrick, P. (2015). A review of the existing and alternative methods for greener nitrogen fixation. *Chem. Eng. Process. Process Intensif.* 90: 24–33.

**28** Vervloessem, E., Gorbanev, Y., Nikiforov, A. et al. (2022). Sustainable $NO_x$ production from air in pulsed plasma: elucidating the chemistry behind the low energy consumption. *Green Chem.* 24 (2): 916–929.

**29** Wang, M., Khan, M.A., Mohsin, I. et al. (2021). Can sustainable ammonia synthesis pathways compete with fossil-fuel based Haber–Bosch processes? *Energy Environ. Sci.* 14 (5): 2535–2548.

**30** Rouwenhorst, K.H.R. and Lefferts, L. (2020). Feasibility study of plasma-catalytic ammonia synthesis for energy storage applications. *Catal.* 10 (9): 999.

**31** Chauhan, R. and Srivastava, V.C. (2020). Electrochemical denitrification of highly contaminated actual nitrate wastewater by $Ti/RuO_2$ anode and iron cathode. *Chem. Eng. J.* 386: 122065.

**32** Su, L., Li, K., Zhang, H. et al. (2017). Electrochemical nitrate reduction by using a novel $Co_3O_4/Ti$ cathode. *Water Res.* 120: 1–11.

**33** Nguyen, T.T.P., Do, B.K.D., Bui, N.N. et al. (2013). Selectiveness of copper and polypyrrole modified copper electrodes for nitrate electroreduction: a comparative study and application in ground water. *ECS Trans.* 53 (16): 41–52.

**34** Schumann, U. and Huntrieser, H. (2007). The global lightning-induced nitrogen oxides source. *Atmos. Chem. Phys.* 7 (14): 3823–3907.

**35** Liu, W., Xia, M., Zhao, C. et al. (2024). Efficient ammonia synthesis from the air using tandem non-thermal plasma and electrocatalysis at ambient conditions. *Nat. Commun.* 15 (1): 3524.

**36** Sapountzi, F.M., Gracia, J.M., Weststrate, C.J. et al. (2017). Electrocatalysts for the generation of hydrogen, oxygen and synthesis gas. *Prog. Energy Combust. Sci.* 58: 1–35.

**37** Chatenet, M., Pollet, B.G., Dekel, D.R. et al. (2022). Water electrolysis: from textbook knowledge to the latest scientific strategies and industrial developments. *Chem. Soc. Rev.* 51 (11): 4583–4762.

**38** Inta, H.R., Dhanabal, D., Markandaraj, S.S., and Shanmugam, S. (2023). Recent advances in electrocatalytic $NO_x$ reduction into ammonia. *EES Catal.* 1 (5): 645–664.

**39** Xiong, Y., Wang, Y., Zhou, J. et al. (2024). Electrochemical nitrate reduction: ammonia synthesis and the beyond. *Adv. Mater.* 36 (17): 2304021.

**40** Zhang, G., Li, X., Chen, K. et al. (2023). Tandem electrocatalytic nitrate reduction to ammonia on MBenes. *Angew. Chem. Int. Ed.* 62 (13).

**41** Wu, A., Lv, J., Xuan, X. et al. (2023). Electrocatalytic disproportionation of nitric oxide toward efficient nitrogen fixation. *Adv. Energy Mater.* 13 (14): 1–9.

**42** Li, J., Zhan, G., Yang, J. et al. (2020). Efficient ammonia electrosynthesis from nitrate on strained ruthenium nanoclusters. *J. Am. Chem. Soc.* 142 (15): 7036–7046.

**43** Zhang, L., Liang, J., Wang, Y. et al. (2021). High-performance electrochemical NO reduction into $NH_3$ by $MoS_2$ nanosheet. *Angew. Chem. Int. Ed.* 60 (48): 25263–25268.

**44** Zhang, W., Qin, X., Wei, T. et al. (2023). Single atomic cerium sites anchored on nitrogen-doped hollow carbon spheres for highly selective electroreduction of nitric oxide to ammonia. *J. Colloid Interface Sci.* 638: 650–657.

**45** Qi, D., Lv, F., Wei, T. et al. (2022). High-efficiency electrocatalytic NO reduction to $NH_3$ by nanoporous VN. *Nano Res. Energy* 1 (2): 1–7.

**46** Kim, D.Y., Shin, D., Heo, J. et al. (2020). Unveiling electrode-electrolyte design-based NO reduction for $NH_3$ synthesis. *ACS Energy Lett.* 5 (11): 3647–3656.

**47** Chen, K., Wang, G., Guo, Y. et al. (2023). Iridium single-atom catalyst for highly efficient NO electroreduction to $NH_3$. *Nano Res* 16 (7): 8737–8742.

**48** Wang, Y., Zhou, W., Jia, R. et al. (2020). Unveiling the activity origin of a copper-based electrocatalyst for selective nitrate reduction to ammonia. *Angew. Chem. Int. Ed.* 59 (13): 5350–5354.

**49** Shao, J., Jing, H., Wei, P. et al. (2023). Electrochemical synthesis of ammonia from nitric oxide using a copper–tin alloy catalyst. *Nat. Energy* 8 (11): 1273–1283.

**50** Bunea, S., Clemens, K., and Urakawa, A. (2022). Electrified conversion of contaminated water to value: selective conversion of aqueous nitrate to ammonia in a polymer electrolyte membrane cell. *ChemSusChem* 15 (2): e202102180.

**51** Li, S., Han, D., Jiang, G. et al. (2023). Proton exchange membrane electrode assembly for ammonia electrosynthesis from nitrate. *ACS Appl Energy Mater* 6 (9): 5067–5073.

**52** Cheon, S., Kim, W.J., Kim, D.Y. et al. (2022). Electro-synthesis of ammonia from dilute nitric oxide on a gas diffusion electrode. *ACS Energy Lett.* 7 (3): 958–965.

**53** Kim, H.W., Jin, J., Kim, M. et al. (2024). Direct recovery of electro-synthesized ammonia from low-concentration nitric oxide using pulse electrodeposited Cu/C catalyst in a catholyte-free system. *Chem. Eng. J.* 485: 150048.

**54** Bunea, S., Coppens, M., and Urakawa, A. (2023). Electrochemical conversion of NO to $NH_3$ in a PEM cell. *ACS Catal.* 13 (17): 11345–11351.

**55** Ramu, A.G., Renukadevi, R., Silambarasan, P. et al. (2023). Discovering a catholyte free design for gas phase electrocatalytic NO gas reduction to $NH_3$ at room temperature. *J. Environ. Chem. Eng.* 11 (5): 110751.

**56** Kwon, Y.I., Kim, S.K., Kim, Y.B. et al. (2021). Nitric oxide utilization for ammonia production using solid electrolysis cell at atmospheric pressure. *ACS Energy Lett.* 6 (12): 4165–4172.

**57** Ding, J., Li, W., Zhang, H. et al. (2024). A cascade jet plasma oxidation—electroreduction system using Pd-Ni dual-site catalyst for sustainable ammonia production from air. *Adv. Funct. Mater.* 2410768: 1–11.

**58** Zheng, J., Zhang, H., Lv, J. et al. (2023). Enhanced $NH_3$ synthesis from air in a plasma tandem-electrocatalysis system using plasma-engraved N-doped defective $MoS_2$. *JACS Au* 3 (5): 1328–1336.

**59** Li, L., Tang, C., Cui, X. et al. (2021). Efficient nitrogen fixation to ammonia through integration of plasma oxidation with electrocatalytic reduction. *Angew. Chem.* 133 (25): 14250–14256.

**60** Li, W., Zhang, S., Ding, J. et al. (2023). Sustainable nitrogen fixation to produce ammonia by electroreduction of plasma-generated nitrite. *ACS Sustain. Chem. Eng.* 11 (3): 1168–1177.

**61** Sun, J., Zhou, R., Hong, J. et al. (2024). Sustainable ammonia production via nanosecond-pulsed plasma oxidation and electrocatalytic reduction. *Appl. Catal. B Environ.* 342: 123426.

**62** Cui, Y., Yang, H., Dai, C. et al. (2022). Coupling of $LaFeO_3$-plasma catalysis and Cu +/$Cu_0$ electrocatalysis for direct ammonia synthesis from air. *Ind. Eng. Chem. Res.* 61 (14): 4816–4823.

**63** Wu, A., Yang, J., Xu, B. et al. (2021). Direct ammonia synthesis from the air via gliding arc plasma integrated with single atom electrocatalysis. *Appl. Catal. B Environ.* 299 (July): 120667.

**64** Liu, Y., Ma, J., Huang, S. et al. (2023). Highly dispersed copper-iron nanoalloy enhanced electrocatalytic reduction coupled with plasma oxidation for ammonia synthesis from ubiquitous air and water. *Nano Energy* 117: 108840.

**65** Hawtof, R., Ghosh, S., Guarr, E. et al. (2019). Catalyst-free, highly selective synthesis of ammonia from nitrogen and water by a plasma electrolytic system. *Sci. Adv.* 5: eaat5778.

**66** Signorelli, S., Möller, M.N., Coitiño, E.L., and Denicola, A. (2011). Nitrogen dioxide solubility and permeation in lipid membranes. *Arch. Biochem. Biophys.* 512 (2): 190–196.

**67** Ko, B.H., Hasa, B., Shin, H. et al. (2022). Electrochemical reduction of gaseous nitrogen oxides on transition metals at ambient conditions. *J. Am. Chem. Soc.* 144 (3): 1258–1266.

**68** Guerra, V., Silva, T., Pinhão, N., Guaitella, O., Guerra-Garcia, C., Peeters, F.J.J., Tsampas, M.N., and Van De Sanden, M.C.M. (2022) Plasmas for in situ resource utilization on mars: fuels, life support, and agriculture. *J. Appl. Phys.*, 132 (7). 070902-1 to 070902-19

**69** Pandiyan, A., Kyriakou, V., Neagu, D. et al. (2022). $CO_2$ conversion via coupled plasma-electrolysis process. *J $CO_2$ Util* 57: 101904.

**70** Perez-Carbajo, J., Matito-Martos, I., Balestra, S.R.G. et al. (2018). Zeolites for $CO_2$-$CO$-$O_2$ separation to obtain $CO_2$-neutral fuels. *ACS Appl. Mater. Interfaces* 10 (24): 20512–20520.

**71** Luna-Triguero, A., Vicent-Luna, J.M., Jansman, M.J. et al. (2021). Enhancing separation efficiency in European syngas industry by using zeolites. *Catal. Today* 362: 113–121.

**72** Patel, H., Sharma, R.K., Kyriakou, V. et al. (2019). Plasma-activated electrolysis for cogeneration of nitric oxide and hydrogen from water and nitrogen. *ACS Energy Lett.* 4 (9): 2091–2095.

**73** Lin, C., Chen, X., Wang, L. et al. (2024). Electrocatalytic conversion of nitrate into ammonia through heterogeneous catalysis of $NiMoO_4$ and $Cu/Cu_2O$. *Adv. Funct. Mater.* 34 (36): 2401287.

**74** Kormányos, A., Szirmai, A., Endrődi, B., and Janáky, C. (2024). Stable operation of paired $CO_2$ reduction/glycerol oxidation at high current density. *ACS Catal.* 14 (9): 6503–6512.

# 14

# Analytical Techniques for Plasma Catalysis

*Christopher Hardacre[1], Sarayute Chansai[1], and Shanshan Xu[2]*

[1] *Department of Chemical Engineering, School of Engineering, University of Manchester, Manchester, UK*
[2] *Department of Materials, School of Natural Sciences, University of Manchester, Manchester, UK*

## 14.1 Optical Spectroscopy

### 14.1.1 Introduction

Optical spectroscopy is a fundamental diagnostic technique and considered a powerful tool for analyzing plasma chemistry and parameters, providing critical insights into their composition, temperature, and reaction dynamics and kinetics. The measurement and detection of the light emitted, absorbed, or scattered by atoms, ions, and molecules in the plasma enable the identification of important ground-state, metastable-state, and/or excited-state species. In this chapter, the two commonly used techniques, i.e. optical emission spectroscopy (OES) and tunable diode laser absorption spectroscopy (TDLAS), are represented and compared.

OES technique is an analytical technique widely used for the analysis of light that is emitted from the plasma-assisted processes [1–11]. It is essential for the analysis of the plasma-generated gas-phase species, for example, which determine the surface phenomena occurred under the plasma discharge. OES is a powerful and useful tool to investigate the fundamental understanding of mechanistic aspects driven by plasma processes. One significant advantage is that the OES technique is relatively simple and robust to operate and can be used for characterizing plasma chemistry *in situ* for catalytic reactions, and monitoring and controlling some plasma-assisted processes in a real-time measurement for a range of plasma processes [12–14].

*Plasma-Assisted Nitrogen Fixation for Sustainable Process Industries*, First Edition. Edited by Volker Hessel, Annemie Bogaerts, Gabriele Centi, Evgeny Rebrov, and Nguyen Van Duc Long.

Published 2026 by John Wiley & Sons, Inc.

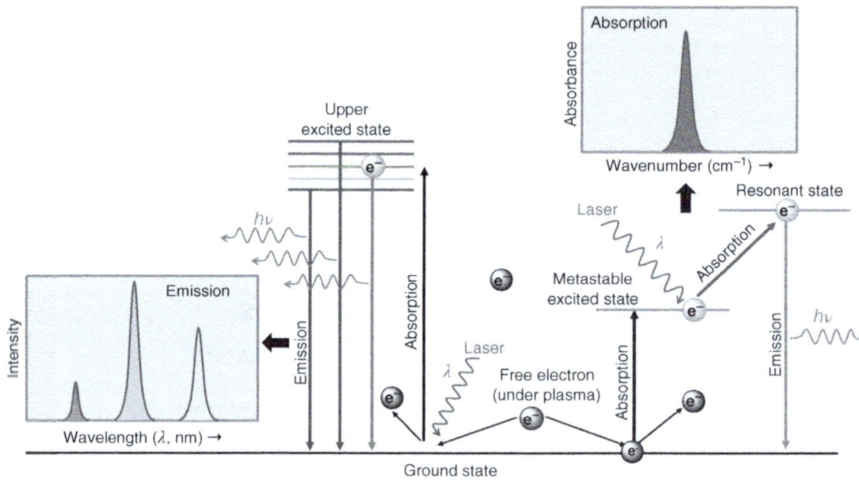


**Figure 14.1** A simple schematic representation of transitions and energy levels involved in absorption and emission spectroscopy.

During the plasma processes, the gas-phase species collide with energetic electrons ejected from an atom's outer shell under electrical plasma discharge, promoting the formation of excited electronic states of ionized atoms and molecules, and radicals (Figure 14.1). These excited states of species are unstable and contain orbital vacancies created by ejected electrons, which subsequently relax to the ground electronic state by spontaneous emission. The excess energy is released during the transition of electrons between the two orbitals in the form of spectrally dispersed lights with a fixed wavelength or radiation energy. Each transition taken place produces the individual characteristic spectral line signal that is captured and collected by detectors and then processed to generate a spectrum showing the light intensity peaks vs. their wavelengths, providing the qualitative information about the sample such as specific intermediates created under plasma-assisted processes. This qualitative analysis is very important to study the overall plasma chemistry for comprehensive understanding of the effect of ionized atoms and molecules, and radicals on surface interactions and reaction mechanism under plasma environment. OES is not only a qualitative technique, but it can also be used quantitatively as the intensity or the peak area of the spectral line can provide the semi-quantitative analysis of the species within the analyte using the relative calibration to improve the accuracy. In this case, a fixed and known concentration of internal standard, i.e. atom or molecule, is added to various concentrations of gas of interest under plasma condition and is compared to the emission intensity of gas of interest. The linear relationship between the relative emission intensity and the concentration of gas of interest can then be established. Moreover, the OES technique can be used to monitor the intensity and spectral distribution of a characteristic OES spectrum of plasma-assisted process to probe the plasma and surface temperatures, determined by the vibrational temperature of species and rotational branch of spectral lines [1, 15–20].

The OES technique is an ideal non-intrusive technique that provides the essential analysis of plasma-driven chemistry and processes to identify the reactive species in the ionized gas-phase and surface species such as rotationally, vibrationally, and electronically excited radicals, ions, and molecules. Most recently, Van Duc long et al. [9] summarized and critically reviewed the practical applications of OES for plasma-assisted ammonia synthesis and nitrogen fixation and addressed the importance of understanding of the plasma-driven reactive species to gain an insight into the reaction mechanism and plasma physics, including electrons, vibrational, and rotational temperature, density, and electric field. However, in this section, we focus on a general overview of a simple use of OES technique to study the plasma-assisted ammonia synthesis using the fundamental knowledge between heterogeneous catalysis and plasma chemistry to obtain valuable insights into the reaction mechanisms, which can assist in catalyst designs and other developments in the field of catalysis.

Similar to the OES technique, TDLAS is also a non-intrusive optical technique widely used for gas analysis and investigation of plasmas chemistry and parameters [21–27]. In principle, TDLAS is based on the laser absorption spectroscopic technique, involving the tuning of a diode laser to a specific wavelength that corresponds to the absorption line of particular gas molecules such as $NH_3$, NO, $NO_2$, $CO_2$, CO, $H_2O$, $O_2$, $CH_4$, and $C_2H_4$, which can be detected *in situ* with high selectivity and sensitivity. Each type of gas has a unique absorption spectrum. As the laser light passes through the plasma reactor, the gas molecules absorb light at these specific (characteristic) wavelengths, yielding qualitative information. Importantly, quantitative analysis can also be carried out and the concentration of gas species can be determined with high accuracy by measuring the amount of light absorbed by gas molecules. TDLAS technique is widely used for the optimization of plasma processes by highly specific, accurate, and real-time monitoring of species densities.

Under plasma conditions, the impact of free energetic electrons on the gas molecules results in the transitions between ground states and excited states. Due to different impact energies of the electrons, gas molecules are excited to different levels (Figure 14.1). One of which is metastable excited state, which generally has longer lifetime than the upper excited state. Under these circumstances, the metastable excited state of gas molecules can be probed by an absorption with a specific wavelength tuned by diode laser module.

TDLAS technique offers some advantages over OES, in which it only measures the excited gas molecules and provides the detailed qualitative information. The high sensitivity and selectivity of TDLAS technique can detect and measure low concentration of plasma-activated gas molecules at metastable state under plasma conditions and ground-state gas species for general gas analysis and gas-sensing technology. Due to the high wavelength resolution obtainable by using a diode laser, the interference from other gases is significantly minimized or negligible, allowing the TDLAS technique to be used with long path lengths and low gas concentrations, which are suitable for remote gas sensing. However, there are some limitations of TDLAS technique, in which it is only capable of measuring the concentration of one gas species at a time. In addition, each laser has a limited range of wavelengths, meaning that multiple lasers may be needed if multiple gases are to be measured.

### 14.1.2 Experimental Setup

OES measurements are typically carried out using a fiber optical cable with an optical probe inserted in the plasma reactor or placed perpendicularly to the quartz plasma reactor (Figure 14.2). The distance between the probe and the quartz tube can be varied to obtain high spectral sensitivity and accuracy. A fiber-optic cable is usually connected to the spectrometer equipped with a CCD (charge-coupled device) array as a detector capable of measuring a wide range of electromagnetic spectrum in the combined regions of ultraviolet (UV), visible (Vis), and near infrared (NIR) with wavelengths between 190 and 1,100 nm.

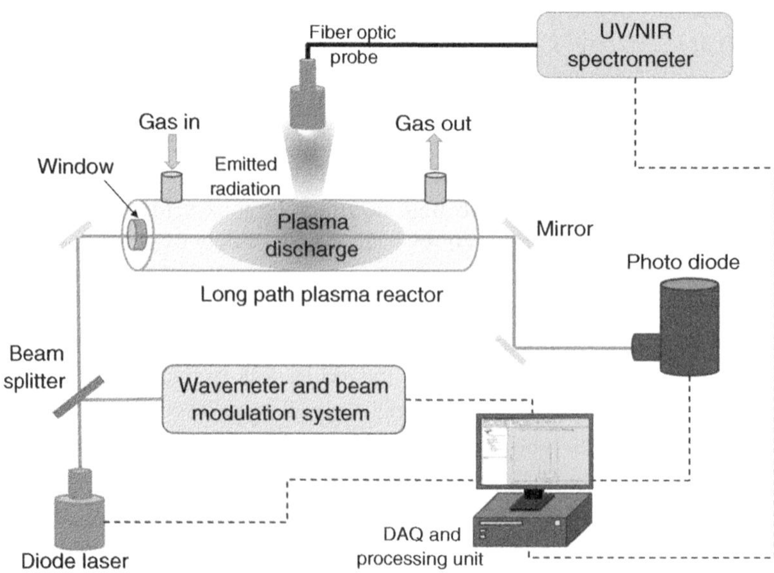


**Figure 14.2** A schematic representation of a typical plasma-assisted catalytic reactor coupled with OES and TDLAS spectroscopic techniques.

As noted above, the TDLAS technique works by directing a laser beam through a gas sample at a specific wavelength that is "tuned" to that of the target compound by adjusting the temperature and bias current. There are five major components for TDLAS setup. First, *tunable diode laser* is the heart of the TDLAS technique, in which the diode laser is precisely tuned to a specific wavelength corresponding to the absorption line of the gas molecule of interest. The tunability to obtain a narrow linewidth makes it suitable for high-resolution spectroscopy. Second, *wavemeter and beam modulation systems* are often used to enhance the signal-to-noise ratio (SNR) by techniques such as wavelength modulation spectroscopy (WMS) or frequency modulation spectroscopy (FMS) to modulate the laser beam at a specific wavelength or frequency. Third, *plasma gas cell or reactor* is generally a chamber where the laser light passes through interact with the gas molecules under plasma conditions. The design of gas cell is varied such as open-path, multi-pass, and single-pass gas cells and can depend on the application and required sensitivity. Fourth, *photodiode or photodetector* measures the reduction in light intensity after passing through the plasma gas cell due to absorption. The change in light intensity is then converted into an electrical signal that can be processed and analyzed. Importantly, the use of high-sensitivity photodetector is essential for detecting extremely weak light signal. Lastly, *data acquisition and processing unit* collects the data and electrical signals from the photodetector and processes them for further analysis. Advanced algorithms and software are used to translate the data into readable digital information.

### 14.1.3 OES Spectrum and Interpretation for Ammonia Synthesis

In this section, typical OES spectra and characteristic features obtained under plasma-assisted ammonia synthesis conditions are examined and utilized the fundamental understanding from thermal heterogeneous catalysis and plasma catalysis to generate and gain insights into the possible chemical reactions occurring during the plasma-assisted ammonia synthesis. Regardless of the $N_2 + H_2$ reaction conditions and the types of plasma used, the OES spectrum shown in Figure 14.3 reveals a considerable number of optical emission peaks for hydrogen atoms and nitrogen atoms and molecules.

Of particular interest is the OES spectrum of the nitrogen-containing species, which can be divided into three distinctive regions [9, 11, 28]:

i) the second positive system (SPS) between 300 and 400 nm,
ii) the first negative system (FNS) between 400 and 500 nm, and
iii) the first positive system (FPS) above 500 nm.

The presence of OES peaks at *ca.* 358 nm in the SPS region and at *ca.* 662 nm in FPS region is generally indicative of the electronic excitation of $N_2$ molecules

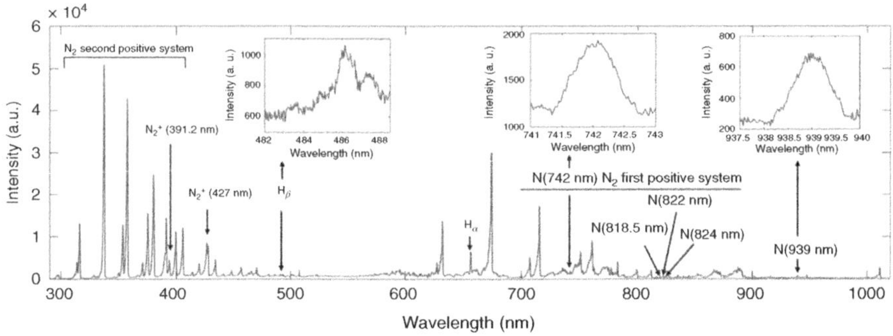


**Figure 14.3** A typical OES spectrum obtained under catalytic ammonia synthesis conditions ($N_2$:$H_2$, 1:3) over glass beads and dielectric barrier discharge plasma conditions at 16 kV. *Source:* [7] / American Chemical Society.

under plasma conditions and associated with the electronic transitions $C^3\Pi \rightarrow B^3\Pi$ and $B^3\Pi_g \rightarrow A^3\Pi_u$, respectively. Interestingly, the detection of OES peaks at *ca.* 391 and 427 nm in FNS region indicates the ionization by electron impact of $N_2$ molecules to form $N_2{}^+$ species. The appearance of the OES peaks at *ca.* 486 and 656 nm, due to the dissociation of $H_2$ molecules, is generally assigned to atomic $H_\beta$ and $H_\alpha$, respectively. Not only are atomic hydrogen detected but also several emission lines at *ca.* 742, 818, 822, 824, 939, and 1,011 nm, that belong to atomic N species present under plasma conditions. It is so important to note that the clear observation of $NH^*$ species by OES technique is often challenging under $N_2 + H_2$ plasma conditions because of the complexity of several OES peaks and bands derived from nitrogen atoms and molecules as well as one of the electronic transitions of $N_2$ molecules, which yields a strong OES peak at *ca.* 337 nm in the SPS. However, the unresolved OES shoulder at *ca.* 336 nm has been assigned as $NH^*$ species due to the electronic transition of $A^3\Pi \rightarrow X^3\Sigma$ [4, 9, 11, 28, 29].

Overall, the OES spectrum directly provides a clear evidence of the formation of several reactive gas-phase species under plasma reaction conditions and the plasma processes derived from the interpretation of OES spectrum can be summarized as follows:

Electron-impact ionization:

$$N_2\,(g) + e^- \rightarrow N_2^+\,(g) + e^- \quad (14.1)$$

$$H_2\,(g) + e^- \rightarrow H_2^+\,(g) + e^- \quad (14.2)$$

Electron-impact dissociation:

$$N_2\,(g) + e^- \rightarrow 2N^{\bullet}\,(g) + e^- \quad (14.3)$$

$$H_2\,(g) + e^- \rightarrow 2H^{\bullet}\,(g) + e^- \quad (14.4)$$

Electron-impact excitation:

$$N_2\,(g) + e^- \rightarrow N_2^*(g) + e^- \tag{14.5}$$

$$H_2\,(g) + e^- \rightarrow H_2^*\,(g) + e^- \tag{14.6}$$

These electron-impact processes drive and facilitate the ammonia synthesis via further inelastic collisions with gas-phase and adsorbed surface species, resulting in the formation of active surface intermediates and the acceleration of several possible surface reactions under plasma conditions to promote the ammonia synthesis reaction. These possible chemical reactions are proposed based on the OES analysis [7, 9, 28] and are specifically categorized as follows with the adsorption sites for surface species denoted as "s":

Nitrogen adsorption:

$$N_2\,(g) + 2s \leftrightarrows 2N\,(ads) \tag{14.7}$$

$$N^\bullet\,(g) + s \leftrightarrows N\,(ads) \tag{14.8}$$

Hydrogen adsorption:

$$H_2\,(g) + 2s \leftrightarrows 2H\,(ads) \tag{14.9}$$

$$H^\bullet\,(g) + s \leftrightarrows H\,(ads) \tag{14.10}$$

$NH_x$ adsorption:

$$NH^\bullet\,(g) + s \leftrightarrows NH\,(ads) \tag{14.11}$$

$$NH_2^\bullet\,(g) + s \leftrightarrows NH_2\,(ads) \tag{14.12}$$

Surface hydrogenation reactions:

$$N\,(ads) + H\,(ads) \leftrightarrows NH\,(ads) + s \tag{14.13}$$

$$NH\,(ads) + H\,(ads) \leftrightarrows NH_2\,(ads) \tag{14.14}$$

$$NH_2\,(ads) + H\,(ads) \leftrightarrows NH_3\,(ads) + s \tag{14.15}$$

Eley-Rideal reactions:

$$N\,(ads) + H^\bullet\,(g) \leftrightarrows NH\,(ads) \tag{14.16}$$

$$NH\,(ads) + H^\bullet\,(g) \leftrightarrows NH_2\,(ads) \tag{14.17}$$

$$NH_2\,(ads) + H^\bullet\,(g) \leftrightarrows NH_3\,(ads) \tag{14.18}$$

$$H\,(ads) + N^\bullet\,(g) \leftrightarrows NH\,(ads) \tag{14.19}$$

$$H\,(ads) + NH^\bullet\,(g) \leftrightarrows NH_2\,(ads) \tag{14.20}$$

$$H\,(ads) + NH_2^\bullet\,(g) \leftrightarrows NH_3\,(ads) \tag{14.21}$$

Ammonia desorption:

$$NH_3\ (ads) + \leftrightarrows NH_3\ (g) + s \tag{14.22}$$

### 14.1.4 OES Analysis and Proposed Reaction Mechanism for Ammonia Synthesis

The plasma-assisted ammonia synthesis has been well-documented and reviewed extensively in the absence and presence of a catalyst [4, 7, 9, 11, 28, 30]. The role of the catalyst is understood to be associated with changing the reaction pathways and reducing the energy requirement to induce the specific reaction to form ammonia compared with the reaction without catalyst under plasma conditions. In general, the energy barriers include the ionization energy, electronic excitation energy, and vibrational and rotational excitation energy of molecular and atomic species that give rise to metastable states of reactive intermediates such as $N^•$, $H^•$, and $NH^•$ radical species to form ammonia.

Inclusively, the proposed reaction mechanisms and pathways of plasma-assisted ammonia synthesis have been comprehensively summarized by Rouwenhorst et al. [30] from a viewpoint of both plasma chemistry and heterogeneous catalysis. First, the reaction mechanism of the *plasma-phase ammonia synthesis* has been proposed to be driven by the interaction between gas-phase radical ($N^•$ and $H^•$) species, which are formed from $N_2$ and $H_2$ dissociation by the plasma processes. Second, the ionization, dissociation, and excitation of both $N_2$ and $H_2$ molecules lead to the specific formation of $NH^•$ radicals and $NH_3$ production away from the catalyst surfaces. In the presence of catalyst, the catalyst surfaces are thought to promote the interaction of radicals such as $N^•$, $H^•$, $NH^•$, and $NH_2{}^•$ species, generated under plasma phase, via adsorption to produce ammonia. The surface-enhanced plasma-driven ammonia synthesis (SEPDAS) significantly involves the reaction between adsorbed $N^•$ and adsorbed $H^•$ radical species, leading to the hydrogenation of $NH_x$ species and the formation of gas-phase $NH_3$ products. Third, the catalyst can also activate the reactant, in which only $H_2$ is dissociated over the active catalyst surfaces, while the dissociation of $N_2$ is limited. In this case, a surface-enhanced semi-catalytic ammonia synthesis is proposed, based on the plasma process in which $N^•$ radical species are created in the plasma phase and then are adsorbed on the catalyst surfaces to react with adsorbed $H^•$ species to form $NH^•$ species and then gas-phase $NH_3$. Finally, where both $H_2$ and $H_2$ molecules are activated and dissociated over the active catalyst surfaces to produce adsorbed $N^•$, $H^•$, $NH^•$, and $NH_2{}^•$ species, this subsequently results in the formation and release of gas-phase $NH_3$. These proposed reaction pathways are named as the plasma-enhanced catalytic ammonia synthesis (PEDAS).

### 14.1.5 OES Analysis for Plasma Dynamics

Not only is OES technique used for studying and identifying the chemical species and reaction, but it can also be used to determine plasma characteristics, e.g. gas temperature ($T_g$), electron density ($n_e$), and electron excitation temperature ($T_{el}$) [1, 5, 15, 16, 19, 20, 31–34]. In general, the gas temperature and the thermal translational energy are correlated as the order of magnitude in the energy separation between rotational levels is considered identical. Under plasma conditions, frequent collisions of gas molecules occur, resulting in a dynamic balance between translational and rotational motions. Therefore, the gas temperature can be determined by the rotational temperature ($T_g = T_{rot}$) from the OES spectra, using the Boltzmann distribution of rotational states, considering the calculation of rotation temperature of diatomic molecules. By fitting and simulating the OES spectrum using either the open-source Python-based MassiveOES software [19, 35], developed by Masaryk University, or the SPECAIR software [15, 33, 34], as a spectroscopic modeling code developed by Stanford University, it is possible to model the OES transitions and extract the rotational and vibrational temperatures of molecules in the plasma discharge. In principle, SPECAIR models the nonequilibrium plasma conditions and uses the Boltzmann distribution for excited states combined with molecular band structure calculations and instrumental broadening effects with a "slit function." The intensity of the emission line associated with the transition from the upper energy levels ($e'$, $v'$, $J'$) to the lower energy level ($e''$, $v''$, $J''$) for the electronic ($e$), vibrational ($v$), and rotational ($J$) levels is expressed as

$$I(\lambda) = (n_{e'v'J'}) \cdot \sum \left(Re^{v'v''}\right)^2 \cdot \frac{S_{J'J''}}{2S' + 1} \cdot (E_{e'v'J'} - E_{e''v''J''})\phi(\lambda - \lambda_0) \quad (14.23)$$

$$n_{e'v'J'} = N \cdot \frac{g_{e'}(2S' + 1) \cdot \exp\left(-\frac{\varepsilon_{el}}{k_B T_{el}} - \frac{\varepsilon_{vib}}{k_B T_{vib}} - \frac{\varepsilon_{rot}}{k_B T_{rot}}\right)'}{\sum_{e'v'J'} \left[g_{e'}(2S' + 1) \cdot \exp\left(-\frac{\varepsilon_{el}}{k_B T_{el}} - \frac{\varepsilon_{vib}}{k_B T_{vib}} - \frac{\varepsilon_{rot}}{k_B T_{rot}}\right)'\right]} \quad (14.24)$$

where

$N$ = total number of molecules
$g_{e'}$ = degeneracy of the electronic level $e'$
$k_B$ = Boltzmann's constant
$\varepsilon_{el}$, $T_{el}$ = electronic energy and temperature for the state, respectively
$\varepsilon_{vib}$, $T_{vib}$ = vibrational energy and temperature for the state, respectively
$\varepsilon_{rot}$, $T_{rot}$ = rotational energy and temperature for the state, respectively
$S$ = spin for the state
$\emptyset(\lambda - \lambda_0)$ = line shape function

Using Eqs. (14.23) and (14.24), SPECAIR simulates the emission intensity derived from a product of the number of molecules ($N$) in the upper state, the population ($n_{e'v'J'}$) in the upper state, the energy difference between the two states ($E_{e'v'J'} - E_{e''v''J''}$), and the probability for spontaneous emission from the upper state to the lower state. The probability constants, $\sum \left(Re^{v'v''}\right)^2$ and $S_{J'J''}$, are not included in the calculation process by SPECAIR. Basically, by providing and adjusting the fitting inputs of the gas temperature ($T_g$) and the electron excitation temperature ($T_{el}$) in Kelvin, SPECAIR then computes and simulates the line intensity vs. wavelength using radiative transition probabilities to match the experimental OES data. In addition, the plasma parameters associated with vibrational–rotational excitation and the ionization and excitation can be determined and extracted. For example, Meehan et al. [21] investigated the fundamental plasma physics of gas heating under 50 Pa and 300 W plasma conditions and compared the gas temperatures obtained by OES techniques for both $O_2$ and $N_2$ (Figure 14.4). The emission band of $O_2$ was selected at *ca.* 760 nm, corresponding to the transition from the long-lived state to the ground state as

$$O_2\left(b^1\Sigma_g^+, v' = 0\right) \rightarrow O_2\left(X^3\Sigma_g^-, v'' = 0\right).$$

As for $N_2$, the emission band of interest is *ca.* 337 nm, associated with the transition $N_2(C)$ to $N_2(B)$ in the SPS as

$$N_2\left(C^3\Pi_u, v' = 0\right) \rightarrow N_2\left(B^3\Pi_g, v'' = 0\right).$$

This transition for $N_2$ is widely used for determining the rotational temperature ($T_{rot} = T_g$) of excited states in nitrogen or nitrogen-containing plasma conditions. Based on the work by Meehan et al. [21], the simulated OES spectra for $O_2$ and $N_2$ yield the gas temperature to be $528 \pm 15$ K and $561 \pm 20$ K, respectively.

In the case of the ammonia synthesis and the catalytic reduction of $NO_x$ to produce $N_2$, the $v = 0$–$2$ rotational band of the SPS $N_2\,(C^3\Pi_u, v' = 0) \rightarrow (B^3\Pi_g, v'' = 2)$ at *ca.* 381 nm can also be used [1, 7, 36]. For example, Stere et al. [36] estimated and reported that the gas temperature of the NTP-assisted SCR of $NO_x$ reduction with *n*-octane and toluene over an $Ag/Al_2O_3$ catalyst by performing the curve fitting of OES spectra of $N_2$ at *ca.* 381 nm. The rotational temperature was extracted from the fitting parameters, giving the corresponding gas temperature of 364 and 375 K at applied peak voltage of 6 and 7 kV. Chen et al. [7] also determined the gas temperature under NTP catalytic ammonia synthesis at various applied voltages over both $SiO_2$ and glass beads to be between 290 and 377 K and between 274 and 336 K, respectively.

In addition to being able to estimate the gas temperature, OES has been used to determine the electron excitation temperature ($T_{el}$), responsible for the ability to excite atoms and gas molecules, and the electron density, representing the negative

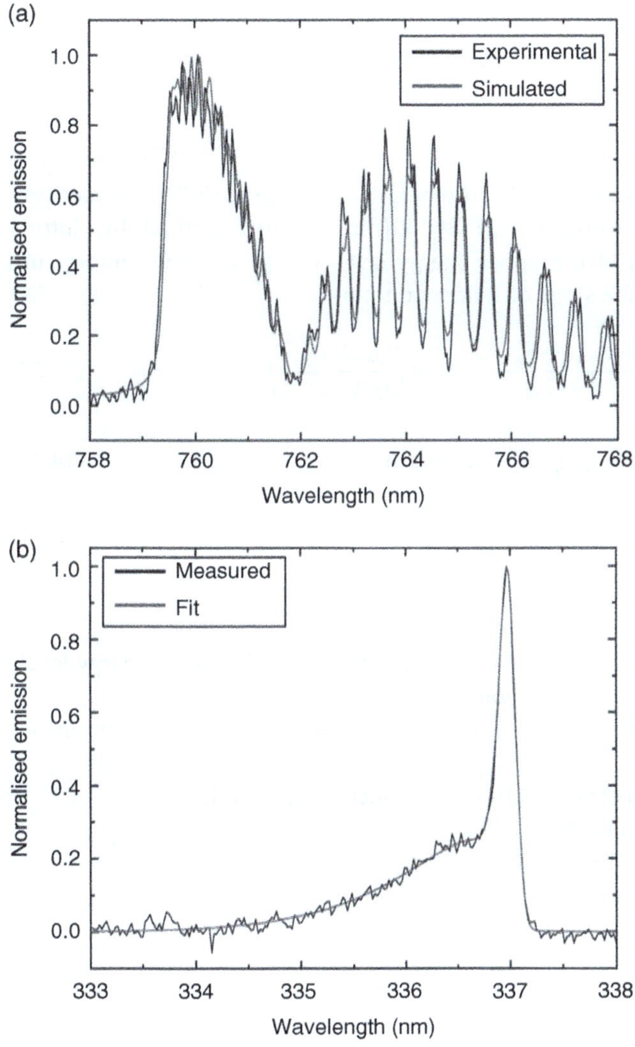


**Figure 14.4** Comparison between simulated and measured OES spectra for (a) $O_2$ and (b) $N_2$ molecules [21].

charge and potential for the interaction of electrons with other molecules. The calculation of electron excitation temperature uses the population of atomic excited states, and the equation derived from Boltzmann distribution with the assumption that the local thermal equilibrium (LTE) or thermodynamic equilibrium in the plasma discharge is reached for the calculation of the electron excitation temperature ($T_{el}$) [5, 16, 17, 37]. The electron temperature can be simply determined by two different methods, i.e. two-line method [5, 8, 18] represented in Eq. (14.25) and Boltzmann plot method [18, 38] expressed in Eq. (14.26). The former uses the ratio of OES intensity between the two emission lines, e.g. between the atomic $H_\alpha$ at *ca.* 656 nm and $H_\beta$ at *ca.* 486 nm or between the ionized molecular $N_2^+$ $\left(B^2\Sigma_u^+, v'=0 \rightarrow X^2\Sigma_g^+, v''=0\right)$ at *ca.* 388 nm $N_2^+$ $\left(B^2\Sigma_u^+, v'=1 \rightarrow X^2\Sigma_g^+, v''=2\right)$ at *ca.* 424 nm. The values for $g$, $A$, and $E'$ can be obtained from the NIST Atomic Spectra Database (currently version 5.12) [39, 40]. In the latter method, by plotting $\ln(I\lambda/Ag)$ as a function of the upper energy level, $E'$, the electron temperature can be extracted from the slope of Boltzmann plot.

$$\ln\left(\frac{I_{at\ \lambda 1}}{I_{at\ \lambda 2}}\right) = -\frac{\left(E'_{at\ \lambda 1} - E'_{at\ \lambda 2}\right)}{k_B T_{el}} + \ln\left(\frac{(gA\lambda_1)_{at\ \lambda 1}}{(gA\lambda_2)_{at\ \lambda 2}}\right) \tag{14.25}$$

$$\ln\left(\frac{I^{\prime\rightarrow\prime\prime}\lambda^{\prime\rightarrow\prime\prime}}{A^{\prime\rightarrow\prime\prime}g'}\right) = -\frac{1}{k_B T_{el}}E' + \text{constant} \tag{14.26}$$

$$\text{Slope} = -\frac{1}{k_B T_{el}} \tag{14.27}$$

where

$I' \rightarrow''$ = emission intensity of transition from upper (′) to lower (″) energy levels
$\lambda' \rightarrow''$ = wavelength of emitted photon (nm)
$A' \rightarrow''$ = Einstein coefficient for spontaneous emission, or transition probability ($s^{-1}$)
$g'$ = degeneracy or statistical weight of the upper energy level
$T_{el}$ = electron temperature (K)
$k_B$ = Boltzmann constant ($eV\,K^{-1}$)
$E'$ = energy of the upper level (eV)
$c$ = speed of light ($m\,s^{-1}$)

As for the electron density ($n_e$, in $cm^{-3}$), the calculation uses the Stark broadening ($\Delta\lambda_{stark}$) of spectral emission lines such as the atomic $H_\alpha$ at *ca.* 656 nm and $H_\beta$ at *ca.* 486 nm. The deconvolution and subtraction between the total full-width half max ($\Delta\lambda_{FWHM}^{Total}$) from the total Gaussian broadening ($\Delta\lambda_{Guass}$) profiles (Doppler and instrumental broadening) yields the total Lorentzian broadening ($\Delta\lambda_{Lorentz}^{Total}$) profiles (natural, van der Waals and resonance broadenings). Then, the Stark broadening and the electron density can be estimated by Eqs. (14.29) and (14.30), respectively [8, 31, 37, 41].

$$\Delta\lambda_{Lorentz}^{Total} = \Delta\lambda_{FWHM}^{Total} - \Delta\lambda_{Guass}^{Total} \tag{14.28}$$

$$\Delta\lambda_{stark} \approx \Delta\lambda_{Lorentz}^{Total} - \Delta\lambda_{vdWaals} - \Delta\lambda_{nat} - \Delta\lambda_{res} \tag{14.29}$$

$$n_e = \left(\frac{\Delta\lambda_{\text{stark}}}{2w_e}\right) \times 10^{16} \tag{14.30}$$

where $w_e$ is the electron-impact parameter [31, 42–44].

### 14.1.6 TDLAS Analysis for Plasma Dynamics and Kinetics

The TDLAS technique offers a powerful diagnostic tool to measure gas temperature ($T_g$) in plasmas by analyzing the absorption of laser light at specific molecular transitions. Additionally, the "HITRAN 2016 database" [45] is a specific version of the HITRAN (High-Resolution Transmission Molecular Absorption) spectral database, which complies important spectroscopic parameters and data used for simulating, modeling, and analyzing of the gas-phase molecular absorption for various gaseous species, e.g. OH, $N_2$, $NH_3$, $H_2O$, $N_2O$, and CO. This database substantially provides an extensive dataset in a wide wavelength range from UV to microwave of molecular transition parameters, including wavelengths, absorption line strengths, temperature dependencies, and pressure broadening coefficients. Essentially, this database enables the modeling and analyzing of plasma parameters.

One of the key plasma diagnostics obtained from the TDLAS technique is the analysis of gas temperature, which is typically determined using the ratio of absorption line intensities from two rotational or vibrational transitions of a molecular species. Since the population of energy levels follows the Boltzmann distribution, the gas temperature can be extracted from the relative absorption strengths of these transitions [25]. Alternatively, Doppler broadening of absorption lines and collisional line shapes due to pressure broadening, which depends on the thermal motion of gas molecules, can also be used for the estimation of gas temperature [21, 22, 25, 26, 46] and the information on species densities and molecular velocity distributions, associated with gas molecules at certain temperatures, for studying the diffusion and energy transfer in nonequilibrium plasmas.

Importantly, TDLAS is a highly sensitive technique due to the absorption of laser light at specific rotational–vibrational transitions, thus facilitating the investigation of the gas-phase kinetics under plasma-activated processes by tracking the temporal evolution and concentration of reactive species of interest in real-time measurement, including the time-dependent densities of electrons, ions, neutral, and excited-state species [21–23, 26, 46]. The kinetic modeling can be studied using a zero-dimensional (0D) plasma kinetics solver, e.g. ZDPlasKin [24, 47–49], which is commonly considered to be a powerful tool used to model plasma chemistry, species evolution, and reaction kinetics in low-temperature plasmas. It is widely applied in plasma discharge modeling, plasma-assisted catalysis, atmospheric plasmas, and combustion studies.

In principle, ZDPlasKin assumes that the plasma environment is homogeneous with the 0D approximation, in which the plasma properties are uniform throughout the reactor. The 0D approximation does not include the diffusion or flow dynamics and the spatial variations in species densities, temperatures, or electric fields. It solves a set of coupled rate equations: (i) the temporal evolution of concentration of plasma species, associated with electron impact reactions, ionization, excitation, recombination, dissociation, and interactions with neutral species and (ii) electron kinetics, associated with electron density. The use of ZDPlasKin with the experimental data from plasma diagnostic techniques such as OES and TDLAS can provide an insight into ionization, dissociation, recombination, and secondary reactions and can help validate and refine plasma reaction mechanisms, enabling optimized plasma processes different applications. Additionally, ZDPlasKin can be well-integrated with BOLSIG+ [24, 47, 49, 50], which is a Boltzmann equation solver based on the electron energy distribution function (EEDF), describing the electron distribution over different energy levels in the plasma environments. This is crucial for both fundamental properties in plasma physics and understanding of electron-impact reactions and energy transfer mechanisms because EEDF determines the reaction rates of electron-impact processes, such as ionization, excitation, and dissociation of gas molecules. Based on the assumption that plasma discharge is under equilibrium, the EEDF uses a Maxwellian distribution, where most electrons have moderate energy and only a few have very high or very low energy.

Although ZDPlasKin solver is the most useful tool for fast plasma kinetic analysis, especially with OES and TDLAS data to validate the species densities and reaction rates, ZDPlasKin cannot resolve the gradients in plasma density, temperature, electric field, or discharge. Moreover, it is not suitable for flowing plasma in large reactors and atmospheric plasma jets where the effect of diffusion and mass transfer on the reaction kinetics can be critical.

## 14.2 Infrared Spectroscopy

### 14.2.1 Introduction

Infrared (IR) spectroscopy is one of the most popular spectroscopic techniques used to determine the chemical functional groups in a sample and has been used extensively for the study of catalytic reactions. In this case, IR has been instrumental in the investigation reaction mechanism to determine the active intermediates for the formation of desirable products. The fundamental understanding of these relationships is one of the key steps to further develop new catalyst designs and formulations and improve the catalytic processes. The advantages include its simple operation, fast data acquisition, chemical sensitivity and specificity, and affordable costs. Typically, the mode of operation for heterogeneous catalysts falls into two main systems.

i) Transmission mode, in which the IR radiation directly passes through a sample. This method is quantitative but requires samples, which are able to be sufficiently transparent, and surface analysis of catalytic reactions is significantly limited.
ii) Reflectance mode, for example, **D**iffuse **R**eflectance **I**nfra**R**ed **F**ourier **T**ransform **S**pectroscopy (DRIFTS), in which the IR radiation is absorbed, scattered with many internal reflections, and reflected off the surface of the sample and is collected by an ellipsoidal mirror and directed to the detector (Figure 14.5a). DRIFTS allows the study of surface phenomena easily (Figure 14.5b); however, the signals are strongly dependent on the nature of the samples, such as the shape, size, roughness and absorption, and hence DRIFTS is difficult to be used quantitatively between samples.

The DRIFTS technique has become widely popular and attractive because the DRIFTS measurements can be performed under realistic reaction conditions (*in situ* and *operando*) and there are no complicated methods for sample preparation. Typically, the background spectrum is recorded before exposing to reaction conditions and then the acquisition of DRIFT spectra of adsorbed species during reaction is recorded by a subtraction from the background spectrum. Most importantly, under catalytic reaction conditions, the *in situ* DRIFTS technique provides each chemical bond (of adsorbed surface species) associated with a characteristic IR frequency that is also affected by the chemical surroundings, for example, a change of catalyst and support materials or reaction conditions can shift the DRIFTS signals from a certain wavenumber, making this technique the most commonly used in the field of heterogeneous catalysis to study and monitor the changes of adsorbed surface species during the catalytic reactions.

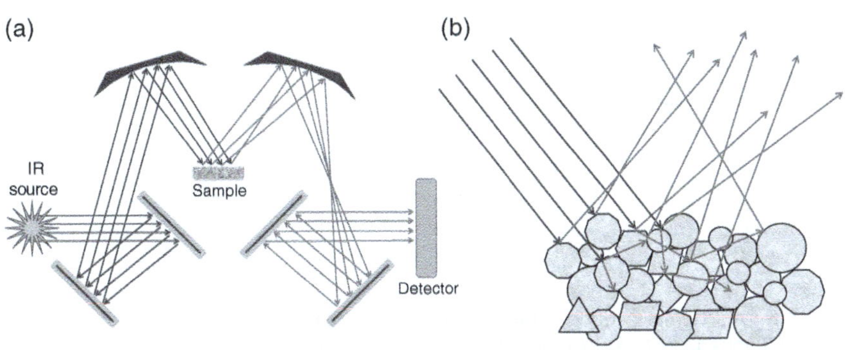


**Figure 14.5** A schematic representation of (a) DRIFTS technique and (b) internal reflectance and diffuse phenomena.

In recent years, in the field of catalysis, there have been a large number of comprehensive review papers and book chapters associated with the *in situ* and *operando* DRIFTS technique with and without complementary techniques such as mass spectrometry (MS), providing the readers the in-depth overview of *in situ* DRIFTS studies on various heterogeneously catalyzed reactions and applications covering the investigation of adsorbed species and intermediates, active sites, solid–gas interfaces/interactions, and the reaction mechanisms. The more recent critical review by Dell'Orco et al. [51] has inclusively reported and summarized the designs and configurations of DRIFTS cells with dielectric barrier discharge (DBD) plasma by exploring the opportunities and the efficacy of using the *in situ* and *operando* DRIFTS techniques with plasma chemistry to study the plasma-generated surface species and active intermediates. However, plasma-coupled transmission IR reactors are not included.

This section covers the IR cell designs for plasma catalysis in both transmission and reflectance sampling modes, and the methodologies for plasma-assisted catalysis applications together with case studies.

### 14.2.2 *In situ* Plasma-IR/DRIFTS Cell Designs and Setups

The early development of the *in situ* IR spectroscopic technique coupled with the DBD plasma discharge was carried out by Rivallan et al. [52], who reported the mechanistic insights into the isopropanol conversion and the gas-phase and adsorbed surface species over alumina under plasma conditions. The transmission IR reactor was designed and made using an outer KBr plate (as well as IR windows) with a grid electrode as dielectric barrier, while using another grid electrode as high-voltage electrode inside the IR cell with catalyst wafer placed perpendicular to the IR beam (Figure 14.6a).

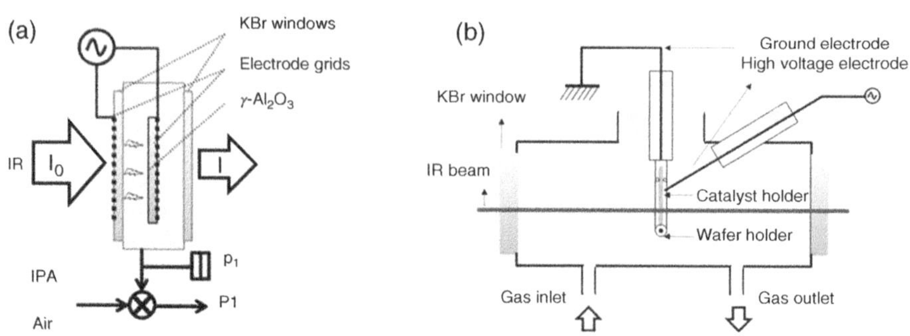


**Figure 14.6** A schematic representation of a DBD plasma transmission IR cell developed. *Source:* (a) Rivallan et al. [52] / John Wiley & Sons and *Source:* (b) Jia and Rousseau [53] / Springer Nature / CC BY 4.0.

Similarly, a new design of DBD plasma discharge incorporated with an IR transmission Pyrex cell, the so-called sorbent-TRACK, was developed and reported by Jia and Rousseau [53] for probing and monitoring the changes of adsorbed species and the surface oxidation of isopropanol and acetone over $CeO_2$ under a direct exposure to plasma discharge as shown in Figure 14.6b. Although the transmission IR technique can be straightforwardly measured and provide both qualitative and quantitative information, the sample preparation for the transmission mode analysis is not as easy because the catalyst sample is usually made as pellet and wafer, which must be transparent for the IR beam to pass through. Also, one of the drawbacks is that this transmission IR reactor has a large dead volume, leading to most of the gases by-passing the catalyst bed.

Alternative to the IR transmission cell, the first report on the use of DRIFTS-MS technique coupled with the DBD–nonthermal plasma (NTP) chemistry for the *in situ* and *operando* investigation of the selective catalytic reduction (SCR) of $NO_x$ with *n*-octane and toluene over Ag-based catalyst was published by Stere et al. [36]. The dome of the DRIFTS cell was simply modified to accommodate a plasma jet, as shown in Figure 14.7a. Therein, the tip of the plasma quartz reactor

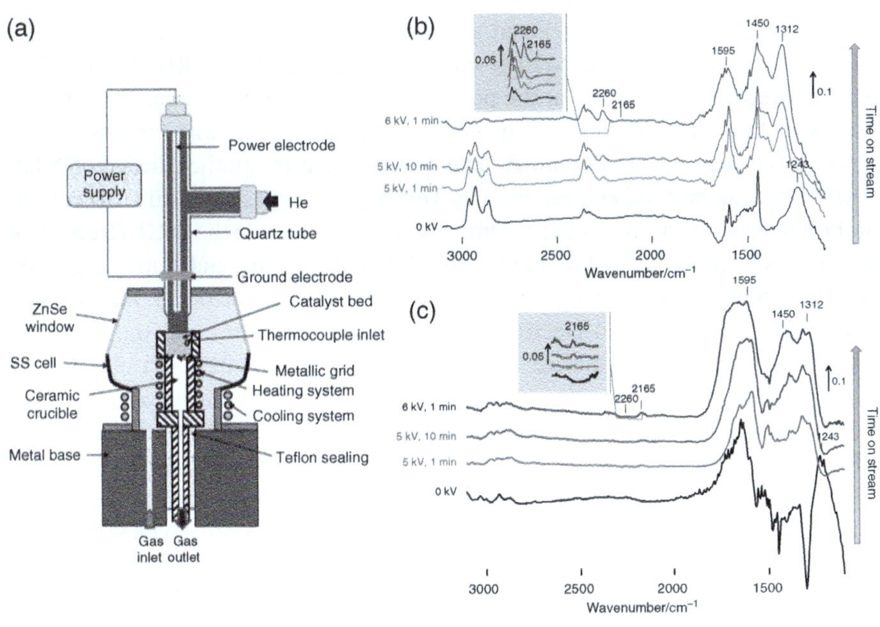


**Figure 14.7** A DRIFTS setup coupled with DBD-NTP reactor above the catalyst bed: (a) a sketch of the modified dome for NTP-DRIFTS system and *in situ* DRIFT spectra under plasma reaction conditions during the SCR of $NO_x$ reaction with (b) *n*-octane and (c) toluene. *Source:* [36] / American Chemical Society.

was carefully positioned to ensure that the tip of the plasma discharge was in contact with the catalyst surfaces.

Subsequently, several developments of *in situ* DRIFTS cells and configurations have been reported [54–59]. Zhang et al. [54] and Zhang et al. [55] designed and developed the DRIFTS reactors with the plasma jets for the catalytic methane decomposition and the catalytic $CO_2$ hydrogenation reaction to methanol, respectively. Recently, Lee et al. [56] developed a multimodel *in situ* and *operando* set of techniques by combining the polarization-modulation infrared reflectance-absorption spectroscopy (PM-IRAS), OES, and MS to investigate the relationship between plasma chemistry and plasma – surface species and reactions. The catalytic methane conversion was used to evaluate the combined analytical system.

Rodrigues et al. [57] developed the NTP-DRIFTS cell, where the catalyst sample was placed on the dielectric ground electrode with the high-voltage electrode pointing directly toward the catalyst bed. The gap between the tip of metal electrode and the dielectric surface was kept fixed at 1 mm. By applying the high-frequency voltage, the catalyst surface was directly exposed to the plasma discharge in the gap between the electrode and the dielectric, allowing the IR beam to probe the *in situ* characterization of surface species under isopropanol and toluene decomposition reactions. Similarly, as reported by Rodrigues et al. [57], Sheng et al. [58] further developed an *in situ* NTP-DRIFTS cell using the needle-type electrode connected to high voltage on the side of the DRIFTS cell, while placing the ground electrode at the periphery of the ceramic crucible. In comparison with the NTP-DRIFTS cell, made by Rodrigues et al., was the position of power electrode and the gas outlet. The development of another design of NTP-DRIFTS reactor was led and reported by Turan et al. [59] as a helical DBD configuration to generate a small-scale plasma discharge in the side the DRIFTS cell. The design was a replica of the Praying Mantis™ DRIFTS reactor and composed of the two electrodes inserted through the dome of the DRIFTS reactor, i.e. high voltage power and ground helical electrodes. The plasma chemistry and the electrical properties of the helical surface DBD were investigated using both air-exposed and dried KBr powder samples without a particular catalytic reaction tested.

Furthermore, Stere et al. [60] has published a newly designed and developed DRIFTS reactor with DBD and NTP incorporated inside the catalyst bed. A type K thermocouple was placed inside the capillary tube and used as a power high-voltage electrode inserted underneath the DRIFTS base, whereas the heating wire was used as ground electrode, which was wrapped around the sample crucible. Once the voltage was applied, the plasma discharge was directly in contact with the catalyst bed, which is similar to most of the DBD plasma plug flow reactors used for catalytic testing. To evaluate this newly designed NTP-DRIFTS cell, plasma-assisted total oxidation of methane on a Pd-based catalyst was carried out and compared with thermal catalysis.

Up to date, there have been various custom-designed configurations of both transmission and DRIFTS reactor coupled with DBD plasma system for *in situ* investigation and characterization of plasma chemistry, plasma–catalyst surface interactions, and plasma-induced surface species and reactions. The publications of various designed NTP-DRIFTS reactors as powerful tools to examine the plasma catalysis and its fundamental understanding of the mechanism of complex plasma catalytic reactions are the gateway to the new development of catalyst and plasma process designs and the significant improvement of the activity, the productivity, the selectivity, and the energy efficiency of plasma catalysis.

### 14.2.3 In-plasma *in situ* DRIFTS Analysis for $NO_x$ Reduction

Stere et al. [36] developed and used the *in situ* and *operando* DRIFTS-MS technique coupled with the NTP system to investigate and probe the surface species and intermediates for the NTP-assisted SCR of nitrogen oxides ($NO_x$) by *n*-octane and toluene over a 2 wt% $Ag/A_2O_3$ catalyst to produce nitrogen and carbon dioxide. The role of NTP chemistry in enhancing the catalytic performance of NOx reduction and in enabling the different reaction mechanism and pathways to form intermediates and products between octane and toluene reductants was reported.

The analysis of *in situ* NTP-DRIFTS spectra (Figure 14.7b and c) suggested that the mechanism for the SCR of NOx with *n*-octane under plasma reaction conditions is similar to the one found and reported for the thermal reaction with isocyanates (–NCO, 2,260 $cm^{-1}$) being the active intermediates. Stere et al. [36] showed that the plasma studies provide further evidence of this hypothesis as the selectivity toward cyanide (–CN, 2,165 $cm^{-1}$) compared with isocyanate is significantly lower than that reported for the thermal systems. This may also indicate that the major route in the case of *n*-octane is via the organic nitro ($R\text{-}NO_2$) species on reaction of the $NO_x$ with the hydrocarbon. In contrast, when using toluene as a reducing agent, the catalytic $NO_x$ reduction may follow a different pathway under plasma discharge since no isocyanate species were observed. However, it is also possible that the water from toluene oxidation converts isocyanate very rapidly to ammonia in the case of the SCR reaction with toluene. Therefore, the route via the organic nitrites (R-ONO) may be important, leading to the formation of cyanide species. During the formation of $N_2$, the cyanide could be converted to isocyanate, which is rapidly hydrolyzed to ammonia, which then reacts with NO to give $N_2$.

### 14.2.4 Post Plasma *in situ* IR Analysis for Ammonia Synthesis

Winter et al. [61] combined the steady-state flow reactor experiments, the computational kinetic modeling, and the post plasma measurements using the *in situ* IR transmission technique to probe and identify the surface intermediates and to gain insight into surface reaction phenomena influenced by plasma chemistry and heating environment during the catalytic ammonia synthesis. It is worth noting that, after each reaction under low-pressure DBD plasma and ammonia synthesis conditions at elevated temperature, the catalytic chamber was evacuated, and the catalyst bed was lowered into a separate chamber where the IR beam passed through. The IR spectra of spent sample were then collected under vacuum to characterize the stable adsorbed surface species and intermediates as indicator of the reaction mechanism occurring under plasma-assisted ammonia synthesis over Fe- and Ni-supported alumina catalysts.

The post plasma IR analysis (Figure 14.8) revealed several strongly adsorbed N-containing surface species such as $NH_3$ (1,658, 1,620, 1,482, 1,425, and 1,205 $cm^{-1}$), $NH_x$ (1,658, 1,589, 1,482, 1,425, 1,238, and 1,100 $cm^{-1}$), and $N_2H_y$ (1,612, 1,393, 1,373, and 1,190 $cm^{-1}$) species. Winter et al. [61] also reported that the changes in the types of adsorbates over NTP reaction time by shifting from $NH_x$ to product species provided a direct evidence of surface reactions occurring on a catalyst during a plasma-activated ammonia synthesis reaction. It was also reported that the differences between the species adsorbed on Ni/γ-$Al_2O_3$ and Fe/γ-$Al_2O_3$, detected by Fourier Transform infrared (FTIR) measurements and calculated by computational kinetic modeling of N(ads) and H(ads) recombination probabilities, explaining the enhancement in the gas-phase reaction rate over the Ni/γ-$Al_2O_3$ catalyst because the IR spectra showed higher concentration of adsorbed species over the Ni/γ-$Al_2O_3$ than Fe/γ-$Al_2O_3$ catalyst. The formation of plasma-enhanced $NH_x$ adsorbed species was identified as important intermediates as observed by OES technique as well. It was clear that the Ni catalyst under plasma discharge facilitated the formation of adsorbed $NH_x$ species over $N_2H_y$, hence an increase in ammonia production when comparing to the Fe/γ-$Al_2O_3$ catalyst.

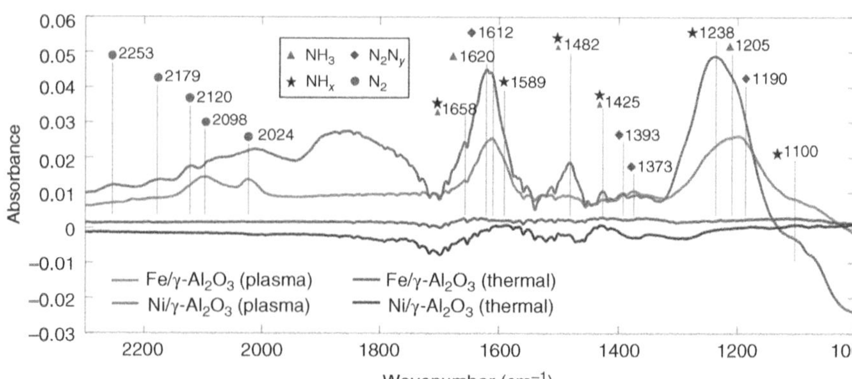


**Figure 14.8** Comparison of adsorbed species detected by the transmission IR technique on Ni and Fe/γ-$Al_2O_3$ catalysts after the plasma-activated reaction of $N_2$ and $H_2$ at 598 K and 12.5 Torr using nonthermal plasma at 18.0 kHz and 1.11 kV for 120 min, and after thermal activation. *Source:* [61] / American Chemical Society.

## 14.3 Summary

In this chapter, the diagnostic analytical tools for fast analysis of plasma chemistry and reactions based on spectroscopic techniques, e.g. OES, TDLAS, and FTIR/DRIFTS, have been summarized for studying a fundamental understanding of nonequilibrium plasma chemistry and kinetics. The identification of plasma species and parameters under transient and/or steady-state conditions is important to gain a crucial insight into any short-lived and plasma-activated species, especially under *in situ* and *operando* conditions where the fast analysis and real-time measurement may provide details of intermediate species formed under plasma reaction conditions. Table 14.1 compares both spatial and temporal resolutions with limitations between the three analytic techniques.

The OES technique has been proven to be one of the powerful tools used to rapidly probe the active species and the fundamental understanding of plasma chemistry and parameters and to obtain an insight into the reaction mechanism of plasma-assisted catalytic reaction. Although this technique has been used extensively, only excited species are detected and the relationship between the optical emission

**Table 14.1** Comparison of spatial and temporal resolutions in OES, TDLAS, and FTIR/DRIFTS.

| Technique | Spatial resolution | Temporal resolution | Limitations |
|---|---|---|---|
| OES | • ~μm – mm (dependent on optics) | • ~ns – ms (fast due to spontaneous emission) | • *Spatial resolution*: limited by spectrometer optics and collection system<br>• *Temporal resolution*: limited by detector speed |
| TDLAS | • ~mm – cm (line-of-sight average, due to laser path length) | • ~μs – ms (fast response depending on tuning speed) | • *Spatial resolution*: limited by the laser path<br>• *Temporal resolution*: limited by laser tuning speed. |
| FTIR/DRIFTS | • ~μm – mm (dependent on optics and aperture) | • ~s – min (slow due to an average of IR signal) | • *Spatial resolution*: detecting chemical information only on surface species<br>• *Temporal resolution*: limited by the IR signal averaging for good sensitivity |

intensity and the concentration of ground electronic-state species is not straightforward. Moreover, the OES spectrum can be complicated when molecular species such as $H_2O$, $N_2$, and $CO_2$ produce continuous spectra that may overlap with atomic spectral lines and when different atoms or molecules emit radiation at nearly the same wavelengths, providing the closely spaced emission spectral lines that may not be resolved by spectrometer. Therefore, the information obtained by OES to investigate the surface interactions and the changes in gas-phase species involving in adsorption, desorption, and surface reactions is often limited. In addition, there is a potential problem associated with the complex algorithms of deconvolution used for extracting the information from the overlapped spectra. However, the best possible solution is to use a high-resolution spectrometer with the Echelle grating to separate the overlapping emission lines and to allow a detection of more individual species in a complex plasma reaction condition.

TDLAS technique offers high sensitivity and selectivity measurements for trace analysis at low concentrations over a narrow spectral range, which can be applied to the plasma gas phase from the UV–Vis to the IR regions. The tunable laser in the Vis to UV range (800–200 nm) and in the IR region enables precise targeting of specific gas absorption lines with minimal or no interferences from the other gases. The high sensitivity of TDLAS with a very fast time response allows for the estimation of the absolute density of both short-lived and long-lived species to be detected as well as plasma parameters and kinetics of gas-phase species involved in plasma-activated processes.

FTIR/DRIFTS technique offers a different approach, in contrast to both OES and TDLAS, using the diffuse and reflectance of IR light to probe ground-state adsorbed species on catalyst surfaces under catalytic reaction conditions for studying and unraveling the surface chemistry and reactions involving not only thermal-activated but also plasma-activated intermediates and other adsorbed species. Although the DRIFTS analysis offers valuable insight into chemical bonding and adsorbed surface compositions, its sensitivity is generally lower compared to OES and TDLAS techniques and the detection of transient species such as short-lived surface intermediates is difficult. Moreover, the interpretation of DRIFTS data can be complex due to the overlapping of spectral features influenced by the repulsive interactions between several adsorbed surface species. To the best of our knowledge, the plasma-assisted catalytic ammonia synthesis has not yet been carried out under plasma reaction conditions using the NTP-DRIFTS technique to probe the plasma-induced surface species and reactions. This chapter can open further opportunities for researchers to develop and use the NTP-DRIFTS technique to gain further insights into the reaction mechanism of ammonia synthesis under plasma conditions.

Overall, each technique brings its own strengths to plasma analysis. Table 14.2 compares various features of the three analytical techniques, depending on the

**Table 14.2** Comparison of different aspects of OES, TDLAS, and FTIR/DRIFTS for plasma reactions.

| Technique | Principle | Analytical type | Key strength | Limitations |
|---|---|---|---|---|
| OES | • Light emission from excited atoms and ions | • Gas phase<br>• Excited state<br>• Qualitative<br>• Semi-quantitative | • Gas temperature<br>• Electron temperature<br>• Electron density<br>• Bulk plasma | • Cannot detect ground state<br>• Calibration is possible, but not simple |
| TDLAS | • NIR/IR absorption at a specific wavelength | • Gas phase<br>• Ground state<br>• Metastable state<br>• Qualitative<br>• Quantitative | • Gas temperature<br>• Gas kinetics<br>• Bulk plasma | • Can only detect one gas species at a time |
| FTIR/DRIFTS | • IR absorption based on diffuse-reflective light | • Adsorbed species<br>• Ground state<br>• Qualitative<br>• Semi-quantitative | • Surface chemistry<br>• Surface reaction<br>• Chemical information | • Cannot detect gas-phase reaction<br>• Only detect IR active species<br>• Calibration is possible, but not simple |

specific analytical tools and purposes, and whether the excited- or ground-state gas-phase compositions, plasma parameters, or surface interactions are of interest. Therefore, combining two or three diagnostic techniques as complimentary analytical tools is essential and can potentially offer a comprehensive and increased understanding of plasma-activated catalytic processes for further development of the catalytic materials and chemical processes. For example, OES and TDLAS techniques can provide a complete understanding of gas-phase plasma chemistry and parameters, while the combined TDLAS and DRIFTS techniques can elucidate and connect the plasma reactions for both gas-phase and surface species to the surface interactions. Importantly, the three-combined OES, TDLAS, and DRIFTS analytical techniques can substantially reveal a full picture of plasma thermodynamic and chemical reaction kinetic, which significantly enhances the understanding of plasma-chemical reactions, plasma-assisted catalysis, and atmospheric plasma processes.

## References


1 Zhang, Q.Y., Shi, D.Q., Xu, W. et al. (2015). Determination of vibrational and rotational temperatures in highly constricted nitrogen plasmas by fitting the second positive system of N2 molecules. *AIP Adv.* 5: 057158. https://doi.org/10.1063/1.4921916.

2 Ribière, M., Eichwald, O., and Yousfi, M. (2020). Microwave absorption and optical emission spectrometry analyses of ambient air plasmas induced by pulsed electron beams. *J. Appl. Phys.* 128: 093304. https://doi.org/10.1063/5.0015482.

3 Zaplotnik, R., Primc, G., and Vesel, A. (2021). Optical emission spectroscopy as a diagnostic tool for characterization of atmospheric plasma jets. *Appl. Sci.* 11: 2275. https://doi.org/10.3390/app11052275.

4 Gorky, F., Best, A., Jasinski, J. et al. (2021). Plasma catalytic ammonia synthesis on Ni nanoparticles: The size effect. *J. Catal.* 393: 369–380. https://doi.org/10.1016/j.jcat.2020.11.030.

5 Rachdi, L., Sushkov, V., and Hofmann, M. (2022). Optical emission spectroscopy diagnostics for plasma parameters investigation in a duo-plasmaline surface-wave sustained discharge. *Spectrochim. Acta B At. Spectrosc.* 194: 106432. https://doi.org/10.1016/j.sab.2022.106432.

6 Liu, Y., Wang, C.-W., Xu, X.-F. et al. (2022). Synergistic effect of Co–Ni bimetal on plasma catalytic ammonia synthesis. *Plasma Chem. Plasma Process.* 42: 267–282. https://doi.org/10.1007/s11090-021-10223-1.

7 Chen, Z., Jaiswal, S., Diallo, A. et al. (2022). Effect of porous catalyst support on plasma-assisted catalysis for ammonia synthesis. *J. Phys. Chem. A* 126: 8741–8752. https://doi.org/10.1021/acs.jpca.2c05023.

8 Zhang, S., Zeng, X., Bai, H. et al. (2022). Optical emission spectroscopy measurement of plasma parameters in a nanosecond pulsed spark discharge for $CO_2/CH_4$ dry reforming. *Spectrochim. Acta A Mol. Biomol. Spectrosc.* 267: 120590. https://doi.org/10.1016/j.saa.2021.120590.

9 Van Duc Long, N., Al-Bared, M., Lin, L. et al. (2022). Understanding plasma-assisted ammonia synthesis via crossing discipline borders of literature: a critical review. *Chem. Eng. Sci.* 263: 118097. https://doi.org/10.1016/j.ces.2022.118097.

10 Liu, T.-W., Gorky, F., Carreon, M.L., and Gómez-Gualdrón, D.A. (2022). Energetics of reaction pathways enabled by N and H radicals during catalytic, plasma-assisted $NH_3$ synthesis. *ACS Sustain. Chem. Eng.* 10: 2034–2051. https://doi.org/10.1021/acssuschemeng.1c05660.

11 Gao, B., Cao, G., Hu, D. et al. (2025). Insight into the effect of support properties on DBD plasma-catalytic $NH_3$ synthesis over Ru-Co bimetallic catalysts. *Fuel* 382: 133802. https://doi.org/10.1016/j.fuel.2024.133802.

12 Mackus, A.J.M., Heil, S.B.S., Langereis, E. et al. (2009). Optical emission spectroscopy as a tool for studying, optimizing, and monitoring plasma-assisted atomic layer deposition processes. *J. Vac. Sci. Technol. A* 28: 77–87. https://doi.org/10.1116/1.3256227.

13 Profijt, H.B., Potts, S.E., van de Sanden, M.C.M., and Kessels, W.M.M. (2011). Plasma-assisted atomic layer deposition: basics, opportunities, and challenges. *J. Vac. Sci. Technol. A* 29: 050801. https://doi.org/10.1116/1.3609974.

14 Hanna, A.R. and Fisher, E.R. (2020). Investigating recent developments and applications of optical plasma spectroscopy: a review. *J. Vac. Sci. Technol. A* 38: 020806. https://doi.org/10.1116/1.5141844.

15 Laux, C.O., Spence, T.G., Kruger, C.H., and Zare, R.N. (2003). Optical diagnostics of atmospheric pressure air plasmas. *Plasma Sources Sci. Technol.* 12: 125. https://doi.org/10.1088/0963-0252/12/2/301.

16 Zhu, X.-M. and Pu, Y.-K. (2008). Using OES to determine electron temperature and density in low-pressure nitrogen and argon plasmas. *Plasma Sources Sci. Technol.* 17: 024002. https://doi.org/10.1088/0963-0252/17/2/024002.

17 Windajanti, J.M., Santjojo, D.J.D.H., and Pamungkas, M.A. (2021). Abdurrouf, determination of electron excitation temperature in an RF-DC hollow cathode nitrogen plasma. *J. Phys. Conf. Ser.* 1825: 012050. https://doi.org/10.1088/1742-6596/1825/1/012050.

18 Turiel, S., Gribov, A., Maler, D., and Krasik, Y.E. (2024). Characterization of the plasma generated by a compact theta pinch. *Plasma* 7: 978–997. https://doi.org/10.3390/plasma7040053.

19 Voráč, J., Synek, P., Procházka, V., and Hoder, T. (2017). State-by-state emission spectra fitting for non-equilibrium plasmas: OH spectra of surface barrier discharge at argon/water interface. *J. Phys. D. Appl. Phys.* 50: 294002. https://doi.org/10.1088/1361-6463/aa7570.

**20** Park, H., You, S.J., and Choe, W. (2010). Correlation between excitation temperature and electron temperature with two groups of electron energy distributions. *Phys. Plasma* 17: 103501. https://doi.org/10.1063/1.3486093.

**21** Meehan, D.N., Niemi, K., and Wagenaars, E. (2020). Gas temperature measurements in a pulsed, low-pressure inductively coupled plasma in oxygen. *Jpn. J. Appl. Phys.* 59: SHHB03. https://doi.org/10.35848/1347-4065/ab7313.

**22** Mignogna, D., Jans, E.R., Raskar, S., and Adamovich, I.V. (2022). Generation and decay of $N_2(A_3\Sigma u+)$ molecules in reacting $CO_2$ and $CH_4$ plasmas. *Plasma Sources Sci. Technol.* 31: 115005. https://doi.org/10.1088/1361-6595/ac99fc.

**23** Winters, C., Eckert, Z., Yin, Z. et al. (2017). Measurements and kinetic modeling of atomic species in fuel-oxidizer mixtures excited by a repetitive nanosecond pulse discharge. *J. Phys. D. Appl. Phys.* 51: 015202. https://doi.org/10.1088/1361-6463/aa9942.

**24** Liu, K., Geng, W., Zhou, X. et al. (2023). Transition mechanisms between selective $O_3$ and $NO_x$ generation modes in atmospheric-pressure plasmas: decoupling specific discharge energy and gas temperature effects. *Plasma Sources Sci. Technol.* 32: 025005. https://doi.org/10.1088/1361-6595/acb814.

**25** Bolshov, M.A., Kuritsyn, Y.A., and Romanovskii, Y.V. (2015). Tunable diode laser spectroscopy as a technique for combustion diagnostics. *Spectrochim. Acta B At. Spectrosc.* 106: 45–66. https://doi.org/10.1016/j.sab.2015.01.010.

**26** O'Connell, D., Gans, T., Crintea, D.L. et al. (2008). Neutral gas depletion mechanisms in dense low-temperature argon plasmas. *J. Phys. D. Appl. Phys.* 41: 035208. https://doi.org/10.1088/0022-3727/41/3/035208.

**27** Stritzke, F., Diemel, O., and Wagner, S. (2015). TDLAS-based $NH_3$ mole fraction measurement for exhaust diagnostics during selective catalytic reduction using a fiber-coupled 2.2-μm DFB diode laser. *Appl. Phys. B Lasers Opt.* 119: 143–152. https://doi.org/10.1007/s00340-015-6073-5.

**28** Wang, Y., Craven, M., Yu, X. et al. (2019). Plasma-enhanced catalytic synthesis of ammonia over a $Ni/Al_2O_3$ catalyst at near-room temperature: Insights into the importance of the catalyst surface on the reaction mechanism. *ACS Catal.* 9: 10780–10793. https://doi.org/10.1021/acscatal.9b02538.

**29** Gershman, S., Fetsch, H., Gorky, F., and Carreon, M.L. (2022). Identifying regimes during plasma catalytic ammonia synthesis. *Plasma Chem. Plasma Process.* 42: 731–757. https://doi.org/10.1007/s11090-022-10258-y.

**30** Rouwenhorst, K.H.R., Kim, H.-H., and Lefferts, L. (2019). Vibrationally excited activation of N2 in plasma-enhanced catalytic ammonia synthesis: a kinetic analysis. *ACS Sustain. Chem. Eng.* 7: 17515–17522. https://doi.org/10.1021/acssuschemeng.9b04997.

**31** Djurović, S. and Konjević, N. (2009). On the use of non-hydrogenic spectral lines for low electron density and high pressure plasma diagnostics. *Plasma Sources Sci. Technol.* 18: 035011. https://doi.org/10.1088/0963-0252/18/3/035011.

**32** Pambudi, D.R.S., Hanif, M.A., Santjojo, D.J.D.H., and Padaga, M.C. (2020). Masruroh, optical emission spectroscopy study of the electron temperature and electron density dependence on the pressure chamber for the carbon deposition produced by argon plasma sputtering. *IOP Conf. Ser. Mater. Sci. Eng.* 833: 012083. https://doi.org/10.1088/1757-899X/833/1/012083.

**33** Savino, L., Cinquegrana, D., French, A. et al. (2022). High resolved optical emission Spectroscopy as accurate physics methodology for plasma freestream temperature characterization. *J. Quant. Spectrosc. Radiat. Transf.* 291: 108323. https://doi.org/10.1016/j.jqsrt.2022.108323.

**34** Verreycken, T., van Gessel, A.F.H., Pageau, A., and Bruggeman, P. (2011). Validation of gas temperature measurements by OES in an atmospheric air glow discharge with water electrode using Rayleigh scattering. *Plasma Sources Sci. Technol.* 20: 024002. https://doi.org/10.1088/0963-0252/20/2/024002.

**35** Patel, R., Peelen, R., van Oijen, J. et al. (2023). How pulse energy affects ignition efficiency of DBD plasma-assisted combustion. *J. Phys. D. Appl. Phys.* 57: 025501. https://doi.org/10.1088/1361-6463/acf942.

**36** Stere, C.E., Adress, W., Burch, R. et al. (2015). Probing a non-thermal plasma activated heterogeneously catalyzed reaction using in situ DRIFTS-MS. *ACS Catal.* 5: 956–964. https://doi.org/10.1021/cs5019265.

**37** Zhu, X.-M., Walsh, J.L., Chen, W.-C., and Pu, Y.-K. (2012). Measurement of the temporal evolution of electron density in a nanosecond pulsed argon microplasma: using both Stark broadening and an OES line-ratio method. *J. Phys. D. Appl. Phys.* 45: 295201. https://doi.org/10.1088/0022-3727/45/29/295201.

**38** Rao, A.P., Gragston, M., Patnaik, A.K. et al. (2019). Measurement of electron density and temperature from laser-induced nitrogen plasma at elevated pressure (1–6 bar). *Opt. Express, OE* 27: 33779–33788. https://doi.org/10.1364/OE.27.033779.

**39** Kramida, A., Ralchenko, Y., and Reader, J. (2009). NIST atomic spectra database (version 5.12), NIST (2009). https://dx.doi.org/10.18434/T4W30F.

**40** Ralchenko, Y. and Kramida, A. (2020). Development of NIST atomic databases and online tools. *Atoms* 8: 56. https://doi.org/10.3390/atoms8030056.

**41** Torres, J., Palomares, J.M., Sola, A. et al. (2007). A Stark broadening method to determine simultaneously the electron temperature and density in high-pressure microwave plasmas. *J. Phys. D. Appl. Phys.* 40: 5929. https://doi.org/10.1088/0022-3727/40/19/023.

**42** Konjević, N. (1999). Plasma broadening and shifting of non-hydrogenic spectral lines: present status and applications. *Phys. Rep.* 316: 339–401. https://doi.org/10.1016/S0370-1573(98)00132-X.

**43** Suchoňová, M., Veis, P., Karhunen, J. et al. (2017). Determination of deuterium depth profiles in fusion-relevant wall materials by nanosecond LIBS. *Nuclear Mater. Energy* 12: 611–616. https://doi.org/10.1016/j.nme.2017.05.013.

44 Hahn, D.W. and Omenetto, N. (2010). Laser-induced breakdown spectroscopy (LIBS), Part I: review of basic diagnostics and plasma—particle interactions: still-challenging issues within the analytical plasma community. *Appl. Spectrosc.* 64: 335A–336A. https://doi.org/10.1366/000370210793561691.

45 Gordon, I.E., Rothman, L.S., Hill, C. et al. (2017). The HITRAN2016 molecular spectroscopic database. *J. Quant. Spectrosc. Radiat. Transf.* 203: 3–69. https://doi.org/10.1016/j.jqsrt.2017.06.038.

46 van der Horst, R.M., Verreycken, T., van Veldhuizen, E.M., and Bruggeman, P.J. (2012). Time-resolved optical emission spectroscopy of nanosecond pulsed discharges in atmospheric-pressure $N_2$ and $N_2/H_2O$ mixtures. *J. Phys. D. Appl. Phys.* 45: 345201. https://doi.org/10.1088/0022-3727/45/34/345201.

47 Pancheshnyi, S., Eismann, B., Hagelaar, G., and Pitchford, L. (2008). ZDPlasKin: a new tool for plasmachemical simulations. *Bull. Am. Phys. Soc.* 53: MWP1.00060.

48 Mao, X., Zhong, H., Liu, N. et al. (2024). Ignition enhancement and NO*x* formation of $NH_3$/air mixtures by non-equilibrium plasma discharge. *Combust. Flame* 259: 113140. https://doi.org/10.1016/j.combustflame.2023.113140.

49 Zhong, H., Mao, X., Liu, N. et al. (2023). Understanding non-equilibrium $N_2O/NO_x$ chemistry in plasma-assisted low-temperature $NH_3$ oxidation. *Combust. Flame* 256: 112948. https://doi.org/10.1016/j.combustflame.2023.112948.

50 Hagelaar, G.J.M. and Pitchford, L.C. (2005). Solving the Boltzmann equation to obtain electron transport coefficients and rate coefficients for fluid models. *Plasma Sources Sci. Technol.* 14: 722. https://doi.org/10.1088/0963-0252/14/4/011.

51 Dell'Orco, S., Leick, N., Alleman, J.L. et al. (2024). Exploring opportunities in operando DRIFTS and complementary techniques for advancing plasma catalysis. *EES. Catal.* 2: 1059–1071. https://doi.org/10.1039/D4EY00088A.

52 Rivallan, M., Fourré, E., Aiello, S. et al. (2012). Insights into the mechanisms of isopropanol conversion on γ-Al $2O_3$ by dielectric barrier discharge. *Plasma Process. Polym.* 9: 850–854. https://doi.org/10.1002/ppap.201200021.

53 Jia, Z. and Rousseau, A. (2016). Sorbent track: quantitative monitoring of adsorbed VOCs under in-situ plasma exposure. *Sci. Rep.* 6: https://doi.org/10.1038/srep31888.

54 Zhang, S., Li, Y., Knoll, A., and Oehrlein, G.S. (2020). Mechanistic aspects of plasma-enhanced catalytic methane decomposition by time-resolved operando diffuse reflectance infrared Fourier transform spectroscopy. *J. Phys. D. Appl. Phys.* 53: 215201. https://doi.org/10.1088/1361-6463/ab795b.

55 Zhang, X., Sun, Z., Shan, Y. et al. (2023). Boosting methanol production via plasma catalytic $CO_2$ hydrogenation over a MnOx/$ZrO_2$ catalyst, Catal. *Sci. Technol.* 13: 2529–2539. https://doi.org/10.1039/D2CY02015G.

56 Lee, G., Go, D.B., and O'Brien, C.P. (2021). Direct observation of plasma-stimulated activation of surface species using multimodal in situ/operando spectroscopy combining polarization-modulation infrared reflection-absorption spectroscopy,

optical emission spectroscopy, and mass spectrometry. *ACS Appl. Mater. Interfaces* 13: 56242–56253. https://doi.org/10.1021/acsami.1c18169.

**57** Rodrigues, A., Tatibouët, J.M., and Fourré, E. (2016). Operando DRIFT spectroscopy characterization of intermediate species on catalysts surface in VOC removal from air by non-thermal plasma assisted catalysis. *Plasma Chem. Plasma Process.* 36: 901–915. https://doi.org/10.1007/s11090-016-9718-1.

**58** Sheng, Z., Kim, H.-H., Yao, S., and Nozaki, T. (2020). Plasma-chemical promotion of catalysis for $CH_4$ dry reforming: unveiling plasma-enabled reaction mechanisms. *Phys. Chem. Chem. Phys.* 22: 19349–19358. https://doi.org/10.1039/D0CP03127E.

**59** Turan, N., Barboun, P.M., Nayak, P.K. et al. (2020). Development of a small-scale helical surface dielectric barrier discharge for characterizing plasma–surface interfaces. *J. Phys. D. Appl. Phys.* 53: 275201. https://doi.org/10.1088/1361-6463/ab8320.

**60** Stere, C., Chansai, S., Gholami, R. et al. (2020). A design of a fixed bed plasma DRIFTS cell for studying the NTP-assisted heterogeneously catalysed reactions. *Cat. Sci. Technol.* 10: 1458–1466. https://doi.org/10.1039/D0CY00036A.

**61** Winter, L.R., Ashford, B., Hong, J. et al. (2020). Identifying surface reaction intermediates in plasma catalytic ammonia synthesis. *ACS Catal.* 10: 14763–14774. https://doi.org/10.1021/acscatal.0c03166.

# 15

# Perspectives in Plasma-Based Nitrogen Fixation for Fertilizer Applications

*Yury Gorbanev*[1,2] *and Annemie Bogaerts*[1,2]

[1] *Research Group PLASMANT, Department of Chemistry, University of Antwerp, Antwerp, Belgium*
[2] *Electrification Institute, University of Antwerp, Antwerp, Belgium*

## 15.1 Nitrogen Compounds Used for Soil Fertilization

As an introductory section, we start with a brief glance at the history and diversity of the nitrogen-containing compounds used for soil fertilization. In the past century, between 1920s and 2020s, the population on Earth has increased approximately 4 times. Besides advances in medicine and the associated longevity of life, an obvious increase in the provision demand could only be met with artificial (synthetic) fertilizers, produced specifically via the Haber-Bosch (HB) process, which was implemented at about the same time in the beginning of the twentieth century (more on this in Section 15.2) [1, 2]. However, before the HB process met its first industrial application, fertilizers were known for a very long time.

In general, elemental nitrogen is known to be essential for sustaining living organisms on the planet, forming the structural basis of amino acids and hence proteins [1]. Plants (e.g. crops) are not merely consumed by the ever-growing population directly but also used as a feed for livestock [3]. The only source of nitrogen for plants is from molecular nitrogen in air that must first be activated through either (i) $N_2$ oxidation into (H)$NO_x$ or (ii) $N_2$ reduction into $NH_3$ [4].

To some extent, the process of nitrogen fixation occurs in nature via abiotic processes (e.g. lightning) as well as biotic ones (e.g. by aquatic and nonaquatic microorganisms). On the basic level, both processes have been known in human history for a very long time, together with their ability to provide the basis for the growth of crops as a food source [5, 6]. However, the amount of nitrogen produced this way is by far not sufficient to sustain the increased demand for crops and hence

*Plasma-Assisted Nitrogen Fixation for Sustainable Process Industries*, First Edition. Edited by Volker Hessel, Annemie Bogaerts, Gabriele Centi, Evgeny Rebrov, and Nguyen Van Duc Long.

Published 2026 by John Wiley & Sons, Inc.

for fertilizers. Specifically, the extent of natural fertilization is usually estimated to be less than half of the total fertilization capacity [7, 8].

Among the two aforementioned routes of $N_2$ fixation, ultimately, the $N_2$ reduction is preferable, primarily because the nitrate fertilizers (produced via $N_2$ oxidation to $NO_3^-$) are reduced inside the plants anyway, yielding $NH_2$ moieties [9]. Thus, introducing N in the already-reduced form ($NH_x$) is beneficial, also because the reduction of nitrate ($NO_3^-$) to nitrite ($NO_2^-$) as the first step of the global reduction process can be detrimental to the proliferation of some, although not all, plants due to the nitrite's toxicity [10, 11].

The pre-HB era saw nitrogen fertilizers mainly obtained via mining of niter ($KNO_3$) and the use of guano. Naturally, the former was a source of oxidized nitrogen with all associated restrictions, while the latter was composed of both reduced and oxidized forms of N together with potassium and phosphate content [2]. Notably, elemental K and P are also fundamental for plant growth but are largely outside the focus of this chapter. With that said, it should be acknowledged that some of the first synthetic fertilizers were the so-called "superphosphates" produced in the mid-nineteenth century by dissolving phosphorites (i.e. mined resources composed of bones and coprolites) in sulfuric acid [12].

Since the 1920s, fertilizers have been largely produced synthetically, and their global annual production has grown from approximately from 160 to 210 Mt between the 2000s and 2020s, of which *ca.* 60% (or approximately 120 Mt in 2020) are represented by nitrogen-containing fertilizers, with phosphate and potassium constituting the remaining 40% [13]. Speaking of synthetic phosphate fertilizers, to date, the monoammonium phosphate (MAP) and diammonium phosphate (DAP) fertilizers are still commercially produced by reacting $NH_3$ with phosphoric acid. However, many other fertilizers are produced as well. The wide variety of synthetic fertilizers employed nowadays ranges from ammonium phosphates or sulfates (where N is present in the cationic form) to nitrates (with N in the anionic form, or both in cationic and anionic form, i.e. $NH_4NO_3$) and even organic molecules such as urea, as shown in Figure 15.1. A combination of urea and ammonium nitrate constitutes the typical fertilizer produced in a combined Haber-Bosch-Ostwald process: a urea/ammonium nitrate (UAN) solution [14]. Importantly, as is clear from Figure 15.1, all modern-day nitrogen fertilizers take their root in the requisite "raw" material: anhydrous ammonia $NH_3$, synthesized catalytically in an HB process.

It is worth mentioning that one of the first industrial processes for fertilizer production, namely, the Birkeland-Eyde process, was in fact plasma based: it utilized hydroelectric energy to drive $N_2$ fixation into $NO_x$ (NO, $NO_2$) using a thermal plasma in air [15]. However, due to the high energy cost (EC) (*ca.* 2.4 MJ $mol^{-1}$ when operating with oxygen-enriched air, and even higher in normal air) and high electricity price, compared to the cheap fossil fuel-based HB process, the Birkeland-Eyde process was abandoned in 1920, and currently, the industry-dominating process is Haber-Bosch(-Ostwald). More details on the HB and Ostwald processes are given in Section 15.2.

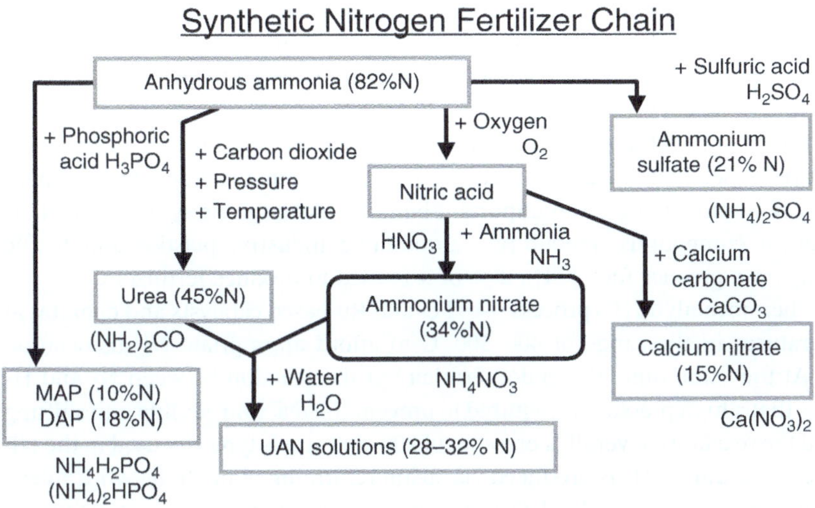


**Figure 15.1** Commonly used fertilizers, their production pathways and conditions, and nitrogen content in percentage by weight. MAP = monoammonium phosphate, DAP = diammonium phosphate, and UAN = urea/ammonium nitrate.

Regarding the nitrogen content specifically, the highest values are associated with two of the most used N-containing fertilizers: urea $(NH_2)_2CO$ and ammonium nitrate $NH_4NO_3$ (45 and 34% N by weight, respectively). Coupled with their high solubility in water, this makes them ideal for direct implementation as fertilizer solutions in farming. However, urea and UAN both release $CO_2$ upon the inherent fertilizer decomposition in the soil [16, 17]. This compromises the decarbonization strategy and reduces the benefits of the highest nitrogen content introduced into the farming land [17]. At the same time, ammonium nitrate has the second-highest N content and is carbon free. Thus, we chose it as the target compound to illustrate the plasma-based approach in fertilizer production. However, before we delve into the plasma field, we should introduce the HB process, combined with the Ostwald process, in a little more detail. For example, its metrics (e.g. EC) are often used as a comparative standard for assessing the viability of a plasma nitrogen fixation setup [18–20].

## 15.2 Fertilizer Production: Haber-Bosch-Ostwald Process and Plasma for Nitrogen Fixation

As mentioned in Section 15.1, the global synthetic fertilizer production reached over two hundred million tons in the early 2020s, very nearly fitting the increasing global demand [13, 21]. The dominant process for synthetic nitrogen fixation is the

HB process, accounting for 29% of the total nitrogen fixation on Earth [4]. The HB process was invented in 1908 and commercialized within merely half a decade in 1913 [1, 22]. In essence, it is a heterogeneous catalysis-based $N_2$ fixation via reduction with $H_2$, which produces annually 150 Mt $NH_3$ [1, 23]. Comparing this number with the 120 Mt yearly production amount of nitrogen-based fertilizers in recent years (see Section 15.1), it is easy to see that despite the many diverse applications of ammonia (e.g. as an important building block for the synthesis of pharmaceutical compounds, metallurgy, automotive industry, plastics and textile industry, as a greener fuel, [24]), 80% of it is used to produce fertilizers.

The thermocatalytic HB process uses Fe- and Ru-based catalysts and operates at temperatures in the range of 400–500 °C to afford appropriate $N_2$ dissociation rates. At the same time, the endothermicity of the reaction between $N_2$ and $H_2$ implies that a high pressure is required to prevent the reaction equilibrium shifting toward the reactants. Overall, a pressure of 200–300 atm is typically used in the HB plants. The required $H_2$ is produced via steam reforming of methane, with subsequent gas separation: CO, $H_2$, $H_2O$, $CO_2$ (from the accompanying reverse water gas shift reaction), and the unreacted $CH_4$. Multiple reforming and gas separation and purification stages are required to yield an $N_2$:$H_2$ mixture of approximately the stoichiometric 1:3 ratio [25, 26]. In the final stage, the gas mixture containing 25 vol% of the produced $NH_3$ is cooled down, compressed, and condensed, affording liquid $NH_3$ as the final product [27]. The EC of the HB process is typically 0.45–0.6 MJ $mol^{-1}$ $NH_3$ produced [28], which is much lower than the EC of nitrogen fixation by plasma, regardless of the pathway – oxidative or reductive (see Section 15.3).

However, $NH_3$ itself is, obviously, not a fertilizer, but merely a precursor for one, as discussed in Section 15.1. Reacting $NH_3$ with $HNO_3$ converts it into a fertilizer on which we focus in this chapter – $NH_4NO_3$. $HNO_3$ required for this is produced by the Ostwald process, created likewise at the beginning of the twentieth century and currently directly linked to the HB process. It consists of several oxidation stages but is practically defined as the thermocatalytic oxidation of $NH_3$ (produced by the HB process) by oxygen over Pt-based catalysts at 900 °C and 8 atm, followed by subsequent oxidation of the formed NO into $NO_2$, with subsequent absorption of $NO_2$ in $H_2O$ forming $HNO_3$ [29, 30]. This $HNO_3$ is further combined with the non-oxidized $NH_3$ from HB to form $NH_4NO_3$.

Thus, it is evident that the process overall consists of a multitude of stages requiring high temperatures and/or pressures. A conceptual representation of the $NH_4NO_3$ production via the Haber-Bosch-Ostwald process is shown in Figure 15.2a. $CH_4$ used as the source of $H_2$ is also the source of energy for this highly energy-intensive process, which uses nearly 2% of the total energy supply worldwide and 3–5% of the globally extracted natural gas [31, 32].

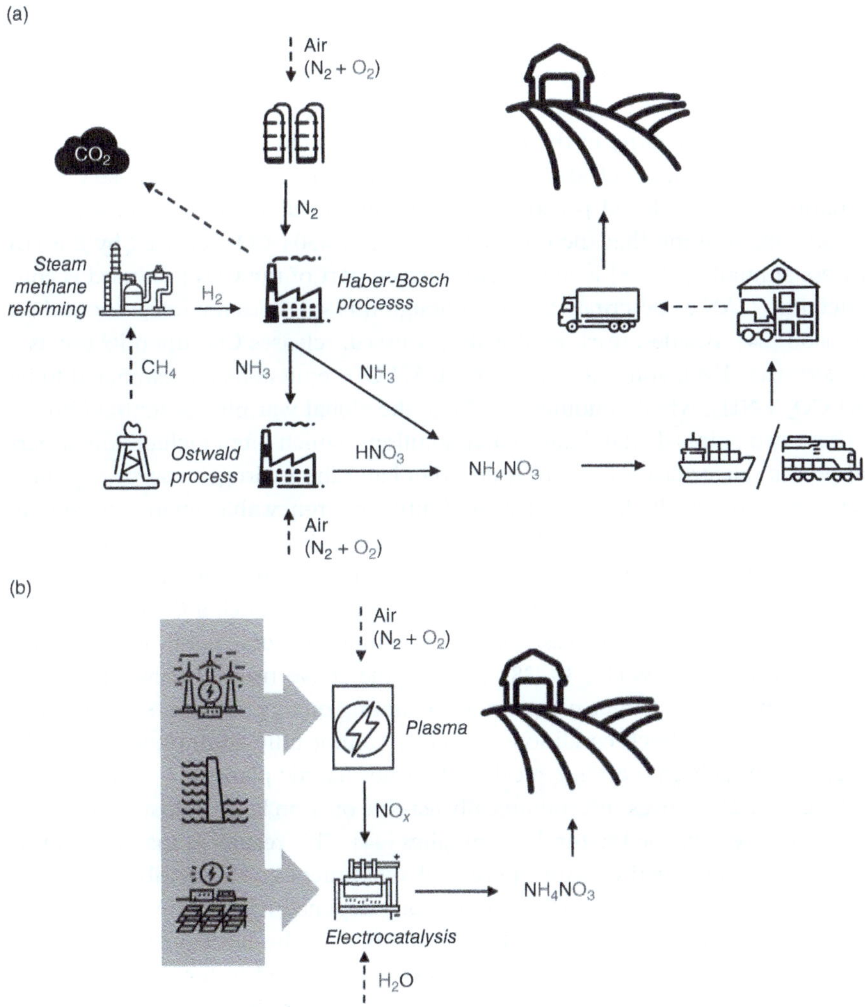


**Figure 15.2** Schematic representation of the fertilizer production and delivery chain for the Haber-Bosch-Ostwald-based (a) and plasma-based (b) $NH_4NO_3$ production. The plasma-based route used as an example here involves electrocatalytic reduction as a quasi-one-pot, two-step process to directly obtain an aqueous solution of $NH_4NO_3$.

In terms of the ECs, this combined process of fertilizer production has been extensively optimized over the century of its existence, with the current values lying in the range of 0.7–0.75 MJ mol$^{-1}$ [15]. However, we infer that although the EC is an important comparison standard, its sole usage is inherently flawed.

For example, taking into account the heavy practical reliance of the process on natural gas means that the high prices and the scarcity of $CH_4$ can cause massive hindrance to the functioning of the Haber-Bosch-Ostwald plants. Especially the immediate scarcity of $CH_4$ due to the unprecedented crisis events in 2020 and 2022 caused the majority of the HB plants in Europe to be halted [33].

Another important metric is the $CO_2$ emission, or the so-called C-factor: the amount of $CO_2$ produced per amount of product in a chemical process [34]. It is worth mentioning that the often-cited value of 450 t $CO_2$ released by the HB process annually [31, 35] is not strictly correct: part of the $CO_2$ produced is converted into UAN as described above, although this still does not offset the carbon emission [24]. Besides, urea, as already discussed, releases $CO_2$ upon its use as a soil fertilizer. Even still, the actual non-UAN $CO_2$ emissions are estimated to be 1.6 t $CO_2$/t $NH_3$, which amounts to 17% of the global warming potential [36].

These two points demand alternative solutions, which often include full or partial process electrification, such as decarbonizing the hydrogen source (e.g. produced electrolytically from water) and utilizing renewable energy to supply power to the HB process [24, 37]. However, the EC of even the $NH_3$ produced in such electrified HB processes is nearly 1.5 times higher than that of the traditional fossil fuel-based process (0.7 vs. 0.45 MJ $mol^{-1}$), reducing the extent of the process optimization advances and approaching the total fossil-based Haber-Bosch-Ostwald process [15, 19, 38]. Besides, the active rise of renewable energy suffers from inherent intermittency of fluctuating energy sources, such as wind and solar [39, 40]. Such conditions currently cannot fully accommodate the HB process with its high inertia, typical for thermocatalytic plants [41–43].

Hence, the HB process is economically feasible only on large industrial scales in regions with established natural gas supplies [44]. This results in massive centralized production, together with the related subsequent costly distribution of the produced $NH_3$ (see Figure 15.2), and a strong dependency on price volatility in the energy market. Furthermore, the distribution of the fertilizer produced via the combined Haber-Bosch-Ostwald process is performed in the most economically feasible form, i.e. not as a UAN solution, but rather as a dry solid salt $NH_4NO_3$, the explosive properties of which have led to multiple lethal accidents during transportation and storage in the past decade alone [45]. The recent geopolitical events also showed that an unstable supply of $CH_4$ (to be used for fertilizer production in a specific region) and the ready-made fertilizer delivery (into a country) create an unsustainable network with unpredictable performance.

The above explains why there is a need for alternative, more sustainable, and decentralized fertilizer production processes. Plasma is one of such alternatives, fully compatible with the intermittence of the renewable energy supplies due to its virtually non-existent process inertia [28, 40]. However, the current strategy does not seek for plasma to compete with the performance metrics of HB, which has been extensively optimized for over a century of its existence. In contrast, at the current stage of development, the use of plasma technology is not plausible for a large-scale production of pure $NH_3$, e.g. for energy purposes (as a hydrogen carrier) or plastics. However, since most of the $NH_3$ produced via HB is used for fertilizers, let us consider this particular application. Figure 15.2b depicts a hypothetical example process of on-site plasma-based nitrogen fixation, where the need for fossil resources, gas reforming and separation stages, as well as the costly and dangerous distribution and storage are completely eliminated, due to the direct production of the $NH_4NO_3$ solution via $NO_x$ generation by plasma coupled with partial electrochemical reduction of the aqueous $NO_x^-$ to $NH_4^+$. Thus, plasma offers an auxiliary technology, which enables a fully decentralized production of fertilizers to be used at the same site, in a form that is also directly applicable for irrigation. Perhaps the most important factor is that there is no fossil dependence: the whole process is entirely electrified and powered with renewable electricity.

## 15.3 Metrics of Various Pathways of Plasma-Based Nitrogen Fixation

As detailed in Chapters 5, 8, 11–13 in this book, nitrogen fixation by plasma can proceed in two directions: oxidation of N into $(H)NO_x$ and reduction into $NH_3$. This can be done in a variety of ways, with different nitrogen-containing feed gas mixtures, e.g. air or pure nitrogen [20], with or without $H_2O$ present in the system [40, 46–48], with $H_2$ as a direct H source or with alternative reducing agents, such as hydrocarbons, or oxidizing agents, such as $CO_2$ [49, 50]. Thus, some of the possible reactive mixtures, which enable plasma-based nitrogen fixation, are as follows: $N_2 + H_2$; $N_2 + O_2$; $N_2 + H_2O$; $N_2 + C_xH_yO_z$ (hydrocarbons and alcohols); and $N_2 + O_2 + H_2O$. Furthermore, the process can proceed only via gas-phase reactions or with catalysts, chemical looping, etc. [51] In turn, based on the position and the role of the catalyst in the overall process, plasma catalysis can also be subdivided into "direct" plasma catalysis (or "in-plasma catalysis") and post-plasma catalysis [51, 52]. A conceptual overview of some of the reactive mixtures is shown in Figure 15.3.

The diversity of the chemical nitrogen fixation pathways is accompanied by the diversity of the applied plasma types. For example, plasma catalysis commonly employed for the direct synthesis of $NH_3$ from $N_2 + H_2$ implies the use of highly non-equilibrium, low-temperature plasma reactors such as dielectric barrier discharge (DBD) plasma, due to the restrictions put by thermal stability of the catalysts [51]. Oxidation of $N_2$ has no such restrictions, and although DBD plasmas

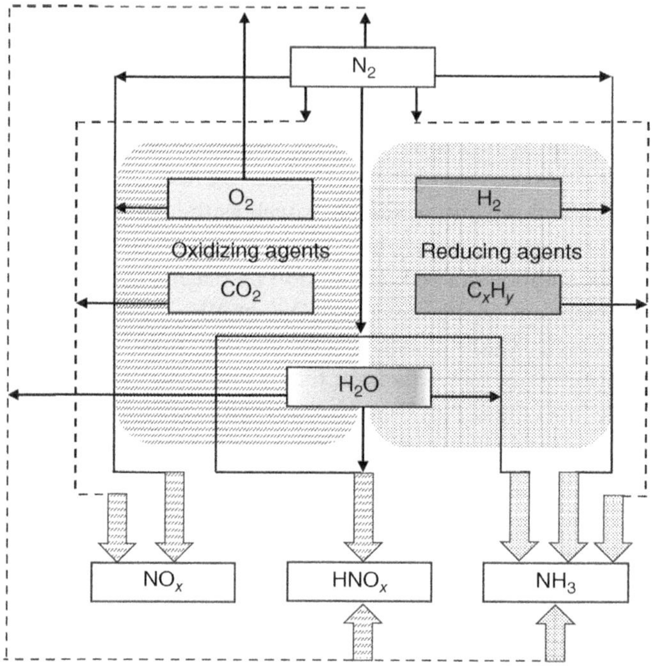


**Figure 15.3** Various pathways of plasma-based nitrogen fixation, based on the reducing/oxidizing agent ($H_2O$ can act as both). Dashed lines mark the less commonly researched pathways.

(often in combinations with catalysis) are also researched in this context [48], "warm" plasmas such as gliding arc (GA), spark, or microwave (MW) plasmas are the main focus of the $NO_x$-oriented studies [20, 53].

It is evident that a clear comparison of this enormous variety of processes requires a common type of metric. Specifically, the EC of the overall process has been shown as a key metric of nitrogen fixation in the overall value chain of fertilizer synthesis [54]. Another important metric to be considered is the production rate (PR): low PR is inefficient even at a very low EC due to the inevitable hidden costs of upscaling or multiplication [55, 56].

Notably, it is not uncommon to see some literature sources mention the concentration of the produced nitrogen fixation products as another key metric. However, this is mostly valid for the formation of (gaseous) $NH_3$, the separation of which from the gas mixture is suggested to be energy efficient only in concentrations above 1 vol% [19]. In the case of $NO_x$ formation and their subsequent absorption in $H_2O$ (as shown in Figure 15.2, and discussed in detail in Section 15.4), low concentrations of $NO_x$ in the post-plasma gas stream do not present a problem. Similarly, because the remaining gas is basically air, no re-circulation of the unconverted substrate is required: "new" air can be fed into the plasma reactor instead, with the remaining air simply exhausted (after $NO_x$ removal via absorption). Hence, we focus on the EC and PR as arguably the most representative metrics [57].

Table 15.1 summarizes the EC and PR values of various pathways of plasma-based nitrogen fixation. The selected references were chosen based on the criterion of the best process metrics: either one of the lowest reported EC or one of the highest reported PR (or the combination).

As evident from the presented data, despite the multiple attempts to gain synergy from plasma catalysis for both nitrogen reduction and nitrogen oxidation, or the use of alternative sources of H ($H_2O$, $C_xH_yO_z$) for $NH_3$ production, the most energy-effective processes, which at the same time possess the highest values of PR, are nitrogen oxidation processes, specifically – a rotating GA and a MW plasma, directly using air (or oxygen-enriched air) as the feed gas (see Table 15.1, $N_2 + O_2$). This is also highlighted in two recent reviews, both of which unambiguously suggest that "warm" plasmas ([gliding] arcs, MW, etc.) outperform other types of plasma in nitrogen fixation [20, 51]. However, despite the low plasma-based EC, MW plasma operation is usually associated with higher total EC (i.e. including the full plug power) due to the use of external cooling equipment, often amounting to, or surpassing, the MW plasma power supply energy uptake [82]. Therefore, (gliding) arc plasmas are (currently) among the most promising tools in energy-efficient plasma-based nitrogen fixation, with the EC values roaming around 2 MJ $mol^{-1}$ and up to 80 g $h^{-1}$ $NO_x$ PR [20, 57].

It is worth mentioning that early plasma studies on nitrogen fixation, conducted in the 1980s, reported $NO_x$ production at very low EC (0.3 MJ $mol^{-1}$), but these results were obtained at very low pressure (0.013 atm) [83], hence not suitable for industry. Moreover, these results were never reproduced since then. Pulsed spark plasmas have shown potential of producing $NO_x$ at a low EC (below 1 MJ $mol^{-1}$), albeit at very limited PRs (up to only 0.3 g $h^{-1}$) (see Table 15.1) [69, 70]. On the other hand, due to the near-equilibrium nature of (gliding) arc and MW plasmas, they cannot reach EC lower than the thermodynamic equilibrium limit in air (i.e. 2.7 MJ $mol^{-1}$) at atmospheric pressure [84]. The reason that some papers in Table 15.1 reported lower EC is due to operation above atmospheric pressure, or with post-plasma quenching, or the use of oxygen-enriched air.

Of course, it must be emphasized that the EC values shown in Table 15.1 correspond to the plasma-deposited power, i.e. the electrical power of the discharge. These values are typically the only ones reported in literature, albeit they do not include the actual power consumed by the plasma power supply from the electrical grid. Actual "plug power" values (which can be up to an order of magnitude higher than the plasma-only power [53, 69]) are required to assess the viability of the processes, and this may change the discussion and conclusions contained in Section 15.3.

**Table 15.1** Typical examples of plasma-based nitrogen fixation processes and their metrics (energy cost, production rate).

| Type of nitrogen fixation | Plasma type | Energy cost ($MJ\ mol^{-1}$) | | Production rate ($mg\ h^{-1}$) | | Employment details |
|---|---|---|---|---|---|---|
| | | $NH_3$ | $NO_x$ | $NH_3$ | $NO_x$ | |
| $N_2 + H_2$ | | | | | | |
| Reduction | DBD [58] | 102 | | 154 | | No catalyst |
| | DBD with catalyst [59] | 437 | — | 8 | — | $Ni/Al_2O_3$ |
| | DBD with catalyst [60] | 74 | — | 26 | — | $Co\text{-}Ni/Al_2O_3$ |
| | DBD with catalyst [61] | 32 | — | 38 | — | $Ru/Al_2O_3$ |
| | DBD with catalyst [62] | 58 | — | 75 | — | Ni-Mg/SBA-15 |
| | DBD with catalyst [63] | 80 | — | 40 | — | $CoCe/Al_2O_3$ |
| $N_2 + H_2O$ | | | | | | |
| Oxidation and/or reduction | DBD jet with $H_2O$ droplets [64] | 7,854 | 3,010 | <1 | 3 | Product accumulation in liquid |
| | GA above $H_2O$ [65] | 2,601 | 1,579 | <1 | 6 | Product accumulation in liquid |
| | Spark jet above water [66] | 52 | 187 | n/a | <1 | Product accumulation in liquid |
| | Spark jet with $H_2O$ vapor [53] | 771 | 73 | <1 | 2 | Products in the gas phase |
| $N_2 + C_xH_yO_z$ | | | | | | |
| Reduction and/or oxidation | Packed bed DBD in $N_2 + CH_4$ [67] | 2.7 | — | 5,210 | — | Borosilicate glass beds |
| | $N_2$ jet above $EtOH/H_2O$ [68] | 352 | 9,534 | 52 | 5 | Products in liquid, 20 vol% aqueous EtOH solution |
| $N_2 + O_2$ | | | | | | |
| | Pulsed AC spark [69] | — | 0.4 | — | 24 | Air ($N_2$:$O_2$ 4:1) |
| | Pulsed DC spark [70] | — | 0.4 | — | 300 | Air ($N_2$:$O_2$ 4:1) |
| | DC glow [71] | — | 2.8 | — | 1,900 | Air ($N_2$:$O_2$ 4:1) |
| | Low-current coaxial plasmatron [72] | — | 3.4 | — | 3,000 | Air ($N_2$:$O_2$ 4:1) |

| | | | | | | |
|---|---|---|---|---|---|---|
| | GA [73] | — | 3.7 | — | 1,600 | $N_2$:$O_2$ 1:1 |
| | Rotating GA [74] | — | 2.5 | | 7,000 | $N_2$:$O_2$ 1:1 |
| | Rotating GA with effusion nozzle [75] | — | 2.1 | — | 7,400 | $N_2$:$O_2$ 1:1 |
| | Rotating GA [57] | — | 2.3 | — | 32,000 | Air ($N_2$:$O_2$ 4:1), 4 atm |
| | Rotating GA [57] | — | 1.8 | — | 68,900 | $N_2$:$O_2$ 1:1, 4 atm |
| | MW [76] | — | 2.0 | — | 86,000 | $N_2$:$O_2$ 1:1 |
| $N_2 + O_2 + H_2O$ | | | | | | |
| Oxidation and reduction | Plasma-electrolytic system [77] | 14.1 | 8.6 | 3 | — | 0.1 M buffer with 500 ppm $NO_2^-$ |
| | Hybrid plasma-electrocatalytic system [78] | 15.5 | 13.7 | 23 | <1 | Cu nanowires catalyst |
| | Hybrid plasma-electrocatalytic system [79] | — | 117 | 3 | 69 | $LaFeO_3$ catalyst, 0.1 M KOH electrolyte |
| | Hybrid plasma-electrocatalytic system [80] | 41 | 40 | 40 | 6 | $Co_3O_4$ nanoparticles catalyst, 0.11 M $NO_x^-$ + 1 M NaOH electrolyte |
| | Hybrid plasma-electrocatalytic system [81] | 3.1 | 2.4 | 3 | 1 | Co Sas/N-C catalyst, 0.1 M KOH electrolyte |

In the section of DBD plasma catalysis for $NH_3$ production from $N_2$ and $H_2$, the PRs are reported in mg per h per g of catalyst: mg $h^{-1}$ $g_{cat}^{-1}$.

## 15.4 Perspectives of $NH_4NO_3$ Production by Plasma-Based Nitrogen Fixation

We will now discuss the potential ways of producing the fertilizer $NH_4NO_3$, which is the focus of the present chapter, with plasma-based nitrogen fixation. Due to the absence (at least to the best of our knowledge) of currently existing commercially implemented processes of such type, the nature of this section is inherently more hypothetical and speculative, although based on the metrics discussed in Section 15.2. Hence, the aim of this section is to discuss the perspectives of plasma-based nitrogen fixation for fertilizer production based on the process performance metrics discussed above.

Let us first consider several pathways of $NH_4NO_3$ production involving plasma-based nitrogen fixation. For the $NH_4NO_3$ production, the process can be subdivided into two steps: (i) $NH_3$ and (ii) $HNO_3$ generation.

### 15.4.1 Pathway Through $NH_3$ Synthesis: Plasma Reduction Followed by Oxidation

Naturally, the most direct alternative to HB is the direct synthesis of $NH_3$ and its subsequent partial oxidation to $HNO_3$, followed by combining $NH_3$ and $HNO_3$ into $NH_4NO_3$ as the ultimate output product (Figure 15.4). As discussed above, the route using C-containing H sources is not feasible in the long term within the new decarbonization policies. Therefore, the reduction agents are limited to $H_2$ and $H_2O$. Furthermore, the choice of plasma type is also strictly limited to highly non-equilibrium, non-thermal plasmas. DBD plasmas satisfy these criteria, but even in combination with catalysts, they typically have a low throughput (and thus low PR) and high EC. The EC is usually in the range of 30–100 MJ $mol^{-1}$ (see Table 15.1), while the estimated requirement for plasma-based nitrogen fixation to be competitive with the industry-dominating Haber-Bosch-Ostwald process is 1–1.5 MJ $mol^{-1}$ [15, 38]. Moreover, the production of $NH_4NO_3$ requires an

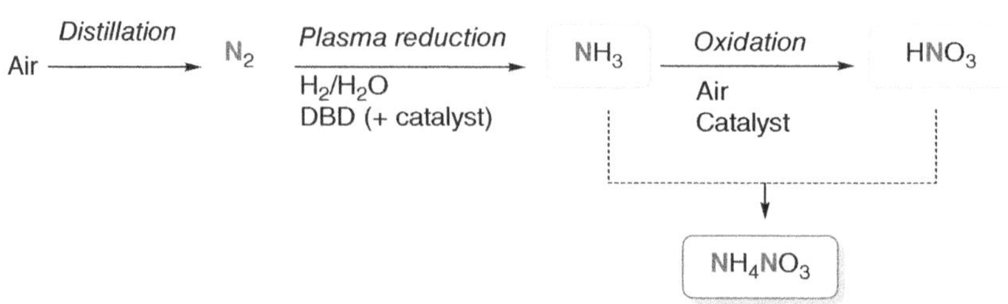


**Figure 15.4** Pathway of $NH_4NO_3$ production via plasma-based nitrogen fixation into $NH_3$ coupled with its subsequent (partial) oxidation to $HNO_3$.

oxidation step, with the initially produced $NH_3$ oxidized into $HNO_3$. Although distillation of $N_2$ from the air is not necessarily needed because $NH_3$ can also be produced by plasma from the air with, e.g. $H_2O$ as the H source [53, 85], the EC and PR values remain far above and below the required levels, respectively. The combination of these factors makes such a route not the most feasible one.

### 15.4.2 Pathway Through both $NH_3$ and $NO_x$ Synthesis: Combined Plasma Reduction and Plasma Oxidation

An alternative route could involve the production of $NH_3$ via plasma nitrogen reduction, while a separate process of plasma nitrogen oxidation is used to produce $NO_x$, with the products of both processes later combined to yield $NH_4NO_3$, as conceptually illustrated in Figure 15.5. Here, the production of $NO_x$ can be done from air in direct plasma $N_2$ oxidation. This oxidative plasma nitrogen fixation can be performed in a variety of ways, which are described in detail in Chapter 6 of this book. It can be implemented using dry air directly in more thermal, "warm"

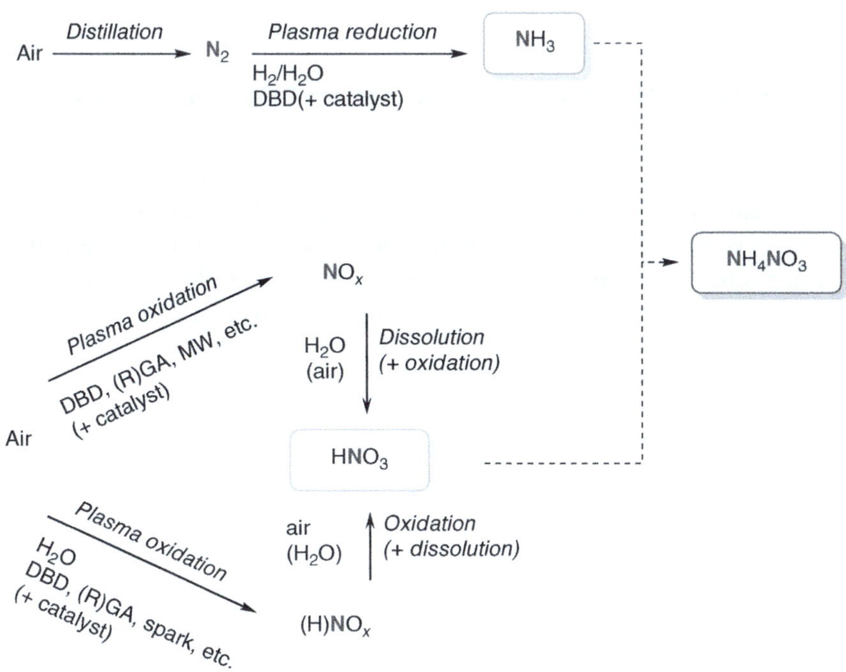


**Figure 15.5** Pathway of $NH_4NO_3$ production via plasma-based nitrogen oxidation into $(H)NO_x$ coupled with plasma-based nitrogen reduction into $NH_3$.

plasmas, which have the lowest EC (2–3 MJ $mol^{-1}$) and the highest PR (tens of g $h^{-1}$ on a laboratory scale) of all reported routes of plasma-based nitrogen fixation (see Table 15.1) [57, 76] Moreover, when using not regular air ($N_2$:$O_2$ of *ca.* 4:1) but oxygen-enriched air (e.g. $N_2$:$O_2$ 1:1), the EC and PR can be further improved and reach values <2 MJ $mol^{-1}$ and ≥ 80 g $h^{-1}$, respectively (see Table 15.1), albeit at a cost required to produce the oxygen-enriched air. The plasma-formed $NO_x$ can be further dissolved in $H_2O$, which, in the presence of unreacted $O_2$ from the air, results ultimately in an $HNO_3$ solution.

Here, we must address another point: plasma-based nitrogen fixation in liquid $H_2O$ systems ($N_2 + H_2O$, $N_2 + O_2 + H_2O$). Although technically $HNO_x$ can be formed directly in the liquid phase when creating direct contact between liquid $H_2O$ and plasma, part of the energy is inevitably lost to the evaporation of $H_2O$ and the quenching of excited $N_2$, which has a negative impact on the EC of the process [48, 86, 87].

Altogether, this route has a higher electrification potential due to the stronger feasibility of using plasma for $N_2$ oxidation than the one described in Figure 15.4. However, the direct plasma-based $N_2$ reduction as part of this route is still a limiting factor.

### 15.4.3 Pathway Through $NO_x$ Synthesis: Plasma Oxidation Followed by Reduction

In our opinion, the most promising plasma-based pathway to $NH_4NO_3$, which does not involve plasma $N_2$ reduction, is shown in Figure 15.6. Here, the only substrate is air, without the need for separation of $O_2$ and $N_2$. Air is used to produce $NO_x$ using warm (or thermal) plasmas, which have significantly higher throughput (or PR) and lower EC than non-thermal (e.g. DBD) plasmas.

After this, the $NO_x$ needs to be reduced to $NH_3$. One of the ways proposed in the literature is thermocatalytic overreduction of $NO_x$ over lean $NO_x$ trap catalysts, commonly used in the automotive industry in the presence of $H_2$, the so-called PNOCRA process (plasma-nitrogen oxidation and catalytic reduction to ammonia) [18, 19]. The advantage of this process is that it does not require a separation step because the catalyst acts as a $NO_x$ adsorber itself. However, it does require producing $H_2$, which is used as a reducing agent, either from $CH_4$ (which, again, compromises decarbonization) or electrolytically from $H_2O$, and an external energy (heat) source to activate the thermocatalytic reaction. A very low EC was reported for such a combined plasma oxidation/thermocatalytic reduction process, i.e. 2.1 MJ $mol^{-1}$, which, to our knowledge, is the lowest EC for decentralized small-scale green $NH_3$ production reported so far, although this EC is accompanied by a low PR (17 mg $h^{-1}$) [18].

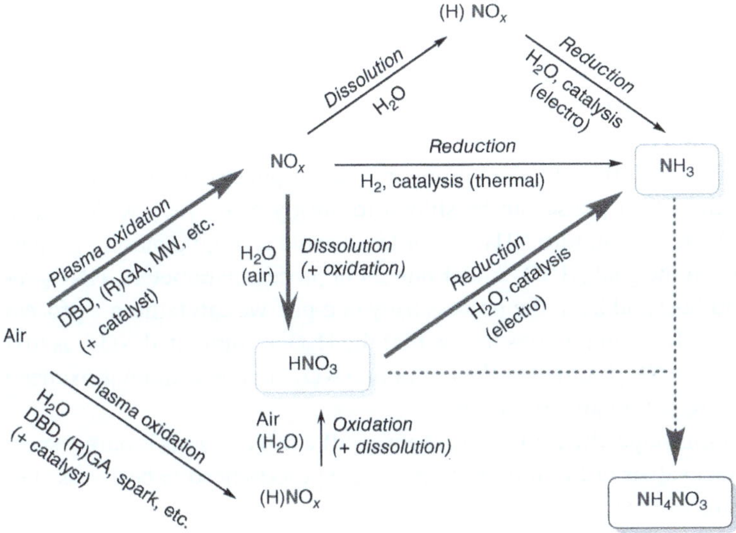


**Figure 15.6** Pathway of $NH_4NO_3$ production via plasma nitrogen oxidation to (H)$NO_x$ coupled with its subsequent (partial) reduction to $NH_3$.

As an alternative to thermocatalytic reduction, electrochemical reduction of the plasma-produced $NO_x$ is another way to $NH_3$. The typical performance of hybrid plasma-electrocatalytic systems is also presented in Table 15.1 (last rows). The EC in such systems is the sum of $N_2$ oxidation and $N_2$ reduction steps, performed in series via plasma and electrochemical approaches, respectively. The electrocatalytic $NO_x$ into $NH_3$ process illustrates some of the best state-of-the-art EC values of plasma-based $NH_3$ production, ranging in general between 3 and 15 MJ $mol^{-1}$ $NH_3$, and reasonable PRs for laboratory-scale systems [88].

Electrochemical reduction does not specifically involve plasma chemistry, and hence it is not discussed in detail in this chapter. We refer to Chapter 13 for more details. However, one important note should be made: although $NO_3^-$ in aqueous solution can be reduced to $NH_3$, the reduction of $NO_2^-$ is more facile [77]. The ratio of $NO_2^-$ to $NO_3^-$ accumulated in an aqueous solution after passing the post-plasma $NO_x$-containing air through it depends on the initial ratio of NO:$NO_2$ produced during $N_2$ oxidation and can be controlled by varying the plasma parameters. This led us to speculatively propose several possible concepts.

In the first scenario, the plasma-produced $NO_x$ mixture is dominated by $NO_2$, which, accompanied by continuous recirculation through $H_2O$ with $O_2$ from air, forms predominantly $NO_3^-$ in the solution [89]. Without any separation or purification steps, this $NO_3^-$ is further partially electrochemically reduced to $NH_4^+$ in the same solution. Such hypothetical one-pot synthesis is perhaps the simplest on-site fertilizer production concept, utilizing only renewable energy (plasma and electrochemical cells can be powered by, e.g. solar panels) and ubiquitous materials such as air and water.

On the other hand, if the $NO_x$-containing plasma output stream has lower fractions of $NO_2$, the liquid phase can be shifted to mostly $NO_2^-$ [90], which has a lower EC for the reduction into $NH_4^+$. In order to avoid changing plasma properties depending on the goal ($HNO_3$ under one set of process parameters, $HNO_2$-to-$NH_3$ under another) and keep the process truly one-pot, we can fathom a process in which the $NO_2$ fraction remains low so that the $H_2O$-accumulated $NO_2^-$ is first partially reduced to $NH_4^+$ and then the remaining $NO_2^-$ in the solution is oxidized to $NO_3^-$ by recirculating air through it.

Of course, at this stage, the comparison between the two concepts is purely speculative: the exact values of EC and PR of the whole process need to be assessed in dedicated future works.

It must be acknowledged that if the ultimate process goal is fertilizers without producing $NH_4NO_3$ specifically, the plasma $N_2$ oxidation step (as in Figures 15.5 and 15.6) with subsequent absorption into an aqueous medium is sufficient to produce nitrates, which already possess fertilizer properties. However, to turn them from acidic solutions into applicable fertilizers, neutralization is still required, e.g. with caustic soda or other chemicals. This, together with the increased N content of $NH_4NO_3$, makes the latter a more desirable product, as was also discussed in Section 15.1.

## 15.5 Plasma-Based Nitrogen Fixation for Reduction of $NH_3$ Emissions and Simultaneous Fertilizer Production

Before we proceed to the final remarks, it must be emphasized that the discussion above concerns complete fertilizer production by plasma from ubiquitous "raw" materials such as air. However, implementations of plasma-based nitrogen fixation with the purpose of producing $NH_4NO_3$ go further beyond that.

We already mentioned that application of urea fertilizers (i.e. urea and UAN) results in $CO_2$ emissions and is thus less optimal than carbon-free fertilizers [91]. However, this is not the only issue of volatilization of fertilizers.

For example, many natural fertilizers (e.g. cattle and pig waste) currently suffer from $NH_3$ loss but can be effectively recycled or improved [92]. The manure contains N (*ca.* 2–8 g kg$^{-1}$) with an $NH_4^+$ content of approximately 70%. However, this ammonia content is largely lost to the atmosphere: $NH_3$ is released over time due to the high pH in the animal waste [93]. This volatilization is estimated in the range of tens of kg per 10,000 $m^2$ of fertilized soil [94]. Industrial-level attempts are made to reduce such $NH_3$ losses by producing $NO_x$ via plasma $N_2$ oxidation from the air, and further applying this $NO_x$ to create $NH_4NO_3$ in the waste slurry [95]. $HNO_3$ produced in the process serves several purposes: it (i) directly reduces the pH by neutralizing the basic environment, (ii) disables the bacteria that cause undesirable $CH_4$ (a greenhouse gas) emissions from waste, and (iii) results in saturating the waste slurry with $NH_4NO_3$. Once again, though, the technology used is "warm" plasma to produce $NO_x$, which makes $NH_4NO_3$ by reacting with the volatile $NH_3$ in the manure.

Another farm-related approach is the removal of $NH_3$ from the stable air, as regulated by the EU law [96]. Currently, this is done, e.g. with acid scrubbers: a commercial acid is dissolved in water, which is used to capture $NH_3$ from air via sprinkling the acid-containing water in a specialized chamber with a high throughput of the stable air [96]. Effectively, this leads not only to the removal of $NH_3$ as per the regulations but also to the accumulation of ammonium-based fertilizers. Typically, $H_2SO_4$ is used, producing $(NH_4)_2SO_4$ as a fertilizer directly in an aqueous solution. The production/distribution costs of the sulfuric acid, and the somewhat limited solubility of $(NH_4)_2SO_4$ in water, put restrictions on the process [97].

A plausible alternative is a process where $NO_x$ is produced by plasma first and then absorbed in water using aqueous absorption columns to yield $HNO_3$. This $HNO_3$ is then used to produce $NH_4NO_3$ while capturing $NH_3$ from the stable air. $NH_4NO_3$ can be concentrated in solutions more than $(NH_4)_2SO_4$ and utilized directly on site [98, 99]. Such a process does not require commercial products with costly delivery, works directly with air and renewable (solar) electricity, and produces fertilizer with a higher N content (see Figure 15.1). Moreover, the process can be time-differentiated in a similar manner to the plasma oxidation/electrocatalytic reduction described above. Plasma production of $NO_x$ and the subsequent accumulation of $HNO_3$ in water can be made virtually time-independent of the $NH_3$ scrubbing process.

## 15.6 Conclusion and Outlook: Challenges and Perspectives of Plasma-Based Nitrogen Fixation

In this chapter, we introduced various types of fertilizers, discussed nitrogen-based fertilizers specifically, and attempted to bridge a gap between the fundamentals of plasma-based nitrogen fixation and producing the actual fertilizers using plasma-enabled processes. So far, the majority of the researchers have focused on improving $NO_x$ and $NH_3$ synthesis separately to achieve the best process performance in terms of lowest EC and highest PR.

While major progress has been made in recent years, especially toward $NO_x$ formation in GA and MW plasmas, it seems not easy to reach EC values below the thermodynamic equilibrium limit (i.e. 2.7 MJ $mol^{-1}$ in air at atmospheric pressure, somewhat lower at elevated pressure or in oxygen-enriched air). This is because, in spite of earlier claims that these warm plasmas promote vibrational excitation (which could lead to more efficient $NO_x$ formation, because the vibrationally excited molecules lower the activation energy of $N_2$ dissociation and of the Zeldovich mechanism, i.e. vibrationally enhanced Zeldovich mechanism), the gas temperature in these warm plasmas is around 3,000 K or more, which also promotes vibrational–translational relaxation, hence depopulating the vibrational levels. Therefore, in practice, the $NO_x$ formation proceeds by the thermal Zeldovich mechanism, which explains why the EC does not drop below the thermodynamic equilibrium limit.

On the other hand, Rouwenhorst et al. estimated that the EC of plasma-based $NO_x$ formation should drop to 1–1.5 MJ $mol^{-1}$ to become competitive with the combined Haber-Bosch-Ostwald process [38]. Several ways for improvement have already been demonstrated, such as operating at enhanced pressure, or in oxygen-enriched air, which both lead to a lower thermodynamic equilibrium limit, but these methods still do not seem to be able to beat the thermodynamics. Hence, fundamental changes in the operating conditions are needed. A further reduction in EC might be possible by post-plasma quenching, i.e. fast cooling, thereby avoiding the back-reactions of NO with N or O atoms, back into $N_2$ and $O_2$. The latter might also be accomplished using scavenging materials for N or O atoms, or by fast oxidation of NO into $NO_2$. Finally, the ultimate aim would still be to exploit vibrational–translational non-equilibrium in atmospheric pressure plasmas, while keeping high $NO_x$ concentrations. We believe detailed self-consistent computational fluid dynamics modeling is needed to better understand how (and whether) such conditions can be reached.

In addition to the need for further reduction of EC, there is a need to increase the PR before the technology can become industrially relevant, i.e. reactor upscaling. Upscaled reactors were proposed for plasma-based NF into $NO_x$ [84]. In a small reactor operating in air, an EC of 2.8 MJ $mol^{-1}$ (hence approaching the minimum thermodynamic equilibrium EC at atmospheric pressure) was obtained for an $NO_x$ concentration of 1.7%, at a flow rate of 20 Ln $min^{-1}$, yielding a PR of 33 g $h^{-1}$. However, the upscaled reactor with the same design yielded much lower $NO_x$ concentrations, due to the higher flow rates, resulting in lower specific energy input values. Practical limitations (of the power supply) made it not possible to operate at higher power, which otherwise would have resulted in higher $NO_x$ concentrations as well. Nevertheless, despite the lower $NO_x$ concentrations reached in this upscaled reactor, the same EC could be obtained. On the other hand, the lower $NO_x$ concentrations resulted in the same PR as in the small reactor, despite the much higher flow rates (up to 300 Ln min$^{-1}$). The low $NO_x$ concentrations in the upscaled reactor resulted from the contracted nature of the plasma arc in the center of the reactor, resulting in a limited fraction of the gas being treated by the plasma, which was indeed identified as the limiting factor for upscaling. However, by changing the reactor configuration into a torch design, the amount of gas treated by the plasma could be enhanced, leading to higher $NO_x$ concentrations, although still much lower than in the small reactor. Nevertheless, due to the higher flow rates, this resulted in a PR of 80 g h$^{-1}$ at an EC of 2.9 MJ mol$^{-1}$. In other words, the PR was much enhanced, at roughly the same EC as the small reactor, demonstrating successful upscaling. This example stresses the challenges (and the importance of reactor design) in upscaling. More in general, such bottlenecks must be solved to achieve a wide application of this technology.

In literature reviews, the cost of electricity is often considered a challenge for the viability of the commercially viable plasma-based nitrogen fixation. We would like to emphasize that in our opinion, plasma-based nitrogen fixation should be based on renewable electricity, to avoid hidden emissions and costs. This brings another important point: the economic considerations of the initial costs of, e.g. solar panels. In general, economically the situation is a lot more complex. The costs of every venture include not only the EC and PR of a process but also the initial costs of the equipment and its wear, followed by regular part replacement [55].

However, these (techno-)economic evaluations, including possible capital costs, etc., are beyond the scope of this chapter and are instead discussed in Chapter 10. Likewise, the environmental impact must be considered, which is discussed in Chapter 9.

Overall, plasma technology enables a large variety of approaches, directions, and possibilities in nitrogen fixation for fertilizer production, making them compatible with electrification and decarbonization policies. The $NH_4NO_3$ synthesis routes discussed here have a solid foundation, being based on the research-relevant scale values of the EC and PR, and could be the focus of future investigations. Beyond this fundamental scientific information, a detailed techno-economic analysis of the proposed approaches can reveal the potential capital costs of the process [54, 100].

## References

1 Erisman, J.W., Sutton, M.A., Galloway, J. et al. (2008). How a century of ammonia synthesis changed the world. *Nat. Geosci.* 1 (10): 636–639. https://doi.org/10.1038/ngeo325.

2 Kasting, J.F. and Siefert, J.L. (2013). The nitrogen fix. *Nature* 501 (7465): 6–6. https://doi.org/10.1038/501006a.
3 Quintero-Herrera, S., Zwolinski, P., Evrard, D. et al. (2022). The role of livestock feed fertilization as an improvement of sustainability in the dairy sector. *Sustain Production Consum.* 31: 448–458. https://doi.org/10.1016/j.spc.2022.03.014.
4 Canfield, D.E., Glazer, A.N., and Falkowski, P.G. (2010). The evolution and future of earth's nitrogen cycle. *Science* 330 (6001): 192–196. https://doi.org/10.1126/science.1186120.
5 Kuypers, M.M.M., Marchant, H.K., and Kartal, B. (2018). The microbial nitrogen-cycling network. *Nat. Rev. Microbiol.* 16 (5): 263–276. https://doi.org/10.1038/nrmicro.2018.9.
6 Ohyama, T. (2014). *Advances in Biology and Ecology of Nitrogen Fixation.* IntechOpen https://doi.org/10.5772/56990.
7 Smil, V. (2002). Nitrogen and food production: proteins for human diets. *AMBIO J. Hum. Environ.* 31 (2): 126–131.
8 Ludemann, C.I., Gruere, A., Heffer, P., and Dobermann, A. (2022). Global data on fertilizer use by crop and by country. *Sci Data* 9 (1): 501. https://doi.org/10.1038/s41597-022-01592-z.
9 Zayed, O., Hewedy, O.A., Abdelmoteleb, A. et al. (2023). Nitrogen journey in plants: from uptake to metabolism, stress response, and microbe interaction. *Biomolecules* 13 (10): 1443.
10 Oke, O.L. (1966). Nitrite toxicity to plants. *Nature* 212 (5061): 528–528. https://doi.org/10.1038/212528a0.
11 Hachiya, T., Ueda, N., Kitagawa, M. et al. (2016). Arabidopsis root-type ferredoxin: NADP(H) oxidoreductase 2 is involved in detoxification of nitrite in roots. *Plant Cell Physiol.* 57 (11): 2440–2450. https://doi.org/10.1093/pcp/pcw158.
12 Russel, D.A. and Williams, G.G. (1977). History of chemical fertilizer development. *Soil Sci. Soc. Am. J.* 41 (2): 260–265. https://doi.org/10.2136/sssaj1977.03615995004100020020x.
13 Menegat, S., Ledo, A., and Tirado, R. (2022). Greenhouse gas emissions from global production and use of nitrogen synthetic fertilisers in agriculture. *Sci. Rep.* 12 (1): 14490. https://doi.org/10.1038/s41598-022-18773-w.
14 Ren, B., Guo, Y., Liu, P. et al. (2021). Effects of urea-ammonium nitrate solution on yield, N2O emission, and nitrogen efficiency of summer maize under integration of water and fertilizer. *Front. Plant Sci.* 12: 700331. https://doi.org/10.3389/fpls.2021.700331.
15 Rouwenhorst, K.H.R., Jardali, F., Bogaerts, A., and Lefferts, L. (2021). From the Birkeland–Eyde process towards energy-efficient plasma-based NOX synthesis: a techno-economic analysis. *Energy Environ. Sci.* 14 (5): 2520–2534. https://doi.org/10.1039/D0EE03763J.
16 Kim, G.W., Alam, M.A., Lee, J.J. et al. (2017). Assessment of direct carbon dioxide emission factor from urea fertilizer in temperate upland soil during warm and cold

cropping season. *Eur. J. Soil Biol.* 83: 76–83. https://doi.org/10.1016/j.ejsobi.2017.10.005.

**17** Ye, X.P. (2024). Decarbonizing nitrogen fertilizer for agriculture with nonthermal plasma technology. *Des. Eng.* 5 (3): 1823–1837.

**18** Hollevoet, L., Vervloessem, E., Gorbanev, Y. et al. (2022). Energy-efficient small-scale ammonia synthesis process with plasma-enabled nitrogen oxidation and catalytic reduction of adsorbed NOx. *ChemSusChem* 15 (10): e202102526. https://doi.org/10.1002/cssc.202102526.

**19** Hollevoet, L., Jardali, F., Gorbanev, Y. et al. (2020). Towards green ammonia synthesis through plasma-driven nitrogen oxidation and catalytic reduction. *Angew. Chem. Int. Ed.* 59 (52): 23825–23829. https://doi.org/10.1002/anie.202011676.

**20** Abdelaziz, A.A., Komuro, A., Teramoto, Y. et al. (2024). Atmospheric-pressure plasmas for NOx production: short review on current status. *Curr. Opin. Green Sustain. Chem.* 50: 100977. https://doi.org/10.1016/j.cogsc.2024.100977.

**21** Gao, Y., Dong, K., and Yue, Y. (2024). Projecting global fertilizer consumption under shared socioeconomic pathway (SSP) scenarios using an approach of ensemble machine learning. *Sci. Total Environ.* 912: 169130. https://doi.org/10.1016/j.scitotenv.2023.169130.

**22** Galloway, J.N. and Cowling, E.B. (2002). Reactive nitrogen and the world: 200 years of change. *AMBIO J. Hum. Environ.* 31 (2): 64–71.

**23** Wang, Y. and Meyer, T.J. (2019). A route to renewable energy triggered by the Haber-Bosch process. *Chem* 5 (3): 496–497. https://doi.org/10.1016/j.chempr.2019.02.021.

**24** Smith, C., Hill, A.K., and Torrente-Murciano, L. (2020). Current and future role of Haber–Bosch ammonia in a carbon-free energy landscape. *Energy Environ. Sci.* 13 (2): 331–344. https://doi.org/10.1039/C9EE02873K.

**25** Cheema, I.I. and Krewer, U. (2018). Operating envelope of Haber–Bosch process design for power-to-ammonia. *RSC Adv.* 8 (61): 34926–34936. https://doi.org/10.1039/C8RA06821F.

**26** Mansourian, R., Mousavi, S.M., and Rahimpour, M.R. (2024). Chapter 4 – Membrane reactors for ammonia synthesis/production. In: *Progresses in Ammonia: Science, Technology and Membranes* (ed. A. Basile and M.R. Rahimpour), 111–130. Elsevier.

**27** Humphreys, J., Lan, R., and Tao, S. (2021). Development and recent progress on ammonia synthesis catalysts for Haber–Bosch process. *Adv. Energy Sustain. Res.* 2 (1): 2000043. https://doi.org/10.1002/aesr.202000043.

**28** Rouwenhorst, K.H.R., Engelmann, Y., van 't Veer, K. et al. (2020). Plasma-driven catalysis: green ammonia synthesis with intermittent electricity. *Green Chem.* 22 (19): 6258–6287. https://doi.org/10.1039/D0GC02058C.

**29** Grande, C.A., Andreassen, K.A., Cavka, J.H. et al. (2018). Process intensification in nitric acid plants by catalytic oxidation of nitric oxide. *Ind. Eng. Chem. Res.* 57 (31): 10180–10186. https://doi.org/10.1021/acs.iecr.8b01483.

**30** Borodin, D., Rahinov, I., Galparsoro, O. et al. (2021). Kinetics of NH3 desorption and diffusion on Pt: implications for the Ostwald process. *J. Am. Chem. Soc.* 143 (43): 18305–18316. https://doi.org/10.1021/jacs.1c09269.

**31** Baltrusaitis, J. (2017). Sustainable ammonia production. *ACS Sustain. Chem. Eng.* 5 (11): 9527–9527. https://doi.org/10.1021/acssuschemeng.7b03719.

**32** Ohyama, T. (2014). *Advances in Biology and Ecology of Nitrogen Fixation*. IntechOpen: Rijeka.

**33** T. Rajabloo Ammonia, production, applications, and the effect of its phase-out; EPOC Belgium, 2022. https://epocbelgium.be/sites/epoc/files/Ammonia%20phase%20out%20consequence%20_%202.pdf.

**34** Voss, B., Andersen, S.I., Taarning, E., and Christensen, C.H. (2009). C factors pinpoint resource utilization in chemical industrial processes. *ChemSusChem* 2 (12): 1152–1162. https://doi.org/10.1002/cssc.200900215.

**35** Nayak-Luke, R.M. and Bañares-Alcántara, R. (2020). Techno-economic viability of islanded green ammonia as a carbon-free energy vector and as a substitute for conventional production. *Energy Environ. Sci.* 13 (9): 2957–2966. https://doi.org/10.1039/D0EE01707H.

**36** Osorio-Tejada, J., Tran, N.N., and Hessel, V. (2022). Techno-environmental assessment of small-scale Haber-Bosch and plasma-assisted ammonia supply chains. *Sci. Total Environ.* 826: 154162. https://doi.org/10.1016/j.scitotenv.2022.154162.

**37** Yüzbaşıoğlu, A.E., Tatarhan, A.H., and Gezerman, A.O. (2021). Decarbonization in ammonia production, new technological methods in industrial scale ammonia production and critical evaluations. *Heliyon* 7 (10): e08257. https://doi.org/10.1016/j.heliyon.2021.e08257.

**38** Rouwenhorst, K.H.R., Jardali, F., Bogaerts, A., and Lefferts, L. (2023). Correction: from the Birkeland–Eyde process towards energy-efficient plasma-based NOX synthesis: a techno-economic analysis. *Energy Environ. Sci.* 16 (12): 6170–6173. https://doi.org/10.1039/D3EE90066E.

**39** Fernández, C.A., Chapman, O., Brown, M.A. et al. (2024). Achieving decentralized, electrified, and decarbonized ammonia production. *Environ. Sci. Technol.* 58 (16): 6964–6977. https://doi.org/10.1021/acs.est.3c10751.

**40** Winter, L.R. and Chen, J.G. (2021). N2 fixation by plasma-activated processes. *Joule* 5 (2): 300–315. https://doi.org/10.1016/j.joule.2020.11.009.

**41** Ghavam, S., Vahdati, M., Wilson, I.A.G., and Styring, P. (2021). Sustainable ammonia production processes. *Front. Energy Res.* 9: 580808. https://doi.org/10.3389/fenrg.2021.580808.

**42** Fahr, S., Schiedeck, M., Schwarzhuber, J. et al. (2023). Design and thermodynamic analysis of a large-scale ammonia reactor for increased load flexibility. *Chem. Eng. J.* 471: 144612. https://doi.org/10.1016/j.cej.2023.144612.

**43** Mallapragada, D.S., Dvorkin, Y., Modestino, M.A. et al. (2023). Decarbonization of the chemical industry through electrification: barriers and opportunities. *Joule* 7 (1): 23–41. https://doi.org/10.1016/j.joule.2022.12.008.

**44** Hochman, G., Goldman, A.S., Felder, F.A. et al. (2020). Potential economic feasibility of direct electrochemical nitrogen reduction as a route to ammonia. *ACS Sustain. Chem. Eng.* 8 (24): 8938–8948. https://doi.org/10.1021/acssuschemeng.0c01206.

**45** Pittman, W., Han, Z., Harding, B. et al. (2014). Lessons to be learned from an analysis of ammonium nitrate disasters in the last 100 years. *J. Hazard. Mater.* 280: 472–477. https://doi.org/10.1016/j.jhazmat.2014.08.037.

**46** Huang, Z., Xiao, A., Liu, D. et al. (2022). Plasma-water-based nitrogen fixation: status, mechanisms, and opportunities. *Plasma Process. Polym.* 19 (4): 2100198. https://doi.org/10.1002/ppap.202100198.

**47** Yoshida, S., Takatsuji, Y., and Haruyama, T. (2024). Green ammonia synthesis technology that does not require H2 gas: reaction technology and prospects for ammonia synthesis using H2O as a direct hydrogen source. *Curr. Opin. Green Sustain. Chem.* 50: 100980. https://doi.org/10.1016/j.cogsc.2024.100980.

**48** Robinson, C. and Stapelmann, K. (2024). Plasma treating water for nitrate based nitrogen fertilizer – A review of recent device designs. *Curr. Opin. Green Sustain. Chem.* 50: 100978. https://doi.org/10.1016/j.cogsc.2024.100978.

**49** Winter, L.R. (2022). Fertilizer and oxygen from Martian atmosphere via microwave plasma. *Chem* 8 (10): 2576–2578. https://doi.org/10.1016/j.chempr.2022.09.018.

**50** Simon, J. and Winter, L.R. (2024). Plasma-activated co-conversion of n2 and c1 gases towards value-added products. *Curr. Opin. Green Sustain. Chem.* 51: 100985. https://doi.org/10.1016/j.cogsc.2024.100985.

**51** Gorbanev, Y., Fedirchyk, I., and Bogaerts, A. (2024). Plasma catalysis in ammonia production and decomposition: use it, or lose it? *Curr. Opin. Green Sustain. Chem.* 47: 100916. https://doi.org/10.1016/j.cogsc.2024.100916.

**52** Hosseini, H. (2023). Dielectric barrier discharge plasma catalysis as an alternative approach for the synthesis of ammonia: a review. *RSC Adv.* 13 (40): 28211–28223. https://doi.org/10.1039/D3RA05580A.

**53** Vervloessem, E., Gromov, M., De Geyter, N. et al. (2023). NH3 and HNOx formation and loss in nitrogen fixation from air with water vapor by nonequilibrium plasma. *ACS Sustain. Chem. Eng.* 11 (10): 4289–4298. https://doi.org/10.1021/acssuschemeng.3c00208.

**54** Manaigo, F., Rouwenhorst, K., Bogaerts, A., and Snyders, R. (2024). Feasibility study of a small-scale fertilizer production facility based on plasma nitrogen fixation. *Energy Convers. Manag.* 302: 118124. https://doi.org/10.1016/j.enconman.2024.118124.

55 O'Modhrain, C., Trenchev, G., Gorbanev, Y., and Bogaerts, A. (2024). Upscaling plasma-based CO2 conversion: case study of a multi-reactor gliding arc plasmatron. *ACS Eng. Au* 4 (3): 333–344. https://doi.org/10.1021/acsengineeringau.3c00067.

56 Nozaki, T., Chen, X., Kim, D.-Y., and Kim, H.-H. (2024). Plasma fluidized beds and their scalability. *Curr. Opin. Green Sustain. Chem.* 51: 100984. https://doi.org/10.1016/j.cogsc.2024.100984.

57 Tsonev, I., O'Modhrain, C., Bogaerts, A., and Gorbanev, Y. (2023). Nitrogen fixation by an arc plasma at elevated pressure to increase the energy efficiency and production rate of NOx. *ACS Sustain. Chem. Eng.* 11 (5): 1888–1897. https://doi.org/10.1021/acssuschemeng.2c06357.

58 Mindong, B., Zhitao, Z., Xiyao, B. et al. (2003). Plasma synthesis of ammonia with a microgap dielectric barrier discharge at ambient pressure. *IEEE Trans. Plasma Sci.* 31 (6): 1285–1291. https://doi.org/10.1109/TPS.2003.818761.

59 Wang, Y., Craven, M., Yu, X. et al. (2019). Plasma-enhanced catalytic synthesis of ammonia over a Ni/$Al_2O_3$ catalyst at near-room temperature: insights into the importance of the catalyst surface on the reaction mechanism. *ACS Catal.* 9 (12): 10780–10793. https://doi.org/10.1021/acscatal.9b02538.

60 Liu, Y., Wang, C.W., Xu, X.F. et al. (2022). Synergistic effect of Co–Ni bimetal on plasma catalytic ammonia synthesis. *Plasma Chem. Plasma Process.* 42 (2): 267–282. https://doi.org/10.1007/s11090-021-10223-1.

61 Li, S., van Raak, T., and Gallucci, F. (2020). Investigating the operation parameters for ammonia synthesis in dielectric barrier discharge reactors. *J. Phys. D. Appl. Phys.* 53 (1): 014008. https://doi.org/10.1088/1361-6463/ab4b37.

62 Li, S., Shao, Y., Chen, H., and Fan, X. (2022). Nonthermal plasma catalytic ammonia synthesis over a Ni catalyst supported on MgO/SBA-15. *Ind. Eng. Chem. Res.* 61 (9): 3292–3302. https://doi.org/10.1021/acs.iecr.1c04968.

63 Ndayirinde, C., Gorbanev, Y., Ciocarlan, R.-G. et al. (2023). Plasma-catalytic ammonia synthesis: packed catalysts act as plasma modifiers. *Catal. Today* 419: 114156. https://doi.org/10.1016/j.cattod.2023.114156.

64 Toth, J.R., Abuyazid, N.H., Lacks, D.J. et al. (2020). A plasma-water droplet reactor for process-intensified, continuous nitrogen fixation at atmospheric pressure. *ACS Sustain. Chem. Eng.* 8 (39): 14845–14854. https://doi.org/10.1021/acssuschemeng.0c04432.

65 Indumathy, B., Ananthanarasimhan, J., Rao, L. et al. (2022). Catalyst-free production of ammonia by means of interaction between a gliding arc plasma and water surface. *J. Phys. D. Appl. Phys.* 55 (39): 395501. https://doi.org/10.1088/1361-6463/ac7b52.

66 Gorbanev, Y., Vervloessem, E., Nikiforov, A., and Bogaerts, A. (2020). Nitrogen fixation with water vapor by nonequilibrium plasma: toward sustainable

ammonia production. *ACS Sustain. Chem. Eng.* 8 (7): 2996–3004. https://doi.org/10.1021/acssuschemeng.9b07849.

**67** Bai, M., Zhang, Z., Bai, M. et al. (2008). Synthesis of ammonia using CH4/N2 plasmas based on micro-gap discharge under environmentally friendly condition. *Plasma Chem. Plasma Process.* 28 (4): 405–414. https://doi.org/10.1007/s11090-008-9132-4.

**68** Kubota, Y., Koga, K., Ohno, M., and Hara, T.T. (2010). Synthesis of ammonia through direct chemical reactions between an atmospheric nitrogen plasma jet and a liquid. *Plasma Fusion Res.* 5: 042–042.

**69** Vervloessem, E., Gorbanev, Y., Nikiforov, A. et al. (2022). Sustainable NOx production from air in pulsed plasma: elucidating the chemistry behind the low energy consumption. *Green Chem.* 24 (2): 916–929. https://doi.org/10.1039/D1GC02762J.

**70** Britun, N., Gamaleev, V., and Hori, M. (2021). Evidence of near-the-limit energy cost NO formation in atmospheric spark discharge. *Plasma Sources Sci. Technol.* 30 (8): 08LT02. https://doi.org/10.1088/1361-6595/ac12bf.

**71** Pei, X., Gidon, D., and Graves, D.B. (2020). Specific energy cost for nitrogen fixation as NOx using DC glow discharge in air. *J. Phys. D. Appl. Phys.* 53 (4): 044002. https://doi.org/10.1088/1361-6463/ab5095.

**72** Korolev, Y.D., Frants, O.B., Landl, N.V., and Suslov, A.I. (2012). Low-current plasmatron as a source of nitrogen oxide molecules. *IEEE Trans. Plasma Sci.* 40 (11): 2837–2842. https://doi.org/10.1109/TPS.2012.2201755.

**73** Wang, W., Patil, B., Heijkers, S. et al. (2017). Nitrogen fixation by gliding arc plasma: better insight by chemical kinetics modelling. *ChemSusChem* 10 (10): 2145–2157. https://doi.org/10.1002/cssc.201700095.

**74** Jardali, F., Van Alphen, S., Creel, J. et al. (2021). NOx production in a rotating gliding arc plasma: potential avenue for sustainable nitrogen fixation. *Green Chem.* 23 (4): 1748–1757. https://doi.org/10.1039/D0GC03521A.

**75** Van Alphen, S., Ahmadi Eshtehardi, H., O'Modhrain, C. et al. (2022). Effusion nozzle for energy-efficient NOx production in a rotating gliding arc plasma reactor. *Chem. Eng. J.* 443: 136529. https://doi.org/10.1016/j.cej.2022.136529.

**76** Kelly, S. and Bogaerts, A. (2021). Nitrogen fixation in an electrode-free microwave plasma. *Joule* 5 (11): 3006–3030. https://doi.org/10.1016/j.joule.2021.09.009.

**77** Luo, Y., Jiang, H., Ding, L.-X. et al. (2023). Selective synthesis of either nitric acid or ammonia from air by electrolyte regulation in a plasma electrolytic system. *ACS Sustain. Chem. Eng.* 11 (32): 11737–11744. https://doi.org/10.1021/acssuschemeng.2c06506.

**78** Sun, J., Alam, D., Daiyan, R. et al. (2021). A hybrid plasma electrocatalytic process for sustainable ammonia production. *Energy Environ. Sci.* 14 (2): 865–872. https://doi.org/10.1039/D0EE03769A.

**79** Cui, Y., Yang, H., Dai, C. et al. (2022). Coupling of $LaFeO_3$–plasma catalysis and $Cu+/Cu_0$ electrocatalysis for direct ammonia synthesis from air. *Ind. Eng. Chem. Res.* 61 (14): 4816–4823. https://doi.org/10.1021/acs.iecr.2c00107.

**80** Meng, Z., Yao, J.-X., Sun, C.-N. et al. (2022). Efficient ammonia production beginning from enhanced air activation. *Adv. Energy Mater.* 12 (38): 2202105. https://doi.org/10.1002/aenm.202202105.

**81** Wu, A., Yang, J., Xu, B. et al. (2021). Direct ammonia synthesis from the air via gliding arc plasma integrated with single atom electrocatalysis. *Appl. Catal. B Environ.* 299: 120667. https://doi.org/10.1016/j.apcatb.2021.120667.

**82** Detz, R.J. and van der Zwaan, B. (2022). Cost projections for microwave plasma CO production using renewable energy. *J. Energy Chem.* 71: 507–513. https://doi.org/10.1016/j.jechem.2022.04.014.

**83** Rusanov, V.D., Fridman, A.A., and Sholin, G.V. (1981). The physics of a chemically active plasma with nonequilibrium vibrational excitation of molecules. *Soviet Phys. Uspekhi* 24 (6): 447. https://doi.org/10.1070/PU1981v024n06ABEH004884.

**84** Tsonev, I., Ahmadi Eshtehardi, H., Delplancke, M.-P., and Bogaerts, A. (2024). Importance of geometric effects in scaling up energy-efficient plasma-based nitrogen fixation. *Sustainable Energy & Fuels* 8 (10): 2191–2209. https://doi.org/10.1039/D3SE01615C.

**85** Haruyama, T., Namise, T., Shimoshimizu, N. et al. (2016). Non-catalyzed one-step synthesis of ammonia from atmospheric air and water. *Green Chem.* 18 (16): 4536–4541. https://doi.org/10.1039/C6GC01560C.

**86** Gromov, M., Kamarinopoulou, N., De Geyter, N. et al. (2022). Plasma-assisted nitrogen fixation: the effect of water presence. *Green Chem.* 24 (24): 9677–9689. https://doi.org/10.1039/D2GC03063B.

**87** Adhami Sayad Mahaleh, M., Narimisa, M., Nikiforov, A. et al. (2023). Nitrogen oxidation in a multi-pin plasma system in the presence and absence of a plasma/liquid interface. *Appl. Sci.* 13 (13): 7619.

**88** Sun, J., Zhou, R., Hong, J. et al. (2024). Sustainable ammonia production via nanosecond-pulsed plasma oxidation and electrocatalytic reduction. *Appl. Catal. B Environ.* 342: 123426. https://doi.org/10.1016/j.apcatb.2023.123426.

**89** Talviste, R., Jõgi, I., Raud, S. et al. (2022). Nitrite and nitrate production by NO and $NO_2$ dissolution in water utilizing plasma jet resembling gas flow pattern. *Plasma Chem. Plasma Process.* 42 (5): 1101–1114. https://doi.org/10.1007/s11090-022-10270-2.

**90** Noori, H., Raud, J., Talviste, R., and Jõgi, I. (2021). Water dissolution of nitrogen oxides produced by ozone oxidation of nitric oxide. *Ozone Sci. Eng.* 43 (3): 284–294. https://doi.org/10.1080/01919512.2020.1839379.

**91** Klimczyk, M., Siczek, A., and Schimmelpfennig, L. (2021). Improving the efficiency of urea-based fertilization leading to reduction in ammonia emission. *Sci. Total Environ.* 771: 145483. https://doi.org/10.1016/j.scitotenv.2021.145483.

**92** Li, T., Wang, Z., Wang, C. et al. (2022). Ammonia volatilization mitigation in crop farming: a review of fertilizer amendment technologies and mechanisms. *Chemosphere* 303: 134944. https://doi.org/10.1016/j.chemosphere.2022.134944.

**93** Font-Palma, C. (2019). Methods for the treatment of cattle manure—a review. *C* 5 (2): 27.

**94** Vilarrasa-Nogué, M., Teira-Esmatges, M.R., González-Llinàs, E. et al. (2020). Ammonia volatilisation from pig slurry and ANS with DMPP applied to Westerwold ryegrass (*Lolium multiflorum* Lam., cv. Trinova) under Mediterranean conditions. *Sci. Total Environ.* 724: 137918. https://doi.org/10.1016/j.scitotenv.2020.137918.

**95** Ingels, R. and Graves, D.B. (2015). Improving the efficiency of organic fertilizer and nitrogen use via air plasma and distributed renewable energy. *Plasma Med.* 5 (2–4): 257–270. https://doi.org/10.1615/PlasmaMed.2016015763.

**96** Melse, R.W. and Ogink, N.W.M. (2005). Air scrubbing techniques for ammonia and odor reduction at livestock operations: review of on-farm research in the Netherlands. *Trans. ASAE* 48 (6): 2303–2313. https://doi.org/10.13031/2013.20094.

**97** Zisopoulos, F.K., van der Goot, A.J., and Boom, R.M. (2018). Exergy destruction in ammonia scrubbers. *Resour. Conserv. Recycl.* 136: 153–165. https://doi.org/10.1016/j.resconrec.2018.04.014.

**98** *Nieuwe luchtwasser filtert stikstof uit stallen en zet die om in waardevolle meststof*, 2024. https://nieuws.kuleuven.be/nl/2024/nieuwe-luchtwasser-filtert-stikstof-uit-stallen-en-zet-die-om-in-waardevolle-meststof (accessed 2024 November 27, 2024).

**99** KU Leuven Luchtwasser kort. https://www.youtube.com/watch?v=oqNWvmlWbqk&list=PLW6FW0nPN9TewT1K-KjmDuhqowcaGs6A9&index=4 (accessed November 27, 2024).

**100** Osorio-Tejada, J.L., Rebrov, E., and Hessel, V. (2023). Internalisation of environmental costs of decentralised nitrogen fertilisers production. *Int. J. Life Cycle Assess.* 28 (11): 1590–1603. https://doi.org/10.1007/s11367-023-02187-5.

# Index



*Plasma-Assisted Nitrogen Fixation for Sustainable Process Industries*, First Edition. Edited by Volker Hessel, Annemie Bogaerts, Gabriele Centi, Evgeny Rebrov, and Nguyen Van Duc Long.

Published 2026 by John Wiley & Sons, Inc.

Printed and bound by CPI Group (UK) Ltd, Croydon, CR0 4YY

31/03/2026

14853484-0001